DATE DUE

CHALLENGED

An Overview of Humanity's Stewardship of Earth

EARTH

World Scientific Lecture Notes in Physics

Stephen F Lincoln

University of Adelaide, Australia

CHALLENGED

An Overview of Humanity's Stewardship of Earth

EARTH

 Imperial College Press

Published by

Imperial College Press
57 Shelton Street
Covent Garden
London WC2H 9HE

Distributed by

World Scientific Publishing Co. Pte. Ltd.
5 Toh Tuck Link, Singapore 596224
USA office: 27 Warren Street, Suite 401-402, Hackensack, NJ 07601
UK office: 57 Shelton Street, Covent Garden, London WC2H 9HE

Library of Congress Cataloging-in-Publication Data
Lincoln, Stephen F.
    Challenged Earth : an overview of humanity's stewardship of Earth / Stephen
  F. Lincoln.
     p. cm.
  Includes bibliographical references and index.
  ISBN-13 978-1-86094-526-7
  ISBN-10 1-86094-526-0
  1. Environmental responsibility. 2. Environmental protection. 3. Sustainable development.
4. Environmental ethics. 5. Environmental management. I. Title.
  GE195.7 .L56 2006
  363'7--dc22

                          2006284933

British Library Cataloguing-in-Publication Data
A catalogue record for this book is available from the British Library.

First published 2006
Reprinted 2008

Printed in Singapore by B & JO Enterprise

ACC Library Services
Austin, Texas

For Christine, my best friend and wife

Preface

This book seeks to present an overview of humanity's interactions with Earth at the beginning of a new century. One of the most striking occurrences of the twentieth century was the growth of the human population from one and a half billion in 1900 to more than six billion in 2000. This was largely achieved through improvements in healthcare and greatly increased food production. However, with these achievements came increased pressure on the environment, the loss of species, depletion of water resources, land degradation, exhaustion of fisheries, pollution, climate change and depletion of the ozone layer. At the beginning of the twenty-first century, all of the continents and oceans have been changed by human activity while population continues to grow and appears set to level out at ten to eleven billion in 2100. Such is the present and likely extent of human induced climate change that it is sometimes said that the warm Holocene epoch, that has lasted for eleven thousand or so years, is being transformed into an Anthropocene epoch. Thus, humans have effectively become the stewards of Earth and its multitudinous lifeforms. The exercise of this stewardship and the future well-being of humanity are intimately intertwined; a realization that is increasingly reflected in international treaties exemplified by the Rio de Janeiro, Montreal and Kyoto Protocols that are concerned with the preservation of biodiversity, protection of the ozone layer and climate change, respectively. Within this context, this book considers life on Earth, population growth, the pressures on water and food supply, genetic engineering, health, energy supply, climate change and the ozone layer in nine chapters. My wife Christine and my colleagues, Ray Choate, David Hayman, Lee Kersten, Bruce May, Derrick Rowley, Andrew Smith and Ching-hwang Yen have collectively read all of the manuscript and provided valuable advice. Nevertheless, any inaccuracies or errors are entirely the responsibility of the author.

Stephen F. Lincoln
The University of Adelaide, 2005

Contents

Chapter 1

The Living Planet

"Man is but earth 'Tis true', but earth is the centre."
John Donne, LXXX Sermons, 1640.

1.1. Life on Earth

Earth, the third planet from the sun, was formed some 4.6 billion years ago [1,2]. The first simple life forms appeared a billion years later and, despite suffering five major extinctions since then, the oceans and landmasses now teem with a huge diversity of animals and plants [3]. From space an obvious sign of life is the seasonal variation of photosynthesis that causes the colours of the continents to change with the great burst of growth of deciduous plants in springtime and the fall of leaves in autumn. A more sophisticated observer might see that the infrared radiation emitted by Earth shows gaps where absorptions by atmospheric water, methane, ozone and carbon dioxide occur as seen in Fig. 1.1 [4]. In contrast, the infrared radiation from Earth's apparently lifeless neighbours, Venus and Mars, the second and fourth planets from the sun, shows a large gap due to absorption by atmospheric carbon dioxide alone. Water is essential to all known forms of life and the presence of ozone indicates that oxygen, from which it is produced by solar radiation, is also present in Earth's atmosphere. Oxygen is produced by cyanobacteria, algae and green plants and is essential to the life of air breathing animals [5]. Conversely, methane is produced from dead animal and plant tissue by anaerobic bacteria that shun oxygen. Thus, the infrared absorptions due to water, ozone and methane may be viewed as the signature of life written large in Earth's atmosphere, a signature eagerly sought by those seeking life on other planets [6]. Meanwhile, humanity signals its presence with a barrage of microwave and radio transmissions and city lights sprinkled across the night side of Earth [7]. However, as population grows and demands for resources increase, a question mark hangs over humanity's tenancy of the planet on which some one hundred billion humans have so far lived [8]. Earth now faces a challenge to the resilience of its biological environment and habitability

1

of a magnitude never before presented by a living species. It is the nature and possible consequences of this challenge that are explored in this book.

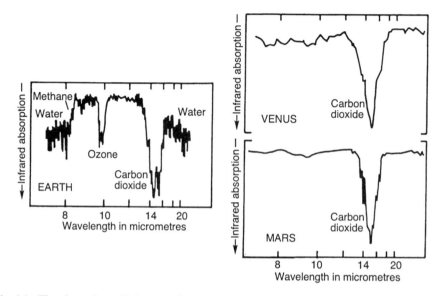

Fig. 1.1. The absorption of infrared radiation emitted by Earth, Venus and Mars by carbon dioxide in their atmospheres. The additional infrared absorptions by water, methane and ozone in Earth's atmosphere indicate the presence of life. Adapted from [4].

1.2. Life's Vicissitudes

A plethora of animal and plant species have appeared on Earth only to be either much diminished or to disappear altogether in five great extinctions [9]. Of all the species that have lived almost all have vanished. The Late Ordovician extinction saw the demise of eighty-five percent of all species 440 million years ago and was followed by the loss of eighty percent of species in the Late Devonian extinction 365 million years ago. The largest extinction, the Permian-Triassic, occurred 250 million years ago with the loss of ninety-six percent of species only to be followed by the loss of seventy-six percent of species 205 million years ago in the Late Triassic extinction. These great extinctions took place over tens of thousands to a million years. They appear to have been caused by asteroid impacts, volcanic eruptions and interrelated changes in climate, sea levels and atmospheric composition that occurred as the continents slowly drifted to their present positions.

The Cretaceous-Tertiary extinction occurred sixty-five million years ago with the loss of eighty percent of all living species including the dinosaurs [10]. This has been attributed to a large asteroid striking Earth at Chicxulub on the coast of the Yucatàn peninsula of Mexico. After the initial massive destruction caused by the impact which may have triggered major volcanic eruptions, the debris, smoke and dust that filled Earth's atmosphere greatly reduced the sunlight reaching the surface and caused a global cooling and reduction in photosynthesis. This was probably followed by a substantial global warming caused by the loading of the atmosphere with carbon dioxide and water from vaporization of carbonate rocks and ocean water caused by the heat of the impact. While these events would have made life very difficult, a 2004 study suggests that this asteroid impact occurred some 300,000 years before the Cretaceous-Tertiary extinction began and that a series of later asteroid impacts may have caused the extinction [11]. However, such is the tenacity of life that the demise of one set of species has invariably been accompanied by the rise of new species. Today, of all of Earth's estimated 100 million or so living species, ranging from single cell organisms to advanced plant and animals, only about 1.5 million have been described [12,13].

Of these multitudinous lifeforms, humans are now dominant and with the birth of Adnan Nevic in Sarajevo on 12 October 1999 the United Nations formally recognized that the human population had reached six billion [14]. This population has arisen from a few thousand African progenitors of some 200,000 years ago and thrives in the warm Holocene epoch that has lasted almost 12,000 years since the retreat of the great ice sheets of the last Ice Age [15]. So large is this population that there remains no part of Earth that has not been changed to some extent as a consequence [16,17]. However, the record of past extinctions holds a salutary message for humanity, a young species by comparison with other species that have inhabited Earth such as the dinosaurs that were dominant for 165 million years. Humans are at once the most intelligent, inventive and, both wittingly and unwittingly, destructive species that Earth has yet seen. It is this last characteristic that is responsible for a sixth great species extinction now underway [18,19]. This holds a great danger for it is possible that a combination of the destruction of complete ecosystems, or biomes, an imbalance in the biological diversity, or biodiversity, of species and a decline in the populations of individual species in the environment may become so great that humanity's future is jeopardized. This could arise through an overwhelming of biological recycling systems, a failure of food supply and the spread of disease. Fortunately, a growing awareness that humanity has become the custodian of the environment shows signs of leading to a more perceptive approach to the health of the planet.

1.3. A New Realization

The slow realization that humanity has a global stewardship role has an interesting history. Humans have always possessed a lively curiosity that was probably a major stimulus for their spread to all of the habitable continents during their nomadic phase. Once largely settled, they retained an urge to explore. Thus, in relatively recent times humans spread across the Pacific to settle the myriad islands of Melanesia, Micronesia, Oceania and Polynesia [20]. The Arabs and Chinese explored the southern oceans in pursuit of trade and the Europeans circumnavigated Earth and settled in lands far from their places of birth. Because of the slowness of their vessels a feeling of the vastness of the oceans and the planet was implanted in the minds of these seafarers and was passed on to humanity at large. However, this was to change slowly from the late nineteenth and early twentieth centuries with the advent of flight by balloon and aircraft that led to a shrinking perception of the size of the planet.

The next major step came with the launch of the first satellite, Sputnik 1, on 4 October 1957, that heralded the beginning of the space age and had millions watching the night sky to marvel at this small artificial moon regularly traversing its Earth orbit [21,22]. Shortly afterwards, on 3 November 1957, Sputnik 2 took the first of Earth's inhabitants, the dog Laika, into orbit. A few years later, on 12 April 1961, Yuri Gagarin in the spacecraft Vostok became the first man to orbit Earth during the Cold War driven space race between the Soviet Union and the United States [23]. The next immense step came with the Apollo 11 mission and the landing on the moon of the first man, Neil Armstrong, on 20 July 1969, who memorably said as he stepped onto the lunar surface: "One small step for (a) man, one giant leap for mankind" [24]. Remarkable as this achievement and subsequent lunar landings were technologically, it is probable that the pictures of Earth from the moon that they provided generated the greatest impact. For the first time all humanity could see Earth as a magnificent translucent blue planet on which are traced the outlines of continents, oceans, rivers and mountain ranges partially covered by clouds suspended in a surprisingly thin envelope of air. This now familiar picture of Earth against the blackness and immensity of space has reinforced a global awareness of the smallness of the planet and the fragility of its habitability.

This raises the question as to whether humanity is intelligent and capable enough to maintain Earth in a sufficiently healthy state to sustain the generations to come [25]. In seeking an answer to this question a wide range of issues is explored in this book. This chapter begins the exploration through a brief examination of the demands placed on Earth by humanity and the perceptions and understandings that were slowly reached in the twentieth century. The chapters that follow contain more detailed considerations of the pressures of population growth, the basic necessities of water and food, a remarkable new understanding of biology, the ever present possibility of disease, the seemingly insatiable demand for energy, and human induced global atmospheric changes and the consequent beginning of climate change.

1.4. Humanity's Footprint

As a walker on a sandy seashore leaves footprints to be washed away by the incoming tide, so humanity leaves ecological footprints on the environment, but so large have they become that nature's regenerative capacity is no longer able to wash them away. And yet for most of humanity these footprints are made unknowingly in seeking to satisfy everyday needs and aspirations. Few would relish being accused of despoiling Earth, yet this despoliation has accelerated with population growth and expectations of higher living standards. The point has now been reached were humanity's impact threatens to impose a decline in the quality of life and possibly to render some now habitable parts of Earth close to uninhabitable because of either land degradation or flooding or both.

Such a dismal prognosis is not new but, probably just in time, an increasing global awareness of the problem may restore a balance between humanity's demands and Earth's regenerative capacity. The nature of the problem is shown by a 1999 study of the ecological footprint of the people of the county of Malmöhus in southern Sweden that is a microcosm of the highly demanding developed world and is illustrated by Fig. 1.2 [26]. The size of this footprint was established by comparing the resources consumed and the waste produced by the people of Malmöhus with the ability of the county's environment to absorb these demands and to completely regenerate while reserving a minimum of twelve percent of the county's area to preserve the biodiversity of nature [27]. In other

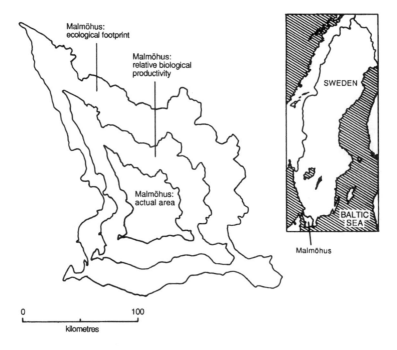

Fig. 1.2. The county of Malmöhus showing its actual area, its biologically productive area on the basis of average global productivity, and its ecological footprint in terms of the consumption and waste production of its population. Adapted from [26].

words, it is an assessment of the ability of Malmöhus to sustain the lifestyle of its people while retaining its animal and plant inhabitants in a healthy state. The footprint includes the consumption of energy, food, clothing and other requirements of the people of Malmöhus and the waste they produce. This waste is mainly carbon dioxide to be absorbed by the growth of permanent forests [28], and nitrates and phosphates from sewerage and agricultural fertilizer runoff to be filtered by wetlands and assimilated by plants and microorganisms that protect fresh water supplies and aquatic ecologies [29]. It also takes into account the balance between imported and exported consumables and the waste produced from them and so incorporates the use of land and ocean beyond Malmöhus' borders.

Malmöhus is fortunate as the 1.2 hectares of available productive land per person are so biologically productive that they are equivalent to 3.4 hectares of land of globally averaged productivity. When twelve percent of this is set aside for maintaining biodiversity, 3.0 hectares of land of globally averaged productivity are available to each person in Malmöhus. However, the Swedish ecological footprint is 7.2 hectares of land of globally averaged productivity so that land elsewhere has to make up the difference of 4.2 hectares for each person in Malmöhus. This difference represents the ecological deficit for Malmöhus as is illustrated in Fig. 1.2. Fortunately for Swedes, 8.2 hectares of land of globally averaged productivity per person are available in Sweden after subtracting twelve percent to preserve biological diversity. However, if every nation adopted the 1997 Swedish lifestyle the capacity of Earth would be exceeded threefold.

Ecological footprint calculations are almost certainly on the low side as accounting for the effect of every human activity on the environment is probably unachievable. Even so, such calculations invariably show the ecological footprints of most nations to be greatly overtaxing the environment's capacity to regenerate as is seen from the selection of nations in Fig. 1.3 [30]. Thus, in 1997 the average Canadian used 7.0 hectares of globally average productive land and 0.7 of a hectare of productive ocean to give a total of 7.7 hectares per person. This compared with an available biological capacity of 9.6 hectares per person to sustain the Canadian lifestyle. By the same calculation, an American required 10.3 hectares whereas there were only 6.7 hectares available per person in the United States that as a consequence had an ecological deficit of 3.6 hectares per person. In contrast, an Indian had a much smaller ecological footprint of 0.8 hectares per person and yet India could provide only 0.5 hectares for each of its people and had an ecological deficit of 0.3 hectares. This compared with an average footprint of 2.8 hectares per person for Earth's entire population in 1997, whereas only 2.0 hectares were available so that there was an ecological deficit of 0.8 hectares per person. Generally, the larger the ecological footprint per person the higher is the standard of living of the nation and its gross domestic product. Thus, Fig. 1.3 largely reflects the relative wealth of nations. Clearly, Earth is ecologically overstretched and as the developing nations seek to attain the living standards of the developed nations and population grows this overstretching will increase unless humanity becomes less demanding and far more efficient at using Earth's resources.

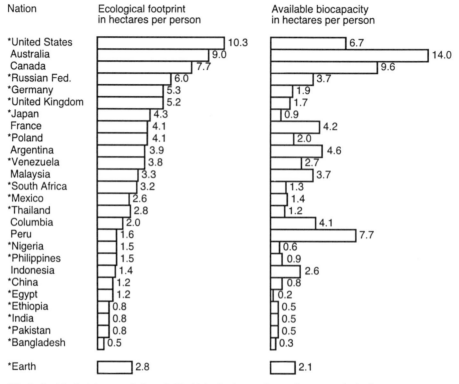

Nation	Ecological footprint in hectares per person	Available biocapacity in hectares per person
*United States	10.3	6.7
Australia	9.0	14.0
Canada	7.7	9.6
*Russian Fed.	6.0	3.7
*Germany	5.3	1.9
*United Kingdom	5.2	1.7
*Japan	4.3	0.9
France	4.1	4.2
*Poland	4.1	2.0
Argentina	3.9	4.6
*Venezuela	3.8	2.7
Malaysia	3.3	3.7
*South Africa	3.2	1.3
*Mexico	2.6	1.4
*Thailand	2.8	1.2
Columbia	2.0	4.1
Peru	1.6	7.7
*Nigeria	1.5	0.6
*Philippines	1.5	0.9
Indonesia	1.4	2.6
*China	1.2	0.8
*Egypt	1.2	0.2
*Ethiopia	0.8	0.5
*India	0.8	0.5
*Pakistan	0.8	0.5
*Bangladesh	0.5	0.3
*Earth	2.8	2.1

*Ecological footprint exceeded available biological capacity to give an ecological deficit at the 1997 standard of living.

Fig. 1.3. The ecological footprints and available biological capacity for a selection of nations. Data from [30].

When the 1997 ecological footprint per person is multiplied by a nation's population a national ecological footprint is obtained as shown in Fig. 1.4 for five of each of the more populous developed and developing nations that together used 46.2 percent of Earth's biological capacity. Thus, the United States and China used the largest portions and Japan used eight times the portion taken by Bangladesh despite having a similar population.

Another assessment of humanity's impact was gained from a 2004 estimate that about twenty percent of the amount of annual plant growth, and through it animal growth, on land and referred to as net primary production was consumed by humans as food and materials [31]. There was a great variation in this consumption from close to zero in sparsely inhabited regions to over 30,000 percent in large urban areas. On a regional basis, Africa consumed 12.40 percent, East Asia, 63.25 percent, South and Central Asia, 80.39 percent, Western Europe, 72.22 percent, North America, 23.69 percent and South and Central America, 6.09 percent. Average consumption per person in the developed nations was almost twice that of the developing nations where eighty-three percent of the population lived. Should the per person consumption of the latter nations rise to match those of the former, global consumption would rise to thirty-five percent of Earth's net primary production on land.

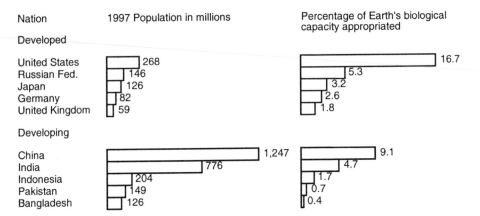

Fig. 1.4. The 1997 appropriations of Earth's biological capacity by ten nations [30].

1.5. Economics and the Environment

Despite humanity's dependency on the life supporting services that the environment performs, these services are generally undervalued because they have always been present and it is assumed that they will continue to be so. Briefly stated, these services include the regulation of climate, the recycling of the nutrients and materials that sustain living organisms, the recycling of fresh water, soil protection, crop pollination and pest control amongst others, and yet these are generally assumed to be "free". Although all of humanity's everyday and economic activities are dependent on these services, economic considerations seldom include them [12,32]. One estimate of the monetary value of these services is gained from the massive and somewhat controversial

Biosphere 2 experiment set up in 1991 in the Sonoran Desert of Arizona to support eight people for two years [33]. This large self-sustaining and closed environment contained some 4,000 plant and animal species in tropical rainforest, marsh, desert, cropland and oceanic ecosystems. Its source of power was sunlight and electricity that drove pumps, sensors, scrubbers and cooling systems to maintain habitability. It cost US$150 million to run for two years that averaged to US$9 million per person for a year. To provide similarly for a six billion human population at this rate would cost US$54 thousand trillion annually, or about 3,000 times Earth's 1997 gross domestic product, an impossibly high figure. Obviously this is an extreme example as such complete life support systems would only become necessary in the event of a total environmental breakdown, a distant possibility that fortunately there still appears to be time to avoid. Evidence that humanity is awakening to such a possibility may be drawn from the internationally enforced prohibitions of the use of the ozone depleting chlorofluorocarbons, or CFCs [34], and the increasingly stringent international sanctions practiced and proposed against global warming carbon dioxide emissions [35,36].

Until quite recently, economists tended not to overtly count the value of environmental assets among a nation's wealth [37]. As a result rivers, aquifers, forests, cropland, fisheries and other natural assets could be exploited beyond their regenerative ability without the consequential impoverishment appearing explicitly in national accounts. In the last decades of the twentieth century that saw the widespread adoption of market economics as the management model for national and international affairs, change arrived slowly with the first attempts to place monetary value on environmental assets. Such valuation is inevitably complex and may vary greatly depending on the perceptions and ambitions of the valuer. Nevertheless, a full valuation of environmental assets is essential if further impoverishment of Earth's biological capacity is to be avoided [38].

In 1997, a detailed attempt to cost the global value of services provided by the environment arrived at the huge value of US$33 trillion that compared with Earth's gross domestic product of US$18 trillion [39]. This assessment subdivided the environment into eight parts, or ecosystems, leaving others uncosted as shown in Fig. 1.5. Inevitably, this analysis has drawn considerable comment and some criticism [40]. Even so, it provides an important new element in the intensifying global argument between those proposing economic development and those opposing it on environmental grounds that often include aesthetically and ethically based assessments that are difficult to quantify in economic terms [18,41]. Nevertheless, it seems likely that the type of injudicious developments of the past, that when fully costed lost more for humanity than they gained, will be increasingly resisted both on environmental and economic grounds.

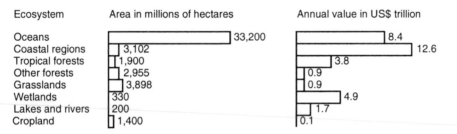

Ecosystem	Area in millions of hectares	Annual value in US$ trillion
Oceans	33,200	8.4
Coastal regions	3,102	12.6
Tropical forests	1,900	3.8
Other forests	2,955	0.9
Grasslands	3,898	0.9
Wetlands	330	4.9
Lakes and rivers	200	1.7
Cropland	1,400	0.1

The total annual economic value of the surface environment, excluding desert, tundra, urban, and ice and rock ecosystems is US$33.3 trillion.

Fig. 1.5. The estimated monetary value of Earth's environment. There is some uncertainty in the range of upper and lower limits in the estimations of the averaged value shown for each ecosystem. For the entire environment, this range was estimated to be US$16-54 trillion with an average value of US$33.3 trillion in 1997 [39].

While estimation of the monetary value of the services provided by the environment as shown in Fig. 1.5 is a major step forward, the practical realization of this worth requires a sophisticated approach. This was shown by the city of New York that draws much of its water from the Catskill Mountains [42]. Until the early 1990s water purification by microorganisms, root systems, and sedimentation in the catchment area cleansed water to the levels required for urban supply. However, gradually increasing levels of sewerage, fertilizers and pesticides in the catchment caused water quality to deteriorate to an unacceptable level. As a consequence, New York was faced with either building a purification plant at a cost of US$6-8 billion with annual running costs of US$300 million, or restoring the Catskill water catchment by buying land in and around it and subsidizing the improvement and building of local sewerage treatment farms at a cost of US$1-1.5 billion. The latter course of action was chosen with an expectation that major savings would accrue. Similar realizations of the economic and environmental advantages of biological water purification have resulted in either the restoration or construction of hundreds of wetlands for this purpose in Europe, North America and elsewhere [43].

As awareness of the environmental damage often caused by commerce and industry grows so does regulation to reduce such damage. In time, this will inevitably lead to companies either having to include in the cost of their products a component to repair the environmental damage the manufacture and use of their products might cause, or to pay taxes to repair such damage, or a combination of these environmental protection measures. This is already the case for acid rain-causing sulfur dioxide in developed nations where the acidity of rain has fallen as a result [44,45]. Accompanying such changes are likely to be accelerating changes in consumer choice made either on the economic grounds of the increasing cost of environmentally damaging products or on the basis of choosing to protect the environment.

1.6. Changing Earth

Most of humanity has seen wilderness cleared for agricultural use, forests felled for timber, suburbs encroaching onto farmland and other changes in the landscape. Frequently accompanied by a decrease in wildlife, these changes in the environment are the most immediately obvious ones that humanity has wrought as population and economic activity have increased. However, in the latter decades of the twentieth century and at the beginning of the twenty-first century, it is probable that the more subtly detectable depletion of the ozone layer and increasing levels of carbon dioxide in the atmosphere and their consequences have been the most publicly discussed of humanity's impacts on Earth. In particular, the issue of global warming has attracted great attention [35,36].

Evidence that Earth is warming has come from a range of sources as exemplified by underground temperatures measured from 616 boreholes on every continent except Antarctica [46]. This study shows that Earth has slowly warmed by about 1°C over the past 500 years with the greatest increase in the rate of warming occurring in the twentieth century. This coincides with the increased carbon dioxide content of the atmosphere, caused by fossil fuel burning, cement production and land clearing, retaining increasing amounts of

Earth's infrared radiation in an enhanced greenhouse effect. Other studies that deduce the increase in Earth's temperature by a range of methods show that the twentieth century was the warmest over the past 400 to 1,000 years with its last decade being particularly warm [47].

While the eleven year solar cycle causes variations in the amount of sunlight reaching Earth, as does the ash and smoke from volcanic eruptions, there appear to be no natural phenomena that can completely account for the rise in Earth's temperature during the twentieth century. As ice sheets and glaciers melt and water expands with rising temperature as the oceans warm, flooding of low lying land is inevitable together with climate changes due to the increased warmth of the atmosphere and the greater amount of water vapour that it retains [48,49]. These potential changes have greatly focused public and governmental attention on humanity's effects on Earth particularly among the sixty percent of humanity that lives in low lying coastal areas.

While the present level of public funding of environmental research and protection is unprecedented, concern about environmental issues is not a recent thing. Human induced environmental degradation began thousands of years ago as evidenced by some of the stark landscape of modern Greece where grazing, tree felling and arable farming destabilized soil on hillsides led to extensive erosion [50]. Aristotle commented on the conversion of the Argive marshes to agricultural use and fertile land around Mycenae becoming dry and barren [51]:

"In the time of the Trojan wars the Argive was marshy and could only support a small population, whereas the land of Mycenae was in good condition (and for this reason Mycenae was the superior). But now the opposite is the case ... the land of Mycenae has become ... dry and barren, while Argive land that was formerly barren has now become fruitful. Now the same process that has taken place in this small district must be supposed to be going on over whole countries and on a large scale."

The origins of the particularly strong concern about humanity's impact on Earth in the western developed nations at the end of the twentieth century can be traced back to the seventeenth and eighteenth centuries when European nations became alarmed at the environmental effects of their exploitation of newly colonized lands. As a consequence they sought to conserve large tracts of their new possessions by establishing forest and other reserves in places as far apart as Africa, Australia, India, Mauritius, North America, Southeast Asia and the West Indies [52].

Sometimes an individual can powerfully focus attention on an environmental issue. This was the case in 1962 when Rachel Carson's book, Silent Spring, drew attention to the disappearance of songbirds in many regions of the United States because of the use of DDT (dichlorodiphenyltrichloroethane) and other organochlorine insecticides [53]. These insecticides were intended to protect crops from insect pests but had the effect of destroying most insects, be they pest, spectator or benefactor, and thereby diminishing the food supplies for songbirds. Subsequently, it was found that the accumulation of pesticides in raptors at the top of the food chain, such as eagles, hawks and kestrels, caused the shells of their eggs to become thin and break in the nest with a consequent decline in successful breeding. While the benefits of the use of organochlorine insecticides were substantial in terms of crop protection and controlling mosquitoes, flies and lice that spread malaria, typhus and other diseases, their persistence in the environment inevitably caused them to enter the wildlife and human food chain. As a consequence their use in the developed world is now greatly restricted and their use by developing nations is much decreased.

Nevertheless, great environmental changes have occurred because of population increase and the concomitant demands on resources and the accompanying land clearing, forestry, grazing, urbanization, mining, trawling, dredging and related activities. The recycling of carbon, nitrogen, phosphorus and sulfur between air, water and soil, and plants and animals has been changed through fossil fuel burning and the use of agricultural fertilizers. And these activities have caused the loss of animals and plants through habitat changes, agriculture and fishing practices and the release of chemicals into the environment [19,53-55].

1.7. An Ecological Accounting

Almost three quarters of Earth's habitable land are either partially or totally dominated by humanity leaving a quarter available as wilderness according to a 1994 report as shown in Fig. 1.6 [56]. Of the total area of the continents some 162.1 million square kilometres, a little over half, is undisturbed by humans. But this includes uninhabitable rocky, desert and ice covered land that supports little life. Europe is the most human dominated habitable continent while Australasia and South America are the least dominated. Within this global picture enormous variability occurs among countries. This is exemplified by the contrast between the United Kingdom, where almost all land is habitable and less than one percent is undisturbed by humans, and Australia where only twenty percent of land is habitable and is human dominated, the remainder being either desert or arid scrub. However, even in the sparsely populated areas humanity has left its mark as is discussed below.

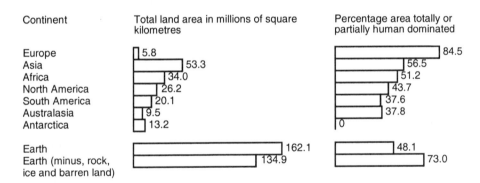

Fig. 1.6. The total area of the continents and the percentage areas under total or partially human domination in 1994. Data from [56].

The study of Earth and its animals and plants has been a human interest for thousands of years. However, as population has grown the regions of Earth unaffected by human activities have become vanishingly small and the opportunities to study a pristine part of nature have all but disappeared. This has induced a gradual change in attitude towards the concept of managing the planet, notably among ecologists. Richard Gallagher and Betsy Carpenter succinctly expressed this changing attitude in their introduction to a special report entitled

"Human-Dominated Ecosystems" in the 25 July 1997 issue of the journal Science [57] where they wrote:

> "Ecologists traditionally have sought to study pristine ecosystems to try to get at the workings of nature without the confounding influence of human activity. But that approach is collapsing in the wake of scientists' realization that there are no places left on Earth that don't fall under humanity's shadow. Furthermore, many scientists now believe that eventually all ecosystems will have to be managed to one extent or another, and, to do this well, managers will need sound advice."

Almost eight years earlier another journal, Scientific American, devoted an entire issue to the subject of "Managing Planet Earth" that examined the prospects for sustainable development of Earth's finite resources and fragile environment [58]. Thus, although Fig. 1.6 shows that 51.9 percent of Earth's land area is not directly disturbed by humanity, it is not insulated from the increasing levels of carbon dioxide, methane, oxides of nitrogen and sulfur, and a variety of other gases generated by human activities that are steadily changing the atmosphere. Nor is there protection from water and wind borne contaminants. Nevertheless, the relatively undisturbed areas of Earth provide refuge for much wildlife including seasonal migratory insects and birds that find themselves increasingly at risk as they either migrate to or traverse the human dominated land around them [59].

A more specific catalogue of the changes effected by humanity, largely since the beginning of the twentieth century, appears in Fig. 1.7. Humanity's impact on the atmosphere through increases in carbon dioxide and methane levels and decreases in stratospheric ozone levels changes has a global reach [60-62]. As population has grown the total area of land in agricultural use increased 466 percent from 1700 to 1980 and was accompanied by a clearing of forty percent of tropical forests [63]. While the rate of this increase slowed from 1960 onwards, food production increased dramatically through the introduction of high yielding crops and an intensification of use of pesticides and fertilizers in the "Green Revolution" [64]. The extent of industrial nitrogen fixation to produce nitrogenous fertilizers now exceeds the amount of nitrogen fixed biologically and by lightening.

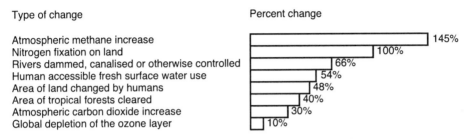

Type of change   Percent change

Atmospheric methane increase	145%
Nitrogen fixation on land	100%
Rivers dammed, canalised or otherwise controlled	66%
Human accessible fresh surface water use	54%
Area of land changed by humans	48%
Area of tropical forests cleared	40%
Atmospheric carbon dioxide increase	30%
Global depletion of the ozone layer	10%

Fig. 1.7. Human induced global changes in the environment. Based on [17,59-66].

Without the greatly increased use of nitrogenous fertilizers and the corresponding increase in the productivity of cropland, some 1.5 billion people would have insufficient food to live [65]. However, the drainage of excess fertilizers into rivers, lakes and oceans, frequently leads to excessive algal and plant growth that unbalances aquatic systems with deleterious consequences. Humanity now uses at least fifty-four percent of accessible fresh water [66] and in an effort to conserve this vital resource, to generate electricity and to prevent flooding, the flow of many of Earth's major rivers is controlled through dams, levees and similar constructions. As a consequence, some major rivers such as the Colorado, Ganges and Nile now deliver very little water to the ocean. While this selection of global changes is not comprehensive it does show that the domination of the environment by humanity is massive.

1.8. Biodiversity: How Many and How Much?

The great variety, or biodiversity, of life on Earth, exists at three levels. The first level is the ecosystem, or biome, such as coral reefs, rainforests, savannahs and deserts where the most biologically prolific and abundant are generally the most vulnerable to human impact. The second level contains all of the species such as bacteria, algae, insects, birds, fish, humans, shrubs and trees that make up the biome, and the third level consists of the diversity of genes in the genome of each species. Biodiversity has probably reached one of its highest levels ever, but unfortunately understanding and knowledge of biodiversity is inadequate at the very time when concern about humanity's impact on biodiversity is increasing.

While estimates of the number of living species are as high as 100 million of which 1.5 million or so have been identified, it might be thought that all of the large living species would be known by now, but even this is probably not so. A new species of large mammal was discovered on average every three years in the last half of the twentieth century and a new large marine animal every five years [67]. Occasionally, new trees were discovered as exemplified by the Wollemi pine, *Wollemia nobilis*, in Australia in 1994 [68]. Across the whole spectrum of life about 300 new species were found each day. Some of these newly found species exist under extreme conditions as exemplified by invertebrate species living in the superheated water of deep ocean hydrothermal vents [69] and the SLiMEs (subsurface lithoautotropic microbial ecosystems), communities of bacteria and fungi that live up to three kilometres beneath the surface in the pores of igneous rock [70].

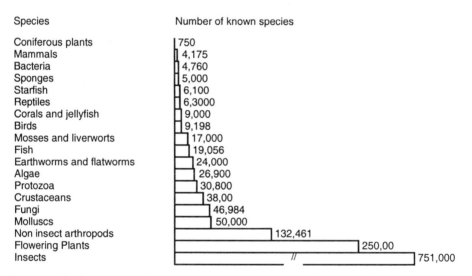

Species	Number of known species
Coniferous plants	750
Mammals	4,175
Bacteria	4,760
Sponges	5,000
Starfish	6,100
Reptiles	6,3000
Corals and jellyfish	9,000
Birds	9,198
Mosses and liverworts	17,000
Fish	19,056
Earthworms and flatworms	24,000
Algae	26,900
Protozoa	30,800
Crustaceans	38,00
Fungi	46,984
Molluscs	50,000
Non insect arthropods	132,461
Flowering Plants	250,00
Insects	751,000

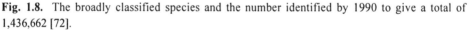

Fig. 1.8. The broadly classified species and the number identified by 1990 to give a total of 1,436,662 [72].

Generally, humanity is more aware of the larger animals and plants because of their visibility and usually expresses more concern about their conservation than for that of beetles, fungi and microorganisms. However, it is insects that make up one half of all known living species while the 4,000 or so mammals represent only a quarter of a percent as Fig. 1.8 shows [71,72]. Although the variety of animals greatly outweighs that of plants, it is plants that make up the vastly larger part of the mass of living tissue. It is also plants that through photosynthesis and recycling carbon dioxide, oxygen, nitrogen, phosphorus, sulfur and water, in collaboration with soil bacteria and fungi, integrate life into the colossal recyclings, or biogeochemical cycles, that support animal and plant life.

The size of this process is seen from the huge reservoir of carbon contained in plants and animals shown in Fig. 1.9 [73]. The rainforests are by far the most biologically productive biomes, which explains the growing concern accompanying their depletion. Roughly one tenth of the carbon in the biomes is recycled annually through photosynthesis and plant growth totals about sixty billion tonnes as seen in Fig. 1.10 [74]. This plant growth represents the

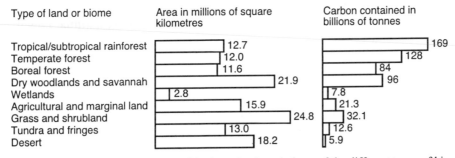

Fig. 1.9. The amounts of carbon stored in the animals and plants of the different types of biomes total 556.7 billion tonnes. The amount of carbon stored in kilograms per square metre ranges from 13.5 through 10.5, 7.2, 4.4, 2.7, 1.4, 1.4, 0.97 to 0.32 as the figure is descended from the prolific life of the rainforests to the sparseness of the deserts. Data from [73].

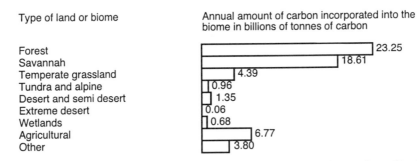

Fig. 1.10. The annual amount of carbon incorporated into living tissue in a variety of land types through photosynthesis and plant growth totals 59.86 billion tonnes according to [74].

difference between the amount of carbon assimilated through photosynthesis and the amount released through respiration and is approximately matched by the amount of carbon released through decay as discussed in section 8.4. A single rainforest tree pumps some 10,000 tonnes, or cubic metres, of water into the atmosphere over a hundred year lifespan [18]. While the huge magnitudes of the quantities listed in Figs. 1.9 and 1.10 are very impressive, so is humanity's capacity to greatly alter a biome and thereby its vital recycling role.

1.9. The Sixth Extinction: The Loss of Biomes

Earth is presently undergoing a major loss of biodiversity that is often referred to as a sixth extinction by analogy to the five great species extinctions preceding it [19,75]. This loss began thousands of years ago as humans became sufficiently organized to dominate and change their environment. With the growth of this organization the loss of biodiversity accelerated to reach an extraordinary level at the beginning of the twenty-first century. Humanity is now so dominant that the destruction of small biomes such as wetlands through drainage and mangrove swamps by dredging is commonplace. However, some of the biomes now threatened are so large that their destruction may cause substantial local, and possibly global, climate change as exemplified by the huge Indonesian rainforest fires of 1997.

In August to December 1997, a great smoke cloud spread over much of Southeast Asia presaging the greatest ever air pollution disaster caused by humanity as shown in Fig. 1.11. This arose through a conjunction of the need to accommodate and feed an Indonesian population growing by three million people a year, a desire to increase gross domestic product, a prolonged drought resulting from the El Niño climatic phenomenon and the burning of tens of thousands of square kilometres of rainforest [76]. The resulting smoke cloud reduced visibility to 100 metres and less, enveloped most of Indonesia, all of Malaysia, Singapore and Brunei and the southern tip of Thailand from mid August to December 1997, and caused at least 100,000 people to become ill. The loss of wildlife was colossal.

This disaster was preceded by largely avoidable and extensive forest fires in 1983, 1991 and 1994 in Indonesia when El Niño induced droughts occurred. The imminence of an El Niño event, and the consequent onset of drought in

Indonesia, was forecast from observations of variations in ocean temperatures across the Pacific early in 1997. Yet, despite the obvious danger, fires were

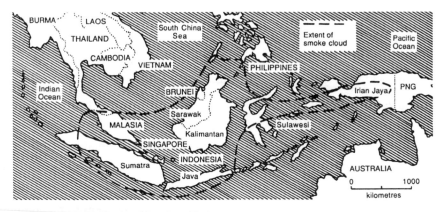

Fig. 1.11. The burning rainforests of Irian Jaya, Java, Kalimantan, Sulawesi and Sumatra generated a huge smoke cloud that thickly cloaked most of Indonesia, all of Malaysia and Brunei and the southern tip of Thailand from mid August to December 1997.

deliberately lit by forestry and plantation companies to clear vast tracts of land through burning to make way for rice paddies and oil palm plantations after they had cleared the valuable timber from the forests. Slash and burn peasant farmers added greatly to the fires that rapidly spread out of control in the drought conditions of 1997. The combination of the cavalier attitudes of the 175 Indonesian and Malaysian forestry companies mainly responsible for the fires with the worst drought in Indonesia for fifty years brought on the predictable disaster [77]. In October 1997, and only after great pressure from governments of nearby nations and wider international pressure, the Indonesian government belatedly withdrew the licenses of twenty-nine forestry and plantation companies who had flouted the weakly enforced environmental laws. It is estimated that in six months the burning forests and peat bogs released one billion tonnes of carbon, or 3.7 billion tonnes of carbon dioxide, into the atmosphere. This compared with the 3.3 billion tonnes of carbon dioxide released by Western Europe annually through fossil fuel burning.

While the Indonesian fires were burning, the American NOAA-12 satellite detected more than 24,000 fires burning in the Amazonian rainforests of Brazil in August and September, some of which spread into Columbia. It is estimated that the 1997 rate of global deforestation was at least 20,000 square kilometres a year, the highest rate since 1988 that was preceded by accelerating and massive deforestation in Amazonia from 1970 onwards. And the burning and deforestation are continuing [78-80]. By 1983, of the 24,500,000 square kilometres of tropical rainforests and seasonal forests that once existed only some 10,000,000 square kilometres remained [81]. About 8,000,000 square kilometres had been converted to agricultural use, 3,000,000 square kilometres to slash and burn agriculture and 3,500,000 square kilometres to pasture. Apart from the loss of the huge forest transpiration that pumps colossal amounts of water into the atmosphere and thereby greatly influences climate and rainfall, it appears that the smoke generated by forest fires causes clouds to form smaller water droplets that are too small to fall as rain and so reduce rainfall in smoke affected areas [82]. This may have contributed to the decrease in tropical rainfall over the last one hundred years.

1.10. The Sixth Extinction: The Loss of Species

The common saying "Dead as a dodo" is a reflection of humanity's acceptance of responsibility for the extinction of other inhabitants of Earth such as the dodo, a flightless bird last seen in Mauritius in the 1670s [52]. Other well-known extinctions are those of the great auk (1844), Steller's sea cow (1768), the sea mink (1880) and the West Indian monk seal (1952) [83]. These are but a few examples of recent species losses whose origins may be traced back thousands of years as humans spread across the continents taking with them fire, exotic animals and plants that supported their hunting and developing agricultural skills and forced other life into retreat [84]. In the Americas fifty-seven large animal species, including mammoths, mastodons, elephants, giant sloths, lions, sabre toothed tigers and glyptodonts, disappeared between ten and twelve thousand years ago and similar losses occurred in Africa, Asia and Europe. In New Zealand half of all the giant bird and insect species were extinct a few hundred years after the Maoris' arrival. In Australia fifty giant marsupials, including giant kangaroos, and animals resembling tapirs, ground sloths, lions, dogs and rhinoceros, were seemingly destroyed by the aborigines during their 50,000 years occupation of the island continent. Globally, about twenty percent of all bird species have disappeared in an extinction that started thousands of years ago as humans hunted the large and easily caught flightless birds. Since 1800, of about 9,600 bird species, seventy-five have become extinct and 1,100 face extinction [85]. Of about 4,300 mammals, sixty have become extinct and 650 face extinction, and of about 4,700 reptiles, twenty have become extinct and 210 face extinction.

These extinctions are symptomatic of the constant drive to increase food supply depleting animal populations either through hunting or habitat destruction. Slowly, humanity is becoming aware of the self-destructive aspects of this process as shown by the rising concern about the relatively recent depletion of fisheries that at one time were thought to be inexhaustible [54]. Although providing about fifteen percent of animal protein for human consumption, few fisheries are now operating at a sustainable level. Directly connected to some of this decline is the global destruction of fifty percent of mangrove swamps that act as fish nurseries and provide coast protection. The loss of the latter capacity is held responsible for increased flooding and loss of life during coastal cyclones [86]. Land clearance and logging have put ten percent of trees, some 8,750 species, at risk of extinction that compares with eleven percent of bird species at risk of extinction, a coincidence that suggests a strong link between loss of habitat and loss of wildlife [87]. There are some bright spots where conservation has saved species from extinction, usually involving large animals. This is exemplified by the biggest of all species, the whale that had become severely depleted by commercial hunting [88]. Starting in the 1970s, most nations banned whale hunting to conserve the species and, although some types of whale may have become extinct, the populations of several members of the whale family have increased to the point where their future appears assured. Some land animals have also been brought back from the verge of extinction, as in the case of the California condor [89], but generally such successes are having only a small effect on the general tide of extinction.

The total recorded loss of species since 1600 is 724, with sixty being lost between 1900 and 1950. This compares with an estimated natural loss rate of one animal and one plant species every 100 to 1000 years. Both the recorded and estimated natural rates of loss of species are small when compared with current extinction rates estimated to be several thousands annually [19]. The reason for this is that the method of estimating this very high species loss rate is not based on definite identification of species lost, most of them small and unnoticed by the casual onlooker, but instead is calculated on the lost of habitat as wilderness is cleared. The tropical rainforests are thought to contain at least sixty percent of Earth's animal and plant species [77,90]. As a consequence concern about the loss of biodiversity heightened with the great acceleration in the clearing of these forests to the point where they covered less than half of their original area by the end of the twentieth century. It is estimated that if the rainforests are reduced to less than ten percent of their original area, as may occur, some 65,000 species, animal and plant, large and small, will be lost [91]. Comparisons of such estimates with the disappearance of species in the fossil record suggest that extinction rates at the end of the twentieth century could be at least 10,000 times greater than the natural loss rate [92]. The fossil record suggests that after past great extinctions it has taken up to ten million years for biodiversity to fully recover, and the same may well be the case for the current human induced extinction quite apart from its impact on future evolutionary patterns [93-95].

Apart from the uncertainty about the number of species that exist, there is also considerable uncertainty about how many of a particular species exists. In some cases these numbers are very small as exemplified by the single known examples of the trees *Diospyros angulata* (Mauritius), *Carpinus putoensis* (China) and *Holmskioldia gigas* (Tanzania) and the three or four examples of *Hibiscus clayi* (Hawaii), *Scalesia atracyloides* (Galapagos Islands) and *Ilex khasiana* (India) [87]. Many large mammals are at risk of extinction because their populations have been greatly reduced and are subject to increasing human competition for living space. This is typified by the three hundred mountain gorillas living in the 330 square kilometre Bwindi Impenetrable Forest rainforest reserve in Uganda that represent half of the remaining population of mountain gorillas [96]. Elsewhere in Africa, the great ape populations are being decimated by hunting and Ebola infection [97]. Another example comes from the ocean where leatherback turtles face extinction as a consequence of being inadvertently caught during commercial fishing [98]. Thus, while 1,367 leatherback females nested at one of their nesting beaches, Playa Grande in Costa Rica, in 1988-89, this number fell steadily until only 117 nested in 1998-1999, and a further fall to fifty was predicted for 2003-2004. In complete contrast, twelve million roundworms, 46,000 small earthworms and their relatives, and a similar number of insects and mites have been found to live under a square metre of Danish pasture [18]. A gram of fertile agricultural soil is capable of supporting over 2.5 million bacteria, 400,000 fungi, 50,000 algae and 30,000 protozoa.

Twenty-five areas, covering twelve percent of Earth's landmasses and spread over every continent except Antarctica, are impressively rich in plant and animal species and have been termed biodiversity hotspots [99]. In 1995, more than 1.1 billion people lived in these areas with an average population density of seventy-three people per square kilometre that compared with an average global population density of forty-two people per square kilometre when ice and rock covered land was excluded. As population grows, so pressure increases on the animals and plants of these biodiversity hotspots, as is the case in the other inhabited areas. It was for this reason that in 1992, at the United Nations Conference on Environment and Development in Rio de Janeiro, over a hundred nations signed the Convention on Biological Diversity to set aside areas to conserve biodiversity [100].

It may be argued that as intensive agriculture, aquaculture and forestry concentrate on those animals and plants that provide for humanity's needs, the loss of other species is inevitable and need not affect humanity's future. However, such an argument is untenable as it neglects the underpinning of the food supply. This is the breeding of extremely productive crops and domestic animals from wild species in parallel with technology that frees the bulk of the population of the developed nations from food production to pursue the myriad other activities that characterize sophisticated societies. At present sixty percent of humanity's food is directly derived from just three crops, maize, rice and wheat that are produced in a massive monocultural agricultural system and were first bred from wild species some 10,000 years ago. The 1997 global wheat crop covered an area of 250 million hectares with at least 500 trillion individual plants [93]. However, during a long breeding program to produce the high yielding modern strains that coincided with the destruction of much of the natural environment with increase in population, many of the wild progenitors of agricultural wheat and other wheat strains have been lost [101].

The narrowness of the genetic origins of major food crops is shown by almost all of the hard red winter wheat varieties grown in the United States originating from only two varieties from Poland and Russia [101]. Almost all soybeans grown in the United States originated from a dozen strains found in a small area of northeast China [102]. Such genetic narrowness places humanity in a vulnerable situation as monocultures are particularly susceptible to the spread of disease because of the closeness of identical plants over extensive areas. This is not a speculative assertion as was shown by the great potato famine in Ireland in the late nineteenth century that arose because of the spread of potato blight from field to field to produce widespread starvation and destitution [103]. It was a number of such disasters that led to the setting up of seed banks to preserve genetic diversity in agriculturally important crops and their progenitors [102].

Unfortunately, seed banks, reserves and similar stratagems cannot protect the vast majority of plant species from which great benefit for humanity flows and from which yet to be discovered benefit may flow. Thus, digitalis, morphine and quinine are directly derived from large plants and the first antibiotic, penicillin, was extracted from a mould [18]. Cyclosporin, now widely used to suppress the immune system and prevent organ transplant rejection, is derived from a similar source, as is gliotoxin that promises applications similar to those of cyclosporin. Humanity has to become reconciled with a dependence on biodiversity for the Earth's habitability and incorporate its protection into everyday agricultural and other activities.

Almost perversely, as this largely unwitting experiment with Earth's biodiversity has been occurring, a deep understanding of biodiversity at the molecular level developed during the twentieth century that appears set to have a substantial effect on humanity, and thereby Earth, in diverse ways in the twenty-first century. This understanding is based on the double helical molecule deoxyribonucleic acid, DNA, that contains the genome of each living organism and its genetic code. As other species and their genomes are disappearing, humanity celebrated the publication of the near complete sequence of the human genome in 2004 that promises to lead to the prevention and cure of previously intractable diseases and increases in the human lifespan [104]. Similarly a growing understanding of the genomics of plants and animals has already resulted in genetically engineered new plants and animals, the production of which, while the subject of much debate, may result in considerable benefit to humanity [105]. This "new biology" is explored in more detail in Chapter 5.

1.11. The Human Condition

Great as is humanity's dominance of Earth, and impressive as were the advances in scientific endeavour and the benefits arising from them during the twentieth century, the beginning of the twenty-first century finds the major part of humanity still preoccupied with gaining sustenance and shelter and the avoidance of disease and death. This is despite life expectancy in the developed nations becoming unprecedently long. Globally a million or so people die each week while about 2.5 million are born, many of whom soon join the ranks of the dying through war, famine and disease, especially in sub-Saharan Africa.

Humans have a unique ability to threaten their own survival through systematic war. The twentieth century saw the most technologically advanced nations engage in the First and Second World Wars of 1914-1918 and 1939-1945 that were hugely destructive of life and economically devastating. During the Cold War of the latter part of the twentieth century, the possibility of global nuclear war posed the greatest ever threat to humanity's tenancy of Earth. At its height every human was threatened with a nuclear equivalent of three tonnes of the high explosive trinitrotoluene, TNT [106]. It is probable that the fear of the awesome destructive power of nuclear weaponry acted as the dominant deterrent to a major conflict between the massively armed Cold War opponents. Fortunately, the threat of global nuclear war has receded and nuclear armouries are now much reduced. Nevertheless, conventional wars, large and small, continue to be waged as a testimony to human aggressiveness.

Devastating as war is, famine and disease claimed far more lives during the twentieth century and are likely to do so in the twenty-first century. The major diseases leading to death globally are diseases that infrequently lead to death in the developed world. The exception is the new disease, autoimmune deficiency syndrome, AIDS, that is associated with the human immune deficiency virus, HIV, that lowers the body's ability to resist disease. The 1996 World Health Report of the World Health Organization listed the ten diseases with the biggest death toll in 1995 as shown in Fig. 1.12 [107]. The vast majority of the seventeen million deaths from these diseases occurred among the children of developing nations and compared with a total global death toll of fifty-two million.

Disease   Number of deaths in millions in 1995

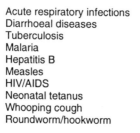
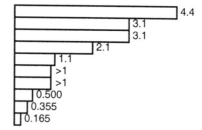

Acute respiratory infections   4.4
Diarrhoeal diseases   3.1
Tuberculosis   3.1
Malaria   2.1
Hepatitis B   1.1
Measles   >1
HIV/AIDS   >1
Neonatal tetanus   0.500
Whooping cough   0.355
Roundworm/hookworm   0.165

Fig. 1.12. The ten biggest killer infectious diseases in 1995 according to the World Health Organization [107].

By 1998, HIV/AIDS had become the major cause of death in sub-Saharan Africa with an annual death toll of 2.2 million that compared with 200,000 lives lost in all of the African conflicts in that year, and 23.3 million people were infected that compared with a global total of 33.4 million [108]. So devastating is the impact of HIV/AIDS on economic, social and political stability in sub-Saharan Africa that in January 2000, the United Nations Security Council declared the pandemic a threat to world stability. However, by the end of 2004 the global total of HIV/AIDS sufferers had risen to 39.4 million of which 25.4 million were in sub-Saharan Africa [109]. Some 3.1 million people died of HIV/AIDS in 2004 and there were 4.9 million new infections of which 3.1 million were in sub-Saharan Africa, 890,000 in South and Southeast Asia, 290,000 in East Asia, 240,000 in South and Central America, and 210,000 in Eastern Europe and Central Asia. Given the large populations of these regions, they could soon rival sub-Saharan Africa in the total number of infections, the global peak of which appears to be years away.

1.12. Natural Disasters

Earth is a restless planet and every so often natural occurrences such as extreme weather events or movement of tectonic plates either directly or indirectly result in disaster. So great can be the loss of life and economic activity from such disasters that the last decade of the twentieth century was declared the Decade for Natural Disaster Reduction by the United Nations [110]. The hazards specifically considered were earthquakes, windstorms, tsunamis, floods, landslides, volcanic eruptions, wildfires, grasshopper and locust infestations, and drought and desertification. To gain an idea of the magnitude that such disasters can assume several examples are now briefly discussed.

On 15 June 1991, the six centuries dormant Philippines volcano, Pinatubo, erupted spectacularly [111]. This, the largest volcanic eruption of the twentieth century, is variously estimated to have ejected four to eight cubic kilometres of dust and ash and twenty million tonnes of sulfur dioxide into the atmosphere, much of it entering the stratosphere. The immediate effect was to smother a huge tract of the surrounding countryside in metres thick layers of dust and ash that collapsed buildings, choked water courses and displaced hundreds of thousands of people from farms and villages. As the eruption rumbled on, much of the dust and sulfur dioxide was washed out of the atmosphere by rain. However, in the stratosphere, far above the rainstorms, the fine dust, together with a sulfuric acid aerosol arising from reaction of sulfur dioxide with atmospheric moisture, formed a girdle around Earth above the equator that reflected sunlight back into space to cause a slight global cooling. While the effect of the dust decreased rapidly as it fell back to the surface, the cooling effect of the sulfuric acid aerosol lessened more gradually over three or so years as it slowly re-entered the lower atmosphere. On the ground in the immediate vicinity of Pinatubo, that was still rumbling in 1995, great viscous streams of magma, or lahars, continued to flow from the caldera. Although the loss of life was small, the loss of livelihood was enormous. Unlike the situation with Pinatubo, often many people die in volcanic eruptions and it is thought that 274,443 people died in a variety of ways as a consequence of 304 volcanic eruptions during the sixteenth to twentieth centuries [112]

Early in the morning of 17 January 1995, an earthquake measuring 6.9-7.2 on the Richter scale struck the Japanese harbour city of Kobe killing over 6,000 people and rendering 310,000 homeless as it destroyed freeways and water, gas and electricity supplies [113]. This earthquake was Japan's second most devastating after the 1923 Tokyo earthquake that killed 142,000 people. Japan continuously experiences earth tremors because it lies close to the junction of the Philippine and Eurasian tectonic plates that grind against each other along with the jostling of all of the other tectonic plates that make up Earth's surface [1]. Every so often this produces a major earthquake to which cities at the edge of tectonic plates are particularly vulnerable. Thus, on the eastern side of the Pacific

Ocean, Los Angeles and San Francisco frequently experience tremors as the Pacific plate grinds against the North American plate along the San Andreas fault [114]. Similar movements occur elsewhere, as exemplified by that along the North Anatolian fault line that resulted in a 7.4 Richter scale earthquake and the loss of more than 15,000 lives in the Turkish city of Izmit on the morning of 17 August 1999 [115]. This was followed by the 8 Richter scale Bhuj-Anjar-Bhachau earthquake in western India that took 30,000 lives on the morning of 26 January 2001, and another earthquake at Bam in Iran took 50,000 lives on the morning of 26 December 2003 [116,117].

Earthquakes are hazards whose danger can be lessened by building to withstand their force, recognizing of early warning signs and having well prepared emergency services. Even so, the huge waves, or tsunamis, caused by earthquakes beneath the ocean can strike with little warning, as was the case in Papua New Guinea in the evening 17 July 1998 [118]. There, a 7.1 Richter scale earthquake thirty kilometres offshore produced a series of waves in rapid succession, some up to fifteen metres high. These completely overwhelmed the low lying coastal villages of Sissano, Arop, Warapu and others taking more than 2,200 lives and obliterating not only houses, but also stripping most vegetation from the coastline. Tsunamis are much more powerful than the surface waves generated by storms because they reach to the ocean floor and carry huge volumes of water with them on fronts of hundreds of kilometres at up to 800 kilometres an hour over thousands of kilometres. Although tsunamis may be scarcely noticed in the deep ocean, their effect as they pile up on a shoreline is devastating.

Most tsunamis occur in the Pacific Ocean, ten of which took more than 4,000 lives from 1990 to 2000. Tragically, this familiarity with tsunamis proved no preparation for the huge tsunami that killed some 300,000 people living in Indian Ocean coastal areas on 26 December 2004 [119]. Jostling between the Indian and Burma tectonic plates caused a 9.3 Richter scale earthquake that caused some parts of the seafloor to drop by 2.5 metres and other parts to rise five metres over an area of 25,000 square kilometres 200 kilometres west of Sumatra. This generated a fast moving half metre high wave 100 kilometres wide that piled up on the coast of the Indonesian province of Aceh to produce fifteen metre high waves that took at least 230,000 lives with further heavy loss of life in India, Sri

Lanka, Thailand, Myanmar (Burma) and the Maldives. So massive was the tsunami that it killed many on the eastern side of the Indian Ocean in Somalia and other East African nations. In addition to the huge loss of life and property damage of many US$ billion, complete ecosystems were stripped away in low lying areas such that restoration and re-establishment of livelihoods will take decades to achieve. If an early warning ocean floor tsunami sensor system had been in place in the Indian Ocean, similar to that in the Pacific Ocean, it is probable that the loss of life would have been much less. Tragically, the tsunami was followed on 28 March 2005 by a magnitude 8.7 Richter scale earthquake on the seabed close to the December earthquake and a further 2,000 lives were lost on the island of Nias [120].

On 29 October 1998, a weakening hurricane, Mitch, swept in from the Caribbean to bring high winds and torrential rain to most of Honduras and large parts of Nicaragua, El Salvador and Guatemala [121]. Entire villages were obliterated by surging streams, rivers of mud and landslides to leave up to 24,000 people dead or missing and causing damage in excess of US$5 billion. Over 600 millimetres of rain fell in the Honduran mountains in a single day. This, combined with the extensive land clearance for cattle grazing that accelerated water runoff and destabilized the soil, made major flooding, huge soil erosion and landslides inevitable. In the wake of Mitch, the sodden countryside was ideal for the proliferation of water borne diseases such as cholera of which there were more than 30,000 cases. Diseases transmitted by mosquitoes, that thrive under such waterlogged conditions, resulted in more than 30,000 cases of malaria and more than a thousand cases of dengue fever. This is a typical pattern of extreme weather and land clearing combining to produce disasters large and small. Extensive clearance of rainforest by farmers and loggers in Brazil, Indonesia and Thailand, areas also subject to torrential rain, can be expected to cause losses of life and homes similar to those experienced in Central America.

A very large extreme weather death toll arose when a cyclone struck Bangladesh on the night of 29 April 1991 [122]. Winds in excess of 225 kilometres per hour whipped up seven metre waves in the Bay of Bengal and spread devastation and flooding in coastal regions to kill more than 130,000 and injure over 450,000. More than 850,000 houses were destroyed, 440,000 cattle drowned and 63,000 hectares of crops were lost.

1.13. Unnatural Disasters

Humanity is quite capable of generating its own disasters that show signs of becoming bigger and more deadly as the scale of technology grows. The worst industrial accident to date occurred in India just after midnight on 3 December 1984 when forty tonnes of highly toxic methyl isocyanate escaped from the Union Carbide pesticide plant in Bhopal [123]. The vaporized chemical enveloped a wide area killing over 2,800 people and injuring some 200,000 others, many seriously and many permanently. This provides an example of multiple aspects of human aspirations leading to unforeseen and tragic consequences. Here the manufacture of pesticides to protect crops to feed a burgeoning population of a developing nation seeking to industrialize, perhaps without adequate safeguards to control a major chemical plant operated by a large multinational company seeking to increase its global market share, resulted in a chain of events leading to disaster.

However, developed nations are not immune to such disasters that may present themselves in a variety of forms. Although seldom costly in human life, an all too frequent accident is the spillage of huge quantities of oil from stricken tankers into the ocean and usually onto coastlines with consequent great loss of wildlife and the destruction of fisheries. One of the biggest such spills was that from the super tanker Exxon Valdez that ran aground in Prince William Sound in Alaska on 24 March 1989 spilling forty-two thousand cubic metres of crude oil into the ocean and onto the surrounding coastline [124]. While no human lives were lost through the spill, the large loss of fish, birds, marine mammals and plankton constituted an extensive and long lasting damaging effect on the ecology of Prince William Sound and the rich fishing industry that it supported.

In the early morning of 26 April 1986, the worst ever nuclear accident occurred at Chernobyl in Ukraine [125]. The inappropriate operation of an obsolete RBKN nuclear reactor resulted in a massive explosion, reactor core meltdown and the spread of a large cloud of radioactive fission products over most of Europe and Scandinavia within a few days. Within a few weeks, radioactive fallout was detected as far afield as Japan and the United States. The release of radioactivity was 200 times greater than that produced by the nuclear weapons dropped on Hiroshima and Nagasaki combined [106].

Although the immediate death toll from the explosion was small, several thousand people died from radiation sickness and related diseases in the ten years following the explosion according to some estimates. Many of these deaths occurred among the large numbers of people drafted into the reactor site and environs in an attempt to stop the release of huge amounts of radioactive material that belched from the crippled reactor for ten days and to collect and bury radioactive debris. More deaths also occurred among the populations of Ukraine, Belarus and Russia closest to Chernobyl where millions were exposed to fallout. A thirty kilometre radius zone around Chernobyl became so radioactive that the 135,000 people living in it were evacuated immediately and are unlikely to be allowed to return in the foreseeable future. Since then many more have been evacuated from other contaminated areas to bring the total to more than 200,000 evacuees.

Twenty major radioactive fission and decay products were in the vast radioactive cloud emitted by the reactor. Among them were iodine-131 with a half-life of eight days, cesium-134 and cesium-137 with half-lives of two and thirty years, respectively, strontium-90 with a half-life of thirty years, and several isotopes of plutonium with half-lives varying from thirteen to 24,3600 years. Apart from the deaths from radiation sickness, the most obvious health effect of this contamination was that the incidence of thyroid cancer in children in Gomel in Belarus became much greater than normal. This was caused by radiation damage from iodine-131 that accumulates in the thyroid gland. An increased rate of sperm and foetal abnormalities in the Chernobyl region was also attributed to the accident.

In the immediate aftermath of the accident, contaminated agricultural produce in Austria, Belarus, Finland, Georgia, Germany, Hungary, Norway, Poland, Russia, Sweden, Ukraine and the United Kingdom had to be destroyed, and restrictions on agricultural produce still applied in 2000 even in some of the more distant countries. Water from the Chernobyl area drains into the Pripyat, a tributary of the Dnieper that flows into the Black Sea, and spreads the longer lived radioactive contamination further afield.

These three accidents stemmed largely from humanity seeking to supply its basic needs of sustenance and energy and to increase living standards. This activity is unlikely to diminish in the twenty-first century and, given that very often such activities are carried out at the minimum safeguard level thought necessary, it seems inevitable that similar accidents will occur in the future. The effects of natural and unnatural disasters are placed in perspective in Fig. 1.13 that shows the size of areas affected and the time for the environment to recover [126,127]. At first sight it seems a little odd that intensive agriculture should be identified as an unnatural disaster. However, such agriculture can lead to loss of soil fertility and soil loss through erosion, both of which result in reduced crop yield and sometimes in crop failure, quite apart from loss of biodiversity. Often intensive agriculture leads to further problems in the form of soil salinization through excessive irrigation, and eutrophication of inland and coastal waters as a result of fertilizer runoff. The magnitude of the nuclear accident shown in Fig. 1.13 is based on Chernobyl. The asteroid and comet impact magnitude shows an envelope that incorporates the Tunguska impact of 1804 and larger impacts that humanity would survive, but does not extend to the magnitude of the Yucàtan impact of sixty-five million years ago that humanity would survive with great difficulty [10,11].

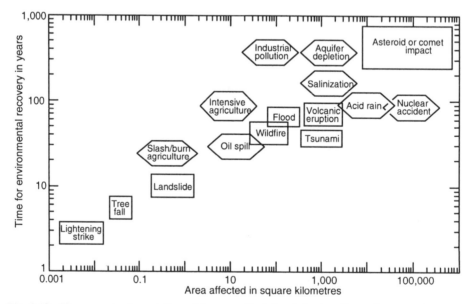

Fig. 1.13. The approximate variation of the area affected and the time for environmental recovery characterizing a selection of natural and human caused environmental impacts represented by rectangles and lozenges, respectively. Constructed from Chernobyl data and those in references [10,11,126].

1.14. Challenged Earth

While Earth could be said to be challenged by humanity to continue to provide the underpinnings of life, the reality is that it is humanity that faces many challenges to maintain Earth as a beautiful and hospitable planet for the generations to come. While inexorable geological and climate changes will occur as Earth continues along its planetary evolution most such changes are likely to be slow on the human timescale and therefore fairly readily adjusted to as long as the Holocene epoch persists. The challenge is to ensure that events are not precipitated that damage Earth's habitability to the point where humanity's existence becomes marginal, possibly to the point of extinction. Should this occur, Earth will still appear as a magnificent blue planet, but something extraordinary will have been either much diminished or lost, humanity, the highest known form of life, and with it much of its companion animal and plant life. Fortunately, it is probable that humanity is sufficiently intelligent and resilient to avoid such a catastrophe. Even so, some of the challenges are massive as exemplified by those posed by the sheer size of the human population and its growth as discussed in Chapter 2.

References

1. C. J. Allègre and S. H. Schneider, The evolution of Earth, *Sci. Am.*, **1994**, *271*, Oct., 44.
2. S. Bowler, Formation of Earth, *New Sci.*, *Inside Science 96*, **1996**, Dec. 14, 1.
3. a) L. E. Orgel, The origin of life on Earth, *Sci. Am.*, **1994**, *271*, Oct., 53. b) S. J. Gould, The evolution of life on Earth, *ibid.*, 63. c) M. J. Benton, Diversification and extinction in the history of life, *Science*, **1995**, *268*, 52.
4. J. R. P. Angel and N. J. Woolf, Searching for life on other planets, *Sci. Am.*, **1996**, *274*, April, 46.
5. a) V. A. Szalai and G. W. Brudvig, How plants produce dioxygen, *Am. Sci.*, **1998**, *86*, 542. b) J. Copley, The story of O, *Nature*, **2001**, *410*, 862. c) G. C. Dismukes, Splitting water, *Science*, **2001**, *292*, 447. d) J. F. Kasting, The rise of atmospheric oxygen, *Science*, **2001**, *293*, 819. e) J. F. Kasting and J. L. Siefert, Life and the evolution of Earth's atmosphere, *Science*, **2002**, *296*, 1066.
6. a) C. Sagan, The search for extraterrestrial life, *Sci. Am.*, **1994**, *271*, Oct., 71. b) J. C. Tarter and C. F. Chyba, Is there life elsewhere in the universe? *Sci. Am.*, **1999**, *281*, Dec., 80. c) I. Crawford, Where are they? *Sci. Am.*, **2000**, *283*, July, 29. d) J. L. Bada, State-of-the-art instruments for detecting extraterrestrial life, *Proc. Natl. Acad. Sci. USA*, **2001**, *98*, 797. e) C. F. Chyba and C. B. Phillips, Possible ecosystems and the search for life on Europa, *ibid.*, 801. f) N. R. Pace, The universal nature of biochemistry, *ibid.*, 805. g) J. I. Lunine, The

occurrence of Jovian planets and the habitability of planetary systems, *ibid.*, 809. h) M. Rees, Our greatest quest, *New Sci.*, **2003**, 12 July, 24.

7.   a) N. Henbest, The dark side of the Earth, *New Sci.*, **1989**, 8 April, 42. b) C. N. H. Doll, J.-P. Muller and C. D. Elvidge, Night-time imagery as a tool for global mapping of socioeconomic parameters and greenhouse gas emissions, *Ambio*, **2000**, *29*, 157. c) M. Murphy, Revealing the dark side of light, *Chem. Indust.*, **2000**, 20 Oct., 627.

8.   C. Haub, How many people have ever lived on Earth? *Popn. Today,* **1995**, Feb., 4.

9.   a) D. H. Erwin, The mother of mass extinctions, *Sci. Am.*, **1996**, *263*, July, 56. b) G. Vines, Mass extinctions, *New Sci.*, *Inside Science 126*, **1999**, 11 Dec., 1. c) R. A. Kerr, Whiff of gas points to mass extinction, *Science*, **2001**, *291*, 1469. d) L. Becker *et al.*, Impact event at the Permian-Triassic boundary: Evidence from extraterrestrial noble gases in fullerenes, *ibid.*, 1530. e) R. Dagani, Buckyballs mark mass extinction, *Chem. Eng. News*, **2001**, *79*, 26 Feb., 9. f) S. Simpson, Deeper impact, *Sci. Am.*, **2001**, *284*, May, 13.

10.   a) C. C. Swisher *et al.*, Coeval ^{40}Ar/^{39}Ar Ages of 65 million years ago from Chicxulub crater melt rock and Cretaceous-Tertiary boundary tektites, *Science*, **1992**, *257*, 954. b) T. Gehrels, Collisions with comets and asteroids, *Sci. Am.*, **1996**, *274*, March, 34. c) R. A. Kerr, Cores document ancient catastrophe, *Science*, **1997**, *275*, 1265. d) C. R. Marshall, Mass extinction probed, *Science*, **1998**, *392*, 17.

11.   G. Keller *et al.*, Chixulub impact predates the K-T boundary mass extinction, *Proc. Natl. Acad. Sci. USA*, **2004**, *101*, 3753.

12.   E. O. Wilson, *The Diversity of Life*, Norton, New York, USA, 1992.

13.   a) R. M. May, How many species inhabit the Earth? *Sci. Am.*, **1992**, *267*, Oct., 18. b) S. L. Pimm *et al.*, The future of biodiversity, *Science*, **1995**, *269*, 347. c) M. Brookes, The species enigma, *New Sci.*, *Inside Science 111*, **1998**, 13 June, 1. d) V. Morell, The variety of life, *Natl. Geog.*, **1999**, *195*, Feb., 7. e) A. Purvis and A. Hector, Getting the measure of biodiversity, *Nature*, **2000**, *405*, 212.

14.   Editorial, Is six billion a reason to celebrate? *Nature Med.*, **1999**, *5*, 1215.

15.   a) A. C. Wilson and R. L. Cann, The recent African genesis of humans, *Sci. Am.*, **1992**, *266*, April, 22. b) A. G. Thorpe and M. H. Wolpoff, The multiregional evolution of humans, *Sci. Am.*, **1992**, *266*, April, 28. c) S. B. Hedge, A start for population genomics, *Nature*, **2000**, *408*, 652. d) L. Dayton, The man from down under, *New Sci.*, **2001**, 13 Jan., 6. e) C. Holden, Oldest human DNA reveals Aussie oddity, *Science*, **2001**, *291*, 230. f) E. Pennisi, Skull study targets Africa-only origins, *ibid.*, 231. g) A. Gibbons, Modern men trace ancestry to African migrants, *ibid.*, 1051. h) Y. Ke *et al.*, African origin of modern humans in East Asia: a tale of 12,000 Y chromosomes, *ibid.*, 1151. i) M. Balter, Anthropologists duel over modern human origins, *ibid.*, 1728. j) R. L. Cann, Tangled genetic routes, *Nature* 2002, *416*, 32. k) A. R. Templeton, Out of Africa again and again, *ibid.*, 45. l) M.-C. King and A. G. Motulsky, Mapping human history, *Science*, **2002**, *298*, 2342.

16.   a) W. C. Clark, Managing planet Earth, *Sci. Am.*, **1989**, *261*, Sept., 19. b) J. Lubchenco, Entering the century of the environment, *Science*, **1998**, *279*, 491.

17.   a) P. M. Vitousek *et al.*, Human domination of Earth's ecosystems, *Science*, **1997**, *277*, 494. b) C. B. Field, Sharing the garden, *Science*, **2001**, *294*, 2490. c) S. Rojstaczer, S. M. Sterling and N. J. Moore, Human appropriation of photosynthesis products, *ibid.*, 2549.

18.   P. R. Ehrlich and A. H. Ehrlich, The value of biodiversity, *Ambio*, **1992**, *21*, 219.

19. a) E. O. Wilson, Threats to biodiversity, *Sci. Am.*, **1989**, *261*, Sept., 60. b) R. Leakey and R. Lewin, *Sixth Extinction: Biodiversity and Its Survival*, Weidenfield and Nicolson, London, UK, 1996. c) M. Holdgate, The ecological significance of biological diversity, *Ambio*, **1996**, *25*, 409. d) T. F. Flannery, Debating extinction, *Science*, **1999**, *283*, 182. e) V. Morell, The sixth extinction, *Natl. Geog.*, **1999**, *195*, Feb., 43. f) F. S. Chapin *et al.*, Consequences of changing biodiversity, *Nature*, **2000**, *405*, 234. g) K. Christen, Biodiversity at the crossroads, *Env. Sci. Technol.*, **2000**, *34*, 123 A. h) E. O. Edwards, Vanishing before our eyes, *Time, Special Edn., Earth Day 2000*, April-May, **2000**. i) B. A. Stein, A fragile cornucopia. Assessing the status of U.S. biodiversity, *Environment*, **2001**, 43, Sept., 13. j) M. Jenkins, Prospects for biodiversity, *Science*, **2003**, *302*, 1175.

20. a) P. Bellwood, The Austronesian dispersal and the origin of languages, *Sci. Am.*, **1991**, *265*, July, 70, b) P. Kirch, *On the Road of the Winds: An Archaeological History of the Pacific Islands Before European Contact*, University of California Press, Berkeley, USA, 2000. c) A. Gibbons, The peopling of the Pacific, *Science*, **2001**, *291*, 1735.

21. a) *Anon.*, The artificial Earth satellite, *Nature*, **1957**, *180*, 734. b) Staff of The Royal Aircraft Establishment, Farnborough, Observations on the orbit of the first Russian earth satellite, *ibid.*, 937. c) F. L. Whipple and J. A. Hynek, Observation of Satellite 1, *Sci. Am.*, **1957**, *197*, Dec., 37.

22. a) *Anon.*, The second artificial Earth satellite, *Nature*, **1957**, *180*, 931. b) H. S. W. Massey, Sputnik II, *New Sci.*, **1957**, 7 Nov., 14. c) *Anon.*, On to the Moon, *Sci. Am.*, **1957**, *197*, Dec., 58.

23. a) *Anon.*, The first manned earth-satellite, 1961μ, *Nature*, **1961**, *190*, 307. b) *Anon.*, Cosmonaut, *Sci. Am.*, **1961**, *204*, May, 74. b) J. M. Logsdon and A. Dupas, Was the race to the moon real? *Sci. Am.*, **1994**, *270*, June, 216.

24. a) P. Stubbs, Bridgehead in space, *New Sci.*, **1969**, 17 July, 114. b) W. Hess, R. Kovach *et al.*, The exploration of the Moon, *Sci. Am.*, **1969**, *221*, Oct., 55. c) D. H. Harland, *Exploring the Moon: The Apollo Expeditions*, Springer, Berlin, Germany, 1999.

25. a) R. W. Kates, Sustaining life on the earth, *Sci. Am.*, **1994**, *271*, Oct., 92. b) J. E. Cohen, Population growth and Earth's human carrying capacity, *Science*, **1995**, *269*, 341. c) J. Bongaarts, Population pressure and the world food supply system in the developing world, *Popn. Dev. Rev.*, **1996**, *22*, 483.

26. M. Wackernagel, L. Lewan and C. Borgström Hansson, Evaluating the use of natural capital with the ecological footprint. Applications in Sweden and subregions, *Ambio*, **1999**, *28*, 604.

27. To preserve biodiversity it is recommended that twelve percent of land area be set aside by the World Commission on Environment and Development, in *Our Common Future*, Oxford University Press, Oxford, UK, 1987.

28. a) D. Schimel *et al.*, Contribution of increasing CO_2 and climate to carbon storage by ecosystems in the United States, *Science*, **1999**, *287*, 2004. b) M. Battl *et al.*, Global carbon sinks and their variability inferred from atmospheric O_2 and $\delta^{13}C$, *Science*, **2000**, *287*, 2467. c) J. Grace and M. Rayment, Respiration in the balance, *Nature*, **2000**, *404*, 819. d) R. Valentini *et al.*, Respiration as the main determinant of carbon balance in European forests, *ibid.*, 861.

29. a) B. Moss, A land awash with nutrients - the problem of eutrofication, *Chem. Indust.*, **1996**, 3 June, 407. b) D. Malakoff, Death by suffocation in the Gulf of Mexico, *Science*, **1998**,

281, 190. c) S. W. Nixon, Enriching the sea to death, *Sci. Am. Quart., The Oceans*, **1998**, *9*, Fall, 48. d) M. Schmiedeskamp, Getting the nutrients out, *ibid.*, 50.

30.   a) M. Wackernagel *et al.*, National natural capital accounting with ecological footprint concept, *Ecol. Econ.*, **1999**, *29*, 359. b) M. Wackernagel *et al.*, Tracking the ecological overshoot of the human economy, *Proc. Natl. Acad. Sci. USA*, **2002**, *99*, 9266.

31.   M. L. Imhoff *et al.*, Global patterns in human consumption of net primary production, *Nature*, **2004**, *429*, 870.

32.   a) N. Myers, Environmental services of biodiversity, *Proc. Natl. Acad. Sci. USA*, **1996**, *93*, 2764.

33.   a) J. Avise, The real message of Biosphere 2, *Cons. Biol.*, **1994**, *8*, 327. b) J. E. Cohen and D. Tilman, Biosphere 2 and biodiversity: The lessons so far, *Science*, **1996**, *274*, 1150. c) G. Walker, Secrets from another Earth, *New Sci.*, **1996**, 18 May, 31. d) J. Mervis, Bye, bye, Biosphere 2, *Science*, **2003**, *302*, 2053.

34.   a) V. Morell, Ozone destroying chlorine tops out, *Science*, **1996**, *271*, 32. b) S. A. Montzka *et al.*, Decline in the tropospheric abundance of halogen from halocarbons: implications for stratospheric ozone depletion, *Science*, **1996**, *272*, 1318. c) D. J. Wuebbles and J. M. Calm, An environmental rational for retention of endangering chemicals, *Science*, **1997**, *278*, 1090.

35.   a) J. Houghton, *Global Warming. The Complete Briefing*, 2nd edn., Cambridge University Press, Cambridge, UK, 1997. b) B. Hileman, Global climate change, *Chem. Eng. News*, **1997**, *75*, 17 Nov., 8. c) T. R. Karl, N. Nicholls and J. Gregory, The coming climate, *Sci. Am.*, **1997**, *276*, May, 54. d) T. R. Karl and K. E. Trenberth, The human impact on climate, *Sci. Am.*, **1999**, *281*, Dec., 62. e) B. Hileman, Case grows for climate change, *Chem. Eng. News*, **1999**, *77*, 9 August, 16. f) D. Sarewitz and R. Pielke, Breaking the global-warming gridlock, *The Atlantic Monthly*, **2000**, July, 54.

36.   a) D. Schneider, The rising seas, *Sci. Am. Quart., The Oceans*, **1998**, *9*, Fall, 28. b) M. Mann, Lessons for a new millennium, *Science*, **2000**, *289*, 253. c) T. J. Crowley, Causes of climate change over the past 1000 years, *Science*, **2000**, *289*, 270.

37.   a) R. Repetto, Accounting for environmental assets, *Sci. Am.*, **1992**, *266*, June, 64. b) R. Repetto, Earth in the balance sheet: Incorporating natural resources in national income accounts, *Environment*, **1992**, *34*, Sept., 12. c) D. Pearce *et al.*, Debt and the environment,. *Sci. Am.*, **1995**, *272*, July, 28.

38.   a) G. Daily, ed., *Nature's Services: Societal Dependence on Natural Ecosystems*, Island Press, Washington, DC, USA, **1997**. b) D. Pimentel *et al.*, Economic and environmental benefits of biodiversity, *BioScience*, **1997**, *47*, Dec., 747. c) N. E. Bockstael, *et al.*, On measuring economic values of nature, *Env. Sci. Technol.*, **2000**, *34*, 1384. d) A. Balmford *et al.*, Economic reasons for conserving wild nature, *Science*, **2002**, *297*, 950.

39.   R. Costanza *et al.*, The value of the world's ecosystem services and natural capital, *Nature*, **1997**, *387*, 253.

40.   a) S. L. Pimm, The value of everything, *Nature*, **1997**, *387*, 231. b) W. Roush, Putting a price tag on nature's bounty, *Science*, **1997**, *276*, 1029. c) E. Masood and L. Garwin, Costing the Earth: when ecology meets economics, *Nature*, **1998**, *395*, 426. d) D. Fullerton and R. Stavins, How economists see the environment, *Nature*, **1998**, *395*, 433.

41.   a) J. Gray, Protect and survive, *New Sci.*, **1998**, 15 August, 48. b) G. C. Daily *et al.*, The value of Nature and the nature of value, *Science*, **2000**, *289*, 395.

42.   a) G. Chichilnisky and G. Heal, Economic returns from the biosphere, *Science*, **1998**, *391*, 629. b) R. H. Platt, P. K. Barten and M. J. Pfeffer, A full clean glass? Managing New York City's watersheds, *Environment*, **2000**, *42*, June, 8. c) L. J. Ehlers, M. J. Pfeffer and C. R. O'Melia, Making watershed management work, *Env. Sci. Technol.*, **2000**, *34*, 464 A.

43.   S. Cole, The emergence of treatment wetlands, *Env. Sci. Technol.*, **1998**, *32*, 18 A.

44.   a) R. A. Frosch and N. E. Gallopoulos, Strategies for manufacturing, *Sci. Am.*, **1989**, *261*, Sept., 94. b) B. R. Allenby and D. J. Richards, eds., *The Greening of Industrial Ecosystems*, National Academy Press, Washington, DC, USA, 1994. c) R. Ayers and U. Simonis, eds., *Industrial Metabolism: Restructuring for Sustainable Development*, United Nations University Press, Florence, Italy, 1994. d) R. A. Frosch, *Sci. Am.*, **1995**, *273*, Sept., 144. e) M. Burke, *Env. Sci. Technol.*, **1997**, *31*, 84 A.

45.   a) R. A. Kerr, Acid rain control: Success on the cheap, *Science*, **1998**, *282*, 1024. b) D. Munton, Dispelling the myths of the acid rain story, *Environment*, **1998**, *40*, July/Aug., 4. c) A. Jenkins, End of the acid reign, *Nature*, **1999**, *401*, 537. d) D. Burtraw and E. Mansur, Environmental effects of SO_2 trading and banking, *Env. Sci. Technol.*, **1999**, *33*, 3489.

46.   a) J. T. Overpeck, The hole record, *Nature*, **2000**, *403*, 714. b) S. Huang, H. N. Pollack and P.-Y. Shen, Temperature trends over the past five centuries reconstructed from borehole temperatures, *Nature*, **2000**, *403*, 756.

47.   a) J. Overpeck, Arctic environmental change of the last four centuries, *Science*, **1997**, *278*, 1251. b) P. D. Jones *et al.*, High resolution paleoclimatic records for the last millennium: interpretation, intergration and comparison with general circulation model control-run temperatures, *Holocene*, **1998**, *8*, 455. c) M. E. Mann, R. S. Bradley and M. K. Hughes, Global-scale temperature patterns and climate forcing over the past six centuries, *Nature*, **1998**, *392*, 779. d) P. Jones, It was the best of times, it was the worst of times, *Science*, **1998**, *280*, 544. e) K. R. Briffa and T. J. Osborn, Seeing the wood from the trees, *Science*, **1999**, *284*, 926. f) R. A. Kerr, Draft report affirms human influence, *Science*, **2000**, *288*, 589.

48.   a) R. A. Kerr, Globe's "missing warming" found in the Ocean, *Science*, **2000**, *287*, 2126. b) S. Levitus *et al.*, Warming of the world ocean, *Science*, **2000**, *287*, 2225. c) D. Dahle-Jensen, The Greenland ice sheet reacts, *Science*, **2000**, *289*, 404.

49.   a) J. D. Milliman, J. D. Broadus and F. Gable, Environmental and economic implications of rising sea-level and subsiding deltas: the Nile and Bengal examples, *Ambio*, **1989**, *18*, 340. b) J. Lewis, Sea-level rise: some implications for Tuvalu, *ibid.*, 458.

50.   C. N. Runnels, Environmental degradation in Ancient Greece, *Sci. Am.*, **1995**, *272*, March, 72.

51.   Aristotle, *Meteorologica*, Book 1, Chapter 14.

52.   R. H. Grove, Origins of Western environmentalism, *Sci. Am.*, **1992**, *267*, July, 22.

53.   a) R. Carlson, *Silent Spring*, Houghton-Mifflin, Boston, USA, **1962**. b) B. G. Loganathan and K. Kannan, Global organochlorine contamination trends: an overview, *Ambio*, **1994**, *23*, 187. c) J. R. Krebs *et al.*, The second silent spring, *Nature*, **1999**, *400*, 611.

54.   D. Malakoff, Extinction on the high seas, *Science*, **1997**, *277*, 686.

55.   a) P. A. Matson *et al.*, Agricultural intensification and ecosystem properties, *Science*, **1997**, *277*, 504. b) L. W. Botsford, J. C. Castilla and C. H. Peterson, The management of fisheries and marine ecosystems, *ibid.*, 509. c) J. B. Hughes, G. C. Daily and P. R. Ehrlich, Population diversity: its extent and extinction, *Science*, **1997**, *278*, 689.

56.   L. Hannah *et al.*, A preliminary inventory of human disturbance of world ecosystems, *Ambio*, **1994**, *23*, 246.

57.   R. Gallagher and B. Carpenter, Human dominated ecosystems, *Science*, **1997**, *277*, 485.

58.   Managing Planet Earth, *Sci. Am.*, **1989**, *261*, Sept.

59.   a) J. Terborgh, Why American songbirds are vanishing, *Sci. Am.*, **1992**, *266*, May, 56. b) S. K. Robinson *et al.*, Regional forest fragmentation and the nesting success of migratory birds, *Science*, **1995**, *267*, 198. c) R. A. Askins, Hostile landscapes and the decline of migratory songbirds, *Science*, **1995**, *267*, 1956. d) L. Tangley, The case of the missing migrants, *Science*, **1996**, *274*, 1299.

60.   a) R. G. Prinn, The interactive atmosphere: global atmospheric-biospheric chemistry, *Ambio*, **1994**, *23*, 50. b) P. J. Crutzen, On the role of CH_4 in atmospheric chemistry: sources, sinks and possible reductions in anthropogenic sources, *Ambio*, **1995**, *24*, 52.

61.   a) P. P. Tans and P. S. Baldwin, Climate change and carbon dioxide, *Ambio*, **1995**, *24*, 376. b) C. D. Keeling *et al.*, Interannual extremes in the rate of rise of atmospheric carbon dioxide since 1980, *Nature*, **1995**, *375*, 666. c) J. T. Houghton and L. G. Meira Filho *et al.*, eds., *Climate Change 1995: The Science of Climate Change*, IPCC, Cambridge University Press, Cambridge, UK, 1996. d) J. T. Houghton *et al.*, eds, *Climate Change 2001: The Scientific Basis*, IPCC, Cambridge University Press, Cambridge, UK, 2001.

62.   a) F. S. Rowland, Stratospheric ozone depletion by chlorofluorocarbons (Nobel Lecture), *Angew. Chem. Int. Edn.*, **1996**, *35*, 1786. b) G. Walker, The hole story, *New Sci.*, **2000**, 25 March, 24.

63.   a) H. Whittaker and G. E. Likens, Primary production: the biosphere and man, *Hum. Ecol.*, **1973**, *1*, 357. b) E. O. Wilson, News and comment, *Science*, **1986**, *234*, 14.

64.   a) W. B. Meyer and B. L. Turner, Human population growth and global land use/cover change, *Annu. Rev. Ecol. Syst.*, **1992**, *23*, 39. b) R. L. Naylor, Energy and resource constraints on intensive agricultural production, *Annu. Rev. Energy Environ.*, **1996**, *21*, 99. c) R. L. Naylor, W. Falcon and E. Savaleta, Variability and growth in grain yields, 1950-94: does the record point to greater instability? *Popn.. Dev. Rev*, **1997**, *23*, 41.

65.   a) V. Smil, Global population and the nitrogen cycle, *Sci. Am.*, **1997**, *277*, July, 58. b) V. Smil, Detonator of the population explosion, *Nature*, **1999**, *400*, 415.

66.   a) S. L. Postel, G. C. Daily and P. R. Ehrlich, Human appropriation of renewable fresh water, *Science*, **1996**, *271*, 785. b) J. N. Abramovitz, *Imperilled Waters, Impoverished Future: The Decline of Freshwater Ecosystems*, Worldwatch Institute, Washington, DC, USA, 1996.

67.   a) R. H. Pine, New mammals not so seldom, *Nature*, **1994**, *368*, 593. b) C. G. M. Paxton, A cumulative species description curve for large open water marine animals, *J. Marine Biol. Assoc.*, **1998**, *78*, 1389.

68.   J. Woodford, *The Wollemi Pine*, Text Publishing, Melbourne, Australia, 2000.

69.   J. F. Grassle, Species diversity in deep-sea communities, *Trends Ecol. Evol.*, **1989**, *4*, 12.

70.   J. K. Fredrickson and T. C. Onstott, Microbes deep inside the Earth, *Sci. Am.*, **1996**, *275*, Oct., 42.

71.   R. M. May, How many species? *Phil. Trans. Roy. Soc. B*, **1990**, *330*, 293.

72.   J. A. McNeely *et al.*, *Conserving the World's Biological Diversity*, International Union for Conservation of Natural Resources, World Resources Institute, Conservation International World Wildlife Fund-US and the World Bank, Gland, Switzerland, 1990.

73. J. S. Olson, J. A. Watts and L. J. Allison, *Carbon in Live Vegetation of Major World Ecosystems*, Oak Ridge National Laboratory Publication, Oak Ridge, USA, 1997.

74. G. L. Ajtay, P. Ketner and P. Duvigneaud, *Terrestrial primary production and phytomass* in *The Global Carbon Cycle*, B. Bolin *et al.*, eds, Wiley, Chichester, UK, 1979.

75. a) O. Tickell, We'll all go together when we go, *New Sci.*, **1996**, 2 March, 42. b) M. J. Novacek and E. E. Cleland, The current biodiversity event: scenarios for mitigation and recovery, *Proc. Natl. Acad. Sci. USA*, **2001**, *98*, 5466. c) D. S. Woodruff, Declines of biomes and biotas and the future of evolution, *ibid.*, 5471. d) N. Knowlton, The future of coral reefs, *ibid.*, 5419. e) S. L. Pimm *et al.*, Can we defy nature's end? *Science*, **2001**, *293*, 2207.

76. a) F. Pearce, Incendiary policies, *New Sci.*, **1997**, 4 Oct., 4. b) D. Swinbanks, Forest fires cause pollution crisis in Asia, *Nature*, **1997**, *389*, 321. c) R. Showstack, Scientists assess impact of Indonesian fires, *EOS*, **1997**, 4 Nov., 493. d) N. Mawdsley, Indonesia aflame, *New Sci.*, **1997**, 25 Oct., 51. e) M. Brauer and J. Hisham-Hashim, Fires in Indonesia: Crisis and reaction, *Env. Sci. Technol.*, **1998**, *33*, 404 A. f) C. A. Legg and Y. Laumonier, Fires in Indonesia, 1997: a remote sensing perspective, *Ambio*, **1999**, *28*, 479. g) M. S. Hamilton, M. O. Miller and A. Whitehouse, Continuing fire threat in Southeast Asia, *Env. Sci. Technol.*, **2000**, *34*, 82 A. h) F. Siegert *et al.*, Increased damage from fires in logged forests during droughts caused by El Niño, *Nature*, **2001**, *414*, 437. i) P. Jepson *et al.*, The end of Indonesia's lowland forests? *Science*, **2001**, *292*, 859. j) D. Schimel and D. Baker, The wildfire factor, *Nature*, **2002**, *420*, 29. k) S. E. Page *et al.*, The amount of carbon released from peat and forest fires in Indonesia during 1997, *ibid.*, 61.

77. F. Pearce, Indonesia's inferno will make us all sweat, *New Sci.*, **1997**, 18 Oct., 22.

78. a) J.-P. Malingreau and C. J. Tucker, Large-scale deforestation in the southeastern Amazon basin of Brazil, *Ambio*, **1988**, *17*, 49. b) W. F. Laurence *et al.*, The future of the Brazilian Amazon, *Science*, **2001**, *291*, 438.

79. F. Pearce, Scorched earth, *New Sci.*, **1997**, 11 Oct., 10.

80. a) F. Pearce, Playing with fire, *New Sci.*, **1998**, 21 March, 37. b) F. Pearce, Beyond hope, *ibid.*, 31 Oct., 24. c) J. G. Goldammer, Forests on fire, *Science*, **1999**, *284*, 1782. d) F. Pearce, Logged out, *New Sci.*, **2002**, 2 March, 6.

81. E. Salati and P. B. Vose, Depletion of tropical rainforests, *Ambio*, **1983**, 12, 67.

82. a) R. Adler, All dried up. Forest fires are upsetting the global weather machine, *New Sci.*, **1999**, 16 October, 15. b) O. B. Toon, How pollution suppresses rain, *Science*, **2000**, *287*, 1763.

83. J. T. Carlton *et al.*, Historical extinctions in the sea, *Annu. Rev. Ecol. Syst.*, **1999**, *30*, 515.

84. a) D. W. Steadman, Prehistoric extinctions of Pacific island birds: biodiversity meets zooarchaeology, *Science*, **1995**, *267*, 1123. b) J. Diamond, Blitzkrieg against the moas, *Science*, **2000**, *287*, 2170. c) R. N. Holdaway and C. Jacomb, Rapid extinction of the moas (Aves: Dinornithiformes): model, test and implication, *Science*, **2000**, *287*, 2250. d) L. Deighton, Mass extinctions pinned on ice age hunters, *Science*, **2001**, *292*, 1819. e) J. Alroy, A multispecies overkill simulation of the end-Pleistocene megafaunal mass extinction, *ibid.*, **2001**, *292*, 1893.

85. S. L. Pimm, *Threatened species around the world*, World Book Multimedia Encyclopaedia, Millennium 2000 (http//www.worldbook.com/fun/wbla/earth/html/ed11.htm)

86.   a) World Resources Institute, *World Resources 1996-1997*, Oxford University Press, New York, USA, 1996. b) F. Pearce, An unnatural disaster, *New Sci.*, **1999**, 6 Nov., 12.

87.   a) N. Williams, Study finds 10% of tree species under threat, *Science*, **1998**, *281*, 1426. b) O. Tickell, Trees on the verge of mass extinction, *New Sci.*, **1998**, 29 Aug., 11. c) J. Josephson, Going, going, gone? Plant species extinction in the 21st century, *Env. Sci. Technol.*, **2000**, *34*, 130 A.

88.   L. R. Gerber, D. P. DeMaster and S. P. Roberts, Measuring success in conservation, *Am. Sci.*, **2000**, *88*, 317.

89.   a) J. P. Cohn, Saving the California condor, *BioScience*, **1999**, *49*, Nov., 864. b) M. Kaplan, Plight of the condor, *New Sci.*, **2002**, 5 Oct., 34.

90.   J.-P. Malingreau and C. J. Tucker, Large-scale deforestation in the southeastern Amazon basin of Brazil, *Ambio*, **1988**, *17*, 49.

91.   a) V. H. Heywood and S. N. Stuart, Species extinction in tropical forests, in *Tropical Deforestation and Species Extinction*, T. C. Whitmore and J. H. Taylor, eds., Chapman and Hall, London, UK, 1992. b) N. C. A. Pitman and P. M. Jørgensen, Estimating the size of the world's threatened flora, *Science*, **2002**, *298*, 989.

92.   R. M. May, J. H. Lawton and N. E. Stork, Assessing extinction rates, in *Extinction Rates*, J. H. Lawton and R. M. May, eds., Oxford University Press, Oxford, UK, 1995.

93.   N. Myers, Mass extinction and evolution, *Science*, **1997**, *278*, 587.

94.   a) N. Myers and A. H. Knoll, The biotic crisis and the future of evolution, *Proc. Natl. Acad. Sci. USA*, **2001**, *98*, 5389. b) D. Jablonski, Lessons from the past: evolutionary impacts of mass extinctions, *ibid.*, 5393. c) D. H. Erwin, Lessons from the past: biotic recoveries from mass extinctions, *ibid.*, 5399. d) M. L. Rosenzweig, Loss of speciation rate will impoverish future diversity, *ibid.*, 5404. e) D. Tilman and C. Lehman, Human-caused environmental change: impacts on plant diversity and evolution, *ibid.*, 5433. f) H. A. Mooney and E. E. Cleland, The evolutionary impact of invasive species, *ibid.*, 5446. g) D. Western, Human-modified ecosystems and future evolution, *ibid.*, 5458. h) P. Erlich, Intervening in evolution: ethics and actions, *ibid.*, 5477.

95.   a) D. Erwin, Palaeontology: life's downs and ups, *Nature*, **2000**, *404*, 129. b) J. W. Kirchmer and A. Weil, Delayed biological recovery from extinctions throughout the fossil record, *Nature*, **2000**, *404*, 177.

96.   a) R. Nowak, Uganda enlists locals in the battle to save the gorillas, *Science*, **1995**, *267*, 1761. b) J. Bohannon, An 11th-hour rescue for Great Apes? *Science*, **2002**, *297*, 2203.

97.   a) J. Whitfield, Ape populations decimated by hunting and Ebola virus, *Nature*, **2003**, *422*, 551. b) P. D. Walsh *et al.*, Catastrophic ape decline in western equatorial Africa, *ibid.*, 611.

98.   J. R. Spotila *et al.*, Pacific leatherback turtle faces extinction, *Nature*, **2000**, *405*, 529.

99.   a) N. Myers, Threatened biotas: hotspots in tropical forests, *Environmentalist*, **1988**, *8*, 178. b) R. P. Cincotta, J. Wisnewski and R. Engelman, Human population in the biodiversity hotspots, *Nature*, **2000**, *404*, 990. c) A. Balmford *et al.*, Conservation conflicts across Africa, *Science*, **2001**, *291*, 2616. d) B. Holmes, Should we seek to save Earth's iconic hotspots? *New Sci.*, **2005**, 5 Feb., 10.

100.   a) United Nations, *Report of the United Nations Conference on Environment and Development*, Rio de Janeiro, 3-14 June 1992, United Nations, New York, USA, 1992. b) L. Glowka, F. Burhenne Guilmin and H, Synge, *A Guide to the Convention on Biodiversity*, IUCN, Gland, Switzerland, 1994.

101.   N. Myers, Population and biodiversity, *Ambio*, **1995**, *24*, 56.

102.   S. D. Tanksley and S. R. McCouch, Seed banks and molecular maps: unlocking genetic potential from the wild, *Science*, **1997**, *277*, 1063.

103.   a) R. Edwards, Tomorrow's bitter harvest, *New Sci.*, **1996**, 17 August, 14. b) R. Nelson, The blight is back, *Sci. Am.*, **1998**, *278*, June, 18.

104.   a) L. D. Stein, End of the beginning, *Nature*, **2004**, *431*, 915. b) International Human Genome Sequencing Consortium, Finishing the euchromic sequence of the human genome, *ibid.*, 931.

105.   M. J. Reiss and R. Straughan, *Improving Nature? The Science and Ethics of Genetic Engineering*, Cambridge University Press, Cambridge, UK, **1996**.

106.   a) F. Barnaby, The effects of a global nuclear war: the arsenals, *Ambio*, **1982**, *11*, 76. b) F. Barnaby *et al.*, Reference scenario: how a nuclear war might be fought, *ibid.*, 94.

107.   a) M. Day, Scourge of infection kills Third World's young, *New Sci.*, **1996**, 25 May, 6. b) C. Holden, Ominous trends for infectious diseases, *Science* **1996**, *272*, 1269.

108.   a) M. Balter, On World AIDS Day, a shadow looms over Southern Africa, *Science*, **1998**, *282*, 1790. b) K. Birmingham, UN acknowledges HIV/AIDS as a threat to world peace, *Nature Med.*, **2000**, *6*, 117.

109.   *AIDS Epidemic Update, December 2004*, UNAIDS/WHO, Geneva, Switzerland., 2004.

110.   a) F. Press and R. M. Hamilton, Mitigating natural disasters, *Science*, **1999**, *284*, 1927. b) W. D. Iwan, *et al.*, Mitigation emerges as major strategy for reducing losses caused by natural disasters. *Science*, **1999**, *284*, 1943.

111.   a) G. Brasseur and C. Granier, Mount Pinatubo aerosols, chlorfluorocarbons, and ozone depletion, *Science*, **1992**, *257*, 1239. b) J. Horgan, Volcanic disruption, *Sci. Am.*, **1992**, *266*, March, 16. c) D. Schneider, A river (of mud) still runs through it, *Sci. Am.*, **1995**, *273*, July, 17. d) M. R. Carroll, Volcanic sulphur in the balance, *Nature*, **1997**, *389*, 543. e) V. Kress, Magma mixing as a source for Pinatubo sulphur, *Nature*, **1997**, *389*, 591. f) C. G. Newhall, J. A. Power and R. S. Punongbayan, "To make grow", *Science*, **2002**, *295*, 1241.

112.   T. Simkin, L. Siebert and R. Blong, Volcanic fatalities - lessons from the historical record, *Science*, **2001**, *291*, 255.

113.   a) D. Normile, Quake builds case for strong codes, *Science*, **1995**, *267*, 444. b) U. Tsunogai and H. Wakita, Precursory chemical changes in ground water: Kobe earthquake, *Science*, **1995**, *269*, 61. c) E. M. Noam and H. Sato, Kobe's lesson: dial 711 for "open" emergency communications, *Science*, **1996**, *274*, 739.

114.   Scientist of the U.S. Geological Survey and the Southern California Earthquake Center, The magnitude 6.7 Northridge, California, Earthquake of 17 January 1994, *Science*, **1994**, *266*, 389.

115.   a) T. Appenzeller, In Turkey, havoc from a falling-domino effect, *Science*, **1999**, *285*, 1334. b) A. Barka, The 17 August 1999 Izmit earthquake, *ibid.*, 1858. c) S. Leavy, Stress test, *Sci. Am.*, **1999**, *281*, Dec., 16.

116.   R. Kerr, Tectonics, design combine for India disaster - more coming, *Science*, **2001**, *291*, 802.

117.   H. K. Gupta *et al.*, The deadliest intraplate earthquake, *Science*, **2001**, *291*, 2101. b) A. Lawler, *Science*, Earthquake allows rare glimpse into Bam's past – and future, **2004**, *303*, 1463.

118. a) F. I. Gonzalez, Tsunami, *Sci Am.*, **1999**, *280*, May, 44. b) R. Koenig, Researchers target deadly tsunamis, *Science*, **2001**, *293*, 1251.

119. a) Y. Bhattacharjee, In wake of disaster, scientists seek out clues to prevention, *Science*, **2005**, *307*, 22. b) E. Kintisch, U.S. Clamor grows for global network of ocean sensors, *ibid.*, 191. c) R.A. Kerr, Failure to gauge the quake crippled the warning effort, *ibid.*, 201. d) E. Kintisch, Global tsunami warning system takes shape, *ibid.*, 331. e) R. Stone, A race to beat the odds, *ibid.*, 502. f) J. Hogan and E. Young, Will we be ready for the next one? *New Sci.*, **2005**, 15 Jan., 12. g) F. Pearce and B. Holmes, The impact will last decades, *ibid.*, 14. h) R. Nowak, Reconstructing a most deadly wave, *ibid.*, 16. i) K. Davis, Anatomy of a quake, *ibid.*, 17. j) J. McCloskey, S. S. Nalbant and S. Steacy, Earthquake risk from co-seismic stress, *Nature,* **2005**, *434*, 291. k) K. Sieh, What happened and what's next? *Ibid.*, 573. l) S. Stein and E. A. Okal, Speed and size of the Sumatra earthquake, *ibid.*, 581. m) S. Ni, H. Kanamori and D. Helmberger, Energy radiation from the Sumatra earthquake, *ibid.*, 582. n) M. Mukerjee, The scarred Earth, *Sci. Am.*, **2005**, *292*, March, 8.

120. a) M. Hopkin, Indonesia spared a tsunami as disaster quake strikes, *Nature*, **2005**, *434*, 547. b) R. A Kerr, Model shows islands muted tsunami after latest Indonesian earthquake, *Science*, **2005**, *308*, 341

121. a) J. Copley, Recipe for disaster: Why a weakening hurricane wrought such havoc, *New Sci.*, **1998**, 14 Nov., 5. b) P. R. Epstein, Climate and health, *Science*, **1999**, *285*, 347.

122. O. Sattaur, Counting the cost of catastrophe, *New Sci.*, **1991**, 29 June, 13.

123. a) A. Vaidyanathan, A sorry technological tale from India, *New Sci.*, **1985**, 21 Feb., 36. b) D. MacKenzie, Design failings that caused disaster, *New Sci.*, **1985**, 28 March, 3. c) A. Rosencranz, Bhopal, transnational corporations, and hazardous technologies, *Ambio*, **1988**, *17*, 336. d) B. Allen, Bhopal: 15 years on, *Green Chem.*, **2000**, G56. e) C. Crabb, Revisiting the Bhopal tragedy, *Science*, **2004**, *306*, 1679. f) J.-F. Tremblay, Bhopal today, *Chem. Eng. News*, **2005**, *83*, 24 Jan., 28.

124. a) M. Barinaga, Shipwreck fouls the water, *Nature* **1989**, *338*, 451. b) L. Dayton, Scientists converge on stricken sound in Alaska, *New Sci.*, **1989**, 8 April, 20. c) P. Coles, Fisheries first to suffer, *Nature*, **1989**, *338*, 533. d) J. F. Piatt and C. J. Lensink, Exxon Valdez bird toll, *Nature*, **1989**, *342*, 865. e) T. A. Birkland, In the wake of the Exxon Valdez, *Environment*, **1998**, 40, Sept., 7. f) F. Pearce, In the thick of it, *New Sci.*, **2001**, 5 May, 4.

125. a) V. Rich, An ill wind from Chernobyl, *New Sci.*, **1991**, 20 April, 20. b) M. Bojcun, The legacy of Chernobyl, *New Sci.*, **1991**, 20 April, 24. c) R. Edwards, Chernobyl floods put millions at risk, *New Sci.*, **1996**, 23 March, 4. d) M. Freemantle, Ten years after Chernobyl consequences are still emerging, *Chem. Eng. News*, **1996**, *74*, 29 April, 18. e) Y. M. Shcherbak, Ten years of the Chornobyl era, *Sci. Am.*, **1996**, *274*, April, 32. f) T. J. Smith *et al.*, *Nature*, Chernobyl's legacy in food and water, **2000**, *405*, 141. g) R. Stone, Living in the shadow of Chornobyl, *Science*, **2001**, *292*, 420.

126. a) W. R. Jordan, M. E. Gilpin and J. D. Aber, *Restoration Ecology*, Cambridge University Press, Cambridge, UK, 1987. b) A. P. Dobson, A. D. Bradshaw and A. J. M. Baker, Hopes for the future: restoration ecology and conservation biology, *Science*, **1997**, *277*, 515.

127. a) F. Pearce, Reap what you sow, *New Sci.*, **2001**, 21 April, 11. b) D. Tilman *et al.*, Forecasting agriculturally driven global environmental change, *Science*, **2001**, *292*, 281.

Chapter 2

The Human Population

"Poverty, environment and population can no longer be dealt with, or
even thought of, as separate issues."

Gro Harlem Bruntland, Norwegian prime minister, at the United Nations
Conference on Environment and Development, The "Earth Summit", Rio
de Janeiro, June 1992.

2.1. The Pressure of Population

On 19 February 2005, at 07:54 GMT, the human population of Earth was
estimated to be 6,419,676,868 by the United States Census Bureau that updates
its estimate every second [1]. According to these statistics, the population is
increasing by about seventy-three million a year as 130 million children replace
the fifty-eight million who die. Thus, the number of children born is roughly
equal to six times the population of Australia, twice that of the United Kingdom,
similar to that of Japan and half that of the United States. Such population
growth is startling and is increasingly occupying the thoughts of national
governments and the United Nations. There is a fast growing awareness that a
limit exists for the capacity of Earth to support such a burgeoning population and
its aspirations towards higher living standards although there is considerable
debate as to what this limit is.

Concerns about Earth's fast increasing population and its real and potential
environmental consequences resulted in the United Nations Conference on
Environment and Development, more popularly known as the "Earth Summit",
that took place in Rio de Janeiro, Brazil, in June of 1992. This wide ranging
conference produced a treaty seeking to prevent potentially dangerous climate
changes by controlling greenhouse gas emissions resulting from human, or
anthropogenic, activities that contribute to a global temperature increase. It also
produced a second treaty seeking to protect the biodiversity of plants and animals
that is increasingly threatened. While these treaties were signed by 153 nations,
agreement among the diversity of signing nations was incomplete [2,3]. A more

recent conference, The Third Conference of Parties to the Framework Convention on Climate Change, in Kyoto, Japan, in December 1997, and subsequent conferences sought to establish legally binding national limits for greenhouse gas emissions, particularly that of carbon dioxide, under the Kyoto Protocol [4]. Inevitably, the divergent views expressed on the proposed limits reflected the differing natures of the economies and the stages of development of the nations attending the conference. However, the Kyoto Protocol was ratified by 141 nations upon whom it became legally binding on 16 February 2005. Irrespective of the various views on the effectiveness of this protocol, there can be little doubt that the growth of atmospheric carbon dioxide levels has produced a powerful focus on the pressures placed on Earth by the activities of a rapidly growing human population.

Concerns about population growth are as longstanding as recorded history. Babylonian (about 1600 BC) and Greek (776-580 BC) historians commented on overpopulation and similar concerns have been regularly expressed up to the present time [5,6]. However, the human population has never been so immense as it is today, and the speed and quantity of information exchange have become so great with the growth of global communications that news of famine, flood, epidemic and other threatening events becomes general knowledge within hours. A combination of these and related factors has generated a concern that Earth may not adequately sustain this increasing population and that resources will be exploited to such an extent that the habitability of Earth will suffer a major decline as a consequence of land degradation, deforestation, climate change and pandemic disease [6-16]. This, in turn, is predicted by some to lead to conflict as nations increasingly go to war to either defend their own resources or to take those of others [17]. Already there are signs that developments leading towards these chilling scenarios may be under way. However, the overall picture is complex and predictions of national or global catastrophe are recurrent themes in most civilizations. Accordingly, this chapter examines the nature of the human population and predictions about its growth in the twenty-first century to set the scene for discussion of the many other interrelated factors that affect humanity's present and future well-being in the chapters that follow.

2.2. Where People Live

Humans have spread to permanently populate all of Earth's continents except Antarctica. According to the United Nations 2003 estimates, the majority of people, some 3,823 million, lived in Asia, while 851 million lived in Africa, 726 million lived in Europe, 543 million lived in South and Central America, 326 million lived in North America and 32 million lived in Oceania [18]. The major portion of the human population was in the developing nations of the southern landmasses while the population of the developed nations was largely concentrated in the north. A selection of the populations of some of the larger nations from these two groups and a third group, previously called the Soviet bloc, but now called the transitional economies as they undergo change from centrally controlled to free market economic policies, is shown in Fig. 2.1. It is

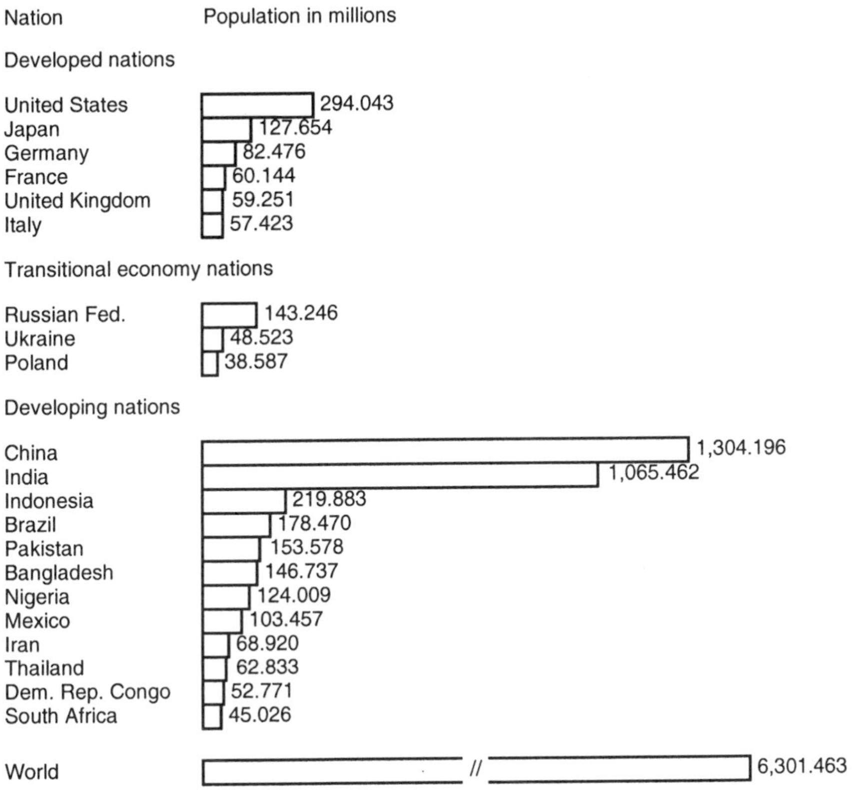

Nation   Population in millions

Developed nations

United States	294.043
Japan	127.654
Germany	82.476
France	60.144
United Kingdom	59.251
Italy	57.423

Transitional economy nations

Russian Fed.	143.246
Ukraine	48.523
Poland	38.587

Developing nations

China	1,304.196
India	1,065.462
Indonesia	219.883
Brazil	178.470
Pakistan	153.578
Bangladesh	146.737
Nigeria	124.009
Mexico	103.457
Iran	68.920
Thailand	62.833
Dem. Rep. Congo	52.771
South Africa	45.026

| World | 6,301.463 |

Fig. 2.1. The 2003 populations of some of the more populous nations [18].

seen that China's population represented 20.7 percent of Earth's 2003 population of 6,301 million followed by India with 16.9 percent and the United States with 4.7 percent. The impacts of such huge populations on humanity's future are likely to be multifold, the foremost of which is Earth's ability to sustain the fast growing global population, for it is this that most affects all life on the planet.

2.3. Carrying Capacity

Ecologists use the concept of the "carrying capacity" of a habitat for a particular animal as a guide to survival of that animal [5,6,10,15,16,19]. In simple terms, carrying capacity is a measure of the ability of a habitat to sustain the living requirements of an animal population without impairment of the habitat's ability to regenerate continually and completely. Should the animal population exceed this capacity, habitat degradation and starvation and depopulation will follow. However, estimation of carrying capacity is not quite this simple for the human population as humans have an amazing ability to both adapt to their habitat and to adapt it to their needs. As a consequence estimations of Earth's human carrying capacity vary greatly.

An early expression of Earth's carrying capacity was expressed by the Englishman Thomas Malthus in 1789 in "An Essay on The Principle of Population", who wrote:

> "The happiness of a country does not depend, absolutely, upon its poverty or its riches, upon its youth or its age, upon its being thinly or fully inhabited, but upon the rapidity with which it is increasing, upon the degree in which the yearly increase of food approaches to the yearly increase of an unrestricted population."

He was not optimistic about the ability of humans to increase food production, and predicted that population growth would rapidly exceed food supply. Fortunately, this prediction has proved incorrect for the last two hundred years during which time food supply has matched or bettered demand although local climate variations, distribution shortcomings and political malevolence or ineptitude have often generated devastating famines [11,19,20]. Nevertheless,

Malthus' ideas continue to be influential in population fora and there is a pervasive concern that Earth's human carrying capacity will be exceeded in the not too distant future unless some very perceptive and innovative action is taken.

Almost two hundred years after Malthus, in 1968, the American, Garrett. Hardin, wrote his essay, "The Tragedy of the Commons", in which he discussed the increasing strain placed by a growing human population on Earth's finite resources and thereby the sustainability of reasonable living standards [21]. The "commons" discussed by Hardin are resources such as rivers, lakes and oceans and the fish in them, forests and their wood, and the air to breath to which individual access is largely unrestrained and are available to be exploited by those in a position to do so irrespective of the effect on humanity at large. He painted a bleak picture of uncontrolled and increasing exploitation of these resources leading to their exhaustion, breakdown and pollution unless access to them was carefully controlled. Hardin's ideas have been influential in resource, environmental and sociological discussions and, largely coincidentally, access to the "commons" has been restricted through fishery exclusion zones, forestry protection measures, atmospheric pollution controls and other resource conservation and protection measures [22].

As knowledge of Earth has grown at an ever increasing rate during the twentieth century and into the beginning of the twenty-first century, it seems reasonable to assume that an accurate estimation of the number of humans that Earth can support should be possible. Indeed, many such estimates of Earth's human carrying capacity have been made but they vary greatly as is seen in Fig. 2.2. Upon reflection, this variation is not too surprising as any estimate must be based on the living standards and quality of life expected. Once these are decided, the estimation of carrying capacity then varies according to another estimation, that of Earth's ability to sustainably supply the required amount of water, food, shelter and energy. Superimposed on these requirements is the extent to which the preservation of the other inhabitants of Earth, animals and plants, and their habitats, wilderness and recreation areas, and the myriad other factors that determine the habitability of Earth are rated in importance. As a consequence, upper estimates of Earth's carrying capacity range from a low one billion at 1970 United States levels of affluence, estimated by H. R. Hulett [23], to a high 1,022 billion estimated by C. T. De Wit on the basis of Earth's

maximum photosynthetic ability to produce food [24]. This latter estimate is 162
times higher than Earth's 2003 human population and, if evenly spread over all
land, about 162.1 million square kilometres when Antarctica is included, would
result in a population density of 6,300 people per square kilometre. While
making a similar estimation of 1,000 billion as the upper limit based on energy
considerations, C. Marchetti suggested that two thirds of such a population could

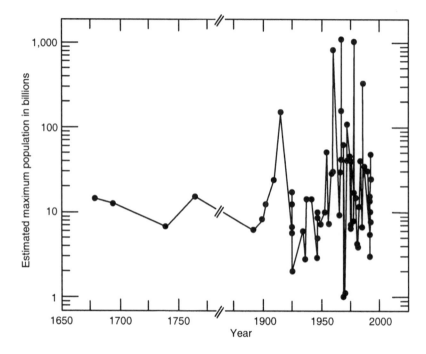

Fig. 2.2. Upper estimations of the number of people that Earth can sustain. Adapted from [5,6].
Reference [5] contains references to all of the estimates shown in the figure.

be accommodated in floating cities at sea [25]. These two very high upper
estimates are unrealistic in terms of Earth remaining a desirable place to live, as
the authors indicate. However, when contrasted with Hulett's estimate of one
billion, they illustrate the great differences generated depending on the criteria
used to make estimations of Earth's carrying capacity [26]. Inevitably, debate
about Earth's carrying capacity heightens interest in the current human
population and its growth in the twenty-first century.

2.4. The Growth of Population

According to a United Nations 1997 estimate, Earth's population was expected to have grown from a billion in 1804 to eleven billion in 2093 as shown in Fig. 2.3 [27]. While the growth from one to two billion took 123 years, the addition of each further billion accelerated rapidly until six billion was achieved [28], but it was predicted that the approach to eleven billion in 2093 would slow considerably suggesting that Earth's population may stabilize at around that number [27,29]. This prediction assumed a continuation of the 1997 global trends in birth rates and death rates until 2093. While the prediction of a slowing of Earth's population growth to stabilize in 2093 is encouraging, the predicted population at stabilization of eleven billion is not far short of being double Earth's 1997 population. Later United Nations analyses, based on 2000 and 2002 trends, suggested that the rate of increase of population was slowing and predicted 2050 populations of 9.3 and 8.9 billion, respectively [18]. However, unless a combination of technological, economic, political and cultural development can be used to alleviate the stress that much of humanity and the biosphere show signs of suffering at the current much lower population level, a levelling out of population of ten to eleven million by the end of the twenty-first century could render Earth a very difficult place to live for much of humanity.

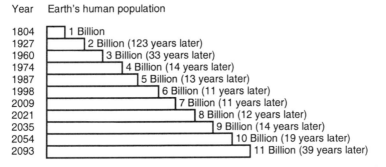

Year   Earth's human population

1804 — 1 Billion
1927 — 2 Billion (123 years later)
1960 — 3 Billion (33 years later)
1974 — 4 Billion (14 years later)
1987 — 5 Billion (13 years later)
1998 — 6 Billion (11 years later)
2009 — 7 Billion (11 years later)
2021 — 8 Billion (12 years later)
2035 — 9 Billion (14 years later)
2054 — 10 Billion (19 years later)
2093 — 11 Billion (39 years later)

Fig. 2.3. Earth's actual and estimated population increase from 1804 to 2093. Data from UN 1997 estimates [27].

Nutritional levels and disease were major factors determining population growth prior to the first half of the twentieth century, but with the general improvement in food supply and healthcare, albeit unevenly achieved on a global scale, human fertility has become the dominant factor determining population growth. During the last two thousand years global population growth has increased greatly from an average of about 0.04% per year from AD 1 to 1650 to a peak of 2.1% from 1965 to 1970 and its decline to 1.35% in 1996 and beyond [1,15]. Birth rates vary widely from nation to nation and generally as life expectancy and living standards improve so smaller families are favoured, a trend first seen in North America and Europe in the nineteenth century. There are signs that this trend is becoming global but it will take some time before this stabilizes Earth's population. At an average of 2.11 children born to every woman that allows for children dying before reaching puberty, the global population will stabilize and below this birth rate will decline. However, population stabilization or decline will not occur until seven to eight decades later when most of those alive when the 2.11 or lower birth rate was achieved have died. This is because at the time when the birth rate falls to 2.11 a very large proportion of the female population will either be in childhood or of childbearing age. On current predictions, Japan, that achieved this birth rate in the 1950s, will experience either a levelling off or a decrease in population from early in the twenty-first century onwards as seen in Fig. 2.4 [30,31]. Europe appears likely to reach a similar state soon afterwards [32].

In contrast, the least affluent section of humanity in sub-Saharan Africa is unlikely to reach a stable population until well into the twenty-third century, although the AIDS/HIV pandemic may alter this prediction [33]. Asia and South and Central America will probably achieve stationary populations towards the end of that century. Among individual nations, India with a 2000-2005 birth rate of 3.01 could reach a stable population of 1.5 billion by the end of the twenty-first century and become the most populous nation. China, acutely aware of its burgeoning population, introduced a closely monitored one child family program, with some exclusions for minority ethnic groups, and in 2000-2005 had a birth rate of 1.83. While this program has been criticized on the grounds of restriction of civil liberties, the alternative situation where a rapidly growing population overwhelmed an already scarcely adequate food supply would be a disaster. In contrast, Pakistan with a 2000-2005 birth rate of 5.08 appears set to stabilize its population at 350 million that its present agricultural productivity seems unlikely to be able to sustain. While such a birth rate was once typical of developing nations, those of Pakistan's neighbours, India and Bangladesh had fallen to 3.01 and 3.46, respectively, in 2000-2005 as an increasing awareness of the disadvantages of overpopulation resulted in major efforts to decrease birth rates. At the family level, such population control strategies may result in child abandonment and infanticide in cultures where sons are preferred to daughters as is discussed in section 2.11. Nigeria and Ethiopia are also expected to stabilize their populations at 350 million and Iran, Mexico, the Philippines, Vietnam and the Democratic Republic of Congo will probably reach more than 150 million before their populations stabilize.

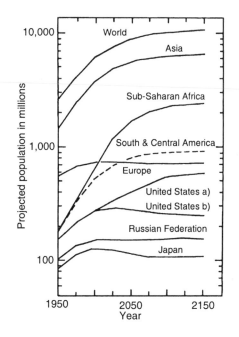

Fig. 2.4. Actual and predicted human populations. Adapted from [30]. Curves US (a) and US (b) represent different assumptions about birth rate and immigration. The population change for South and Central America is shown as a broken curve to distinguish it from adjacent curves.

The above predictions are made on the basis of current birth rate trends. When other factors are taken into account population predictions can vary substantially. This is shown by the United States, a nation that presently accepts large numbers of immigrants. On the basis of current birth rates and immigration rates, the population of the United States could reach more than 500 million by the twenty-second century, whereas a decrease in both birth rate and immigration could result in a population stabilizing at around 250 million as shown by curves US a) and US b), respectively, in Fig. 2.4. In such immigrant nations, substantial differences may arise in birth rates between ethnic groups as in the United States where birth rates of non whites and Hispanic Americans are well above replacement rate while those of white Americans are below replacement rate.

It is obvious that prediction of changes in population is beset with assumptions about birth rates, death rates and migration rates. The impact of such assumptions show very clearly in the predictions of human population change made by Wolfgang Lutz and colleagues in 1996 and shown in Fig. 2.5 [34]. These demographers calculated population changes on the basis of combinations of low, central and high birth rates and death rates that change smoothly from 1995 onwards. For Western Europe the 1995 birth rate of 1.67 was assumed to change to 1.39, 1.89 and 2.39 in 2080-2085 in the low, central

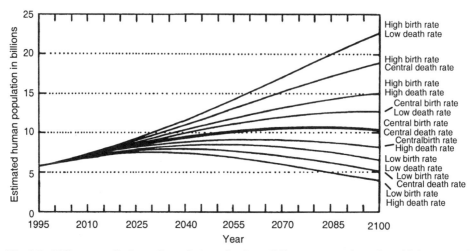

Fig. 2.5. Different predictions of population arise from different assumptions about birth rates and death rates. Adapted from [34].

and high birth rate scenarios. In sub-Saharan Africa the 1995 birth rate of 6.18 was assumed to change to 1.44, 1.94 and 2.44 in 2080-2085 in the low, central and high birth rate scenarios. Death rates were calculated on the basis of life expectancy at birth. Thus, in Western Europe male life expectancy was expected to change from 72.1 in 1995 to 75.9, 83.4 and 90.9 in 2080-2085 for low, central and high death rates, and the corresponding changes in female life expectancy were from 78.6 in 1995 to 82.4, 91.1 and 99.9 in 2080-2085. In sub-Saharan Africa, male life expectancy was expected to change from 50.6 in 1995 to 43.1, 58.1 and 73.1 in 2080-2085 for the low, central and high scenarios, and the corresponding changes in female life expectancy were from 53.9 in 1995 to 46.4, 62.7 and 78.9 in 2080-2085. Analogous changes were assumed for the populations of the other eleven regions into which Earth was subdivided for the purpose of the calculations. A seemingly most probable annual migration of 1.7 million people from the developing to the developed nations was also assumed to occur.

According to this analysis, the most likely scenario was that where central birth rate and death rate applied as indicated by the bold curve in Fig. 2.5. This predicted that population would increase from 5.7 billion in 1995 to 7.9 billion in 2020, and 9.9 billion in 2050, to reach a maximum of 10.6 billion in 2080, and 10.4 billion 2100 as population growth entered a slow decline. This is quite similar to the other predictions that have already been discussed. However, Fig. 2.5 also shows some other population change possibilities that illustrate the effect of changes in assumed birth rate and death rate. In the extreme case of high birth rate and low death rate, population would climb to 22.7 billion by 2100. Such a growth in population would almost certainly lead to great suffering through famine, environmental degradation and wars fought over resources. At the other extreme of low birth rate and high death rate, a maximum population of about 7.35 billion was predicted for 2030 to be followed by a population decline to 3.9 billion by 2100. Although this extreme scenario would lessen the impact on Earth's carrying capacity, this desirable objective could only be achieved at the cost of great suffering as indicated by the high death rate. Such a situation now prevails in sub-Saharan Africa due to AIDS/HIV [33]. A lessening of suffering was offered by the low birth rate and low death rate scenario where the population reached a maximum of 8.5 million by 2050 before declining to 6.5 billion by 2100. Under this circumstance Earth's resources would be severely stretched at a maximum population of 8.5 billion but, providing that this stress could be kept within tolerable bounds, the population decline thereafter offered some relief.

Inevitably, population predictions are subject to substantial uncertainty and this must be recognized when discussing them [35,36]. Thus, a 2001 prediction that there is an eighty-five percent chance that Earth's population will stop growing before 2100 has to be viewed in the context that by the same estimate there is a sixty percent chance that Earth's population will not exceed ten billion before 2100 and a fifteen percent chance that it will be less than six billion. Such predictions depend very much on the statistical methods used and can result in substantially different predictions as shown by one estimate that there is a ninety-five percent chance that in 2050 Earth's population will be between 6.6 and 11.4 billion, whereas another estimate indicates that there is a ninety-five percent chance that it will be between 7.9 and 10.9 billion.

2.5. Birth Rates and Population Change

To this point, discussion has largely dwelt on global population issues. It is now appropriate to look at some of the differences between nations. The birth rates and population changes in the selection of nations discussed in section 2.2 show marked differences as is seen in Fig. 2.6. Apart from the United States, all of the developed and transitional economy nations have birth rates well below replacement level and show population changes ranging from an increase of 1.03 percent for the United States to a decrease of 0.78 percent for Ukraine. The relatively high population growth for the United States reflects its large intake of immigrants while the declining population of Ukraine is typical of six of the seven European nations of the former Soviet Union. The Japanese birth rate and population changes differ little from those of West European nations and emphasize the close correlation between low birth rates and developed nation status. It is anticipated that over the next century most nations will complete similar demographic transitions to the stage where their birth rates and death rates are similar so that either population increase will slow markedly, or population decrease will set in, or a stable population will be achieved depending on the particular nation concerned [37,38].

There are also considerable variations in birth rate and population change among the developing nations. While birth rates in China and Thailand are below replacement level, high birth rates and population growth rates warn of difficult futures for Pakistan, Nigeria and the Democratic Republic of Congo unless these issues are urgently addressed.

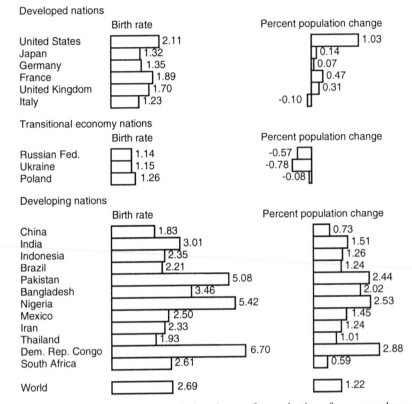

Fig. 2.6. Birth rates and percentage population changes for a selection of more populous nations, 2000-2005 [18].

2.6. Wealth and Birth Rate

It has been seen that the wealthy developed nations have birth rates well below replacement levels in most cases and it was once thought that achievement of developed nation status was the central factor in stabilizing population. The variation of birth rate with per capita gross domestic product, GDP [39], shown in Fig. 2.7, indicates that there is a strong correlation between a developed nation being wealthy and possessing a low birth rate. However, wealth is not the only prerequisite for achieving a birth rate below replacement levels as shown by China and Thailand. The transitional economy nations also have low birth rates despite being less wealthy, but to some extent their gross domestic products are misleading due to changed circumstances. Because of the rigidly controlled currency exchange rates and the shelter from competition provided by the centrally directed national economies of the former Soviet Bloc, exposure to the global market economy accompanying the collapse of communism resulted in great falls in their currency exchange rates and dislocation of their economies. Nevertheless, these nations are technologically sophisticated and their wealth should increase quite rapidly as they adjust to their new economic environments.

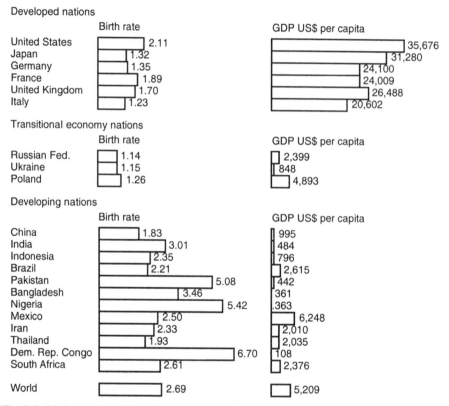

Fig. 2.7. Birth rates for 2000-2005 and 2002 gross domestic products (GDP) of a selection of more populous nations. Data from [18,39].

2.7. Education and Birth Rate

There is strong evidence that birth rate declines as education becomes available to all, but a detailed analysis of this linkage is complex. It is almost invariably the case that the economies of highly educated nations rely more on sophisticated technology and provision of a multitude of services than on manual labour and that both male and female lifestyle and career choices are diverse. While in the developed and transitional economy nations females are generally required to receive the same primary and secondary education as males and are expected to have equal access to tertiary education, this is not invariably the case in developing nations as Fig. 2.8 shows [40]. Only thirty-five percent of Pakistani females attended primary school in 1995 compared with seventy percent of males, and the situation was similar in Bangladesh [41]. This is reflected in Fig. 2.8 in the 2000 relative illiteracy rates for these nations. While, the illiteracy rates are lower for the other nations, in most cases illiteracy is higher

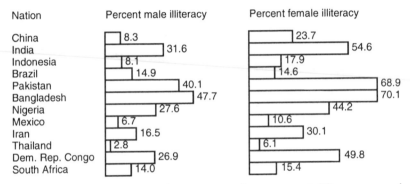

Fig. 2.8. Estimated percentage illiteracy rates of persons aged fifteen years and above in a selection of the more populous developing nations in 2000 [40].

for females because of a bias against them in access to education, a continuation of which in later secondary and tertiary education results in a relatively low proportion of educated females. Apart from the injustice of this situation, there appears to be a significant resultant impact on the birth rate of nations where females are poorly educated. The coincidence of the low birth rates and high educational attainment of females in the developed and transitional economy nations suggests that female control of fertility has at least a major, and probably a dominant, influence on population levels. Quite apart from the right of females to choose the time and frequency of their childbearing, there appears to be a strong correlation between career and lifestyle choices by both males and females and the low birth rates in developed nations. In some such cases birth rates have dropped substantially below replacement rate as exemplified by Germany and Italy where the 2000-2005 birth rates were 1.35 and 1.23, respectively.

Challenged Earth

2.8. Contraception and Birth Rate

In developing nations the strong correlation between the increased use of
contraception and the decline in birth rate is seen in Fig. 2.9 [18,27]. Intuitively,
this is hardly surprising, but it is a vital correlation if Earth's population is to be
held at a sustainable level. It also illustrates the importance of education in
fertility control in achieving this objective. The birth rates in many developing
nations have plummeted by a third to a half since the mid 1960s, but the pattern
has differed from that of the developed nations where a slow decrease in birth

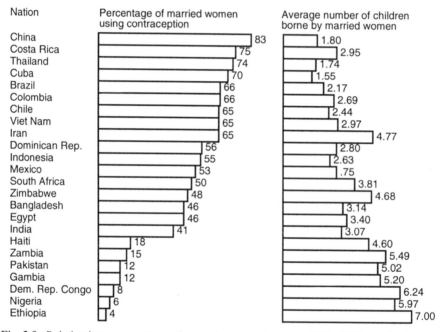

Fig. 2.9. Relation between contraceptive use by married women (1991-1995) and birth rate (1995-2000) in some developing nations [18,27].

rates over several decades occurred only after economic growth brought
improved healthcare and education. It appears that the educational effect of the
mass media and better contraceptive methods available to family planning
programs since the mid 1960s have been major influences on the rapid decline in
birth rate even in those developing nations where improvement in overall living
conditions and educational opportunities has been slow [42].

Generally, the major declines in birth rates have occurred in Asian, South American and Caribbean nations [43]. Until quite recently, birth rate declines have been much less in sub-Saharan African nations, but it now appears that their birth rates are beginning a substantial decline. Apart from differences in family structures and the age at which the first child is born, the additional reasons for these differences appear to be the differing rates at which radio and television have become available as means of communication and the variation in the willingness and abilities of governments to either introduce their own family planning programs or to accept international aid in such programs. The mix of usage of abstinence, rhythm method, condoms, contraceptive pills, intrauterine devices, contraceptive injections, abortion and sterilization in fertility control also appears to vary substantially among nations.

2.9. The Ageing Population

While there is a wide perception that population growth should be slowed, its achievement comes at a cost. As populations stabilize, substantial changes in demography, or population structure, will occur. Already Earth's population is growing older as survival rates for children and woman of childbearing age have generally increased with improvement in healthcare and better nutrition [31,44]. This has been especially noticeable in the developed nations where life expectancy has increased substantially beyond the age of sixty-five since the mid 1960s until 1990, as is shown in Fig. 2.10 [45], and has continued to increase. Thus, in Japan life expectancy for males increased from sixty-five to seventy-six and for females from seventy to eighty-two in the years 1960 to 1990, and represents one of the greater improvements in life expectancy [45,46]. Globally, however, there are great variations in life expectancy. Thus, while the combined life expectancy for males and females born in the period 2000-2005 for the more populous developed nations: the United States (77.1), Japan (81.6), Germany (78.3), France (79.0), the United Kingdom (78.2) and Italy (78.7), shown in years in brackets, were high, those of the larger transitional economy nations; the Russian Federation (66.8), Ukraine (69.7) and Poland (73.9) were significantly less [18].

62 *Challenged Earth*

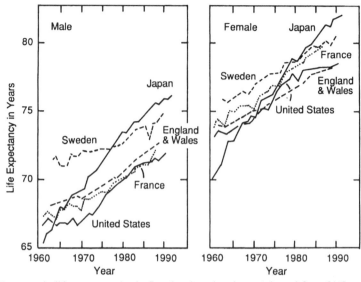

Fig. 2.10. Increase in life expectancies in five developed nations. Adapted from [45].

Currently, life expectancies in the some of the more populous developing nations are similar to those of the transitional economy nations as exemplified by China (71.0), India (63.9), Indonesia, (66.8), Brazil (68.1), Pakistan (61.0), Bangladesh (61.4), Mexico (73.4), Iran (70.3) and Thailand (69.3). However, life expectancies in some of the larger African nations are much less as shown by Nigeria (51.5), the Democratic Republic of Congo (41.8) and South Africa (47.7), substantially because of the HIV/AIDS pandemic in sub-Saharan Africa. Generally, where public commitment to education, healthcare and nutrition have been given reasonable priority in developing nations, male and female life expectancies exceed seventy years, as exemplified by Sri Lanka (72.6) and the Indian state of Kerala [47]. The combination of the differences in birth rate and life expectancy in the more populous nations of Fig. 2.7 has resulted in the proportion of the population under fifteen years of age being considerably less than those over sixty in the developed nations whereas the reverse is the case in the developing nations. The consequence is that the time taken for the developing nations' populations to stabilize will be substantially longer than is the case for the developed nations. Nevertheless, the proportion of elderly people will increase as the global birth rate falls, as shown in Fig. 2.11, and the demographies of the nations will become similar [48].

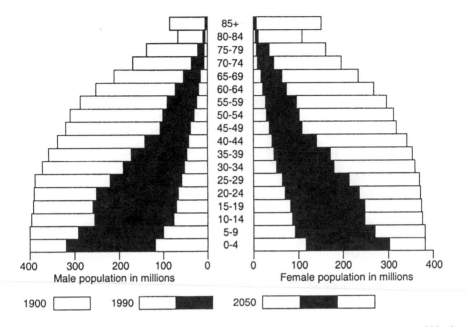

Fig. 2.11. The change in age structure as Earth's population stabilizes. Until about 1900, the population age structure was a narrow pyramid. This structure became broader and more rectilinear in 1990 as the proportion of people attaining greater age increased in the growing population. The predicted population for 2050 shows a significant increase in the number of people in the middle and older age groups. Adapted from [48].

While the stabilizing of population underlying Earth's changing population structure is encouraging, the ageing of populations will place an increasing burden on the generations to follow. This arises from the increase in disability that accompanies ageing coupled with increased longevity. Thus, in a study of the change in the United States population from 1900 to 1990 it was found that the proportion of the 1990 population exceeding eighty-five years was much greater than that of the 1900 population. However, despite the healthier lifestyles enjoyed by the 1990 population, they still suffered a high disability level beyond the age of eighty-five [49]. This entails a requirement for a higher level of aged care coincident with a decline in the proportion of the population of working age and presents a problem not only for the United States, but also for all other nations with similar age distributions. A possible solution for nations with ageing populations facing difficulties with labour shortages and in caring for the aged may be to encourage the migration of young people from developing nations whose populations are still growing [50]. However, the developing nations will

Challenged Earth

face similar problems as their population growth levels out [51], and it is
inevitable that humanity will have to adjust to an older population.

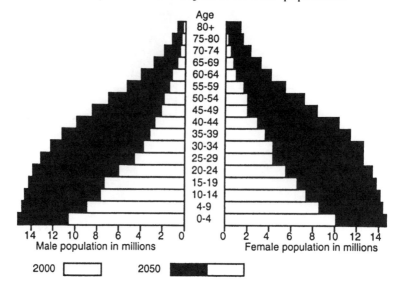

Fig. 2.12. The 2000 and anticipated 2050 population profiles for Nigeria Adapted from [52].

The developing nations show pyramidal population profiles because of their
high birth rates and relatively low life expectancies by comparison with the
developed nations [52]. The 2000 population profile of Nigeria shown in Fig.
2.12 typifies such nations, although its population will age as birth rate declines
and life expectancy increases. Even by 2050 it is anticipated that half of
Nigeria's population will be aged thirty and under. However, the 2050 profile
shown does not take into account the effects of the AIDS/HIV pandemic on
Nigeria, an aspect that is discussed below for Botswana. The population profiles
of developed nations such as Italy are quite different as is seen from Fig. 2.13
[52]. The 2000 population profile for Italy shows a distinct bulge in the twenty to
sixty-five age group, and the expected 2050 profile shows over half of the
population to be aged forty-five or greater.

The devastating effect of the AIDS/HIV pandemic on the 2020 population
profile of Botswana is shown in Fig. 2.14 [52]. It typifies a number of sub-
Saharan African nations where the pandemic shows little sign of slackening.

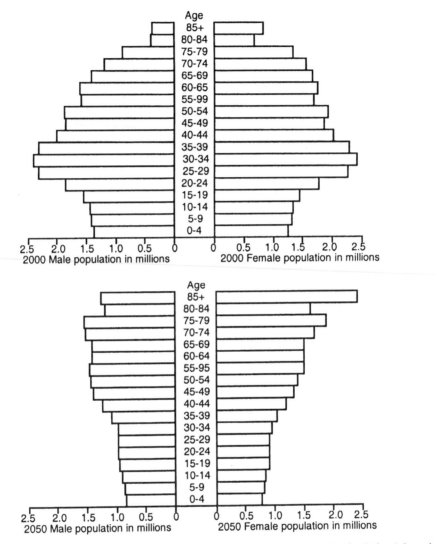

Fig. 2.13. The 2000 (upper) and anticipated 2050 (lower) population profiles for Italy. Adapted from [52].

Challenged Earth

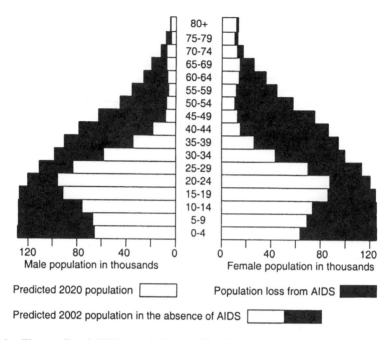

Fig. 2.14. The predicted 2020 population profile for Botswana showing the effect of the AIDS/HIV pandemic. Adapted from [52].

2.10. Life Expectancy Decline in the Transitional Economy Nations

In the Russian Federation a very unusual decline in life expectancy from 64.3 in 1985-1990 to 58.0 in 1995-2000 for males, and from 74.3 in 1985-1990 to 71.5 in 1995-2000 for females occurred [27,53]. Similar declines in life expectancy occurred in Belarus, Estonia, Latvia, Lithuania and Ukraine that were also European states in the dissolved Soviet Union, with only Moldova not showing this trend. Life expectancy increases in other European states that were formerly part of the Soviet Bloc: Bulgaria, Czech Republic, Hungary, Poland, Romania and Slovakia have largely ceased. These unfavourable changes in life expectancies do not appear to be a consequence of changes in statistical compilation since the disintegration of the Soviet Bloc in 1989. To some extent this is borne out by comparison with the eight Asian states of the former Soviet Union: Armenia, Azerbaijan, Georgia, Kazkhakstan, Kyrgyzstan, Tajikistan, Turkmenistan and Uzbekistan that showed modest increases in life expectancy during recent decades with male life expectancy in 1995-2000 ranging from 68.5

in Georgia to 61.2 in Turkmenistan, and female life expectancy ranging from 76.7 in Georgia to 68.0 in Turkmenistan [27].

The decline of life expectancy for males, in particular, in the Russian Federation is disturbingly large and in terms of life expectancy it is now on a par with some of the less fortunate developing nations. In the five year period 1990-1994 the decline in male life expectancy is estimated to have accounted for 1.3 million extra male deaths over those expected if life expectancy had stayed constant, and over half of these deaths were in the age group twenty-five to fifty-nine. Even in a nation of more than 100 million people, this is a very large number of extra male deaths and is similar in magnitude to that which might be anticipated from a substantial conventional war. The major causes of these excess deaths were heart attacks, strokes, alcohol poisoning, homicide and suicide. The situation has not always been this depressing. In 1965, life expectancy in the Soviet Union was similar to that of Japan and the United States, but thereafter increases in life expectancy slowed compared with those in developed nations as communism and the Soviet economy crumbled, and from the mid 1980s male life expectancy declined rapidly with a slower decline occurring for females. It seems that this was accompanied by the adoption of increasingly unhealthy lifestyles characterized by heavy smoking and alcohol consumption, poor diet and lack of exercise [53,54]. A substantial increase in the number of families living in poverty from two percent in 1987 to fifty percent, or seventy-four million people in 1995, coincident with the change from a centrally controlled economy towards a free market economy, exacerbated this situation. This was accompanied by an upsurge in homicide so that statistically the Russian Federation became on a par with South Africa as one of the most violent nations.

When the decline of life expectancy in the Russian Federation is coupled with a low birth rate, the decrease of this nation's population by 0.31 percent, or 460,000 people, each year in 1995-2000 was to be expected. With the exception of Moldova, the seven European states of the former Soviet Union suffered annual population declines totalling 711,000 in 1995-2000, with Estonia, Latvia and Lithuania showing high migration rates, largely of Russians to the Russian Federation.

2.11. The Missing Females

Close inspection of present day populations reveals geographic anomalies in the ratio of males to females. Given similar healthcare and nutrition, medical evidence indicates that females tend to live longer than males and generally female life expectancy exceeds that of the male globally [27]. Although, males outnumber females at conception and birth by a ratio of about 1.06 globally, higher male infant mortality lowered this ratio to 1.01 in 2003 [18]. In that year, the ratio of males to females in the larger developed nations was: United States was 0.97, Japan, 0.96, Germany, 0.96, France, 0.95, the United Kingdom, 0.95, and Italy, 0.94; and in the transitional economy nations it was: Russian Federation, 0.88, Ukraine, 0.87 and Poland, 0.94. While, this ratio is decreased to some extent by the death of males in wars, the contrast with the ratios for some developing nations is particularly marked. For the larger developing nations the ratios were: China, 1.06, India, 1.06, Indonesia, 1.00, Brazil, 0.97, Pakistan, 1.05, Bangladesh, 1.05, Nigeria, 1.01, Mexico, 0.96, Iran, 1.03, Thailand, 0.96, Democratic Republic of Congo, 0.98, and South Africa, 0.96. If a ratio of 0.98, which is probably a little low, is taken as the benchmark, it appears that many millions of females are missing in those countries showing a ratio greater than this. On this basis, China, India, Pakistan, Bangladesh and Iran contributed about thirty-eight and thirty-two, five, three and one million of the missing females, respectively. Yet in the period 1995-2000 the life expectancies of females exceeded that of males in these five nations by 2.8, 0.6, 2.2, 0.1 and 1.5 years, respectively. Thus, a bias appears to exist against females at birth and this possibility is now examined.

Largely because of its huge population and its large number of "missing females", China has attracted especially close attention from demographers. The reported ratio of males to females born in China in 1989 was 1.049 for the first born and 1.209, 1.246 and 1.317 for the second, third and fourth born, with a ratio of 1.138 for all births, and in some localities ratios as high as 1.45 were reported [55]. As there is no evidence for such a bias towards males with successive births having a biological origin, the source of the deficit in female births lies elsewhere [56]. A combination of China's "one child per family" policy and a preference for male children in rural areas, in particular, are the probable sources of the female deficit. This deficit is thought to arise through failure to report many female births, a higher incidence of abortion of female foetuses where prenatal sex identification is available, abandonment of female children and infanticide. In India the survival rate of female children is also substantially less than that of males, particularly in northern India. Generally, in India the bias against the survival of female children is found to be much diminished as female literacy increases, coincident with a decrease in birth rate [57]. While, it is probable that the female deficits in some of the other developing nations arise from different combinations of the factors producing the female deficits in China and India, the inescapable conclusion is that the lives of female children are inexcusably undervalued in those nations showing a deficit in their female population.

2.12. Urbanization

Coincident with population growth there has been a global trend for people to move to the cities in a process called urbanization. In the developed nations this trend has been driven by factors such as improved employment prospects and preferred lifestyles. In the developing nations, these factors are also important, but the much harsher factors of land dispossession, environmental degradation and civil war have often been dominant. According to United Nations estimates, forty-five percent of Earth's population were urban dwellers in 1994, and this was predicted to increase to fifty percent in 2005 and over sixty percent in 2025 [34]. In the developed nations seventy-five percent of the population were urban while only thirty-seven percent were urban in the developing nations. Globally the urban population was growing at a rate of 2.5 percent annually to accommodate about sixty-one million additional people in 1994, while the rural growth rate was only 0.8 percent to accommodate twenty-five million additional people annually.

This movement of population to the cities resulted in 408 cities of one million inhabitants or more in 2003 [58]. Among them are many "megacities" whose populations are very large irrespective of variations in the way that their boundaries are defined. Such megacities are exemplified by the twenty largest in 2003 headed by Tokyo with 35.0 million people, followed by Mexico City (18.7

million), New York (18.3 million), São Paulo (17.9 million), Mumbai (17.4 million), Delhi (14.1 million), Calcutta (13.8 million), Buenos Aires (13.0 million), Shanghai (12.8 million), Jakarta (12.3 million), Los Angeles (12.0), Dhaka (11.6 million), Osaka-Kobe (11.2 million), Rio de Janeiro (11.2 million), Karachi (11.1 million), Beijing (10.8 million), Cairo (10.8 million), Moscow (10.5 million), Manila (10.4 million) and Lagos (6.0 million).

2.13. Migration

Humans have been a migratory species since their appearance on Earth as shown by their spread to every habitable region, and they remain so today. The factors that drive migrations today, starvation, poverty, environmental disasters, persecution and war and, at a milder level, the search for a better lifestyle have always driven migrations. The only difference is that today knowledge of other places to live is more widespread and travel is much easier. In the nineteenth and twentieth centuries a great proportion of migrants was European. This emigration from Europe to the "New World" amounted to more than forty million in the period 1850-1913 although many of these migrants ultimately returned home. Nearly two thirds of migrants headed for the United States with the remainder being absorbed by Australia, Canada, New Zealand and South America [59]. The attraction of these destinations was probably a mixture of the notion of less constraining societies in lands as yet incompletely settled according to European perceptions (although this was unlikely to have been a perception shared by the indigenous peoples of these lands), and the knowledge that Europeans already dominated these destinations.

Another large migrant group over the last two centuries has been the Chinese, of which there were about thirty-seven million living in 136 countries outside the Peoples Republic of China and Taiwan in 1990 [60]. Twenty percent of these "overseas Chinese" were in Indonesia, sixteen percent in Thailand, fifteen percent in Hong Kong, fifteen percent in Malaysia, six percent in Singapore, five percent in Vietnam, four percent in the United States, four percent in Myanmar (Burma), two percent in the Philippines, two percent in Canada, with the remainder in a wide range of other countries.

As recently as the early 1960's the major proportion of intercontinental migrants was still Europeans moving largely to Australasia and North America. Since that time migration from the developing world to the developed world and migration within the developing world has assumed increased importance. Because of a combination of the complexity of assessing the net population change for either a given nation or geographic region, and also because of slowness in publication of this material, one of the most complete and recent sets of global numbers available refers to the period 1985-1989 [27,61]. No attempt at detailed analysis of the flow of migrants is made here; rather the intention is to indicate the magnitude of the numbers of migrants and their origins.

A considerable flow of migrants within the developed world occurred in 1985-1989 so that North America experienced an annual maximum net gain of 125,854 people, Australasia, 47,305 and Western Europe (Belgium, France, Germany, the Netherlands, Sweden and the United Kingdom), 303,315 [27]. For the same three regions, the annual maximum net gain in migrants arriving from the developing world was 830,160, 103,865 and 496,102, to give a total annual maximum net intake of migrants of 956,014, 151,170 and 799,412 that sum to 1,906,601 migrants granted residency. A recent projection indicates that this flow of migrants from the developing nations to the developed nations is likely to continue and will amount to two million annually in the period 2000-2050 [18]. The major recipients of these immigrants on an annual basis are likely to be the United States (1,1000,000 net immigrants), Germany (211,000), Canada (173,000), the United Kingdom (136,000) and Australia (83,000). The nations expected to experience an annual loss of people as migrants are: China (303,000), Mexico (267,000), India (222,000), the Philippines (184,000) and Indonesia (180,000). Another large flow of two to three million workers largely from India, Pakistan and Southeast Asia to the oil rich states of the Persian Gulf occurs annually, but these workers are seldom granted right of residency [62]. The number of migratory people legally on the move is substantially less than the number of would be migrants. This is shown by the illegal migration of Mexicans into the United States that has been estimated to be as high as one million people annually. On the other side of the Atlantic, it is estimated that at least twelve million illegal immigrants live in Western Europe.

A new classification, "environmental refugee", has come into use and describes persons forced to leave their land because it can no longer support them for a wide variety of reasons [63]. An example of this type of refugee is the Marsh Arabs of Iraq who fell into disfavour with the Iraqi government during the Gulf War. In 1992, the villages of the Marsh Arabs were attacked by the Iraqi military and the rivers feeding the 700,000 hectare Qurnah marshes that sustained them were diverted. As the marshes dried out, the Marsh Arabs' source of food and shelter vanished and they were forced to disperse, about 100,000 of them to Iran and about 200,000 within Iraq itself. It has been estimated that some twenty-five million environmental refugees currently exist in the developing nations [64]. This could rise to 200 million by the middle of the twenty-first century if the present predictions of climate change and sea level rises occur. The humanitarian aspects of such a vast number of environmental refugees are appalling and the political implications daunting.

2.14. The Future

After this brief reconnaissance of Earth's population, three dominant conclusions about the future appear to be justified. First, the human population will grow substantially and could reach ten to eleven billion by 2100 before a slow decline sets in. Second, the population will grow older and greatly increase the resources required for care of the aged. Third, the proportion of the population in the developing nations will increase substantially relative to the developed nations. While these conclusions are based on a continuation of the trends discussed above, it should be borne in mind that people and nations change their minds and in doing so may change these trends. These conclusions also lead naturally to thoughts about the supply of water, food and energy available to sustain Earth's growing population, the likely effect of disease on population and the effect of population growth on the habitability of Earth. These aspects are explored in the chapters that follow.

References

1.   The United States Bureau of the Census regularly updates its population estimates by the second on its "World POPclock" that may be accessed through URL http://www.census.

gov/ main/ www/popclock.html. This website also provides other extensive global population statistics.

2. The full text of the Rio Declaration appears in *Popn. Dev. Rev.*, **1992**, *18*, 571.

3. M. Holloway, Population pressure. The road from Rio is paved with factions, *Sci. Am.*, **1992**, *267*, Sept., 36.

4. a) B. Bolin, The Kyoto negotiations on climate change: A science perspective, *Science*, **1998**, *279*, 330. b) T. Blundell, After the Hague? *Chem. Indust.*, **2001**, 2 Jan., 11. c) F. Pearce, A real roasting, *New Sci.*, **2001**, 7 April, 11. d) B. Hileman, Agreement is reached in Bonn, *Chem. Eng. News*, **2001**, *79*, 30 July, 13. e) *Anon.*, Climate respite? *New Sci.*, **2005**, 19 Feb., 4.

5. J. E. Cohen, *How Many People can the Earth Support?* Norton, New York, USA, 1995.

6. J. E. Cohen, Population growth and Earth's human carrying capacity, *Science*, **1995**, *269*, 341.

7. N. Keyfitz, The growing human population, *Sci. Am.*, **1989**, *261*, Sept., 71.

8. E. O. Wilson, Threats to biodiversity, *Sci. Am.*, **1989**, *261*, Sept., 60.

9. J. Bongaarts, Population growth and global warming, *Popn. Dev. Rev.*, **1992**, *18*, 299.

10. G. C. Daily and P. R. Ehrlich, Population, sustainability, and Earth's carrying capacity, *BioScience*, **1992**, *42*, Nov., 761.

11. A. Sen, The economics of life and death, *Sci. Am.*, **1993**, *268*, May, 18.

12. J. Bongaarts, Can the growing human population feed itself? *Sci. Am.*, **1994**, *270*, March, 19.

13. H. W. Kendall and D. Pimentel, Constraints on the expansion of the global food supply, *Ambio*, **1994**, *23*, 198.

14. P. S. Dasgupta, Population, poverty and the local environment, *Sci. Am.*, **1995**, *272*, Feb., 26.

15. K. Arrow *et al.*, Economic growth, carrying capacity, and the environment, *Science*, **1995**, *268*, 520.

16. J. Bongaarts, Population pressure and the world food supply system in the developing world, *Popn. Dev. Rev.*, **1996**, *22*, 483.

17. T. F. Homer-Dixon, J. H. Boutwell and G. W. Rathjens, Environmental change and violent conflict, *Sci. Am.*, **1993**, *268*, Feb., 38.

18. *World Population Prospects. The 2002 Revision*, The Population Division, United Nations, New York, USA, 2003. While the United Nations continuously updates statistics on population and many other aspects of human activities, it is an enormous task and, as a consequence, incomplete statistics are available for the most recent years. Most of the statistics used in this chapter are from this reference and [27].

19. P. R. Ehrlich, A. H. Ehrlich and G. C. Daily, Food security, population, and environment, *Popn. Dev. Rev.*, **1993**, *19*, 1.

20. a) M. Livi-Bacchi, On the costs of collectivisation in the Soviet Union, *Popn. Dev. Rev.*, **1993**, *19*, 743. b) D. Goodkind and L. West, The North Korean famine and its demographic impact, *ibid.*, **2001**, *27*, 219.

21. a) G. Hardin, The tragedy of the commons, *Science*, **1968**, *162*, 1243. b) G. Hardin, Extensions of "The tragedy of the Commons", *Science*, **1998**, *280*, 1682.

22. J. Burger and M. Gochfeld, The tragedy of the commons, *Environment*, **1998**, *40*, Dec., 4.

23. H. R. Hulett, Optimum world population, *BioScience*, **1970**, *20*, Feb., 160.

24. C. T. de Wit, Photosynthesis: Its relationship to overpopulation, in *Harvesting The Sun: Photosynthesis in Plant Life*, A. San Pietro, F. A. Greer, and T. J. Army, eds., Academic Press, New York, USA, 1967.

25. C. Marchetti, 10^{12}: A check on the Earth-carrying capacity for man, *Energy*, **1979**, *4*, 1107.

26. A. S. Moffat, Ecologists look at the big picture, *Science*, **1996**, *273*, 1490.

27. *World Population Prospects*, The Population Division, United Nations, New York, USA, 1997.

28. Editorial, Is six billion a reason to celebrate? *Nature Med.*, **1999**, *5*, 1215.

29. F. Pearce, Population bombshell, *New Sci.*, *Inside Science 112*, **1998**, 11 July, 1.

30. R. Doyle, Global fertility and population, *Sci. Am.*, **1997**, *276*, March, 20.

31. a) H.-P. Kohler, F. C. Billari and J. A. Ortega, The emergence of lowest-low fertility in Europe during the 1990s, *Popn. Dev. Rev.*, **2002**, *28*, 641. b) W. Lutz, B. C. O'Neill and S. Scherbov, Europe's population at a turning point, *Science*, **2003**, *299*, 1991.
32. J. E. Cohen, Human population: The next half century, *Science*, **2003**, *302*, 1172.
33. a) B. Schwartländer *et al.*, AIDS in a new millennium, *Science*, **2000**, *289*, 64. b) United Nations 1998 revision of the world population estimates and projections. *The Demographic Impact of HIV/AIDS*, Population Division, New York, USA, 1998. http://www.popin.org/pop1988/6.htm. c) U.S. Census Bureau, International Population Reports WP/02-2. *The Aids Pandemic in the 21st Century*, U.S. Government Printing Office, Washington, DC, USA, 2004. d) *AIDS Epidemic Update, December 2004*, UNAIDS/WHO, Geneva, Switzerland, 2004. e) *The World Health Report: 2004. Changing History*, WHO, Geneva, Switzerland, 2004.
34. W. Lutz *et al.*, World population scenarios for the 21st century, in *The Future Population of The World: What Can We Assume Today?* W. Lutz, ed., Earthscan, London, UK, 1996, p. 361.
35. W. Lutz, W. Sanderson, and S. Scherbov, Doubling of world population unlikely, *Nature*, **1997**, *387*, 803.
36. a) J. Bongaarts and R. A. Bulatao, eds., *Beyond Six Billion: Forecasting the World's Population*, National Academy Press, Washington, DC, USA, 2000. b) N. Keilman, Uncertain population forecasts, *Nature*, **2001**, *412*, 490. c) W. Lutz, W. Sanderson and S. Scherbov, The end of world population growth, *ibid.*, p. 543.
37. a) J. Bongaarts, Demographic consequences of declining fertility, *Science*, **1998**, *282*, 419. b) F. Pearce, Mamma mia, *New Sci.*, **2002**, 20 July, 38.
38. a) J. Bongaarts and R. A. Bulatao, Completing the demographic transition, *Popn. Dev. Rev.*, **1999**, *25*, 515. b) J. Vallin, The end of the demographic transition: relief or concern? *Popn. Dev. Rev.*, **2002**, *28*, 105.
39. National Accounts Main Aggregates Database, United Nations Statistics Division, New York, USA, 2004. http://unstats.un.org/unsd/snaama/glossResults.asp?Id=5.
40. United Nations Statistic Division, *Indicators on Literacy*, New York, USA, 2000.
41. F. Pearce, The rise and rise of Pakistan's people, *New Sci.*, **1996**, 27 April, 7.
42. J. C. Caldwell and P. Caldwell, High fertility in sub-Saharan Africa, *Sci. Am.*, **1990**, *262*, May, 82.
43. a) B. Robey, S. O. Rutstein and L. Morris, The fertility decline in developing countries, *Sci. Am.*, **1993**, *269*, Dec., 30. b) P. S. Dasgupta, Population, poverty and the local environment, *Sci. Am.*, **1995**, *272*, Feb., 26. c) J. Cleland, A regional review of fertility trends in developing countries, in *The Future Population of The World: What Can We Assume Today?* W. Lutz, ed., Earthscan, London, UK, 1996, p.47. d) C. F. Westoff, Reproductive preferences and future fertility in developing countries, *ibid.*, p. 73. e) M. B. Concepcion, Population policies and family-planning in south-east Asia, *ibid.*, p. 88. f) R. Doyle, World birth-control use, *Sci. Am.*, **1996**, *275*, Sept., 18. g) R. Doyle, Global fertility and population, *Sci. Am.*, **1997**, *276*, March, 20. h) M. Potts, The unmet need for family planning, *Sci. Am.*, **2000**, *282*, Jan., 70.
44. a) R. Doyle, The future of the old, *Sci. Am.*, **1998**, *278*, May, 17. b) F. Pearce, The rise of the wrinklies, *New Sci.*, **1998**, 20 June, 51.
45. S. Oshima, Japan: feeling the strains of an ageing population, *Science*, **1996**, *273*, 44.
46. a) N. Ogawa and R. D. Retherford, The resumption of fertility decline in Japan: 1973-92, *Popn. Dev. Rev.*, **1993**, *19*, 703. b) S. Horiuchi, Greater lifetime expectations, *Nature*, **2000**, *405*, 744.
47. A. Sen, The economics of life and death, *Sci. Am.*, **1993**, *268*, May, 18.
48. S. J. Olshansky, B. A. Carnes and C. K. Cassel, The ageing of the human species, *Sci. Am.*, **1993**, *268*, April, 18.
49. W. Roush, Live long and prosper, *Science*, **1996**, *273*, 42.

50. *Replacement Migration: Is it a Solution to Declining Ageing Populations?* United Nations Population Division, Department of Economic and Social Affairs, New York, USA, 2000. http://www.un.org/esa/population/migration.htm.

51. C. Holden, New populations of old add to poor nations' burdens, *Science*, **1996**, *273*, 46.

52. United States Census Bureau, *World Population Profile 2000*, United States Government Printing Office, Washington, DC, 2000.

53. a) L. C. Chen, F. Wittgenstein and E. McKeon, The upsurge of mortality in Russia: Causes and policy implications, *Popn. Dev. Rev.*, **1996**, *22*, 517. b) T. Lincoln, Death and the demon drink in Russia, *Nature*, **1997**, *388*, 723.

54. R. Stone, Stress: the invisible hand in Eastern Europe's death rates, *Science*, **2000**, *288*, 1732.

55. a) T. H. Hull, Recent trends in sex ratios at birth in China, *Popn. Dev. Rev.*, **1990**, *16*, 63. b) S. Johansson and O. Nygren, The missing girls of China: a new demographic account, *Popn. Dev. Rev.*, **1991**, *17*, 35. c) A. J. Coale, Excess female mortality and the balance of the sexes in the population: An estimate of the number of "missing females", *ibid.*, 517.

56. a) Z. Yi et al., Causes and implications of the recent increase in the reported sex ratio at birth in China, *Popn. Dev. Rev.* **1992**, *18*, 283. b) K. Johnson, The politics of the revival of infant abandonment in China, with special reference to Hunan, *Popn. Dev. Rev.*, **1996**, *22*, 77. c) D. Walfish, National count reveals major societal changes, *Science*, **2001**, *292*, 1823. d) C. Junhong, Prenatal sex determination and sex-selective abortion in rural central China, *Popn. Dev. Rev.*, **2001**, *27*, 259.

57. a) M. Murthi, A.-C. Guio and J. Drèze, Mortality, fertility, and gender bias in India: district-level analysis, *Popn. Dev. Rev.*, **1995**, *21*, 745. b) P. Mayer, India's falling sex ratios, *Popn. Dev. Rev.*, **1999**, *25*, 323. c) F. Arnold, S. Kishor and T. K. Roy, Sex-selective abortions in India, *Popn. Dev. Rev.*, **2002**, *28*, 759.

58. *World Urbanization Prospects: The 2003 Revision*, Population Division, United Nations, New York, USA, 2003.

59. T. J. Hatton and J. G. Williamson, What drove mass migrations from Europe in the late nineteenth century? *Popn. Dev. Rev.*, **1994**, *20*, 533.

60. D. L. Poston, M. X. Mao and M.-Y. Yu, The global distribution of the overseas Chinese around 1990, *Popn. Dev. Rev.*, **1994**, *20*, 631.

61. a) H. Zlotnik, Migration to and from developing regions: A review of past trends, in *The Future Population of The World: What Can We Assume Today?* W. Lutz, ed., Earthscan, London, UK, 1996, p. 299.

62. S. Öberg, Spatial and economic factors in future south-north migration, in *The Future Population of The World: What Can We Assume Today?* W. Lutz, ed., Earthscan, London, UK, 1996, p. 336.

63. N. Fell, Outcasts from Eden, *New Sci.*, **1996**, 31 August, 27.

64. Myers and J. Kent, *Environmental Exodus: An Emergent Crisis in the Global Arena, Climate Institute*, Washington DC, USA, 1995.

Chapter 3

Water: The Vital Resource

"Water is the earth's most distinctive constituent. It set the stage for
the evolution of life and is an essential ingredient of all life today; it
may well be the most precious resource the earth provides to
humankind."

J. W. Maurits la Rivière in Scientific American, September, 1989.

3.1. The Water Planet

When viewed from space, one of the striking impressions of Earth is the vast
area of the oceans that cover seventy percent of the planet's 510 million square
kilometre surface. This abundant water probably arrived in a bombardment of icy
comets over a period of six hundred million years beginning fifty million years
after Earth's formation 4.6 billion years ago, as did that of Venus and Mars [1,2].
However, the later two planets are closer to and further from the sun than is
Earth, respectively, and this has profound consequences. As the sun grew
brighter with its stellar evolution, the water of Venus evaporated and split into
oxygen and hydrogen in the intensifying ultraviolet radiation. Hydrogen, the
lighter gas, escaped into space to leave the hot planet without water. Thus, the
inner boundary of the habitable zone around the sun, where liquid water and the
possibility of life exist on planets, moved outwards beyond Venus whose average
temperature is 420°C. As the sun continues to grow Earth will also lose its water
but that is some billion years in the future. Further out, frozen Mars is close to
the outer boundary of the habitable zone but has insufficient carbon dioxide in its
atmosphere to produce a warming greenhouse effect as strong as that of Earth.
While no liquid water has so far been found on Mars, whose average temperature
is −50°C, there are strong indications that the planetary surface has been eroded
by water. Evidence has been found of frozen water at the poles where it probably
exists as sub-surface glaciers and as fine grains of ice mixed with dust. Thus,
among the planets of the solar system Earth is unique with its plentiful liquid

water that is essential to its myriad lifeforms, the only ones presently known in the universe [3,4].

Most of Earth's 1.39 billion cubic kilometres of water are in the oceans that fill the depressions in the planetary surface to an average depth of 3,800 metres, although some of the ocean trenches are much deeper. This water is continually recycled to the atmosphere through evaporation and transpiration by plants and is returned to the surface as rain and snow. As it flows into aquifers, rivers, lakes and oceans, water leaches minerals from soil and rocks with the consequence that 96.54 percent of all water is so salty that it can neither quench humanity's thirst nor nurture most plants, the foundation of humanity's food supply [5]. Although ocean fish are an important source of food, the oceans contain only seven percent of all animal species while inland waters contain twelve percent with the remaining eighty-one percent living on land. Similarly, the much greater proportion of plants lives on land. Thus, humanity and most other lifeforms are dependent on fresh water, the availability of which is the most critical factor affecting the diversity of life on Earth.

Of Earth's fresh water, 69.6 percent is frozen in glaciers, permanent snow and permafrost as is seen in Fig. 3.1, with some ninety percent of this being in Antarctica. Most of the residual 10,530,000 cubic kilometres is ground water in subterranean aquifers whose replenishment is often so slow that water drawn from them is an exhaustible resource rather like oil. This leaves only 91,000 cubic kilometres of fresh water in lakes, 11,470 cubic kilometres in swamps, 16,500 cubic kilometres of soil moisture and 2,120 cubic kilometres in rivers as renewable surface fresh water, just 0.00874 percent of Earth's water. The lower atmosphere contains about 12,900 cubic kilometres of fresh water as invisible vapour and in clouds that are completely recycled every nine days or so. Finally, all of biology, including humans, contains another 1,120 cubic kilometres of continuously recycled water.

3.2. Water and Life

The vast majority of living organisms consist of at least sixty percent water. Humans, in common with other mammals, are about ninety percent water by weight at birth. Although this proportion decreases with age, adults are still about sixty-five percent water. Bones have the least amount of water at twenty-two percent, brain and muscle are seventy-five percent water and kidneys and blood are eighty-three percent water. Because water is continuously lost from the body in digestive, purification, repair and thermostatting processes it must be constantly replenished. Adults consume up to ten litres of water a day in food and drink and return it to the environment through exhaled breath, perspiration and excretion. The extent of this recycling process varies greatly with temperature and climate and most humans directly exposed to the elements drink much more in summer than in winter. This is because the cooling effect of water evaporating from the skin as perspiration is an important part of the body's thermostatting system. In hot arid climates adults perspire as much as ten litres of water a day while in cooler conditions perspiration rates are much less.

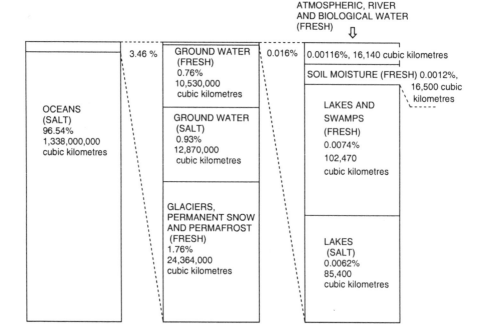

Fig. 3.1. Earth's water resources. Of the total 1,385,984,510 cubic kilometres of water, 96.54% or 1,338,000,000 cubic kilometres fill the oceans, leaving only 3.46% to the landmasses. Of this 34,617,000 cubic kilometres are fresh water of which only 135,110 cubic kilometres are rapidly recycled and available for human use (0.00974% of Earth's total water). Data from [4].

Plants show a similarly great dependence on water. Herbaceous plants are ninety percent water by weight and yet they retain only about one percent of the water they absorb from the soil. Most of the remainder is lost to the atmosphere by transpiration from their leaves that is part of the process whereby nutrients are pumped to the plant's extremities and the plant is thermostatted. In this way plants collectively act as a huge pump transferring water from the soil to the atmosphere and are a major part of a great fresh water purification system that also profoundly affects climate; something to be reflected upon as the clearing of Earth's forests continues.

A more detailed analysis shows that, apart from sustaining life, water is part of life itself. On a molecular scale, the origin of all living tissue is the reaction of water with carbon dioxide to produce carbohydrates through photosynthesis that is powered by sunlight in the green leaves of plants. Subsequently, biochemical processes transform carbohydrates into cellulose, the largest biological reservoir of carbon and the major structural component of plants, and the myriads of other molecules found in living organisms. These include deoxyribonucleic acid, DNA, the blueprint for all living organisms as is discussed in more detail in Chapter 5. Plants are eaten by herbivores that in turn are eaten by carnivores so that all living tissue has its origins in photosynthesis.

Water has still other essential roles in life at the molecular level that warrant exploration. It is formed from the most plentiful atom in the universe, hydrogen (H), and the most abundant atom on Earth, oxygen (O), in a ratio of two to one and has the molecular formula H_2O. This triatomic molecule is bent so that the angle made by the hydrogen atoms bound on either side of the oxygen atom is 105°. However, liquid water does not behave as a collection of small independent molecules. Rather, it acts as a collection of molecules that are linked together. That this is so can be seen when a water droplet falls onto glass. Instead of spreading thinly over the glass surface the droplet forms an almost hemispherical shape because of this linking of the water molecules. Inside the droplet, water molecules move rapidly in all directions, but for fractions of a second a hydrogen atom of one water molecule links to the oxygen atom of another to form a hydrogen bond and create an extensive flickering array of such bonds that largely retains the shape of the droplet.

Hydrogen bonding is one of water's most important characteristics in participating in life [6]. It is only too clear from oil spills that oil does not mix with water, yet part of the despoiling effect of oil floating on water acts out an underlying principle of the chemistry of life. Water cannot form hydrogen bonds with oil and so, while continuing to form hydrogen bonds with itself, it excludes the lighter oil that floats on the surface. Molecules that cannot form hydrogen bonds with water are "hydrophobic", or water hating, while those that can form hydrogen bonds with water are "hydrophilic", or water loving. Most biologically important large molecules such as DNA and proteins are composed of parts that are hydrophobic and parts that are hydrophilic. Because of this, they twist themselves into shapes where their hydrophilic parts turn outward and hydrogen bond with water while their hydrophobic parts turn inwards away from the water. Often they form hydrogen bonds within their own structures. In this way they assume shapes influenced by water that are also those necessary for their biological functions.

The structures of the many types of cells among the ten trillion or so that compose a human, together with bacterial and plant cells, are largely dependent on hydrophobic and hydrophilic interactions with water. Specialized long molecules with oil-like tails, called lipids, form layers in water lined up with their hydrophilic heads side by side and their hydrophobic tails side by side [7]. The heads hydrogen bond with water while the tails do not. Because of this, two of the lipid layers arrange themselves tail-to-tail to form a lipid bilayer, or membrane, where both surfaces exposed to water are made up of hydrophilic lipid heads. Biological cell membranes are composed of similar lipid bilayers that form envelopes that largely define the shape and size of each cell as seen in Fig. 3.2 [8]. Inside animal and plant cells most of the DNA is contained within a nucleus that is itself contained by a lipid membrane. Suspended in the water inside the cell are the many other structures and molecules essential to the functioning of the cell. There are channels in the cell membrane through which water and other molecules can pass in either direction in a controlled manner.

Most mammalian cells are rather like slightly floppy water filled balloons, but they are afforded some rigidity by an internal cytoskeleton composed of proteins. Plant cells lack this cytoskeleton. Their rigidity depends on the combined effects of a cellulose wall outside the cell membrane and water pressure inside the cell that adds to cell rigidity rather like the way air in a balloon gives it rigidity. The wilting of plants that occurs when they are short of water, a consequence of the loss of water from cells and with it rigidity, and the quite rapid revival of the plant on watering and restoration of cell rigidity illustrates the important structural role of water in plant cells. The effects of dehydration on humans are not so physically obvious, but a deficiency of water has a deleterious effect on the functioning of human cells as on plant cells with the result that humans become weak and hallucinate with the onset of dehydration. Clearly, the functioning and construction of animal and plant cells are much more complex than described in this short discussion, but it does give a brief overview of the intimate involvement of water in life. (DNA and biological cells are considered in more detail in Chapter 5.)

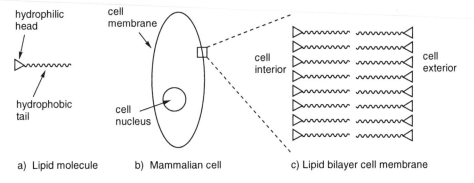

Fig. 3.2. a) A simplified view of a lipid molecule showing its hydrophilic head and hydrophobic tail. b) A simplified view of a mammalian cell showing the cell membrane encapsulating its watery content that includes the nucleus where most of the DNA resides. The cytoskeleton is not shown. c) An expanded simplified view of the cell membrane showing the tail-to-tail arrangement of the lipid bilayer where one surface of lipid heads faces inwards to the cell interior and the other surface of lipid heads faces outwards to the cell exterior.

There are other aspects of water that make Earth habitable. One comes to the fore with water vapour in the atmosphere where it is the most important of the greenhouse gases and, together with carbon dioxide and methane, absorbs much of Earth's reradiated infrared energy to retain the warmth that makes Earth habitable as discussed in more detail in Chapter 8. Another occurs in the troposphere where solar ultraviolet radiation splits a hydrogen atom off the water molecule to produce the hydroxyl radical ($OH^\bullet$). The reactive hydroxyl radical continuously converts naturally produced methane (CH_4), methyl chloride

(CH_3Cl) and similar gases to carbon dioxide, water and other molecules that dissolve in rain and are washed from the atmosphere. If these gases did not undergo this conversion they would build up to high concentrations and have major climate modifying effects. When this action of the hydroxyl radical is considered together with the washing out effect of rain dissolving atmospheric gases such as sulfur trioxide (SO_3) and the oxides of nitrogen (N_2O, NO, NO_2), the important air purifying role of water becomes clear.

A third aspect concerns the commonplace knowledge that ice floats. When water cools it contracts, as do most other liquids. However, water is unusual for at 4°C it reaches its maximum density and below this temperature it expands until at 0°C it becomes less dense than liquid water on which it floats. In ice, water molecules are fixed in position through each hydrogen atom bonding to the oxygen atom of an adjacent water molecule so that each oxygen atom is at the centre of four hydrogen atoms at the apices of a triangular pyramid. This beautiful and repeating pattern results in water molecules being held further apart in ice than they are in liquid water so that they occupy more space and ice is less dense than liquid water. If ice was heavier than liquid water, rivers, lakes and oceans would freeze from the bottom up to become solid with the result that Earth's climate would be very different and life may have evolved differently.

3.3. The Renewal of Fresh Water: The Hydrological Cycle

Although only a small amount of Earth's water is fresh, it is continually replenished by rain and snow that may either fall during most months of the year, or may arrive largely as an annual event such as a monsoon, or may only arrive on rare occasions in arid regions. This massive solar powered replenishment of fresh water is one of the great recycling processes that sustains life on Earth and is illustrated by Fig. 3.3. Water vapour evaporating from the landmasses and oceans, together with that transpired from the leaves of plants, amounts to a total of 577,000 cubic kilometres of fresh water entering the atmosphere annually. Of this, 119,000 cubic kilometres fall on land as rain and snow. Simultaneously, about 47,000 cubic kilometres flow back to the oceans largely as river water (42,570 cubic kilometres) with smaller amounts reaching the ocean as icebergs breaking away from coastal glaciers (2,230 cubic kilometres) and as water percolating through soil and rock (2,200 cubic kilometres). While the magnitude of this fresh water replenishment is enormous, it is the relatively limited availability of fresh water that represents the major factor controlling the upper limit on Earth's human population and the sustainability of living standards [9-11].

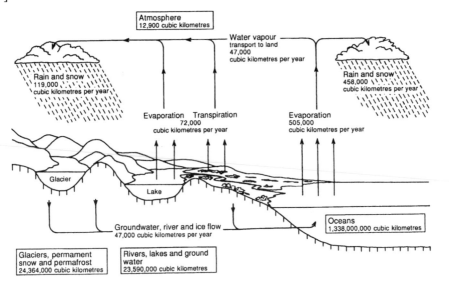

Fig. 3.3. A simplified illustration of the annual global recycling of water, or hydrological cycle, based on [4]. The boxed quantities represent the average amount of water in each area. The unboxed quantities represent average annual water flow in each area.

3.4. Humanity's Use of Fresh Water

Of the 119,000 cubic kilometres of fresh water that fall on land as rain and snow annually, 72,000 cubic kilometres return to the atmosphere through evaporation and transpiration by plants as they absorb water through their roots and release it to the atmosphere through their leaves. Many of these plants are either food crops, are crops that produce oil or fibres, or are trees grown for timber and amenity areas such as gardens and parks. Some 18,200 cubic kilometres, or 25.3 percent, of the water returned to the atmosphere through evaporation and transpiration from the landmasses does so through crop production, as shown in Fig. 3.4 [11]. For every tonne of water returned to the atmosphere in this way about two kilograms of plant material are produced. This estimate accounts for different growth rates of plant species and variations in soil fertility and leads to the conclusion that humanity produces 40.6 billion tonnes of plant material annually, as is also shown in Fig. 3.4. This represents thirty-one percent of the 132 billion tonnes of the total plant material produced globally each year, of which the major part is wild plants. Of the remaining fresh water, only that flowing in rivers is realistically accessible to humanity in significant quantities

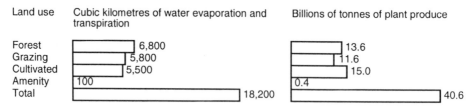

Fig. 3.4. Annual global water usage through evaporation and transpiration in plant production for human use. Data from [11].

Global estimations of water quantities are subject to some uncertainty and, depending on the method of calculation, some variations in these estimates arise. Thus, using a lower estimate of 40,700 cubic kilometres of river water than that discussed in section 3.3, it is estimated that 20,426 cubic kilometres flow back to the oceans as untapped floodwater and 7,774 cubic kilometres flow along rivers in inaccessible terrain [11,12]. This leaves 12,500 cubic kilometres, or thirty-one percent, of the water flow available for human use. By far the larger proportion of this water, 2,880 cubic kilometres, is taken from rivers and lakes for agricultural use in irrigating 250 million hectares, or 2.5 million square kilometres, of land but about a third is not consumed and passes back into the runoff flow to the oceans as shown in Fig. 3.5. Although industry withdraws the next largest amount, 975 cubic kilometres, for cooling and cleansing, a high proportion is discharged back into the runoff flow, often in a polluted state. Municipal use withdraws 300 cubic kilometres of fresh water for household supply and sewerage treatment and about one sixth of this water re-enters the runoff flow. There are about 5,500 cubic kilometres of water stored in reservoirs of which as much as 2,000 cubic kilometres is unavailable in the dead space of the reservoirs that cannot be readily drained. (Some five percent of reservoir water is lost through evaporation.) The remaining 2,350 cubic kilometres of fresh water runoff flows in canals, rivers and lakes where it is used for recreational, conservation, transportation and waste absorption. The latter usage includes not only the carriage of treated and untreated sewerage and other anthropogenic effluents, but also that of naturally occurring salts, fertilizers and pesticides that rain washes from agricultural land into rivers and lakes and thence into the oceans. When this great cleansing process is overloaded the resulting pollution can change the biology of natural waterways sometimes to the point of disaster as is discussed below. The 6,780 cubic kilometre sum of these withdrawals of fresh water amounts to seventeen percent of the total runoff, or fifty-four percent of the accessible runoff.

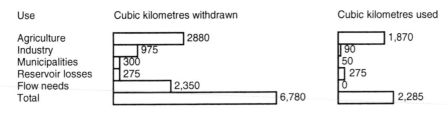

Fig. 3.5. Annual global withdrawal and usage of runoff water in 1990 based on [11].

Looking to the future, it appears that access to fresh water runoff could be increased to 13,700 cubic kilometres by 2025, largely by building dams. However, on the basis of the estimated 2025 population, it is likely that human withdrawals of fresh water could amount to 10,000 cubic kilometres or seventy-three percent of this increased accessible fresh water if consumption per person remains at the 1990 value. While Earth appears to be capable of supplying sufficient fresh water for humanity in the short term, its distribution does not place fresh water in sufficient quantities in the areas of greatest population growth. And it is population growth that poses the biggest threat to the adequate availability of fresh water despite the effects of significant climate change arising from global warming and the accompanying changes in global rainfall patterns [13]. Consequently, it is inevitable that the proportion of water recycled for a variety of uses will have to increase [14].

3.5. The Global Distribution of Fresh Water

Rain and snow falls show great seasonal and geographical variations that combine with patterns of river flow to produce regions where plant growth varies from the profusion of the rainforests to the scarcity of the deserts. Rainfall variation ranges from that of the Atacama Desert in Chile, where no rain falls for years in succession, to that of Mount Waialeale on the Hawaiian island of Kauai where 11.5 metres of rain has fallen in a single year [5]. A broad measure of the fresh water available to the continents is the runoff entering the oceans shown in Fig. 3.6. This water runoff is much less than that of the rain and snowfall of 119,000 cubic kilometres (Fig. 3.3), a difference that arises from the evaporation of water and its transpiration by plants. The exception to this is Antarctica where it is so cold that the amount of evaporation and transpiration is negligible.

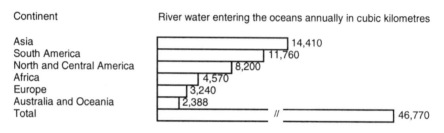

Continent   River water entering the oceans annually in cubic kilometres

Asia   14,410
South America   11,760
North and Central America   8,200
Africa   4,570
Europe   3,240
Australia and Oceania   2,388
Total   //   46,770

Fig. 3.6. Annual water runoff by continent based on [4].

Within geographical regions great differences emerge in the availability of water as in Africa were very little rain falls in the Sahara Desert that covers much of the north of the continent. Most rain falls in sub-Saharan Africa where the bulk of the African population lives. Similarly, Australia, that is of similar size to the continental United States, has an annual river runoff of only 348 cubic kilometres while the neighbouring islands of Oceania, that include New Zealand, Polynesia and Melanesia, have a much smaller surface area and produce an annual river runoff of 2,040 cubic kilometres. Locally, seasonal river runoff may both nurture and destroy life as do the massive spring flows of the Brahmaputra and Ganges generated by the melting snow of the Himalayas. These rivers regularly inundate Bangladesh and parts of India. Yet, without the huge amount of sediment dropped by these rivers on their way to the Bay of Bengal most of Bangladesh and much of the Indian states of Bihar, Uttar Pradesh and West Bengal would be branches of the Indian Ocean [15].

While these regional differences have a great effect on local vegetation and animal life, they do not become meaningful for humanity until populations are taken into account. Then, great and critical differences emerge as is seen from Fig. 3.7 that shows the renewable fresh water available to selected nations in total and on a per person basis. It is on the latter amount that nations may be classified as either water rich, sufficient or poor. In practice all fresh water is not accessible, as discussed earlier, and so Fig. 3.7 overstates the availability of fresh water to a varying extent for each nation [16,17]. However, the very precarious nature of water supply for some nations and the potential for conflict over shared water resources is readily apparent from Fig. 3.7 [18].

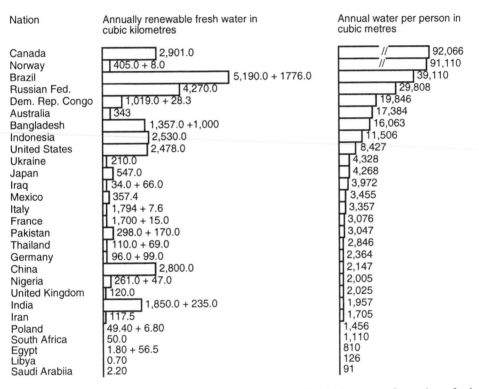

Nation	Annually renewable fresh water in cubic kilometres	Annual water per person in cubic metres
Canada	2,901.0	92,066
Norway	405.0 + 8.0	91,110
Brazil	5,190.0 + 1776.0	39,110
Russian Fed.	4,270.0	29,808
Dem. Rep. Congo	1,019.0 + 28.3	19,846
Australia	343	17,384
Bangladesh	1,357.0 +1,000	16,063
Indonesia	2,530.0	11,506
United States	2,478.0	8,427
Ukraine	210.0	4,328
Japan	547.0	4,268
Iraq	34.0 + 66.0	3,972
Mexico	357.4	3,455
Italy	1,794 + 7.6	3,357
France	1,700 + 15.0	3,076
Pakistan	298.0 + 170.0	3,047
Thailand	110.0 + 69.0	2,846
Germany	96.0 + 99.0	2,364
China	2,800.0	2,147
Nigeria	261.0 + 47.0	2,005
United Kingdom	120.0	2,025
India	1,850.0 + 235.0	1,957
Iran	117.5	1,705
Poland	49.40 + 6.80	1,456
South Africa	50.0	1,110
Egypt	1.80 + 56.5	810
Libya	0.70	126
Saudi Arabiia	2.20	91

Fig. 3.7. Fresh water resources. The first figure for renewable fresh water refers only to fresh water within a nation's boundaries. Where a second figure appears this refers to river water from other nations. For other nations, the single figure refers to the sum of water from both sources. Fresh water values are from [16] and the per person values are calculated on the basis of 2003 populations [17].

The bulk of Egypt's annual renewable water, 56.5 cubic kilometres, flows down the Nile from eight other nations to be supplemented by a meagre 1.80 cubic kilometres of Egyptian rainfall to sustain a rapidly growing population. So far conflict has not erupted over the Nile, but given the desire of the source nations to develop their economies and the inevitable increase in demand for water that this will entail, perceptive diplomacy will be required to manage the flow of the Nile and its sources amicably [19]. Similarly, Bangladesh, India and Pakistan share the waters of the Ganges and Indus in often uneasy partnerships [20]. Iraq is dependent for sixty-six percent of its renewable water on the Tigris and Euphrates that have their headwaters in Turkey and flow through Syria before entering Iraq [21]. China's building of the Gonguoqiao, Manwan, Dachaoshan, Nuozhadu, Gonlamba and Mengsong dams on the Mekong before it flows through Myanmar (Burma), Thailand, Laos and Cambodia and Vietnam to the South China Sea may be detrimentally affecting the river fisheries of these countries [22]. In Europe the Rhine flows through several nations and disputes have arisen over pollution of the river. In total, over 300 river basins are shared by two or more nations and a similar number of international treaties exist to lessen the likelihood of conflict over water [23].

Other nations do not have the luxury of engaging in disputes over shared renewable water resources as for them such resources barely exist. Instead, they must supplement their low rainfalls with ground water drawn from largely exhaustible aquifers. These water poor nations occur largely in North Africa, the Middle East and to a lesser extent in Central Asia. Two examples, Libya and Saudi Arabia, that both draw massive amounts of water from aquifers appear in Fig. 3.7 [24,25]. Australia, that for its land area has a small rainfall, draws water from Earth's largest aquifer, the Great Artesian Basin underlying 1,711,000 square kilometres of eastern Australia and containing 8,700 cubic kilometres of water [26]. The United States draws water from another huge aquifer, the Ogallala aquifer underlying the states of Colorado, Kansas, Nebraska, New Mexico, Oklahoma, South Dakota, Texas and Wyoming, but as the aquifer has declined the cost of pumping water from it has increased to such an extent that irrigation has steadily diminished since the 1970s [5,27].

3.6. Rivers and Lakes

Where significant rainfall occurs the land is threaded with streams and rivers that channel surface water into lakes and oceans. Rivers are very dynamic and, on a global average their water is completely changed every fourteen days or so. The water flow in a river can be enormous as is shown by Earth's second longest river, the Amazon, that on average carries 6,930 cubic kilometres of fresh water, or 14.7 percent of Earth's total annual runoff, to the South Atlantic as seen in Fig. 3.8. So great is this flow that the ocean surface water is fresh many kilometres out from the Amazon's mouth. In contrast, the longest river, the Nile only carries 202 cubic kilometres of fresh water and much of this is diverted before it reaches Egypt where further diversions occur so that only a small part of the river's flow reaches the Mediterranean. This difference arises from the great dissimilarities in the rainfall and size of the Amazon and Nile basins whose locations are shown in Fig. 3.9 together with those of other major rivers [28]. Similar variations in flow are seen for the other rivers appearing in Fig. 3.8.

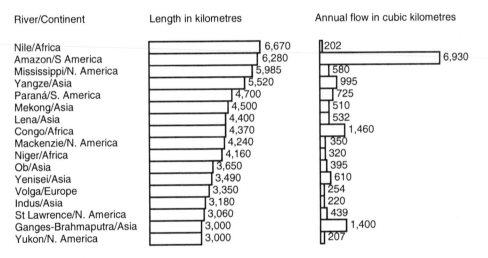

Fig. 3.8. The longest rivers and their average flow, or runoff. Data from [4].

Most fresh water lakes are found in regions covered by ice during the last Ice Age, near glaciers and where large tectonic fractures are found. The largest lake complex is formed by the Great Lakes that straddle the border between the United States and Canada. Lakes Superior, Huron, Michigan, Erie and Ontario together have a surface area of 245,280 square kilometres and hold 22,115 cubic kilometres of fresh water as is shown in Fig. 3.10. However, the 1,741 metres deep Lake Baikal in Central Asia holds the single largest volume of fresh water at 23,000 cubic kilometres.

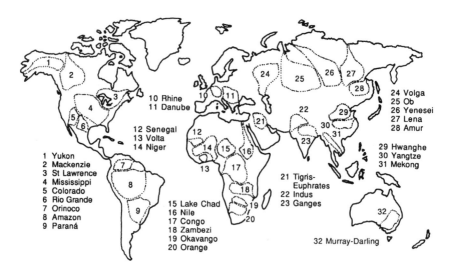

Fig. 3.9. Earth's largest river basins. Adapted from [28].

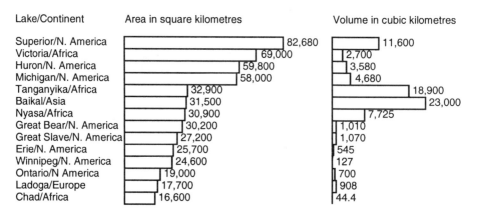

Fig. 3.10. Earth's largest fresh water lakes by area and volume. Data from [5].

Because the flow of water through lakes from which rivers flow to the ocean is often slow they are very vulnerable to human activities despite their often great size. Nevertheless, they have frequently been used as dumping grounds for sewage and industrial effluent in developed nations until quite recently. A prime example off this was the pollution of the Great Lakes to the point where their ecology was greatly damaged, fish caught in them were unfit for eating and in many places swimming was banned [29,30]. Largely landlocked seas such as the Baltic have been subject to similar pollution [29,31]. Fortunately, more stringent regulations now govern human activities associated with the Great Lakes and the Baltic Sea and their water quality and ecology are recovering, albeit slowly.

3.7. The Aral Sea: A Major Environmental Disaster

Lakes that have no exit to the ocean and whose water inflow is entirely balanced by evaporation are very vulnerable to human interference as they do not experience the flushing process that carries dissolved salts and pollutants to the ocean. They are exemplified by the Aral Sea [32,33], the Caspian Sea [34] and Lake Chad [35]. The Aral Sea, with an original area of 69,500 square kilometres, lies in Southwest Kazakhstan and northwest Uzbekistan near the Caspian Sea in Central Asia as shown in Fig. 3.11. It was once Earth's fourth largest inland sea in area behind the Caspian Sea, Lake Superior and Lake Victoria but is now much diminished as a consequence of one of the greatest environmental disasters generated by humans. Prior to the diversion of their water for irrigation, the River Syr flowed through Kazakhstan into the northern part of the Aral Sea and

Fig. 3.11. The Aral Sea and its 1961 and 1992 shorelines. Based on [32]

the River Amu passed through Turkmenistan and Uzbekistan into the southern part. There is no outflow of water and, until the 1950's, water lost by evaporation was annually replaced by the inflow of fifty to sixty cubic kilometres of water from the Amu and Syr that rise amidst the melting snow of the Pamir and Tien Shan mountains on the borders of Afghanistan and China, respectively. The annual flows of the Amu and Syr rivers are seventy-three and thirty-seven cubic kilometres as they leave the mountains, but they lose half of this water through evaporation and percolation into porous soil as they cross the deserts of Central Asia.

In 1900, five percent of the flow of the Amu and Syr where diverted for irrigation. This increased in the 1920s, and from the 1960s increasingly large amounts of water were diverted for irrigation to boost cotton production under the centralized control of the Soviet Union. This resulted in the amount of cotton produced in Uzbekistan approaching the huge production of the United States. The largest of the irrigation canals, the 1,200 kilometre long Karakum Canal, annually diverts fifteen cubic kilometres of water from the Amu to Tukmenistan, but up to a third drains away through the sandy soil of the canal channel. By 1989 virtually no river water flowed into the Aral Sea with the result that its area decreased from 67,900 square kilometres in 1960 to 37,000 square kilometres in 1989, its volume decreased from 1,090 to 340 cubic kilometres and its depth decreased by 14.3 metres [32]. This caused the Aral Sea to separate into two, the northern "Lesser Sea" with an area of about 3,500 square kilometres and the southern "Greater Sea" with an area of 33,500 square kilometres. In the mid 1990s dams were built to prevent water flow from the "Lesser Sea" to the "Greater Sea". Subsequently, the latter continued to contract and split into an eastern and western remnant by 2002, both of which may have evaporated completely within fifteen years. Thus, all that will be left of the Aral Sea will probably be the northern fragment, a small salt lake of only 3,000 or so square kilometres area, sustained by a small annual rainfall of 156 millimetres.

The depletion of the Aral Sea was initially offset economically to some extent by the increased crop production from the resulting 8.5 million hectares of irrigated land. However, the irrigation water became increasingly saline and the water table rose so that forty-five percent, or 7,000 square kilometres of Uzbekistan's agricultural land became salinized and of much decreased fertility. By 1999 cotton yields had dropped to 200 tonnes per square kilometre, much less than in other major cotton producing areas. Similar soil salinization has occurred in the neighbouring republics. Such salinization is a major hazard in dry Central

Asia where evaporation is high and rainfall is insufficient to leach salts from the topsoil. To add to these woes, the contraction of the Aral Sea has exposed more than 35,000 square kilometres of seabed caked with salt mixed with fertilizer and pesticides carried from the irrigated areas by the periodic small flows from the Amu and Syr rivers. It is possible that toxic waste from nearby military bases and the Baikonour space centre are also present. The estimated fifteen to seventy-five million tonnes of salt and dust blown annually over the surrounding region exacts a high price. Each year, some 500 kilograms per hectare fall in solid form and dissolved in rain on the Amu delta at the southern end of the Aral Sea where it has greatly reduced soil fertility. Similar contamination of agricultural land occurs in central Uzbekistan.

The ecological consequences and the interrelated economic effects of the drying of the Aral Sea are devastating. As water salinity increased with decreasing volume, fish species decreased from more than twenty-four to none by 1999. The fishery that once reported annual catches of 40,000-50,000 tonnes closed with the loss of 60,000 jobs. The canneries that once processed the fish are now many kilometres from the shores of the shrunken Aral Sea and sometimes process fish from the Atlantic, Arctic and Pacific Oceans. The 550,000 hectares of reedbeds that originally fringed the coast have shrunk to 20,000 hectares causing a major decline in the cellulose and packaging industry. This, together with the loss of other habitat, coincides with a decline in bird species from 319 to 168 and in mammalian species from seventy to thirty. Coincident with the decline of the Aral Sea, the local population has experienced an increase in hepatitis, typhus, throat cancer, liver disease, birth defects and respiratory problems. This is attributed to an increased bacterial and chemical load of the Amu and Syr from which drinking water is drawn and also to the airborne pesticides and other chemicals and dust blowing from the exposed seabed. It also appears that the decreased evaporation from the diminished Aral Sea may be changing the climate of Central Asia.

With the disintegration of the Soviet Union, the five central Asian republics gained their independence in August 1991 and, in the face of the tensions existing between them over water sharing, began to consider better water and ecological management practices in the Aral Sea region [36]. The World Bank and the United Nations in concert with the Kuwait Fund and the Kreditanstalt fur

Wiedenaufbau have made grants and loans of hundreds of US$ millions designed to begin the reversal of the health and environmental damage caused by the degradation of the Aral Sea. These commitments were made against a background of proposals for massive projects to restore some water flow to the Aral Sea. The Sibaral project envisaged the diversion of water from Siberian rivers 2,500 kilometres to the Aral Sea, but was shelved in 1986 in the face of environmental concerns about the effect of this diversion on western Siberia and the unsettled political situation in the Soviet Union. Since then, the Russian Federation has shown little interest in diverting water from its territory to the independent republics of Central Asia. In the 1990's a proposal to pump water from the Caspian Sea, that is below sea level, along a 450 kilometre canal to the Aral Sea, that is above sea level, gained prominence. However, there is little real prospect of returning the Aral Sea to its former state in the medium term, if ever.

Ironically, there is now a fear that great a surge of water could flow down the Amu river to flood much of the area to the south of the Aral Sea as a consequence of an earthquake in the Pamir mountains [37]. High in the Pamirs, the dam holding back Lake Sarez that formed when the river Murgab was blocked by a landslide after an earthquake in 1911 is becoming unstable. Should the dam burst, seventeen cubic kilometres of water would roar down the narrow Murgab gorge to flood 50,000 square kilometres where some five million people live in parts of Tajikistan, Afghanistan, Uzbekistan and Turkmenistan. Much of this water would find its way to the Aral Sea, but would provide only a temporary relief to its long term loss of water gained at the probable cost of many thousands of lives lost in the path of the flood.

Unfortunately, the lessons of the Aral Sea appear to be unheeded in Africa where a plan to divert a large part of the flow of the Okavango river is proposed from time to time. The landlocked Okavango delta in Botswana is famous for its prolific wildlife and animal migrations that are dependent on water from seasonal rains in neighbouring Angola. In the wet season, this annual inflow of water into the Okavango floods the inland delta and attracts vast herds of herbivorous animals and birds and their predators until the water slowly evaporates and the dry season returns. This seasonal cycle is extremely important to the 60,000 people who live on the fringe of the delta and are dependent on the seasonal flow of the Okavango for their farming and recharging of aquifers. It is similarly important to wildlife and the tourism it generates.

To the south of Angola and the west of Botswana lies Namibia whose border is skirted by the Okavango. During a period of water shortage in 1996 the Namibians proposed to build a pipeline to carry water from the Okavango to Grootfontein and Windhoek as part of a plan that would divert 120 million cubic metres or more of river water early in the twenty-first century [38]. Should this occur, the water flow to the Okavango delta would be severely depleted and the wildlife dependent on the delta, the people living on its fringes and the Botswanan economy would be hard hit as the delta experienced a much decreased wet season and possibly a continuous dry season.

3.8. Rivers and Dams

Rain falls in seasonal cycles and in many regions of Earth the availability of fresh water goes through corresponding cycles of excess and scarcity. It is obviously desirable to even out this variation and since the advent of agricultural settlement humans have sought to store water to see them and their crops through times of scarcity. With growing engineering prowess, diversion of water for agriculture made possible the feeding of otherwise unsustainable populations in Egypt, Mesopotamia and parts of the Indian subcontinent several thousand years ago. However, it was not until the twentieth century that the control of water runoff by the building of large dams on major rivers began to change the face of the planet [39]. There are now more than 45,000 large dams, defined as being greater than fifteen metres in height from foundations to the top, holding back the water of Earth's rivers together with many more smaller dams [40,41]. The largest dams are classified as major dams of which there are about 350. They must satisfy at least one of the criteria: height of at least 150 metres, construction material volume of at least fifteen million cubic metres, reservoir volume of at least twenty-five cubic kilometres, or hydroelectricity generating capacity of at least 1,000 megawatts. The volumes of the reservoirs and the hydroelectric power generated can be huge as is seen from Figs. 3.12 and 3.13 [42].

Challenged Earth

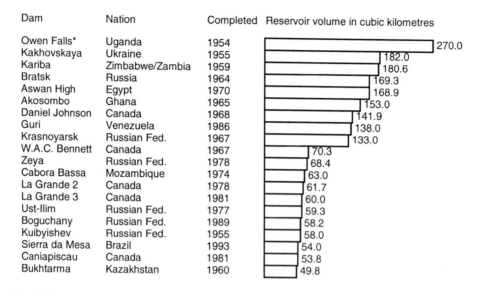

Dam	Nation	Completed	Reservoir volume in cubic kilometres
Owen Falls*	Uganda	1954	270.0
Kakhovskaya	Ukraine	1955	182.0
Kariba	Zimbabwe/Zambia	1959	180.6
Bratsk	Russia	1964	169.3
Aswan High	Egypt	1970	168.9
Akosombo	Ghana	1965	153.0
Daniel Johnson	Canada	1968	141.9
Guri	Venezuela	1986	138.0
Krasnoyarsk	Russian Fed.	1967	133.0
W.A.C. Bennett	Canada	1967	70.3
Zeya	Russian Fed.	1978	68.4
Cabora Bassa	Mozambique	1974	63.0
La Grande 2	Canada	1978	61.7
La Grande 3	Canada	1981	60.0
Ust-Ilim	Russian Fed.	1977	59.3
Boguchany	Russian Fed.	1989	58.2
Kuibyishev	Russian Fed.	1955	58.0
Sierra da Mesa	Brazil	1993	54.0
Caniapiscau	Canada	1981	53.8
Bukhtarma	Kazakhstan	1960	49.8

Fig. 3.12. The dams forming the largest reservoirs by volume [42]. *This represents the additional volume added to Lake Victoria that has a natural volume of 2,430 cubic kilometres.

Dam	Nation	Power generated in megawatts
Three Gorges	China	18,200
Itaipú	Brazil/Paraguay	12,600
Guri	Venezuela	10,300
Sayano-Shushensk	Russian Fed.	6,400
Grand Coulee	United States	6,180
Krasnoyarsk	Russia	6,000
Churchill Falls	Canada	5,428
La Grande 2	Canada	5,328
Bratsk	Russian Fed.	4,500
Ust-Ilim	Russian Fed.	4,320
Tucuruí	Brazil	3,960
Ilha Solteira	Brazil	3,200
Tarbela	Pakistan	3,046
Gezhouba	China	2,715
Nurek	Tadjikistan	2,700
Mica	Canada	2,660
La Grande 4	Canada	2,650
Volgograd	Russia	2,563
Paulo Alfonso IV	Brazil	2,460
Cabora Bassa	Mozambique	2,425

Fig. 3.13. The dams producing the largest amounts of hydroelectricity [42].

The reservoirs behind dams now hold more than 10,000 cubic kilometres of water, roughly five times greater than the water in Earth's rivers. So great is the weight of water in some reservoirs that it has induced earthquakes, and the redistribution of the total weight of reservoir water may be causing a very small change in the tilt of Earth's axis and its rotation about it [43]. Many rivers have become little more than staircases of dams with little free water flow remaining. Very few of the main channels of the major rivers do not have dams across them [44]. Globally, at least 400,000 square kilometres of land have been flooded in the filling of reservoirs behind dams, an area as large as California. In the United States reservoirs cover an area equal to that of the states of New Hampshire and Vermont together [45]. The largest reservoir in area, 5,500 square kilometres, is behind the Akosonbo Dam on the Volta and covers four percent of Ghana. At least thirty million, and probably sixty million, people have been displaced from their land by the filling of the reservoirs behind dams that now cover 0.3 percent of Earth's surface. In many cases the land covered was either fertile farmland or marshes and forests that sustained prolific wildlife.

Dams cause great changes in the environment through which the dammed river flows, changes in river and delta structures and ecologies, and changes in the ocean off the river delta [46,47]. Although some of these changes, such as the flooded river valley behind the dam and the changes in the seasonal river flow are readily seen, other more subtle effects were not at first appreciated [48]. Among such effects, it has been found that fish in reservoirs often have elevated toxic dimethylmercury levels that render them unsafe for eating [49]. Mercury and other salts occur naturally in soil and are carried in runoff water into rivers and lakes whence they find their way to the ocean. When river flows are interrupted behind dams a combination of sedimentation and evaporation concentrates mercury salts in the reservoir. Bacteria then metabolise the mercury salts to dimethylmercury that accumulates in fish higher up the food chain to a greater extent than is the case in free flowing water.

The many effects of building a dam are now widely understood by dam proponents and opponents alike, but they often weigh their relative importance differently. In November 2000, the World Commission on Dams concluded that often the beneficial effects of dams were outweighed by their detrimental effects when the degradation and fragmentation of sixty percent of Earth's rivers, the resulting environmental damage and the displacement of people from their land were considered [41,50]. As a consequence international funding for large dam building appeared to be drying up in 2000, but by 2003 the World Bank was once again looking favourably on renewed funding for dams, especially in

Africa. While the hope that hydroelectric power would offset some of the increase in atmospheric carbon dioxide resulting from fossil fuel burning has been realised to some extent, considerable releases of carbon dioxide occurs due to the breaking down of dead vegetation by aerobic bacteria in the oxygenated upper water held behind a dam [51]. Methane, a much more potent greenhouse gas, is produced by the similar activities of anaerobic bacteria in the deoxygenated deep water. The seasonal and usage variation of dam water levels alternatively drowns and deprives vegetation of water so that the amount of these greenhouse gases produced through decay is substantial.

Rivers frequently cross international boundaries and downstream nations are very vulnerable to the effects of the damming of a river by an upstream nation. There are many rivers where this situation arises and some of the major examples are discussed here. The Nile, on which arid Egypt depends, passes through eight other nations. The Euphrates and the Tigris, on which Iraq depends, rise in Turkey, and the Ganges that rises in the Himalayas and supplies much of Bangladesh's water, first passes through India. Each of these rivers has dams on them and others are proposed. Understandably, concerns over the control of their water and the silt and nutrients that it carries are the causes of tension and potential conflict. However, before some of these rivers and their dams are further explored, the effects of the dams on the Colorado and Columbia rivers of the western United States are first considered for it was there that the building of large dams began.

3.9. Dams on the Colorado and Columbia Rivers

One of the earliest and most famous of the major dams is the Hoover Dam completed in 1936 on the Colorado River in the United States [52] (Fig. 3.14). This 221 metre high arabesque of concrete arcs across the Colorado between the Nevada and Utah banks to retain the Lake Mead reservoir of thirty-five cubic kilometres capacity and heralded the era of great dam construction. It is a powerful sixty million tonne concrete symbol of humanity's ability to control rivers. The exhilaration of the realization of this ability, coupled with perceived economic benefit, resulted in a huge program of dam building in the United States. Before the Hoover Dam was completed, construction of even bigger

dams, the Shasta Dam on the Sacramento in California and the Grand Coulee
Dam on the Columbia in Washington State, began. These were followed by the
Bonneville Dam, also on the Columbia, and other large dams further to the east
on the Missouri: the Garrison, Oahe, Fort Peck and Fort Randal Dams. During
the Second World War, much of the hydroelectricity of the western dams was
used in the production of aluminium for aircraft production and later for the

Fig. 3.14. The Colorado basin and its large dams. The Colorado rises in the Rocky Mountains and
flows 2,400 kilometres to the Gulf of California during which it falls 3,000 metres. It drains an area
of 620,000 square kilometres.

production of uranium-235 and plutonium-239 for nuclear weapons. The
optimism that greeted the end of the war in the United States, coupled with
unprecedented prosperity, obscured some of the concerns about the
environmental effects of large dams. Later, increasingly detailed studies showed

that the economic benefits of dam building had to be offset against significant economic loss in other areas and environmental damage.

A river does not undergo the massive change caused by a dam without unwelcome consequences, some of which are readily foreseen and others that are not immediately obvious but which are well understood. Apart from the interruption of seasonal changes in the water level below the dam, the banks and bed of the river usually undergo deleterious changes. Large dams trap ninety percent or more of the sediment carried by the river so that as the reservoir silts up, the clear water released by the dam often cuts away the softer and looser material of the river bed and banks below the dam without the compensating dropping of sediment that occurred prior to the building of the dam. Nine years after the completion of the Hoover Dam, the Colorado had carried away 110 million cubic metres of material from the first 145 kilometres of its course below the dam. This scouring cut the river banks back by up to fifteen metres annually in some places. In addition, the riverbed was lowered by as much as four metres in places to produce a rocky bottom from which the gravel and sand, in which insects, molluscs and crustaceans live, were removed. These creatures are part of the food chain for fish and birds. Apart from this effect, the lowering of the riverbed lowers the level of the water table on either side of the river so that vegetation sometimes dries out. In 1996 and earlier, controlled flooding of the Grand Canyon from the Glen Canyon Dam showed that the Colorado river beaches could be restored to some extent and that reproduction of variable flooding as experienced with a wild river might diminish the ecological damage caused by large dams [53]. However, such flooding has so far not helped the recovery of marine life in the Colorado delta.

There are six large dams on the Colorado: the Granby, Glen Canyon, Hoover, Parker, Davis and Imperial dams, and fourteen others on its tributaries. Their total storage capacity is 107 cubic kilometres. One third of the total flow of the Colorado evaporates annually, much more than would be lost from the free flowing river whose surface area was much less than that of the reservoirs behind the dams. This evaporation concentrates the salts dissolved in the water that is used up to eighteen times for irrigation as it passes down the river and percolates back carrying salts and agrochemicals leached from the irrigated soil with it. The salinity of the water at the Imperial Dam just north of the Mexican border increased by fifteen percent from 1969 to 1990 and was expected to have increased by fifty-three percent by 2000 to a level 2.4 times the United States' maximum permissible level for drinking water. This poses an increasing problem for Los Angeles, a megacity in thirsty California that has increasing difficulty in meeting its growing water demand [54]. This high salinity has also caused a great decrease in crop yields on fields irrigated with Colorado water in Mexico and continues to cause difficulties between Mexico and the United States despite agreements on maximum permissible salinity levels. Even so, such is the demand for water from the Colorado than in some years its flow into the Gulf of California is reduced to a trickle. The consequence is that the once fertile Colorado delta that teemed with wildlife and was annually replenished with river borne silt and nutrients is now eroded and semi-arid, and its once flourishing offshore fishery is greatly depleted.

Another major western United States river, the Columbia that rises in Canada, crosses into Washington State and forms the border between that state and Oregon for the final third of its journey, has been dammed to the extent that only 170 kilometres of its 2,000 kilometres length flows unimpeded [55]. There are eleven large dams on the Columbia, starting with the Grand Coulee Dam, 1,500 metres across and 168 metres high, and finishing with the Bonneville Dam before the Pacific is reached. Its major tributary, the Snake, is also dammed in several places. The water from the reservoirs is used for hydroelectricity generation and irrigation. As with the Colorado, the economic benefits of damming the Columbia are partially offset by deleterious effects. Of these effects, the disruption and decline of the wild salmon population that depends on access to the Pacific for its breeding cycle has attracted great attention [56]. Now the United States, once the major proponent of dam building, has removed some of its less well planned dams. However, as energy demand increases together with water demand (notably in the semi-arid west) in the twenty-first century, pressures to resume large dam construction are expected to grow.

3.10. Salt in the South: The Murray-Darling River Basin

Australia is Earth's driest inhabited continent and yet it is a major food exporter. This achievement has been aided by extensive irrigation that has resulted in rising water tables bringing salt to the surface so that many previously fertile areas are now shimmering white deserts studded with dead trees. In the Murray-Darling basin in southeastern Australia 672,000 hectares were salt affected in 1996 and at least another 5,000,000 hectares are at risk [57]. Many communities draw their water from the Murray, Darling, Murrumbidgee and Lachlan rivers as they flow south and west as seen in Fig. 3.15. Extensive irrigation has caused the

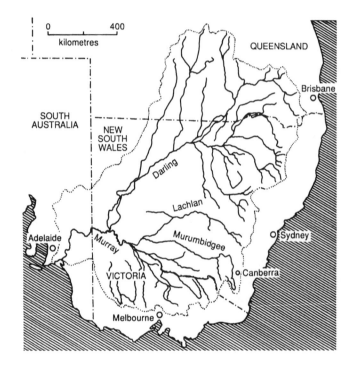

Fig. 3.15. The southward and westward river flow in the Murray-Darling basin.

salinity of the Murray, in particular, to rise precipitately and its flow to be much decreased. In its lower reaches where water is drawn off to supply Adelaide, a city of a million people, salinity is very high and if current agricultural practice continues the water will become too salty to drink within fifty years. In some years the river flow is so low that the mouth of the Murray, where it reaches the Southern Ocean, closes. A significant proportion of the irrigated crops grown in New South Wales and Queensland is thirsty cotton and rice and unless they are replaced by less water demanding crops salinization will increase. However, this problem is even more serious in the great wheat belt of Western Australia. By

1996, 1,804,00 hectares of agricultural land were badly salt affected and will probably take several generations to reclaim, and another 4,305,000 hectares are at risk of salinization if farming practices are not changed.

Australia is but one of many nations facing the loss of cropland through salinization. In the United States more than twenty percent of irrigated land is subject to salinization. Globally, an additional million hectares becomes affected by salinization annually, including those irrigated with water from the Nile, Euphrates, Tigris and Yangtze rivers discussed below.

3.11. River in the Desert: The Nile

Earth's longest river, the 6,825 kilometre long Nile, flows northward through the forbidding Sahara desert to its delta on the Mediterranean coast and has sustained Egyptian civilizations for 6,000 years [58]. Their agriculture and that of the modern state of Egypt have always been completely dependent on irrigation from the Nile and as a consequence this is the most continuously studied and monitored river on Earth. From about 3,000 years ago the seasonal highs and lows of the Nile were recorded by cutting groves in stone walls set in the river bank, and the longest continuous record at the Roda nilometer dates from AD 641 [59]. Yet very little of the Nile's water originates in Egypt but, instead, arises from the rivers of Burundi, the Democratic Republic of Congo, Ethiopia, Kenya, Rwanda, Sudan, Tanzania and Uganda, as seen in Fig. 3.16 [59-61].

The White Nile starts its northward journey as a small stream, the Kagera, in Burundi and skirts the border of Rwanda before flowing through Tanzania into Lake Victoria. From there it passes through Uganda before entering successively Lake Kyoga and Lake Moboto and an immense 135,000 square kilometre marshland, the Sudd, in southern Sudan. There, about twenty cubic kilometres a year, half of the inflowing water, are lost through a combination of evaporation and transpiration from plants, before the White Nile emerges to continue its northward journey. It is planned to lessen this loss by increasing the water flow by channelling the White Nile through the Jonglei Canal that is as yet incomplete largely because of continued armed conflicts between the Christians of southern Sudan and the Moslems to the north. If this project reaches fruition,

Challenged Earth

Fig. 3.16. The Nile's northward journey to the Mediterranean. Based on [60,61].

it will greatly increase the flow of the White Nile, but it will also alter the
ecology of the Sudd on which its prolific wildlife depends. It may also alter the
rainfall patterns of the surrounding area. These potential effects of completing
the Jonglei Canal represent but one example of the consequences of attempting
to alter the flow of water to Egypt. The White Nile is joined near Khartoum in
Sudan by the Blue Nile flowing out of Lake Tana in Ethiopia to form the Nile
proper that traverses Egypt from south to north to its fertile Mediterranean delta,
largely built from silt from the Ethiopian highlands.

Eighty percent of the Nile water entering Egypt comes from summer rain falling in the comparatively small area of the Ethiopian mountains and enters the Nile mainly through the Blue Nile and to a lesser extent through the Atbara [60]. Because of this, the Nile has historically exhibited a very seasonal flow so that prior to the construction of the Aswan dams, the Nile delta experienced major flooding between July and October [59]. The present day delta began to form about 7,500 years ago as global warming and the rise in the level of the Mediterranean accompanying the end of the last Ice Age slowed and the Nile deposited 100 million tonnes of silt annually [62]. The fertility of the delta sustained one of Earth's great civilizations, Pharoic Egypt, from about five thousand years ago. And from those early times to the beginning of the twentieth century the delta grew and the silt and nutrient rich water of the flooding Nile annually replenished its fertility.

The Nile delta has been continuously changed since first settlement. From about two thousand years ago drainage of delta wetlands and irrigation of drier parts intensified and the Rosetta and Damietta channels, through which most Nile water still enters the Mediterranean, were carefully maintained. However, none of this compares with the dramatic changes brought to the Nile delta in the twentieth century. In 1902, the first dam built to mitigate the Nile floods, the Aswan Low Dam, and its subsequent modifications in 1912 and 1934 greatly decreased the amount of silt reaching the delta. As a result, from 1904 onwards erosion at the major Rosetta and Damietta mouths of the Nile reached eighteen to thirty-three metres annually. The completion of the much larger Aswan High Dam in 1964, with the displacement of 100,000 people living in the Wadi Halfa that formed the new reservoir, providing sorely needed hydroelectric power, but further reduced the amount of silt reaching the delta so that erosion at the Rosetta and Damietta mouths rose to 140 to 160 metres annually. The flow of nutrients into the Mediterranean also decreased greatly and the previously prolific sardine fisheries diminished by ninety-five percent after 1964. Meanwhile, behind the wall of the Aswan High Dam, more than 125 million tonnes of silt have been deposited annually at the southern end of Lake Nasser to form what is effectively a new Nile delta twelve kilometres wide, forty metres deep and 200 kilometres long extending to Abu Simel some 1,000 kilometres from the Mediterranean. While the resulting two percent decrease in water storage is of minor concern, changes occurring in the Nile delta north of Cairo, some 22,000 square kilometres in area and representing two thirds of Egypt's cultivable land, are of major concern.

The artificial waterways dug in the Nile delta for transport, drainage and irrigation since the time of the Pharaohs, have been greatly extended to make an extensive network more than 10,000 kilometres in length that only reaches the coast at a few points [63]. Apart from the silt trapped by the Aswan dams, that scoured from the Nile bed north of Aswan settles in the slow moving waters of the artificial waterways, so that very little reaches the coast. Only about thirty-five cubic kilometres of the water released annually from the Aswan High Dam reaches the Nile delta. Evaporation, transpiration by plants, and seepage into underground aquifers accounts for a third of this, another third is trapped in the artificial waterways of the delta, leaving the remaining third to flow into the Mediterranean.

Of Egypt's 1.1 million square kilometres less than five percent is cultivable, most of which lies in the delta, with the remainder extending in narrow strips along the banks of the Nile south of Cairo and near the Bitter Lakes. Egypt's rapidly growing population (64.5 and 71.9 million in 1997 and 2003, respectively) also resides in these cultivated areas and requires more and more land to be irrigated all year round to produce increasing amounts of food. This places a rapidly increasing demand on the Nile for water not only for irrigation dependent agriculture, but also for household, sanitary and industrial use [64]. The overall effect is that agricultural land is suffering from increasing salinity in the absence of the annual flooding with river water that used to wash away the salts left by evaporation.

The annual deposition of silt accompanying the floods prior to the damming of the Nile is no longer available to replenish soil and to compensate for the natural erosion at the coast. Fertilizer runoff, sewerage and industrial pollutants are seriously degrading water quality. In concert with these detrimental effects, the decreased flow of Nile water to the delta threatens the coastal lagoons that supply fifty percent of Egypt's fish catch. Natural processes pose other problems. The levels of the oceans have been slowly rising during the last one hundred years because of the now very slow natural warming process that followed the end of the last Ice Age. This, together with anticipated additional rises due to anthropogenic global warming, could result in rises in the level of the Mediterranean by seventy-nine centimetres by 2050. Natural subsidence of the delta as the soil dries and compacts under its own weight and as water is increasingly withdrawn from wells may amount to sixty-five centimetres to give a total rise of 144 centimetres in the level of the Mediterranean and a loss of a one kilometre band of the delta coastline, or nineteen percent of Egypt's habitable land [65].

3.12. The Twin Rivers: The Euphrates and The Tigris

The Euphrates and Tigris rise scarcely thirty kilometres apart in the high rainfall and mountainous region of Anatolia in southeastern Turkey and follow almost parallel courses to join up in the Shatt al Arab Canal before entering the Persian Gulf as is seen in Fig. 3.17 [28,66]. The Euphrates pursues a tortuous 1,080 kilometre course through Turkey before crossing into Syria for a further 670 kilometres and then flows through Iraq for 950 kilometres. The Tigris flows for 400 kilometres through Turkey, briefly skirting the northeastern corner of Syria before entering Iraq to flow a further 1,500 kilometres to the Persian Gulf. Both rivers carry heavy loads of silt that, once they descend from the plateaus of Syria and northern Iraq, are deposited in the arid Mesopotamian plain. Mesopotamia means "land between the rivers", and the dropping of many millions of tonnes of silt annually by the two rivers has built up a great deposit of fertile soil on which the ancient Sumerian and Babylonian civilizations flourished, and on which modern Iraq depends for much of its food. This great alluvial deposit, 800 kilometres long and 200 kilometres wide, is the shared delta of the Euphrates and Tigris.

In the Mesopotamian plain, agriculture requires extensive irrigation as rainfall seldom exceeds 200 millimetres annually. Because the meandering and silt laden Euphrates and Tigris have build their beds above the level of the surrounding plain, especially in the south, they are confined in their courses by extensive levees. At frequent intervals canals divert water for the extensive irrigation system that incorporates many barrages and sluices to control the river flow. However, because evaporation is high in the hot summer months and much of the irrigation is unsophisticated, at least sixty percent of the cultivated land has become salinized and Iraq is now a major food importer.

Forty years in construction, a 565 kilometres long drainage canal carries saline water drained from land flushed with river water directly to the Persian Gulf through the Basra channel cut to the west of the Shatt al Arab channel. This "third river" lying between the Euphrates and Tigris, from which it is carefully

Fig. 3.17. The southward flow of the Euphrates and Tigris to the Persian Gulf.

kept separate, represents a major effort to reclaim salinized land and to prevent further salinization. However, this scheme also drained much of the extensive Hawr al Hammar marshes, home of the Marsh Arabs and extensive wildlife [66,67]. To some extent this was deliberate and reflected tensions arising from the Gulf War of 1991 and an attempt to deny shelter to government opponents. However, the reduced flow of the Euphrates and Tigris caused by dams in Turkey was also a contributing factor. Coincidentally, drainage of the marshes allowed easier access to the extensive oil deposits thought to lie below them. In 2003, however, the defeat of the Saddam Hussein regime by an American led coalition resulted in the shutting down of pumping stations and the breaching of embankments to allow the restoration of the wetlands to begin.

The control of the headwaters of the Euphrates and Tigris is held by Turkey that thereby controls the future water supply of the downstream states of Syria and Iraq. The natural average annual flow of the Euphrates entering Syria is thirty cubic kilometres and this increases to thirty-two cubic kilometres on entering Iraq after which no significant addition to the flow occurs. The Tigris enters Iraq with a natural average annual flow of twenty-two cubic kilometres that is increased within Iraq by another twenty-seven cubic kilometres by tributaries flowing in from the Zagros Mountains to the east. These natural river flows will be greatly changed by Turkey's Southeast Anatolia Project that incorporates both the Euphrates and the Tigris. The project envisages the construction of several dams to provide 7,700 megawatts of hydroelectric power and sufficient water to irrigate two million hectares of semi-arid land. It also requires the displacement of 250,000 inhabitants of the areas flooded by the dams, a prospect that has led to civil war.

The major Turkish dams are the Keban, the Karakaya and the Ataturk Dam, the largest. The Keban Dam was completed in 1974 and is designed primarily for hydroelectricity generation. It stores thirty cubic kilometres of the water of the Euphrates. The downstream Karakaya Dam was completed in 1988. Further downstream, the massive Ataturk Dam is designed to generate 2,400 megawatts of hydroelectricity and to store up to eighty-two cubic kilometres of water, ten cubic kilometres of which will be provided annually to irrigate 500,000 hectares of land. The magnitude of the control over the Euphrates exercised by Turkey was shown in 1990 when the river flow was stopped for a month causing Syria to suffer a drop in hydroelectricity generation and irrigated crop loss, and Iran to suffer crop loss. This led to major protests from Syria and Iraq. In response, Turkey promised to release a minimum of 15.5 cubic kilometres of Euphrates water to Syria annually. The Southeast Anatolia Project also envisages dams on the Tigris retaining up to seven cubic kilometres of water annually to irrigate 550,000 hectares of land. Thus, the controlled flow of the Euphrates and Tigris out of Turkey will average 30.5 cubic kilometres annually, compared with 52 cubic kilometres before.

Syria, unhappy at the prospect of less water flowing from Turkey, has been active in controlling its own water resources that in turn has led to protests from Iraq. The filling of Lake Assad behind the Tabqa Dam in 1974 substantially decreased the flow of the Euphrates to Iraq that claimed to have suffered a seventy percent crop loss as a result. This led to Iraq sending troops to the Syrian border and a reciprocal response from Syria. A reduced flow of water from Turkey, that may be polluted by fertilizer and pesticide runoff from the extensive Anatolian irrigation areas, imposes a lower than anticipated limit on Syria's hydroelectricity generation and irrigation prospects.

While Syria and Iraq find themselves in difficult situations as downstream states, water rich Turkey has made proposals to supply water from the Seyhan and Ceyhan rivers in southern Turkey. The most ambitious, the Peace Pipeline proposal of 1987, envisaged two pipelines carrying 2.2 cubic kilometres of water annually for domestic use to Syria, Iraq, Jordan, Kuwait, Saudi Arabia and the Gulf States, and possibly Israel. This proposal is not currently under active consideration, but it does highlight the potential for Turkey to become a regulator of water supply in the thirsty Middle East where potential conflict over water access continually simmers. This is exemplified by concerns over access to the relatively small flow of the Jordan River and adjacent aquifers that is a constant undertone to the intermittent disputes between Israel, Jordan and Syria, and on the West Bank between Israelis and Palestinians [68].

3.13. The Three Gorges Dam: The Biggest Dam of All

The largest dam ever built and due for completion in 2009 is designed to control Earth's third longest river, the Yangtze, at a site nearly 4,500 kilometres from its headwaters on the Tibetan plateau as shown in Fig. 3.18 [46,69-71]. This is the Three Gorges Dam at Sandouping, named after the Qutang, Wu and Xiling gorges through which the Yangtze passes between soaring limestone peaks on its way to the Yellow Sea at Shanghai. The dam will be 1.9 kilometres across, 175 metres high, house twenty-six turbines and will be the largest concrete structure on Earth. It will create a reservoir with a surface area of 1,150 square kilometres and a volume of 39.3 cubic kilometres. This enormous reservoir, that started to fill in November 1997 with the closure of the main channel of the Yangtze, will require the resettlement of at least a million people as farmland, towns and villages are flooded. Already, thirteen new towns are being constructed to accommodate some of the displaced population. The drowning of this scenic part of the Yangtze valley will significantly alter the ecology of the river as its current

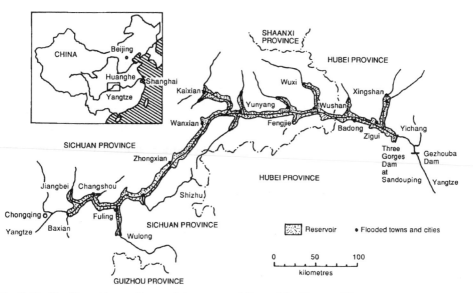

Fig. 3.18. The Three Gorges Dam and the westward flow of the Yangtze River.

slows and its sedimentation patterns change; changes that must be carefully monitored if the problems encountered with earlier Chinese dams are to be avoided. When completed, the dam's turbines will generate 18,200 megawatts of electricity, the largest production of any hydroelectric scheme and roughly equivalent to the output of eighteen nuclear power plants. However, this is not the only purpose of the dam for it is also designed to reduce flooding in the Yangtze floodplain where 400 million people live and sixty-five percent of China's rice is grown. In the past, such floods have drowned hundreds of thousands and destroyed homes and crops on a calamitous scale.

Sun Yat-Sen first proposed the dam in 1919 largely to mitigate the Yangtze floods. Particularly devastating floods drowned 200,000 in 1931 and 1935, and further dam studies were made. In 1954 more floods drowned 30,000 and Mao Tse-tung started the process that, after fluctuating fortunes, led to the decision to build the Three Gorges Dam under the aegis of Li Peng. However, the dam will not eliminate flood risk for in 1954, 1991 and 1998, when many thousands were drowned, most of the floodwater came from tributaries joining the Yangtze east of the dam. The flooding in the Yangtze floodplain invariably results from the failure of the dykes that retain the river and its tributaries in their beds that have built up above the surrounding countryside as they drop some of their heavy load of silt.

The greater the amount of stored water the greater is the magnitude of the potential damage resulting from a failure of its containment. Such a failure occurred at the Banqiao Dam on the Huai, a tributary of the lower Yangtze, in Hunan province in 1975 when a typhoon delivered torrential rain during 5-7 August. This failure released half a cubic kilometre of water that drowned 10,000 people. On a proportional basis, a catastrophic failure of the Three Gorges Dam would drown 800,000 people.

China's second largest river, and Earth's most silt laden, the Huanghe, or the Yellow River, runs to the north of the Yangtze (Fig. 3.18) and has also been the subject of large dam projects [46,70]. The 106 metre high Sanmenxia Dam was built across the last gorge before the Huanghe enters the eastern plains. It was finished in 1960 and was designed to mitigate floods, generate hydroelectric power and provide irrigation water. Some 400,000 people left 66,000 hectares of fertile farmland to make way for the reservoir. However, the designers had not allowed for the 1.5 billion tonnes of silt carried down the Huanghe annually, and by 1964 the reservoir was almost completely filled with fifty million tonnes of silt. This silt accumulation threatened to cause flooding of the ancient capital of Xian. The dam was re-engineered at great expense and the reservoir now holds only about one third of its intended water capacity of eight cubic kilometres. The eight tunnels that once housed turbines are now used to flush silt from the reservoir. Originally designed to produce 1,200 megawatts, the Sanmenxia Dam now produces only 250 megawatts of electricity. Even so, fifteen further dams are planned for the Huanghe, and it is expected that one of these, the 170 metre high Xiaolangi Dam downstream from the Sanmenxia Dam, will have its water storage capacity reduced from 12.6 to five cubic kilometres over ten years as siltation occurs. Nevertheless, it is hoped that together the Sanmenxia and Xiaolangi dams will greatly decrease flooding in the eastern plains where dykes are constantly reinforced and rebuilt to contain the Huanghe that has built its bed up to five metres above the surrounding plains. These floodplain dykes are an integral part of the precarious control of the Huanghe. Their failure would result in catastrophic loss of life as occurred in 1938 when they were deliberately breached by the Chinese army to stop a Japanese advance during the Sino-Japanese War and 300,000 to one million lives were lost [69].

Despite the huge size of the Three Gorges Dam, it represents only a component of a much larger water engineering project commencing in 2003 that will carry fifty cubic kilometres of water annually northwards from the Yangtze to replenish the water starved north [72]. Each of the three components of the project is as large as the Three Gorges Dam. In the east, water will be redirected from Jiangdu, near the mouth of the Yangtze, to Tianjin. More water will be sent northward from a giant reservoir at Danjiangkou in the Yangtze floodplain to Beijing, and in the west water will be diverted from the Yangtze headwaters into the Huanghe. This huge project is China's answer to the ongoing water shortage in the nation's north and its feasibility rests on the huge flow of water in the Yangtze. Eighty percent of China's fresh water is carried by the Yangtze, most of which finds its way to the ocean.

China is not alone in planning such huge water redirections. India has plans to direct the massive flow of the rivers that rise in the Himalayas, and include the Ganges and the Brahmaputra, to the parched south and east of the subcontinent through a network of river linkages, canals, aqueducts and tunnels. Similarly, in North America some in the United States are anxious to divert Canadian river water south to parched Los Angeles and Las Vegas. At present some ten percent of Earth's fresh water is carried by Canadian rivers into the Pacific Ocean. Other massive water redirection projects are being contemplated in Africa, Australia and Spain. While such projects present great engineering challenges, most also involve major disruptions of large human populations and ecologies that at least partially offset the advantages gained from water becoming more plentifully available in the newly supplied areas.

3.14. A River Reclaimed: The Rhine

In 1828, S. T. Coleridge wrote of Köln:

In Köln, a town of monks and bones,
And pavements fang'd with murderous stones
And rags, and hags, and hideous wenches;
I counted two and twenty stenches,
All well defined, and several stinks!
Ye Nymphs that reign o'er sewers and sinks
The river Rhine, it is well known,
Doth wash your city of Cologne;
But tell me, Nymphs, what power divine,
Shall henceforth wash the river Rhine?

and yet he must have been aware that in his native England, London and other cities could match Köln's "stenches" and "stinks" and all of the other deficiencies that he saw. Even so, he was perceptive in seeing that the Rhine could not receive all of Köln's refuse without becoming polluted. Continued such use of the Rhine resulted in the river becoming so polluted in the 1960s and early 1970s that many fish and other aquatic species had disappeared, most of the river became unsafe for swimming, and water quality became so poor that its purification for drinking was severely threatened [73].

The Rhine is a classic example of how a river can be seriously degraded by a dense population and huge industrial and agricultural complexes on its banks using it as an open drain, and then be largely reclaimed through strict environmental controls. From its headwaters in Switzerland, the Rhine flows 1,320 kilometres through Germany, France and Holland with an annual average flow of seventy-five cubic kilometres to the North Sea as shown in Fig. 3.19. Its catchment area is 190,000 square kilometres in area and is inhabited by fifty million people. Eight million of these take their drinking water from the Rhine itself and another ten million take their water from Lake Constance through which it flows. In total, some twenty percent of the Rhine's water is diverted for human and industrial use. Additional to this service, the Rhine carries 150 million tonnes of cargo annually in the 500 barges that ply its navigable 800 kilometres from just below Lake Constance, and provides 8,700,000 megawatt hours of electricity annually from ten hydroelectric plants.

A river put to such extensive use, as is the Rhine, is inevitably prone to pollution. Nevertheless, the political will did not exist to restore the Rhine until the early 1970s when sewerage treatment upgrading began to reduce the river's phosphorus and nitrogen content. This was also assisted by the phasing out of

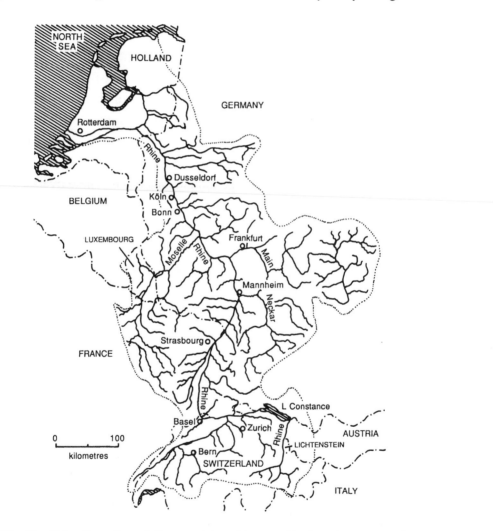

Fig. 3.19. The Rhine flows through four countries with a combined catchment area of some 190,000 square kilometres as shown by the broken line.

phosphate based detergents. Significant nitrogen flows from fertilizer runoff still contribute to its high nitrogen load, but this is expected to decrease with changes in agricultural practice leading to reductions in fertilizer use, and chemical pollution from industry has been progressively decreased [74]. Now, for most of its length from Basel to the North Sea, the oxygen content of the Rhine, a major measure of a river's health, is ninety percent of that expected for a pristine river, aquatic life flourishes and humans have little to fear from consumption of drinking water taken from it after purification.

The restoration of the Rhine provides hope for the restoration of other European rivers, some of which, as exemplified by the Thames, have paralleled the Rhine in greatly improved water quality. Generally, European Community states are required to enforce river water quality regulations as strict as those applied to the Rhine, but to the east great pollution problems still exist. This is notably so with the major rivers Danube, Dnieper, Don, Kuban and a number of smaller rivers that flow through 2.2 million square kilometre catchment areas embracing seventeen nations and 160 million inhabitants. These rivers carry their pollution to the landlocked Black Sea linked to the eastern Mediterranean by the narrow Bosphorus Strait [75]. As a consequence, the Black Sea receives annually 540,000 cubic kilometres of river water loaded with nitrogen, phosphorus and heavy metals together with a plethora of other pollutants. Construction of the Iron Gates Dam in 1970-1972 on the border between the then Yugoslavia and Romania a thousand kilometres upstream has greatly changed the seasonal flow of the Danube that contributes about seventy percent of river water entering the Black Sea. In turn this has markedly decreased the flow of silicates to the Black Sea that are important for the growth of diatomaceous phytoplankton. The combined effect of these changes to the river water entering the Black Sea has lead to extensive algal growth and a great diminution in water oxygen content, or eutrophication, and the demise of the once flourishing fishing industry.

3.15. Water Pollution: A Massive Threat

Water can be described as polluted if it differs from rainwater collected in a pristine environment, but in reality the major portion of water consumed by humans invariably contains salts dissolved from the soil through which it percolates into rivers, lakes and aquifers. Indeed, some of these salts such as those of calcium, magnesium and iron, in particular, are essential to health, and their presence in drinking water is generally beneficial. Sometimes, water from aquifers can be unpalatable and as salty as seawater and is usually called brackish water. Occasionally, the leaching of naturally occurring salts into aquifers is very detrimental to health as is the case in parts of West Bengal and Bangladesh were arsenic salts in well water have caused serious illness for hundreds of thousands and have placed many more at risk [76]. While this is by far the most severe case of such poisoning, arsenic levels in ground water are of concern in at least seventeen countries including Argentina, China and the United States [77]. However, a much wider threat arises from pollution of water by humans.

The most serious component of this pollution arises from bacteria, viruses, protozoa and other organisms from human and animal waste entering drinking water supplies [78]. Mainly in the developing nations of Africa, Asia and Central and South America, a billion people do not have access to safe drinking water and up to three billion have inadequate sanitary facilities. As a result, water borne and sanitation related diseases cause up to five million people to die annually and hundreds of millions more to become ill. Most of these are children under five years of age. The bacterial and viral diseases contracted through drinking contaminated water include cholera, typhus, diarrhoeal illnesses, infectious hepatitis and poliomyelitis. In addition, drinking water may also be contaminated with parasites that cause death and a range of disabilities. The provision of safe drinking water and good sanitation to the poorest one third of humanity who do not have access to it would cost some US$70 billion over ten years, or about one percent of the global military expenditure over the same period. Quite apart from the humanitarian aspects, the removal of this type of pollution, the first of the five major types of water pollution and by far the most lethal identified in a report presented to the United Nations General Assembly in June 1995, might give impetus to tackling all types of water pollution [79].

The other four types of water pollution are just as prevalent in the developed nations as in the developing nations and largely arise from agricultural, industrial and domestic activities. Thus, the second type of pollution is phosphorus and nitrogen from human and animal waste, detergents and agricultural fertilizers that result in excessive algal growth in rivers and lakes. High levels of phosphate and nitrate in water cause excessive fertilization, or eutrophication, of rivers, lakes and coastal areas and encourage the abnormally extensive growth of blue-green algae, or cyanobacteria, often called "algal blooms" [80]. Such algal blooms cut off much of the sunlight that subsurface aquatic plants require and sometimes produce toxins at levels that are life threatening to humans, fish and animals. In fresh water the algal blooms turn water a turbid green, and in marine waters often appear reddish giving rise to the name "red tides". Usually, *Pfiesteria* and similar algae associated with such blooms are toxic to animals, fish, and aquatic plants and can cause shellfish to become toxic to humans.

Sometimes the size of algal blooms is so large that they either greatly or completely deplete oxygen levels in water as they decay; conditions known as hypoxia and anoxia, respectively. This in turn starves other aquatic species of oxygen and can lead to huge fish deaths. Such oxygen depleted areas can be enormous as is now the case in the Gulf Mexico annually, especially during June, July and August when a hypoxic area of up to 18,000 square kilometres forms, the largest found anywhere [81]. This is largely attributed to the fertilizer carried into the gulf from the vast Mississippi watershed that drains forty percent of the area of the contiguous forty-eight states of the United States. The hypoxia occurs in the midst of an important fishery and threatens the economy of a large part of the Gulf of Mexico. Hypoxia occurs naturally in enclosed bays and seas where entering river water is not rapidly mixed as is the case in Chesapeake Bay, the Black Sea and the Baltic Sea. However, the incidence of hypoxia in these and other shallow coastal and estuarine areas seems to be increasing globally, most probably because of increased nutrient levels resulting from human activities.

While some huge fish kills are attributable to fertilizer run off, the origins of others remain uncertain. A typical case was the death of hundreds of millions of pilchards along the entire 5,000 kilometre long southern coast of Australia in 1995 [82]. The kill started in March midway along the coast and rapidly spread east and west coincident in many, but not all, places with a huge rise in the population of the marine alga, *Thalassira*, that secreted a slime that coated the pilchard's gill tissue, thereby depriving them of oxygen and weakening their immune systems.

Excess nitrate in drinking water represents the third type of pollution [74,83]. While nitrate appears to have no directly damaging effect on health, it is converted to nitrite by bacteria in the digestive system [84]. For children up to the age of six this poses a problem as their stomachs produce insufficient acid to restrain bacterial growth to the levels in healthy adults. This exposes the child to higher levels of nitrite that binds to haemoglobin that then loses its role as an oxygen carrier in the blood. At high nitrite levels this results in methaemoglobinemia where children suffer from breathing difficulties, diarrhoea and vomiting and, in extreme cases, unconsciousness and death. This condition is often referred to as the "blue baby" syndrome because symptoms also include a blue colour around the mouth and in the hands and feet. In some cases the bluish coloration may envelope most of the child. Fortunately, this syndrome is effectively treated with vitamin C. Pregnant women consuming high nitrate levels may produce sufficiently high levels of nitrite to induce methaemoglobinemia in their foetuses. Nitrites have also been associated with cancer and other illnesses in adults. As a consequence developed nations have set maximum levels of nitrate considered safe in drinking water (usually fifty milligrams or less per cubic decimetre).

The fourth type of pollution is the more than 100,000 chemicals including persistent and toxic organic pollutants such as polychlorinated biphenyls (PCBs), dichlorodiphenyltrichloroethane (DDT) and similar organochlorine chemicals that are globally spread in water, the environment generally, food and animal and human tissue [85]. While the nature of the effects on health of these chemicals appears to be diverse, an increasing concern has arisen over their hormone mimicking capabilities, particularly their ability to mimic androgens, masculinising hormones, and estrogens, feminising hormones, that control sexual development in foetuses and children [86]. A possible widespread decrease in human sperm count has been attributed to the effects of such hormone mimics [87]. Among wildlife, Floridian alligators suffered from abnormal sexual development after a pesticide spill in Lake Apopka, female fish downstream of chlorine using paper mills have been found to be masculinized, the abnormal development of male genitals by shell fish has been attributed to tributyltin in marine antifouling paints, and reproductive difficulties have arisen in Baltic grey seal with high organochlorine tissue levels. While cause and effect have not been directly linked in all of the above examples, there can be little doubt about the potential effects of the low levels of two hundred or so well characterized drugs, including antibiotics, antidepressants, analgesics, contraceptive agents, chemotherapy drugs and tranquillisers that pass through sewerage treatment plants into the environment [88].

The fifth form of water pollution is that of heavy metals in water of which the best known example is Minamata Bay in Japan into which a chemical company dumped tonnes of mercury from the 1930s through to the 1960s [89]. This was metabolised into dimethylmercury and entered the human food chain through fish and crustaceans so that from the early 1950s people from the Minamata Bay area were stricken with neurological dysfunction, many died, foetuses aborted and children were born with genetic abnormalities as a consequence of seafood being high in mercury. Such mercury poisoning is now known as Minamata disease. Despite this, mercury is used extensively in gold extraction in crudely run mines in Amazonia where some seventy to 170 tonnes annually end up in rivers along with cyanide that is also used in the extraction process [90]. This was also the case in Romania when a tailings dam for a goldmine burst its banks under heavy rain [91]. Apart from mercury, other heavy metals from mineral extraction, industrial use and some agricultural practices have contaminated both water and soil [92]. Formally, chromium, manganese, iron, cobalt, nickel, copper zinc, molybdenum, lead, mercury and cadmium are classified as heavy metals, but because of the relatively high levels of iron and manganese found in soil and the ubiquitous occurrence of iron and zinc in biology, these three metals are often considered less serious pollutants than the others. Cadmium, mercury and lead are very toxic to humans and these, together with chromium, copper, and silver, are toxic to aquatic life. Acidification of natural waters by acid rain is a significant factor in the release of heavy metals and aluminium, from soil and through the corrosion of metal pipes and fittings. A massive global effort is required to tackle these and the other forms of water pollution.

3.16. Acid Rain: Atmospheric Water Pollution

"What have they done to the rain?" was a question asked in a popular song of the 1960's. Although it largely reflected concern about radioactive fallout from atmospheric nuclear tests, it also reflected a growing concern about rain as acidic as vinegar because of atmospheric pollution, and, unsurprisingly, called "acid rain" [93-95]. This concern grew as the prevalence of acid rain increased in the

1970s and 1980s and its deleterious effects became very noticeable through the erosion of stone buildings and statues, unnatural tree deaths and the loss of fish from fresh water lakes. As a consequence, the origins and effects of acid rain became increasingly studied. However, as early as 1852 the Englishman, R. A. Smith, had already found that the acid rain falling on the industrial city of Manchester bleached coloured fabrics, corroded metal surfaces and damaged vegetation [96].

To understand the causes of acid rain it is necessary to look at rain in a pristine environment and to see what is meant by acidity. When water is distilled in a laboratory it is absolutely pure. There is nothing dissolved in it and it is neutral, being neither acid nor alkaline. However, it is quite unlike the rain in a pristine environment that has carbon dioxide dissolved in it that makes it weakly acid. This is because carbon dioxide (CO_2) reacts with water (H_2O) to form carbonic acid (H_2CO_3) that releases a hydrogen ion (H^+) to form a bicarbonate anion (HCO_3^-). It is the released hydrogen ion that makes pristine rainwater naturally weakly acidic. In such rain there is only one hydrogen ion for every 22 million molecules of water (or one gram of hydrogen ions in 398,000 cubic decimetres or kilograms of water) and the rain has a pH of 5.6 on the rather arcane pH scale that is an internationally used measure of acidity. In distilled water there is one gram of hydrogen ions in 10,000,000 cubic decimetres of water and its pH is 7. When there is one gram of hydrogen ions in 1,000,000 cubic decimetres of water the pH is 6 and the acidity has increased tenfold. As the volume of water is decreased in tenfold steps to 100,000, 10,000, 1,000, 100 and 10 cubic decimetres the pH decreases to 5 to 4 to 3 to 2 to 1, respectively, as the acidity increases in tenfold steps. Thus, the pH is equal to the number of zeros after the one in the volume of water in cubic decimetres containing one gram of hydrogen ions. (More succinctly explained, pH is equal to the negative decadic logarithm of the hydrogen ion concentration to a good approximation.) Because pristine rain has a pH of 5.6, it is only rain that has a lower pH and is more acidic, that is called "acid rain". To place this in perspective, battery acid is very acidic and has a pH of 1, lemon juice has a pH of 2.2, vinegar, 2.8, apple juice, 3.1 and milk, 6.6. Water with a pH of 7.0 is neutral and water containing a lower hydrogen ion concentration, is called alkaline as are blood with a pH of 7.4, seawater, 8.1 and ammonia 11.1.

    Rain becomes more acidic than is naturally the case when it dissolves sulfur trioxide (SO_3) that reacts with water to form sulfuric acid (H_2SO_4) that releases two hydrogen ions. Sulfur trioxide results from atmospheric oxygen reacting with sulfur dioxide (SO_2) that is produced from sulfur contained in fossil fuels when they are burnt, and to a lesser extent from the roasting of sulfide ores in metal production. The greater the amount of sulfuric acid dissolved, the more acidic is the rain and the lower is its pH and the damage it causes increases. In addition, oxides of nitrogen produced from fossil fuel burning and from soil bacteria metabolising nitrogenous fertilizers also produce acid rain as they react with atmospheric water to produce nitric acid (HNO_3), to which must be added up to half as much again produced by lightening [97]. Thus, it is not surprising to find that acid rain occurs in industrial areas and their surroundings as is seen for the United States in Fig. 3.20. Rain is usually much more acidic over the more industrial and densely populated northeast, as shown by its lower pH, while less acidic rain falls over the prairie and south western states as shown by higher pH values [98].

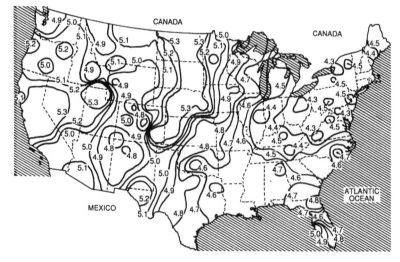

Fig. 3.20. Acid rainfall on the United States in 1997. The number on the inside of each acidity contour line indicates that the pH of rain was of that value or greater while on the other side the pH was less. In four cases in the northeastern states the contour line encloses an area too small to accommodate the pH value and a line indicates that the number shown should be inside the contour. Data from [98].

To place Fig. 3.20 in perspective, although rain with a pH of 4.6 and 4.3 is ten and twenty times more acidic than pristine rain of pH 5.6, short exposure to it does not injure the skin. The rain falling on the United States and Europe is now much less acidic than it was in the 1970s and 1980s when in some areas of particularly high atmospheric pollution rain with pH values of 2.0 to 2.5 fell. One of the effects of such strongly acidic rain can readily be simulated by dropping chalk fragments into vinegar. Chalk is mainly calcium carbonate ($CaCO_3$) and in vinegar bubbles of carbon dioxide to appear as hydrogen ions attack the carbonate to produce water and carbon dioxide. Much of the stone used in buildings and statuary contains calcium carbonate and is eroded at an increasingly rapid rate as the acid rain pH decreases.

While the most readily understood effects of acid rain are seen in the erosion of buildings and statuary and the corrosion of metal surfaces, its effects on the environment are also very serious [99,100]. When acid rain falls on soil high in calcium and magnesium carbonate its acidity is neutralized and its effect on plants and soil microorganisms is less compared with that on soils where carbonates are less plentiful. Even so, calcium and magnesium are leached from the soil and, because these metals are essential to plant growth, vegetation grows less well. Acid rain also releases aluminium from the soil that in turn decreases fine root growth and the uptake of nutrients by plants. When acid rain falls directly on leaves it slowly breaks down their protective layer so that nutrients are leached out. In addition to these individual and directly debilitating effects of acid rain on plants, they combine to reduce plant resistance to disease, fungal and pest attack, and drought and frost. The results have been prominently seen in the forests of Europe, Scandinavia and the northeastern United States and neighbouring parts of Canada where vast areas have been damaged by acid rain. Deciduous trees appear less affected than conifers, but even so deciduous trees show reduced vitality, leaf discoloration, bark damage and crown death. Conifers show all of these effects to a greater extent in addition to substantial root damage. In Western Germany, the Norway spruce, white fir, Scotch pine and beech showed widespread deterioration at the beginning of the 1980s. In Czechoslovakia, Eastern Germany and Poland the situation was even worse. Across the Atlantic, accelerated dying of red spruce on mountaintops of the Adirondacks, the Appalachians, Vermont and New Hampshire in the northeastern United States became evident in the early 1980s. While the solution to this forest destruction is to reduce the sources of acid rain, partial short term remedial action can be taken by fertilizing forests with the nutrients that have been leached from the soil.

Inevitably, acid rain runoff enters rivers and lakes where its effects are often very damaging. If the rivers and lakes are in limestone areas some of the acidity is neutralized while in granite areas little neutralization occurs. The most obvious effect of river and lake acidification is the loss of fish [101]. In the 1980s in the United States at least 212 lakes in the Adirondacks were without fish, almost all trout had disappeared from half of West Virginia's trout streams, and fish were absent from 140 lakes in southeastern Canada. In southern Norway, trout and salmon populations were greatly reduced in major rivers, and in Sweden fish disappeared from 5,000 acidified lakes whose pH had dropped below 5. Another 5,000 Scandinavian lakes with pH values below 6 where beginning to similarly deteriorate. The water in an acid fresh water lake or a river usually appears crystal clear but, rather than indicating water purity, this clarity shows the absence of the phytoplankton on which many fish feed. Apart from this loss of food supply, fish are directly affected themselves. Most fish flourish in fresh water with pH values in the range 6.5 to 9.0, but as the pH drops below 6.5 they progressively disappear. These direct effects of increasing acidity have multiple origins among which are the failure of eggs to hatch and the toxicity of the increased aluminium levels of most acidified lakes [102]. A short term solution to the acidification of lakes is to add either lime (CaO), calcium hydroxide ($Ca(OH)_2$), or limestone, or calcium carbonate ($CaCO_3$), to neutralize the acidity. This also adds calcium to the water, a vital metal for the health of fish, and reduces the level of toxic aluminium.

Unsurprisingly, acid rain is not confined within national borders as the air pollution that causes it is widely transported by the prevailing winds. In 1988-1989 over seventy percent of the atmospheric sulfur falling mainly as acid rain on Austria, Finland, Norway and Sweden came from elsewhere [103]. Similarly, acid rain originating in the northeastern United States falls on southeastern Canada and *vice versa.* Some of the atmospheric sulfuric acid produced in North America is carried across the Atlantic by prevailing westerly winds to add to the acid rain falling on Europe [104]. However, as developing nations industrialize they have added to atmospheric pollution and atmospheric sulfuric acid has built up over China, in particular. The consequence is that a sulfuric acid aerosol produced by human activities now extends continuously from the Pacific off California, across North America and the Atlantic to Europe and northern Africa, and onwards across much of Asia to envelope Japan [105]. In the southern hemisphere, sulfuric acid aerosols are building up over industrializing parts of South America and southern Africa. It is scarcely surprising that uncontrolled exports of pollution between nations causes international irritation, and as early as 1972 the Organization for Economic Cooperation and Development (OECD) set up studies to assess cross border air pollution [106]. Fortunately, the situation is improving with North American and European rain becoming less acid as increasingly effective methods of decreasing sulfur emissions are adopted coincidentally with the imposition of large penalties for atmospheric pollution. Simultaneously, the fitting of catalytic converters to internal combustion engines has decreased the prevalence of oxides of nitrogen [107].

3.17. Water in the Twenty-First Century

If current predictions for a global population approaching nine billion by the middle of the twenty-first century, and perhaps eleven billion by its end, are correct, the fresh water available to each human will be drastically reduced compared with that available in 2000 [108-115]. However, within this picture of scarcity there are at first glance winners and losers as Fig. 3.21 shows [111]. The populations of the developed nations are either growing very slowly, are stationary or are in decline. Generally, these nations are in the temperate regions of North America and Europe and have water supplies that are likely to remain adequate for agricultural, industrial and household use well into the twenty-first century provided that the growing momentum to safeguard water purity and to curb excessive water use is maintained. Many of these nations are major food exporters and those that are not are sufficiently wealthy to import food from elsewhere.

In contrast, the many developing nations that largely occupy either the more arid, the hotter, or the more arid and hotter regions of Earth, will experience a growing water scarcity because of a combination of limited water supplies and growing populations [112]. By 2025, some eighty percent of humanity will live in such nations, many of which are already experiencing water scarcity. It is estimated that the minimum amount of water required for a reasonably nutritional food self-sufficiency and minimal household and industrial needs of

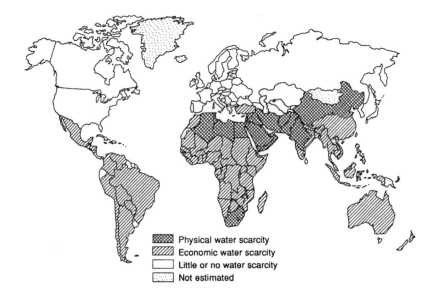

Fig. 3.21. Projected water scarcity in 2025. Some nations are unlikely to experience water scarcity while others are likely to suffer physical water scarcity as their needs increasingly exceed their water resources. Those in between, whose water resources are sufficient to meet their needs but which will require large expenditures to make sufficient water accessible, are described as experiencing economic water scarcity. Adapted from [111].

these nations is nine hundred cubic metres per person annually [109]. On such a basis, it becomes clear that some nations with rapidly growing populations will face a physical water scarcity and will not be self-sufficient in food by 2025. Most nations in the Middle East and North Africa experienced a physical water scarcity in 2000 and by 2025 will be joined by Pakistan, South Africa and large areas of India and China. Their agriculture presently depends greatly on irrigation and, even with major increases in efficiency in irrigation water usage, they will become dependent on food imports to varying degrees, with the possible exception of some parts of Central Asia where increased irrigation efficiency may avert this.

The remaining, mainly developing, nations will suffer an economic form of water scarcity in 2025. Although they presently have sufficient water resources to meet their needs, many will have to increase the availability of their water by twenty-five percent, or more, to maintain food self-sufficiency. This will require the development of large irrigation projects using water from rivers and aquifers, the expense of which may be beyond the capabilities of many nations especially those in sub-Saharan Africa and South America. In summary, by 2025 most of humanity will live in nations experiencing water scarcity and a smaller, but still large, proportion will be dependent on food imports to some extent [113,114]. This is likely to produce new disputes and exacerbate existing ones over water and could fan them into armed conflicts. The developed nations are unlikely to be unaffected by the scarcity experienced by the developing nations, but many of the citizens of the later nations are increasingly seeking refuge in the developed nations. For decades, many millions of such economic refugees have made hazardous journeys to the developed nations instigated by the poor economic futures probable for their nations of origin, at the centre of which is often water scarcity.

3.18. Glimmers of Hope

Despite the rather gloomy forecasts for future water availability, human ingenuity tends to come to the fore in times of crisis and it may be that some of the more extreme effects of water shortage can be averted. At present irrigation produces about forty percent of all food crops and accounts for about sixty-seven percent of water use globally, and in some developing nations accounts for ninety percent of water usage [115]. The extra two billion or so people expected to be on Earth by 2025 will require an additional 790 cubic kilometres of water for irrigation, or about one and a half times the annual flow of the Mississippi, if present farming practices and dietary preferences persist. In many cases the irrigation methods presently used lose much water to evaporation and as runoff. It is estimated that irrigation water usage could probably be reduced by up to fifty percent with the employment of the more efficient irrigation systems that are now available. Dietary preferences for maize fed beef are particularly heavy on water consumption as is apparent from the seventy times more water required to produce a kilogram of maize fed beef than the amount required to produce one kilogram of maize [116]. Nations with a high proportion of meat in their diet could reduce this without significant loss of nutritional value and at the same time either decrease their water usage or increase their food exports.

It is likely that water used in irrigation could be substantially reduced if its real cost was charged to farmers. The price of irrigation water often covers only a fraction of capital and management costs as governments seek to keep the price of food low. However, when the price of irrigation water was raised in Chile the amount of water used decreased by more than twenty percent as farmers adopted less wasteful irrigation methods [117]. On the domestic front, water price increases can substantially decrease wastage, as was the case in Bogor, Indonesia, where price increases were followed by a thirty percent decrease in water use. Globally, large cities are faced with tight water supplies that require innovative thinking to avoid a worsening situation with population growth. Thus, in New York City the replacement of 1.33 million old toilets with modern ones using only a third as much water resulted in the saving of 300,000 cubic metres of water each day [118]. A comparable change occurred in Mexico City through a similar program [114]. While such successes are welcome, most cities lose up to thirty percent of their water through leaky water mains, a loss that could be greatly reduced. Industry has also proved adept at decreasing water usage as shown by the steel industry that in the 1930's used from sixty to eighty tonnes of water to produce a tonne of steel whereas less than six tonnes of water is used now. On a broader scale, Japan used 50,000 tonnes, of water to produce $US1M of commercial output in 1965, while in 1989 only three and a half million tonnes were used. This arose from a combination of increased efficiency of water use and a change in the composition of the commercial output.

While the amount of water used globally continues to increase with population growth, the amount used per person has decreased since 1980. This probably reflects a combination of a growing awareness of water scarcity and price increases, but more will have to be done to minimize the effects of growing demand through population increase. Effort in this direction is reflected in the increasing use of recycled water, particularly for crop irrigation, as exemplified by Israel where seventy percent of municipal water is recycled, mainly for such use [114]. In California, 600 million cubic metres of recycled water were being used annually for irrigation by the mid 1990s, and in Namibia recycled water is used to supplement the household supply by up to thirty percent [114,119]. The purification methods used in reclaiming waste water range from standard sewerage treatment plants to membrane technologies, such as reverse osmosis, microfiltration and ultrafiltration, to distillation [14]. Membrane technologies and distillation are also used to desalinate seawater to provide drinking water. In 2002, about one percent of all drinking water was derived through 12,500 desalination plants notably in the Caribbean, the Mediterranean, the Middle East and increasingly in the United States.

References

1. a) J. F. Kasting, The origins of water on Earth, *Sci. Am. Quart., The Oceans*, **1998**, *9*, Fall, 16. b) A. H. Delsemme, An argument for the cometary origin of the biosphere, *Am. Sci.*, **2001**, *89*, 432.

2. a) J. F. Kasting, O. B. Toon and J. B. Polack, How climate evolved on terrestrial planets, *Sci. Am.*, **1988**, *258*, Feb., 90. b) J. S. Kargel and R. G. Strom, Global climate change on Mars, *Sci. Am.*, **1996**, *275*, Nov., 60. c) K. L. Tanaka, Fountains of youth, *Science*, **2000**, *288*, 2325. d) M. C. Malin and K. S. Edgett, Evidence for recent groundwater seepage and surface runoff on Mars, *Science, ibid.*, 2330. e) T. M. Donahue, Pre-Global Surveyor evidence for Martian ground water, *Proc. Natl. Acad. Sci. USA*, **2001**, *98*, 827. f) V. R. Baker, Water and the Martian landscape, *Nature*, **2001**, *412*, 228. g) R. A. Kerr, Running water eroded a frigid early Mars, *Science*, **2003**, *300*, 1496. h) M. P. Golombek, The surface of Mars: not just dust and rocks, *ibid.*, 2043. i) I. G. Mitrofanov *et al.*, CO_2 snow depth and subsurface water-ice abundance in the northern hemisphere of Mars, *ibid*, 2081. j) V. R. Baker, Icy Martian mysteries, *Nature*, **2003**, *426*, 779. k) J. W. Head *et al.*, Recent ice ages on Mars, *ibid.*, 797. l) D. L. Chandler, Distant shores, *New Sci.*, **2005**, 15 Jan., 30. m) S. C. Solomon *et al.*, New perspectives on ancient Mars, *Science*, **2005**, *307*, 1214. n) D. A. Paige, Ancient Mars: wet in many places, *ibid*, 1575. o) V. R. Baker, Picturing a recently active Mars, *Nature*, **2005**, *434*, 280.

3. J. W. Maurits la Rivière, Threats to the world's water, *Sci. Am.*, **1989**, *261*, Sept., 48.

4. A. Shiklomanov, World fresh water resources in *Water in Crisis: A Guide to the World's Fresh Water Resources*, P. H. Gleick, ed., Oxford University Press, Oxford, UK, 1993, p. 13.

5. P. H. Gleick, An introduction to global fresh water issues, in *Water in Crisis: A Guide to the World's Fresh Water Resources*, P. H. Gleick, ed., Oxford University Press, Oxford, UK, 1993, p. 3.

6. M. Gerstein and M. Levitt, Simulating water and the molecules of life, *Sci. Am.*, **1998**, *279*, Nov., 100.

7. J. Kingsland, Border control, *New Sci., Inside Science 132*, **2000**, 15 July, 1.

8. P. Davis, Control centre, *New Sci., Inside Science 122*, **1999**, 17 July, 1.

9. J. E. Cohen, *How Many People Can The Earth Support?* Norton, New York, USA, 1995.

10. G. K. Heilig, How many people can be fed on Earth? In *The Future Population of The World: What Can We Assume Today?* W. Lutz, ed., Earthscan, London, UK, 1996, p. 196.

11. S. L. Postel, G. C. Daily and P. R. Erhlich, Human appropriation of renewable fresh water, *Science*, **1996**, *271*, 785.

12. B. Holmes, Water, water everywhere..., *New Sci.*, **1996**, 17 Feb., 8.

13. a) C. J. Vörösmarty *et al.*, Global water resources: vulnerability from climate change and population growth, *Science*, **2000**, *289*, 284. b) F. Tao *et al.*, Terrestrial water cycle and the impact of climate change, *Ambio*, **2003**, *32*, 295.

14. a) G. Pinholster, Drinking recycled wastewater, *Env. Sci. Technol.*, **1995**, *29*, 174 A. b) S. Cole, Reclaimed wastewater continues flow toward tap, *ibid.*, **1998**, *32*, 498 A. c) M. B. Brennan, Waterworks, *Chem. Eng. News*, **2001**, 79, April, 32. d) D. Martindale, Sweating the small stuff, *Sci. Am.*, **2001**, *284*, Feb., 38. e) F. Pearce, Sea change for drinking water, *New Sci.*, **2004**, 10 July, 22.

15. a) J. Ives, Floods in Bangladesh: who is to blame? *New Sci.*, **1991**, 13 April, 30. b) F. Pearce, The rivers that won't be tamed, *New Sci.*, **1991**, 13 April, 34.

16. Tables A10 and A16 in *Water in Crisis: A Guide to The World's Fresh Water Resources*, P. H. Gleick, ed., Oxford University Press, Oxford, UK, 1993, pp. 129-133.

17. *World Population Prospects. The 2002 Revision*, The Population Division, United Nations, New York, USA, 2003.

18. a) S. C. McCaffrey, Water, politics and international law, in *Water in Crisis: a Guide to The World's Fresh Water Resources*, P. H. Gleick, ed., Oxford University Press, Oxford, UK, 1993, p. 92. b) A. T. Wolf, Conflict and cooperation along international waterways, *Water Policy*, **1998**, *1*, 251. c) S. L. Postel and A. T. Wolf, Dehydrating conflict, *Foreign Policy*, **2001**, Sept./Oct., 61.

19. P. P. Howell and J. A. Allan, eds., *The Nile. Sharing a Natural Resource*, Cambridge University Press, Cambridge, UK, 1994.

20. a) T. Patel, Bridge over troubled waters, *New Sci.*, **1996**, 30 Nov., 12. b) F. Pearce, Conflict looms over India's river plan, *New Sci.*, **2003**, 1 March, 4.

21. F. Pearce, Water war, *New Sci.*, **2002**, 18 May, 18.

22. a) F. Pearce, China drains life from the Mekong river, *New Sci.*, **2004**, 2 April, 14. b) Z. S. Hogan *et al.*, The imperiled giants of the Mekong, *Am. Sci.*, **2004**, *92*, 228.

23. a) L. Ohlsson, ed., *Hydropolitics. Conflicts over Water as a Development Constraint*, Zed Books, London, UK, 1995. b) Millich and R. G. Varady, Managing transboundary resources. Lessons from river-basin accords, *Environment*, **1998**, *40*, Nov., 10. c) O. Shela, Management of shared water basins: the case of the Zambezi River, *Water Policy*, **2000**, *2*, 65. d) P. Huisman, J. de Jong and K. Wieriks, Transboundary cooperation in shared river basins: experiences from the Rhine, Meuse and North Sea, *ibid.*, 83.

24. F. Pearce, Will Gaddafi's great river run dry? *New Sci.*, **1991**, 7 Sept., 5.

25. A. A. Al-Ibrahim, Excessive use of groundwater resources in Saudi Arabia: impacts and policy options, *Ambio*, **1991**, *20*, 34.

26. Australian Government AGSO Project 24108: Hydrology of the Great Artesian Basin, operating plan 95-96 at http://www.agso.gov.au/95-96/workplan/project/24108.htm.

27. M. Mainguet, *Desertification. Natural background and human mismanagement*, 2nd edn., Springer-Verlag, Berlin, Germany, 1994.

28. M. Newson, *Land and Water Development: River Basin Systems and Their Sustainable Management*, Routledge, London, UK, 1992.

29. a) R. Serafin and J. Zaleski, Baltic Europe, Great Lakes America and ecosystem redevelopment, *Ambio*, **1988**, 17, 99. b) H. J. Harris *et al.*, Importance of the nearshore area for sustainable redevelopment in the Great Lakes with observations on the Baltic Sea, *ibid.*, 112. c) H. A. Regier, *et al.*, Rehabilitative redevelopment of the fish and fisheries of the Baltic Sea and the Great Lakes, *ibid.*, **1988**, 121.

30. a) J. H. Hartig and J. R. Vallentyne, Use of an ecosystem approach to restore degraded areas of the Great Lakes, *Ambio*, **1989**, *14*, 423. b) D. W. Smith, Are PCBs in the Great Lakes approaching a new equilibrium? *Env. Sci. Technol.*, **1995**, *29*, 42 A. c) R. Renner, EPA Great Lakes guidance hits a squall, *ibid.*, 416 A. d) A. Schaefer, Lake Michigan heavily contaminated with PBDEs, *Env. Sci. Technol.*, **2001**, *35*, 139 A. e) J. Pelley, Great Lakes' cleanup too little, too late, *Env. Sci. Technol.*, **2002**, *36*, 446 A.

31. G. R. Francis, Institutions and ecosystem redevelopment with reference to Baltic Europe, *Ambio*, **1988**, 17, 106.

32. a) P. P. Micklin, Desiccation of the Aral Sea: a water management disaster in the Soviet Union, *Science*, **1988**, *241*, 1170. b) V. M. Kotlyakov, The Aral Sea basin. A critical environmental zone, *Environment*, **1991**, Jan./Feb., 4. c) S. Közli, The water and soil crisis in Central Asia - a source of future conflict? ENCOP Occasional Paper 11, Center for Security Policy and Conflict research, Zurich/Swiss Peace Foundation, Zurich/Berne, Switzerland, May, 1994. http://www.fakethz.ch/fsk/encop/11/en11-con.htm. d) R. Stone, Coming to grips with the Aral Sea's grim legacy, *Science*, **1999**, *284*, 30.

33. a) N. Precoda, Requiem for the Aral Sea, *Ambio*, **1991**, *20*, 109. b) S. Postel, *The Last Oasis. Facing Water Scarcity*, Earthscan, London, UK, 1992. c) K. L. Kiessling, Conference on the Aral Sea - Women, children, health and environment, *Ambio*, **1998**, *27*, 560. d) P. Zavialov *et al.*, Hydrographic survey of the dying Aral Sea, *Geophys. Res. Lett.*, **2003**, *30*, 1659. e) N. Jones, "South Aral gone in 15 years", *New Sci.*, **2003**, 19 July, 9.

34. R. Stone, Caspian ecology teeters on the brink, *Science*, **2002**, *295*, 430.

35. a) *Anon.*, African Lake drying up, *Science*, **2001**, *291*, 2081. b) M. T. Coe and J. A. Foley, Human and natural impacts on the water resources of the Lake Chad basin, *J. Geophys. Res.*, **2001**, *106*, 3349.

36. a) *Anon.*, Asian republics agree on joint rescue plan to save the Aral Sea, *Nature*, **1994**, *367*, 206. b) World Bank News Release No 98/1442 ECA, World Bank improves rural water supply and health in the Aral Sea basin in Uzbekistan. http://www.worldbank.org/html/extdr/extdr/extes/wbn0321.htm#water. c) Q. Schiermeier, Ecologists plot to turn the tide for shrinking lake, *Nature*, **2001**, *412*, 756. d) C. Williams, Long time no sea, *New Sci.*, **2003**, 4 Jan., 34. e) C. Pala, To save a vanishing sea, *Science*, **2005**, *307*, 1032.

37. F. Pearce, Hell and high water, *New Sci.*, **1999**, 19 June, 4.

38. a) L. Ramberg, A pipeline from the Okavango River? *Ambio*, **1997**, *26*, 129. b) J. Pallet, ed., *Sharing Water in Southern Africa*, Desert Research Foundation of Namibia, Windhoek, Namibia, 1997, p. 121. c) P. J. Ashton, Water security for multi-national river basin states: The special case of the Okavango River, in Proceedings of the Symposium on *"Water Security for Multi National River Basin States, Opportunity for Development"*, M. Falkenmark and J. Lundquist, eds., Stockholm International Water Institute Report No 8, 2000, p. 110. d) J. M. McCarthy *et al.*, Flooding patterns of the Okavango wetland in Botswana between 1972 and 2000, *Ambio*, **2003**, *32*, 453.

39. a) P. McCully, *Silenced Rivers: The Ecology and Politics of Large Dams*, Zed Books, London, UK, 1996. b) C. J. Vörösmarty and D. Sahagian, Anthropogenic disturbance of the terrestrial water cycle, *BioScience*, **2000**, *50*, Sept., 753. c) J. P. M. Syvitski *et al.*, Impact of Humans on the flux of terrestrial sediment to the global coastal ocean, *Science*, **2005**, *308*, 376.

40. *World Register of Dams 2003*, International Commission on Large Dams, Paris, France, 2003.

41. World Commission on Dams, Dams and Development. *A New Framework for Decision Making*, EarthScan, London, UK, 2000.

42. International Water Power and Dam Construction 1994, *Handbook*, Reed, Sutton, UK, 1994.

43. B. F. Chao, Anthropological impact on global geodynamics due to reservoir water impoundment, *Geophys. Res. Lett.*, **1995**, *22*, 3529

44. C. Nilsson *et al.*, Fragmentation and flow regulation of the world's large river systems,, *Science*, **2005**, *308*, 405.

45. R. S. Devine, The trouble with dams, *Atlantic Monthly*, **1995**, Aug., 64.

46. F. Pearce, A dammed fine mess, *New Sci.*, **1991**, 4 May, 32.

47. a) V. Ittekkot, C. Humborg and P. Schäfer, Hydrological alterations and marine biogeo-chemistry: a silicate issue? *BioScience*, **2000**, *50*, Sept., 776. b) C. Nilsson and K. Berggren, Alterations of riparian ecosystems caused by river regulation, *ibid.*, 783.

48. a) K. Perczel and G. Libik, Environmental effects of the dams system on the Danube at Bös-Nagymaros, *Ambio*, **1989**, *18*, 247. b) J. D. Milliman, Blessed dams or damned dams, *Nature*, **1997**, *386*, 325. c) C. J. Vörösmarty, *et al.*, The storage and ageing of continental runoff in large reservoir systems of the world, *Ambio*, **1997**, *26*, 210. d) F. Pearce, The floodgates open, *New Sci.*, **2000**, 25 March, 16.

49. K. Venkataraman and A. Kumar, A compendium of world wide web sites related to mercury in fish, *Environ. Prog.*, **2000**, *19*, Spring, S11.

50. a) C. Holden, Global commission slams dams, *Science*, **2000**, *290*, 1887. b) M. Scrope, Governments urged to rethink dam projects, *Nature*, **2000**, *408*, 1395. c) F. Pearce, Dismay over call to build more dams, *New Sci.*, **2003**, 22 March, 9. d) F. Pearce, Bleeding the planet dry, *ibid.*, 27.

51. a) V. L. St Louis *et al.*, Reservoir surfaces as sources of greenhouse gases to the atmosphere: a global estimate, *BioScience*, **2000**, *50*, Sept., 766. b) D. Graham-Rowe, Hydro's dirty secret revealed, *New Sci*, **2005**, 26. Feb, 8.

52. a) W. L. Graf, *The Colorado River: Instability and Basin Management*, Association of American Geographers, Washington, DC, USA, 1985. b) R. H. Webb, *Grand Canyon: A Century of Change*, University of Arizona Press, Tucson, USA, 1996.

53. a) M. P. Collier, R. H. Webb and E. D. Andrews, Experimental flooding in Grand Canyon, *Sci. Am.*, **1997**, *276*, Jan., 66. b) M. Kowalewski, Dead delta's former productivity: two trillion shells at the mouth of the Colorado river, *Geology*, **2000**, *28*, 1059. c) J. Pelle, Restoring our rivers, *Env. Sci. Technol.*, **2000**, *34*, 87 A. d) J. P. Cohn, Resurrecting the dammed: a look at Colorado River restoration, *BioScience*, **2001**, *51*, Dec., 998. e) J. W. Jacobs and J. L. Wescoat, Managing river resources: lessons from Glen Canyon, *Environment*, **2002**, *44*, March, 8. f) M. Kaplan, Let the river run, *New Sci.*, **2002**, *28* September, 32. g) K. Powell, Open the floodgates! *Nature*, **2002**, *420*, 356.

54. a) T. Beardsley, Parched policy, *Sci. Am.*, **1991**, *264*, May, 12. b) D. Dziegielewski and D. D. Baumannn. Tapping alternatives: The benefits of managing urban water demands, *Environment*, **1992**, *34*, Nov., 6.

55. a) K. N. Lee, The Columbia river basin: Experiment in sustainability, *Environment*, **1989**, *31*, July/Aug, 6. b) B. Harden, *A River Lost: The Life and Death of the Columbia*, Norton, New York, USA, **1996**.

56. a) K. W. Muckleston, Striking a balance in the Pacific Northwest, *Environment*, **1990**, *32*, Jan./Feb., 10. b) J. M. Volkman, Making room in the Ark: The Endangered Species Act and the Columbia river basin, *Environment*, **1992**, *34*, May, 18. c) J. Pelley, Breaching dams may save salmon, but the science remains an issue, *Env. Sci. Technol.*, **2000**, *34*, 112 A. d) C. C. Mann and M. L. Plummer, Can science rescue salmon? *Science*, **2000**, *289*, 716. e) J. Collie *et al.*, Of salmon and dams, *Science*, **2000**, *290*, 933. f) C. C. Mann and M. L. Plummer, Response, *Science*, **2000**, *290*, 934. g) P. Kareiva, M. Marvier and M. McClure, Recovery and management options for spring/summer Chinook salmon in the Columbia river basin, *Science*, **2000**, *290*, 977. h) D. D. Dauble and D. R. Guest, Comparisons of mainstream spawning habitats for two populations of fall Chinook salmon in the Columbia river basin, *Reg. Rivers Res. Mgt.*, **2000**, *16*, 345. i) P. S. Levin and M. H. Schiewe, Preserving salmon biodiversity, *Am. Sci.*, **2001**, *89*, 220.

57. a) A. Sarre, Monitoring the white death: soil salinity, *Chem. Aust.*, **1999**, 66, 13. b) The salt that won't run to the sea, *The Economist*, **2000**, 5 Feb., 26.

58. A. K. Biswas, Hydrological engineering prior to 600 BC, *Proc. Am. Soc. Civil Eng. J.*, *Hydraulics Div.*, **1967**, *HY5*, 115.

59. T. Evans, History of Nile flows, in *The Nile: Sharing a Natural Resource*, P. P. Howell and J. A. Allan, eds., Cambridge University Press, Cambridge, UK, 1994.

60. J. Hultin, The Nile: source of life, source of conflict, in *Hydropolitics. Conflicts over Water as a Development Constraint*, L. Ohlsson, ed., Zed Books, London, UK, 1995.

61. C. Agnew and E. Anderson, *Water Resources in the Arid Realm*, Routledge, London, UK, 1992.

62. a) D. J. Stanley and A. G. Warne, Nile delta: recent geological evolution and human impact, *Science*, **1993**, *260*, 628. b) D. J. Stanley and A. G. Warne, Worldwide initiation of Holocene marine deltas by deceleration of sea-level rise, *Science*, **1994**, *265*, 228.

63. D. J. Stanley, Nile delta: extreme case of sediment entrapment on a delta plain and consequent coastal land loss, *Marine Geol.*, **1996**, *129*, 189.

64. a) A. K. Biswas, Environmental sustainability of Egyptian agriculture: problems and perspective, *Ambio*, **1993**, *24*, 16. b) D. Conway *et al.*, Future availability of water in Egypt: the interaction of global, regional, and basin scale driving forces in the Nile basin, *Ambio*, **1996**, *25*, 336.

65. J. D. Milliman, J. D. Broadus and F. Gable, Environmental and economic implications of rising sealevel and subsiding deltas: The Nile and Bengal examples, *Ambio*, **1989**, *18*, 340.

66. D. Hillel, The Twin Rivers, in *Rivers of Eden. The struggle for water and the quest for peace in the Middle East*, Ch. 5., Oxford University Press, Oxford, UK, 1994.

67. a) Editorial, Water wars, *New Sci.*, **2001**, 19 May, 3. b) F. Pearce, Iraqi wetlands face total destruction, *ibid.*, 4. c) J. Randerson, Iraquis reclaim their ancient wetlands, *New Sci.*, **2003**, 4 Oct., 6. d) A. Lawler, Reviving Iraq's wetlands, *Science*, **2005**, *307*, 86.

68. a) M. Falkenmark, Middle East hydropolitics: water scarcity and conflicts in the Middle East, *Ambio*, **1989**, *18*, 350. b) F. Pearce, Wells of conflict on the West Bank, *New Sci.*, **1991**, June 1, 26. c) D. Hillel, The River Jordan, in *Rivers of Eden. The struggle for water and the quest for peace in the Middle East*, Ch. 7, Oxford University Press, Oxford, UK, 1994. d) S. Libiszewski, Water disputes in the Jordan basin Region and their role in the resolution of the Arab-Israeli conflict. ENCOP Occasional Paper 13, Center for Security Policy and Conflict Research Zurich/Swiss Peace Foundation, Zurich/Berne, Switzerland, Aug. 1995 (ISBN 3-905641-36-4). http://www.fsk.ethz.ch/encop/13/en13.htm.

69. C. Kwai-Cheong, The Three Gorges Project of China: Resettlement prospects and problems, *Ambio*, **1995**, *24*, 98.

70. A. Zich, China's Three Gorges, *Natl. Geog.*, **1997**, *192*, Sept., 2.

71. a) F. Pearce, The biggest dam in the world, *New Sci.*, **1995**, 28 Jan., 25. b) d) X. Lei, Going against the flow in China, *Science*, **1998**, *280*, 24. c) D. Quing, ed., *The River Dragon has Come! The Three Gorges Dam and the Fate of China's Yangtze River and Its People*, M. E. Sharp, Armonk, NY, USA, 1998. d) J. Wu *et al.*, Three-Gorges Dam – experiment in habitat fragmentation? *Science*, **2003**, *300*, 1239.

72. F. Pearce, Replumbing the planet, *New Sci.*, **2003**, 7 June, 30.

73. K.-G. Malle, Cleaning up the River Rhine, *Sci. Am.*, **1996**, *274*, Jan., 54.

74. N. F. Caraco and J. J. Cole, Human impact on nitrate export: an analysis using major world rivers, *Ambio*, **1999**, *28*, 167.

75. a) L. D. Mee, The Black Sea in crisis: a need for concerted international action, *Ambio*, **1992**, *21*, 278. b) S. Tugrul *et al.*, Changes in the hydrochemistry of the Black Sea inferred from water density profiles, *Nature*, **1992**, *359*, 137. c) V. Fabry, K. Frölich and I. Osvath, Environmental pollution of the Black Sea: a search for answers, *IAEA Bull.*, **1993**, *920*, 20. d) C. Humborg *et al.*, Effect of Danube River dam on Black Sea biogeochemistry and ecosystem structure, *Nature*, **1997**, *386*, 385. e) A. E. Kideys, Fall and rise of the Black Sea ecosystem, *Science*, **2002**, *297*, 1482.

76. a) P. Bagla and J. Kaiser, *Science*, India's spreading health crisis draws global arsenic experts, *Science*, **1996**, *274*, 174. b) R. Nickson *et al.*, Arsenic poisoning of Bangladesh groundwater, *Nature*, **1998**, *395*, 338. c) S. K. Acharyya *et al.*, Arsenic poisoning in the Ganges delta, *Nature*, **1999**, *401*, 545. d) A. M. R. Chowdhur and M. Jakariy, Testing of water for arsenic in Bangladesh, *Science*, **1999**, *284*, 1622. e) A. Coghlan, Bangladeshis sue British geologists for "largest mass poisoning ever", *New Sci.*, **2002**, 7 Sept., 7. f) R. L. Rawls, Tackling arsenic in Bangladesh, *Chem. Eng. News*, **2002**, *80*, Oct., 42. g) F. Pearce and J. Hecht, Flawed water tests put millions at risk, *New Sci.*, **2002**, 16 Nov., 4. h) A. Ananthaswamy, Tainted tube wells pour poison onto food crops, *New Sci.*, **2002**, 7 Dec., 9. i) E. Stokstad, Agricultural pumping linked to arsenic, *Science*, **2002**, *298*, 1535. j) C. F. Harvey *et al.*, Arsenic mobility and groundwater extraction in Bangladesh, *ibid.*, 1602. k) T. Clarke, Delta blues, *Nature*, **2003**, *422*, 254. l) A. M. R. Chowdhury, Arsenic crisis in Bangladesh, *Sci. Am.*, **2004**, *291*, Aug., 70.

77. a) A. H. Welch, *et al.*, Arsenic in ground water supplies of the United States, in *Arsenic Exposure and Health Effects*, W. R. Chappell, C. O. Abernathy and R. L. Cameron, eds., Elsevier New York, USA, 1999. b) W. Lepkowski, Science meets policy in shaping water's future, *Chem. Eng. News*, **1999**, *77*, 6 Dec., 127. c) C. Hogue, A sip of arsenic? *Chem. Eng. News*, **2001**, 79, 21 May, 50. d) D. K. Nordstrom, Worldwide occurrences of arsenic in

ground water, *Science*, **2002**, *296*, 2143. e) F. Pearce, Arsenic's fatal legacy grows, *New Sci.*, **2003**, 9 Aug., 4

78.   a) *Guidelines for Drinking Water Quality*, 2nd, edn., Vols. 1 and 2, World Health Organisation, Geneva, Switzerland, 1993. b) D. Warner, Water policy for safeguarding human health, in *Sustaining Our Waters into the 21st Century*, J. Lundquist and P. Gleick, eds., Background Report to Comprehensive Assessment of the Freshwater Resources of the World. Report presented to the UN General Assembly on 24 June, 1997, (E/CN. 17/1997/9). c) R. Doyle, Access to safe drinking water, *Sci. Am.*, **1997**, *277*, Nov., 19. d) J. Lundqvist, Avert looming hydrocide, *Ambio*, **1998**, *27*, 428.

79.   *Comprehensive Assessment of the Freshwater Resources of the World.* Report presented to the UN General Assembly on 24 June, 1997, (E/CN. 17/1997/9).

80.   a) R. Rosenberg, O. Lindahl and H. Blanck, Silent spring in the sea, *Ambio*, **1988**, *17*, 289. b) R. Dietz, M.-P. Heide-Jørgensen and T. Härkönene, Mass death of Harbor seals (*phoca vitulina*) in Europe, *Ambio*, **1989**, *18*, 258. c) S. Fleischer and L. Stibe, Agriculture kills marine fish in the 1980s. Who is responsible for fish kills in the year 2000? *ibid.*, 347. d) D. M. Anderson, Red tides, *Sci. Am.*, **1994**, *271*, Aug., 52. e) D. K. Atwood *et al.*, eds., Papers from NOAA's nutrient enhanced coastal ocean productivity study - special dedicated issue, *Estuaries*, **1994**, *17*, 729. f) W. W. Carmichael, The toxins of cyanobacteria, *Sci. Am.*, **1994**, *270*, Jan., 64. g) N. F. Caraco, Influence of humans on phosphorus transfers to aquatic systems: a regional scale study using large rivers, in *Phosphorus in the Global Environment: Transfers, Cycles and Management*, H. Tiessen, ed., Wiley, Chichester, UK, 1995. h) L. J. Puckett, Identifying the major sources of nutrient water pollution, *Env. Sci. Technol.*, **1995**, *29*, 408 A. i) F. Pearce, Deadly blooms reach Britain's rivers, *New Sci.*, **1996**, 18 May, 5. j) J. Pelly, Is coastal eutrophication out of control? *Env. Sci. Technol.*, **1997**, *31*, 462 A. k) S. W. Nixon, Enriching the sea to death, *Sci. Am. Quart., The Oceans*, **1998**, *9*, Fall, 48. l) S. Carpenter *et al.*, Nonpoint pollution of surface waters with phosphorus and nitrogen, *Ecol. Appln.*, **1998**, *8*, 559. m) J. Pelly, What is causing toxic algal blooms? *Env. Sci. Technol.*, **1998**, *32*, 26 A. n) N. N. Rabalais, Nitrogen in aquatic systems, *Ambio*, **2002**, *31*, 102.

81.   a) R. E. Turner and N. N. Rabalais, Changes in Mississippi river water quality this century: implications for coastal food webs, *BioScience*, **1991**, *41*, March, 140. b) R. E. Turner and N. N. Rabalais, Coastal eutrophication near the Mississippi river delta, *Nature*, **1994**, *368*, 619. c) N. N. Rabalais *et al.*, Nutrient changes in the Mississippi river and system responses on the adjacent continental shelf, *Estuaries*, **1996**, *19*, 286. d) R. Tyson, "Dead zone" in Gulf of Mexico draws federal - state attention, *Env. Sci. Technol.*, **1997**, *31*, 454 A. e) D. Malakoff, Death by suffocation in the Gulf of Mexico, *Science*, **1998**, *281*, 190. f) Gulf of Mexico hypoxia assessment plan, March 1998, NOAA coastal ocean program office. http://www. cop.noaa.gov/HypoxiaPlan.html. g) D. Ferber, Keeping the Stygian waters at bay, *Science*, **2001**, *291*, 968. h) Comments on g) from, C. S. Snyder, R. W. Howard and D. Winstanley and a response from D. Ferber appear in Hypoxia, fertilizer and the Gulf of Mexico, *Science*, **2001**, *292*, 1485. i) W. J. Mitsch *et al.*, Reducing nitrogen loading to the Gulf of Mexico from the Mississippi river basin: strategies to counter a persistent ecological problem, *BioScience*, **2001**, *51*, May, 373. j) G. F. McIsaac *et al.*, Nitrate flux in the Mississippi river, *Nature*, **2001**, *414*, 166. k) G. F. McIsaac *et al.*, Nitrate flux in the Mississippi river, *ibid.*, 710.

82. G. O'Neill, Ocean anomaly triggers record fish kill, *Science*, **1995**, *268*, 1431.

83. a) B. T. Nolan *et al.*, Risk of nitrate in ground waters of the United States - a national perspective, *Env. Sci. Technol.*, **1997**, *31*, 2229. b) M. Zhang, S. Geng and K. S. Smallwood, Assessing groundwater nitrate contamination for resource and landscape management, *Ambio*, **1998**, *27*, 170.

84. a) A. M. Fan, C. C. Willhite and S. A. Book, Evaluation of the nitrate drinking water standard with reference to infant methemoglobinemia and potential reproductive toxicity, *Reg. Tox. Pharmacol.*, **1987**, *7*, 135. b) C. J. Johnson, Methemoglobinemia: is it coming back to haunt us? *Health Environ. Digest*, **1988**, *1*, 3. c) American Academy of Pediatrics, Infant methemoglobinemia: The role of dietary intake, *Pediatrics*, **1994**, *46*, 475. d) National Research Council, *Nitrate and Nitrite in Drinking Water: A Report of the Board of Environmental Studies and Toxicolgy*, National Academy Press, Washington, DC, USA, 1995. e) A. H. Wolfe and J. Patz, Reactive nitrogen and human health: acute and long-term implications, *Ambio*, **2002**, *31*, 120.

85. a) B. G. Loganathan and K. Kannan, Global organochlorine contamination trends: an overview, *Ambio*, **1994**, *23*, 187. b) J. Falandysz *et al.*, Organochlorine pesticides and polychlorinated biphenyls in cod-liver oils: North Atlantic, Norwegian Sea, North Sea and Baltic Sea, *Ambio*, **1994**, *23*, 288. c) K. A. Kidd *et al.*, High toxaphene concentrations in fish from a subArctic lake, *Science*, **1995**, *269*, 240. d) A. J. Hoffman, An uneasy rebirth at Love Canal, *Environment*, **1995**, *37*, March, 4. e) F. Wania and D Mackay, Tracking the distribution of persistent organic pollutants, *Env. Sci. Technol.*, **1996**, *30*, 390 A. f) F. Pearce, Dirty groundwater runs deep, *New Sci.*, **1996**, 21 Sept., 16. g) F. Pearce, Northern exposure, *New Sci.*, **1997**, 31 May, 25. h) L. Bergström and J. Stenström, Environmental fate of chemicals in soil, *Ambio*, **1998**, *27*, 16.

86. a) *Anon.*, Hormone mimics pose challenges, *Chem. Indust.*, **1996**, 20 May, 364. b) T. Colborn, D. Dumanoski and J. P. Myers, *Our stolen future: are we threatening our fertility, intelligence and survival?* Dutton, New York, USA, 1996. c) T. E. Wiese and W. R. Kelce, An introduction to environmental oestrogens, *Chem. Indust.*, **1997**, 18 Aug., 648. d) P.-E. Olsson, *et al.*, Effects of maternal exposure to estrogen and PCB on different life stages of zebrafish (*Danio rerio*), *Ambio*, **1999**, *28*, 100. e) C. Maczka *et al.*, Evaluating impacts of hormonally active agents in the environment, *Env. Sci. Technol.*, **2000**, *34*, 136 A. f) C. M. Cooney, Science panel confirms low-dose estrogenic effects, *ibid.*, 500 A. g) J. Withgott, Ubiquitous herbicide emasculates frogs, *Science*, **2002**, *296*, 447. h) T. Hayes *et al.*, Feminization of male frogs in the wild, *Nature*, **2002**, *419*, 895. i) R. Renner, Conflict brewing over herbicide's link to frog deformities, *Science*, **2002**, *298*, 938. j) M. Burke, Sex effects, *Chem. Br.*, **2003**, *39*, Jan., 30.

87. N. Bauman, Panic over falling sperm counts may be premature, *New Sci.*, **1996**, 11 May, 10.

88. a) F. Pearce, Something in the water, *New Sci.*, **1999**, 6 March, 19. b) P. S. Zurere, Drugs down the drain, *Chem. Eng. News*, **2000**, *78*, 10 April, 51. c) D. L. Sedlak, J. L. Gray and K. E. Pinkston, Understanding microcontaminants in recycled water, *Env. Sci. Technol.*, **2000**, *34*, 508 A.

89. a) L. J. Goldwater, Mercury in the environment, *Sci. Am.*, **1971**, *224*, May, 15. b) T. Tsubaki and K. Irukayama, *Minamata Disease*, Elsevier, Amsterdam, Netherlands, 1977.

90.   a) J. A. Meech, M. M. Veiga and D. Tromans, Reactivity of mercury from gold mining activities in darkwater ecosystems, *Ambio*, **1998**, *27*, 92. b) J. R. Davée Guimarães *et al.*, Mercury in human and environmental samples from two lakes in Amapá, Brazilian Amazon, *Ambio*, **1999**, *28*, 296. c) L. Maurice-Bourgoin, *et al.*, Mercury pollution in the upper Beni river, Amazonian Basin: Bolivia, *ibid.*, 302. d) F. Pearce, A nightmare revisited: an affliction last seen in the 1950s has struck again, *New Sci.*, **1999**, 6 Feb., 4.

91.   R. Koenig, Wildlife deaths are a grim wake-up call in Eastern Europe, *Science*, **2000**, *287*, 1737.

92.   a) D. R. Smith and A. R. Flegal, Lead in the biosphere: recent trends, *Ambio*, **1995**, *24*, 21. b) R. L. France and J. M. Blais, Lead concentration and stable isotopic evidence for transpolar contamination of plants in the Canadian high Arctic, *Ambio*, **1998**, *27*, 506. c) C. Huamain *et al.*, Heavy metal pollution in soils in China: status and countermeasures, *Ambio*, **1999**, *28*, 130.

93.   a) G. E. Likens and F. H. Bormann, Acid rain: a serious regional environmental problem, *Science*, **1974**, *184*, 1176. b) G. E. Likens *et al.*, Acid rain, *Sci. Am.*, **1979**, *241*, 39. c) B. J. Mason, Acid rain - cause and consequences, *Weather*, **1990**, *45*, 70. d) L. O. Hedin and G. E. Likens, Atmospheric dust and acid rain, *Sci. Am.*, **1996**, *275*, Dec., 58.

94.   D. M. Elsom, *Atmospheric Pollution: A Global Problem*, 2nd edn., Blackwell, Oxford, UK, 1992.

95.   D. Schindler, From acid rain to toxic snow, *Ambio*, **1999**, *28*, 350.

96.   a) R. A. Smith. On the air and rain of Manchester, *Memoirs of the Literary and Philosophical Society of Manchester*, Series 2, **1852**, *10*, 207. b) E. Gorham, Scientific understanding of ecosystem acidification: A historical review, *Ambio*, **1989**, *18*, 150.

97.   a) F. Pearce, Lightening sparks pollution rethink, *New Sci.*, **1997**, 25 Jan., 15. b) C. Price, J. Penner and M. Prather, NO_x from lightening 1. Global distribution based on lightening physics, *J. Geophys. Res.*, **1997**, *102*, 5929. c) C. Price and J. Penner, NO_x from lightening 2. Constraints from the global atmospheric electric circuit, *ibid.*, 5943. d) V. Kiernan, Lightening sharpens acid rain's bite, *New Sci.*, **1997**, 31 May, 17.

98.   a) The National Atmospheric Deposition Program/National Trends Network: http:nadp.sws.uiuc.edu

99.   a) J. Kaiser, Acid rain's dirty business: stealing minerals from soils, *Science*, **1996**, 272, 198. b) G. E. Likens, C. T. Driscoll and D. Buso, Long-term effects of acid rain: response and recovery of a forest ecosystem, *Science*, **1996**, 272, 244. c) C. T. Driscoll *et al.*, Acidic deposition in the Northeastern United States: sources and inputs, ecosystem effects, and management strategies, *BioScience*, **2001**, 51, March, 180. d) K. Krajick, Long-term data show lingering effects from acid rain, *Science*, **2001**, 292, 195.

100. a) B. Ulrich, R. Mayer and P. K. Khanna, Chemical changes due to acid precipitation in a löss derived soil in central Europe, *Soil Sci.*, **1980**, *130*, 193. b) B. Ulrich and J. Pankrath, *Effects of Accumulation of Air Pollutants on Forest Ecosystems*, Reidel, Boston, USA, 1983. c) G. Einbender, *et al.*, The case for immediate controls on acid rain, *Mater. Soc.*, **1982**, *6*, 251. d) A. H. Johnson and T. G. Siccama, Acid deposition and forest decline, *Env. Sci. Technol.*, **1983**, *17*, 249 A. e) B. Prinz, Causes of forest damage in Europe, *Environment*, **1987**, *29*, Nov., 10. f) E. D. Schultz, Air pollution and forest decline in a spruce (*Picea abies*) forest, *Science*, **1989**, *244*, 776. g) P. H. Freer-Smith, Do pollutant-related forest declines threaten the sustainability of forests? *Ambio*, **1998**, *27*, 123. h) C.

Nelleman and J. M. Esser, Crown condition and soil acidification in Norwegian Spruce forests, *ibid.*, 143.

101. a) G. R. Hendrey *et al.*, Some hydrological changes, *Ambio*, **1976**, *5*, 224. b) C. L. Schofield, Acid precipitation: effects on fish, *ibid.*, 228. c) D. W. Schindler, Effects of acid rain on freshwater ecosystems, *Science*, **1988**, *239*, 149. d) A. Henriksen, *et al.*, Northern European lake survey, 1995, *Ambio*, **1998**, *27*, 80. e) B. L. Skjelkvåle and R. F. Wright, Mountain lakes; sensitivity to acid deposition and global climate change, *ibid.*, 280. f) T. Hesthagen, I. H. Sevaldrud and H. M. Berger, Assessment of damage to fish populations in Norwegian lakes due to acidification, *Ambio*, **1999**, *28*, 112.

102. a) R. A. Barnes, Acid rain; media myth or environmental apocalypse? *Catalyst*, **1986**, *2*, 55. b) F. B. Smith, An overview of the acid rain problem, *Meteorol. Mag.*, **1991**, *120*, 77.

103. a) R. A. Barnes, The long range transport of air pollution: a review of European experience, *J. Air Polln. Control Assmt*, **1979**, *29*, 1219. b) J. McCormick, Acid pollution. The international community's continuing struggle, *Environment*, **1998**, *40*, April, 16.

104. B. DeKoker, An acid test, *Sci. Am.*, **1995**, *273*, Oct., 25.

105. a) C. Suplee, Unlocking the climate puzzle, *Natl. Geog.*, **1998**, *193*, May, 38. b) D. G. Streets *et al.*, Energy consumption and acid deposition in northeast Asia, *Ambio*, **1999**, *28*, 135. c) A. McDonald, Combating acid deposition and climate change: Priorities for Asia, *Environment*, **1999**, *41*, April, 4. d) K. E. Wilkening, L. A. Barrie and M. Engle, Trans-Pacific air pollution, *Science*, **2000**, *290*, 65.

106. A. Rosencranz, The acid rain controversy in Europe and North America, *Ambio*, **1986**, *15*, 47.

107. a) R. A. Kerr, Acid rain control: Success on the cheap, *Science*, **1998**, *282*, 1024. b) R. G. Derwent, S. E. Metcalfe and J. D. Whyatt, Environmental benefits of NO_x control in northwestern Europe, *Ambio*, **1998**, *27*, 518. c) D. Munton, Dispelling the myths of the acid rain story, *Environment*, **1998**, *40*, July/Aug., 4. d) A. Jenkins, End of the acid reign, *Nature*, **1999**, *401*, 537. e) N. Carson and D. Munton, Flaws in the conventional wisdom on acid deposition, *Environment*, **2000**, *42*, March, 33. f) J. A. Lynch, V. C. Bowersox and J. W. Grimm, Acid rain reduced in eastern United States, *Env. Sci. Technol.*, **2000**, *34*, 940. g) P. Brimblecombe, Acid drops, *New Sci.*, *Inside Science 150*, **2002**, 25 May, 1.

108. a) G. Fischer and G. Heilig, Population momentum and the demand on land and water resources, *Phil. Trans. Roy. Soc. B*, **1997**, *352*, 869. b) J. S. Wallace and C. H. Batchelor, Managing water resources for crop production, *ibid.*, 937.

109. M. Falkenmark, Meeting water requirements of an expanding world population, *Phil. Trans. Roy. Soc. B*, **1997**, *352*, 929.

110. a) D. Seckler *et al.*, *World water demand and supply, 1990 to 2025: Scenarios and issues*, Research Report 19, International Water Management Institute, Colombo, Sri Lanka, 1998. http://www.cgiar.org.iwmi. b) D. Seckler, R. Barker and U. Amarasinghe, Water scarcity in the twenty-first century, *Int. J. Water Res.*, **1999**, *15*, 29.

111. *Projected water scarcity in 2025*, International Water Management Institute, Colombo, Sri Lanka, 2000. http//www.cgiar.org.iwmi/home /wsmap.htm.

112. a) M. Falkenmark, The massive water scarcity now threatening Africa - why isn't it being addressed? *Ambio*, **1989**, *18*, 112. b) P. J. Ashton, Avoiding conflicts over Africa's water resources, *Ambio*, **2002**, *31*, 236.

113. a) M. Falkenmark *et al.*, Water scarcity as a key factor behind food insecurity: Round table discussion, *Ambio*, **1998**, *27*, 148. b) R. M. Hirsch, T. L. Miller and P. A. Hamilton, Using today's science to plan for tomorrow's water policies, *Environment*, **2001**, *43*, Jan./Feb., 8. c) P. H. Gleick, Making every drop count, *Sci. Am.*, **2001**, *284*, Feb., 29. d) P. H. Gleick, Global water: threats and challenges facing the United States, *Environment*, **2001**, *43*, March, 18. e) K. Brown, Water scarcity: forecasting the future with spotty data, *Science*, **2002**, *297*, 926. f) F. Pearce, Quenching the world's thirst, *New Sci.*, **2002**, 7 Sept., 10. g) H. Hoag, Atlas of a thirsty planet, *Nature*, **2003**, *422*, 252. h) P. H. Gleick, Water use, *Annu. Rev. Energy Environ.*, **2003**, *28*, 275. i) P. H. Gleick, Global freshwater resources: Soft-path solutions for the 21st century, *Science*, **2003**, *302*, 1524. j) O. Schipper, Declining water resources raise food concerns, *Env. Sci. Technol.*, **2003**, *37*, 273 A. k) H. Yang, A water resources threshold and its implications for food security, *ibid.*, 3048.

114. P. H. Gleick, Making every drop count, *Sci. Am.*, **2001**, *284*, Feb., 28.

115. a) S. L. Postel, Water for food production: will there be enough in 2025? *BioScience*, **1998**, *48*, Aug., 629. b) S. Postel, Growing more food with less water, *Sci. Am.*, **2001**, *284*, Feb., 34.

116. D. Pimentel *et al.*, Water resources: agriculture, the environment, and society, *BioScience*, **1997**, *47*, Feb., 97.

117. N. Johnson, C. Revenga and J. Echeverria, Managing water for people and nature, *Science*, **2001**, *292*, 1071.

118. D. Martindale, Leaking away, *Sci. Am.*, **2001**, *284*, Feb., 40.,

119. D. Martindale, Waste not, want not, *Sci. Am.*, **2001**, *284*, Feb., 40.

Chapter 4

Food: Famine and Plenty

"Civilization as it is known today could not have evolved, nor can it survive, without an adequate food supply. Yet food is something that is taken for granted by most world leaders despite the fact that more than half of the population of the world is hungry."

Norman E. Borlaug, pioneer of the Green Revolution and Nobel Peace Laureate, acceptance speech, Oslo, Norway, 11 December, 1970.

"We, the Heads of State and Government, or our representatives, gathered at the World Food Summit at the invitation of the Food and Agricultural Organization of the United Nations, reaffirm the right of everyone to have access to safe and nutritious food, consistent with the right to adequate food and the fundamental right of everyone to be free from hunger."

Rome Declaration on World Food Security, Rome, 13 November, 1996.

4.1. Feeding Humanity

Alternating times of famine and plenty have existed over millennia and continue to this day in the less affluent nations. Some 842 million people were undernourished by the standards of the United Nations Food and Agricultural Organization (FAO) in 2001 as shown in Fig. 4.1 [1]. Millions more suffered from undernutrition during seasonal food scarcity and in times of war. Yet, while the growth of population in the latter half of the twentieth century was without precedent, the global food supply increased apace so that the proportion of humanity suffering from undernutrition decreased from twenty-eight percent in 1981 to seventeen percent in the developing nations in 2001. Food availability rose twenty-seven percent over forty years to reach the 1999-2001 global average of 11,730 kilojoules per person daily. In the developing nations the rise in food availability was thirty-seven percent to 11,200 kilojoules per person daily, with massively populated China contributing greatly to this increase. And the food supply is expected to grow with population for the next twenty years at least in accord with the FAO Rome Declaration on World Food Security and World Food Summit Plan of Action of 1996, although this will require careful planning on a global scale [2-8]. This is a reflection of humanity's ability to refine agriculture to the point where in the United States less than one percent of the population is involved in primary production compared with sixty percent in the early 1890s. Similar changes have occurred in other developed nations. This is exemplified by the growing of wheat in mediaeval France that produced 800 kilograms of grain annually per hectare from 200 kilograms of seed and 500 hours of labour while today northwestern Europe produces 7,500 kilograms of wheat per hectare annually from 150 kilograms of seed and fifteen hours of labour [6]. On a global basis, changes in agriculture have made it possible to feed a single person from the food grown on 2,000 square metres of land whereas ten times that area was required two hundred years ago [8].

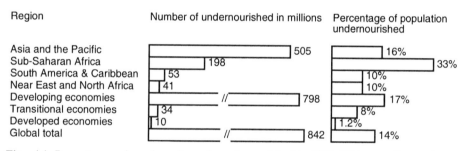

Fig. 4.1. Percentage and number of people undernourished in 1999-2001 according to the FAO [1].

However, the full story is not one of unqualified success for the daily availability of food to a sub-Saharan African has increased only slightly to 9,000 kilojoules from the 8,600 kilojoules of thirty years ago, and generally as the drive to produce more food has progressed so have the strains on the environment whose good health is basic to humanity's well-being. As the land area cleared for agriculture increased during the twentieth century, erosion and soil degradation often followed, although the estimates of its global extent vary greatly with the latest being significantly less than earlier estimates [9-12]. The once seemingly unlimited supply of fish in the oceans has been exploited to the point that within a generation once great fisheries have been exhausted [13]. Some have been diminished by excess fertilizer carried to the oceans by rivers; the very fertilizer that has fuelled the much higher crop productivity that feeds a rapidly increasing population [14,15]. Fortunately, a growing understanding of the ecological balance promises to at least increase efforts to protect the environment from some of the more damaging aspects of food production. At the same time, much of the undernutrition in developing nations arises from sizeable proportions of their populations being either unable to afford to buy or to produce sufficient food [1,9,16].

During the second half of the twentieth century, many nations periodically found it impossible to feed their populations for diverse reasons and received emergency food aid from the United Nations. Thirty-six nations were receiving such aid in July 1998 as shown in Fig. 4.2 [1]. Most of the aided nations suffered either droughts or floods sometimes caused by the periodic El Niño climatic

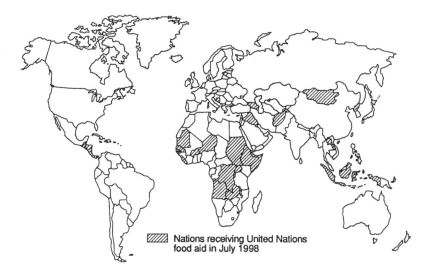

Nations receiving United Nations
food aid in July 1998

Fig. 4.2. The thirty-six nations receiving United Nations food aid in July 1998 [1]. In Central America and the Caribbean, El Salvador, Guatemala, Haiti, Honduras, Nicaragua and Panama received food aid. In East Africa, Burundi, Eritrea, Ethiopia, Rwanda, Somalia, Sudan, Tanzania and Uganda received food aid together with Liberia and Sierra Leone in West Africa. In the Sahel, Burkina Faso, Gambia, Mauritania, Niger and Senegal received food aid, as did the Democratic Republic of Congo in Central Africa and Angola, Mozambique and Zambia in Southern Africa. In Asia, Afghanistan, Indonesia, Iraq, Laos, North Korea and Mongolia received food aid along with Armenia, Azerbaijan, Bosnia-Herzegovina, Georgia and Tajikistan in Eastern Europe and Central Asia.

phenomenon [17]. Others, such as Afghanistan, Angola, Armenia, Azerbaijan, Bosnia-Herzegovina, Ethiopia, Georgia, Iraq, Mozambique, Sierra Leone and

Sudan had their difficulties added to by war while North Korea found its straitened food situation worsened by the restrictions of a centrally controlled economy.

The 1984-1985 Ethiopian famine, in which close to a million people died, shocked humanity and for a time focused attention on the often precarious food supply in the Greater Horn of Africa [18]. Little headway has been made in improving this situation and at the end of 2000 sixteen million people in Ethiopia and neighbouring nations were once again facing a food crisis. These nations are critically dependent on rainfall variations, fertilizer use and seed quality for their agricultural production. This is illustrated by Ethiopia that as a result of a simultaneous improvement in all three of these factors was anticipating a 2004 harvest of cereal and pulse crops of 14.27 million tonnes, a twenty-five percent increase above the average of the previous five years [19]. Nevertheless, it was anticipated that 2.2 million Ethiopians would still need emergency food aid in 2005.

At the beginning of the twenty-first century the challenge is not only to keep the availability of food growing with the human population but also to enable the poorest of the population to either produce or purchase sufficient food. The availability of food can be improved through more effective distribution, reduction in waste and loss to pests and increased productivity through more advanced agricultural practices combined with new ones. One new agricultural practice is the planting of genetically modified food crops that are presently the subject of much debate [20]. On the one hand, this new technology holds the promise of pest resistant crops of greater nutritional value growing on previously unproductive soils in the developing nations. On the other, it has focused attention on the possible detrimental ecological effects of some of the first generation of genetically modified crops, and for the potential for patent rights to put the cost of their seed beyond the reach of farmers in developing nations. The challenge to adequately feed all of humanity is great and it is probable that both pleasant and unpleasant surprises are in store. The sections that follow explore the origins and state of humanity's food supply and give some indications of its probable future.

4.2. Photosynthesis: The Origin of Food

The ultimate source of the amazing global array of foods is the sun that powers the great recycling process that produces plants and animals generation after generation. The annual burst of new green leaves accompanying the increasing warmth of the sun with the arrival of springtime in the boreal temperate lands is probably the most visible manifestation of this great recycling. The green of the leaves arises from a pigment, chlorophyll, that absorbs most of the visible light from the sun except for those wavelengths corresponding to green. It is estimated that a billion tonnes of chlorophyll are produced afresh in green leaves every year to capture the sun's energy through photosynthesis that probably started in cyanobacteria 3.3-3.8 billion years ago [21]. In this process, carbon dioxide and water combine to produce the great array of molecules from which green plants are built. Terrestrial and aquatic plants are the food supply for herbivores that, if unlucky, find themselves food for omnivores and carnivores. Plants also produce oxygen that is essential to the animal metabolism that breaks down food to carbon dioxide and water, in a reversal of the photosynthetic process, and provides energy for physical activity and chemical energy to produce tissue.

The global scale of photosynthesis is massive. Estimates appear to have settled at about 450 billion tonnes of carbon dioxide combining with 184 billion tonnes of water through photosynthesis in green plants to produce 307 billion tonnes of carbohydrate and 327 billion tonnes of oxygen annually. About half of this carbohydrate recombines with oxygen during plant respiration to reverse the photosynthetic process so that the net annual amounts of carbon dioxide and water combined and carbohydrate and oxygen produced are approximately 225, 92, 153 and 164 billion tonnes, respectively [22]. (The recycling of carbon dioxide is discussed in more detail in Chapter 8.) Thus, it is hardly surprising that as Earth's ability to feed the growing human population is subjected to increasing scrutiny, so photosynthesis is under similar scrutiny [23]. (In addition to their photosynthetic activity, green plants also pump into the atmosphere, or transpire, a large part of the 72,000 billion tonnes of water escaping from the land surface through transpiration and evaporation combined in a great water purification process [24].)

In the green leaf, chlorophyll is held in small capsules, or chloroplasts, that are easily seen under a microscope. In the chloroplast, the absorbed sunlight provides the energy to release oxygen from water and also to produce an important energy carrying molecule, adenosine triphosphate. This "light" stage of the photosynthetic process is followed by a "dark" stage where sunlight plays no part. Here, probably the most plentiful enzyme of all, ribulose-1,5-biphosphate carboxylase oxygenase, or rubisco, combines carbon dioxide with ribulose bisphosphate to start the process that produces all green plant derived molecules. However, usually less than one percent of sunlight absorbed by chlorophyll produces plant material. Consequently, if plants could be persuaded to increase the efficiency of photosynthesis to two or three percent much of the concern about humanity's food supply would be alleviated.

The major cause of the inefficiency of photosynthesis arises from the poor performance of rubisco. While rubisco starts the process for the conversion of carbon dioxide into larger molecules in green plants, it also reverses this process through combining with oxygen to produce carbon dioxide from carbohydrates in a process called photorespiration. Fortunately, rubisco in plants such as wheat and rice combines with carbon dioxide about a hundred times more readily than it does with oxygen. However, because the 20.94 percent of oxygen in the atmosphere is much greater than the 0.038 percent of carbon dioxide, fifty percent of the carbon dioxide converted to plant material through photosynthesis reverts to carbon dioxide through photorespiration. When photosynthesis first started 3.3-3.8 billion years ago, Earth's atmosphere contained substantial amounts of carbon dioxide but little oxygen, and photorespiration did not impair the efficiency of photosynthesis. However, as green plants spread, and the amount of atmospheric oxygen grew correspondingly, photorespiration decreased the efficiency of photosynthesis until the current level of oxygen was reached about 2.2 billion years ago [25]. In addition to this drawback, rubisco only reacts two or three times a second that is slow compared with other plant enzymes that carry out their reactions at up to 25,000 times a second. So, despite its immense importance, rubisco is not a very efficient enzyme.

It is here that an opportunity exists for genetic engineers on two fronts [26]. The first arises because the form of rubisco found in red algae is about three times as efficient as that in wheat and rice. The genetic modification of wheat and rice to incorporate the red algal rubisco could allow a faster and more plentiful production of these cereals in principle. The second opportunity arises because wheat and rice, together with ninety-five percent of all green leafed species are C3 plants [27]. These plants are called C3 because they initially produce phosphoglyceric acid, a molecule containing a chain of three carbon atoms, from carbon dioxide. The other five percent of green plants, that include maize, sugarcane and sorghum, are called C4 plants because they initially produce malic and aspartic acids that contain chains of four carbons. In C4 plants, the mesophyll cells surrounding the photosynthesising cells largely exclude oxygen from them and also capture carbon dioxide. Consequently, the photorespiration that lowers the efficiency of photosynthesis is much decreased in contrast to the situation in the C3 plants. Thus, C4 plants photosynthesise more efficiently than C3 plants and there exists the possibility of genetically modifying wheat and rice to become C4 plants, thereby improving their efficiency. However, while such genetic experimentation is feasible, there always exists the possibility that such genetically engineered plants may exhibit other characteristics that either diminish their growth rates or render them unacceptable as food.

4.3. The Need for Food

Whether dining in an elegant Parisian restaurant or eating food from the wild, as hunter-gathers still do, the purpose in both cases is to maintain the body in a healthy state and to provide the energy to carry out a great range of activities [28]. While the hunter-gather is probably more acutely aware of this than most diners in Paris, the daily food requirement is a major way in which Earth's ability to support humanity may be assessed. Most people have a preference for some foods over others on the basis of taste, texture and its compatibility with their digestive systems, and usually eat until their hunger is satisfied, at least in the wealthier nations [29]. Important as these preferences are, they do not lead to a direct measure of the amount and type of food needed for healthy living. And so while an appreciation of cuisine is one of life's pleasures, a more prosaic assessment of humanity's food requirements is needed.

A broad estimate of the nutritional value of food is gained from measuring the amount of energy released when it is burnt. At first sight, this seems an odd measurement of nutrition. However, the heat released in this way, measured in kilojoules, is a direct measure of the energy supplied through food digestion. A human requires between 5,450 to 7,110 kilojoules of food energy a day, depending on age, sex, height and weight, to provide sufficient energy for the heart to beat, the lungs to breath and for the functioning of all of the other processes that maintain a living body without providing energy for any other activities such as walking [30,31]. In other words, this is the energy required to maintain an adult at rest and for a child to grow at rest. Clearly, humanity could not survive in such a state of inactivity and so additional energy has to be acquired. For light activity, food requirement increases to 7,200 to 8,200 kilojoules daily, and for moderate activity from 8,370 to 9,670 kilojoules daily. A man involved in heavy physical work requires 14,600 kilojoules a day. This is the basis used by the FAO to assess humanity's food requirements. When the spread of activities that people engage in is taken into account, it is estimated that the average daily food requirement per person ranges from 10,800 to 12,500 kilojoules if no one is to suffer from undernutrition. On this basis, the FAO considers nations with a food supply in this range to be able to adequately nourish their populations. In those nations that fall below this level, increasingly large proportions of their populations are undernourished as the deficit grows. For a global population of six billion, an adequate food supply should be 65,000 to 75,000 billion kilojoules per day evenly distributed. If, as presently seems probable, Earth's population levels out at about ten to eleven billion at the end of the twenty-first century, some 125,000 billion kilojoules of food daily will be required.

Most nations with an adequate food supply are found in the higher latitudes of the northern hemisphere while most of the hundreds of millions of undernourished people live in developing regions to the south. Here, many nations desperately lacking in nutrition are found, particularly in sub-Saharan Africa, as shown in Fig. 4.3. This illustrates one of the major differences between the developed nations that possess good levels of nutrition, high per capita incomes and birth rates either at or below replacement level, and many of the developing nations where undernutrition is common, per capita income is low and population growth is high. The differences in nutritional levels are also illustrated by Fig. 4.4 that shows one estimate of the populations that Earth could support at various levels of nutrition [32]. These differences arise not only because of difference in wealth and food availability but also because of diet.

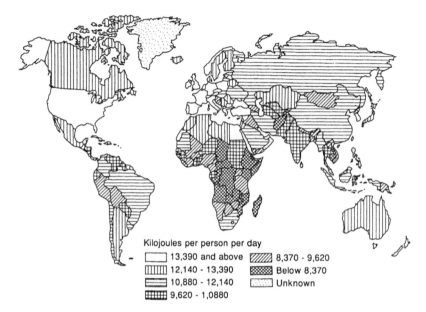

Kilojoules per person per day
☐ 13,390 and above	▨ 8,370 - 9,620
▥ 12,140 - 13,390	▩ Below 8,370
▤ 10,880 - 12,140	▦ Unknown
▤ 9,620 - 1,0880	

Fig. 4.3. Nations classified according to their food availability in kilojoules per person daily in 1998. Based on data in [1].

Nutritional level (1990)   Supportable population in billions

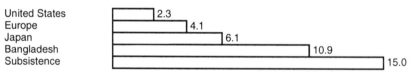

United States	2.3
Europe	4.1
Japan	6.1
Bangladesh	10.9
Subsistence	15.0

Fig. 4.4. The populations that Earth could support at the 1990 level of agricultural production, the dietary preferences and the level of food wastage applying at five different nutritional levels [32]. (The global population was 5.3 billion in 1990.)

During the last thirty years, about forty per cent of all grain has been used as animal feed with fifty percent being used directly as food and the remaining ten percent being retained as seed. In wealthier nations diet contains more animal products as exemplified by Europe where thirty-three percent of food is derived from this source whereas in Africa only six percent is derived from animals. The global diet lies midway between these extremes. While there are some nutritional advantages in animal based foods, the conversion of plant food into animal food is inefficient in terms of bulk and energy value, as shown in Fig. 4.5 [32], which is a major reason for animal based food being less available in developing nations. Even so, there is a growing demand for meat in developing nations. While traditionally many animals have been fed on grass and foliage inedible to humans, there is an increasing tendency to feed animals on grain that could be eaten by humans. This poses a problem for those nations that are already unable to produce sufficient grain to feed their populations [33].

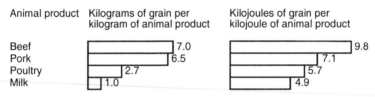

Fig. 4.5. Approximate conversion rates for grain to animal products. Data from [32].

Other factors affecting humanity's food supply are loss of food to rodents, moulds and other pests, loss due to spoilage in processing and waste in preparation and in consumption that is highest in developing nations. Estimates of such losses range from thirty to seventy percent [34]. While much of the post harvest waste can probably be eliminated given the will, farmers fight a continuing battle against new and evolving pests and plant diseases that can result in devastating crop losses. One of the most poignant examples of this was the Irish potato famine caused by potato blight, *Phytophthora infestans*, more commonly known as late blight. It is a fungus that attacks potato crops and, in combination with civil unrest, substantially changed the history of Ireland when in the 1840s the major food, potatoes, was largely destroyed by late blight to the extent that one million people starved to death and one and a half million others emigrated, mainly to the United States [35]. This caused a major depopulation of Ireland that prior to this disaster had a population of eight million. Other parts of Europe also experienced late blight infection but none experienced consequences as devastating as those of Ireland.

The international nature of plant disease is illustrated by the arrival of late blight in Europe, probably from Mexico on a plant specimen. The fungus exists in two strains, A1 and A2, but only A1 appeared in Europe in the 1840s. Fungicides can be used to control late blight and potato varieties resistant to the fungus have been developed. Even so, the appearance of the A2 strain in Europe, and the emergence of more virulent strains of late blight arising from cross-breeding between the A1 and A2 strains were estimated to have reduced global potato crops by fifteen percent in 1998. In the developing countries alone, late blight caused US$2.5 billion in crop losses despite an increase in annual expenditure of US$750 million on fungicides that the new fungal strains were little affected by. These new strains have attacked potato crops in North and South America, Europe, Asia and Africa, sparing only Australia. Fortunately, several new potato varieties, strongly resistant to the new fungal strains, have been bred in Mexico and Peru and their introduction to the potato producing regions should avert further crop loss and lessen fungicide use. In sub-Saharan Africa this could result in potato crops increasing by up to fifty percent.

Major food losses can also result from diseases afflicting poultry and animals raised for food as was the case in the United Kingdom in 2001 when a fast spreading and extensive outbreak of foot and mouth disease resulted in the culling of more than a million cattle, pigs and sheep [36].

4.4. Food and Health

Nutrition and health are inextricably intertwined. Generally, there is a direct connection between a person's height and their body mass that is a good guide to their health [4,30]. A healthy person usually has a body mass index between 18.5 and 25 that is their body mass in kilograms divided by the square of their height in metres. People whose body mass index is above the high end of this range are obese and are predominantly found in developed nations. They suffer from increased risk of stroke, heart disease, diabetes, gout and some forms of cancer that are sometimes called the diseases of affluence. People whose body mass index is below 18.5 are usually suffering from undernutrition and are found predominantly in developing nations.

Nutritional requirements go beyond the energy requirements discussed in section 4.3 and take into account the balance between carbohydrates, proteins, vitamins and other small, but vital, constituents of food required for good health. In developing nations, more than 200 million children suffer from undernutrition caused by a protein deficient diet coupled with infections that increase the need for food while limiting its intake and digestion. Similar undernutrition is also common among pregnant women in developing nations and results in low birth weights, stunted growth and anaemia in their children. Each year thirteen million children under five die as a consequence of undernutrition. For the living, undernutrition has appalling consequences. Vitamin A deficiencies in diet result in half a million children becoming permanently blind each year. For millions, vitamin B, C and D deficiencies cause beri-beri, scurvy and rickets. Iron deficiency causes anaemia in two billion people globally. It is also a major cause of women dying in childbirth in developing nations, lowers learning capacity in children and decreases productivity generally. Yet the situation could have been much worse had agricultural developments not greatly increased the food supply in the latter part of the twentieth century.

Food is often the agent by which disease is transmitted through bacterial or other contamination. Generally, routine hygienic food handling greatly reduces the occurrence of such illness but there are some insidious pathogens found in food that cause deadly diseases as is shown by two examples. The fifth most common cancer, that of the liver, causes tens of thousands of deaths annually with its incidence in developing nations being two to ten times higher than in developed nations. One of its causes is the aflotoxins, some of the most potent hepatocarcinogens known [37]. They are produced by fungi contaminating food and are commonly found in poorly stored groundnuts, maize, rice, cotton seed, dried fruit, spices, vegetable oils, cocoa beans and in milk from cattle fed aflotoxin contaminated feed. As a consequence upper limits are set for the permissible levels of aflotoxins in food.

A second and new disease is a variant of the degenerative and fatal Creutzfeldt-Jacob brain disease, or vCJD, attributed to the eating of beef from cattle infected with bovine spongiform encephalopathy, BSE, also a terminal brain disease commonly known as "mad cow disease" [38]. The first confirmed case of BSE was detected in the United Kingdom in 1986 and an epidemic ensued to afflict a total of 36,680 cattle at its peak in 1992. This disease was attributed to feeding cattle meat and bone meal made from abattoir waste that contained the agent causing scrapie, a neurological disease of sheep that is widespread in the United Kingdom. The banning of this cattle feed decreased the incidence of BSE and five million or so cattle have been slaughtered since 1996 in the European Community in an effort to eliminate BSE. In addition to the occurrence of BSE in British and other European cattle, BSE was subsequently found in North American cattle [39]. The first indication that eating BSE infected beef could cause vCJD in humans emerged in 1996 and since then more than eighty cases have been diagnosed in the United Kingdom alone. The fear is that because of the time taken for the disease to incubate the extent of vCJD infection is unknown and an epidemic may be underway that could kill 136,000 people or more in the United Kingdom [40]. The main symptoms are loss of intellect and control of limbs, and death usually follows within six months. Young and old alike are afflicted by vCJD, which distinguishes it from classical CJD that, although having similar symptoms, is a disease of old age suffered by less than a million people globally each year.

4.5. The Sources of Food

The underpinnings of humanity's food supply are four huge cereal production systems of which major parts lie in the irrigated annually double and triple cropping rice growing areas of tropical and subtropical lowland Asia that produce twenty-five percent of the global rice crop; the irrigated rice and wheat annually double cropping areas of northern India, Pakistan, Nepal and southern China; the largely rained crops of the temperate North American plains that produce more than forty percent of the global maize supply; and the rain fed wheat crops of northwestern and central Europe that account for twenty percent of the global wheat supply.

While Earth's population consumes food derived from some fifty thousand different plants and animals, sixty percent of the global food supply was provided by 2.2 billion tonnes of cereals in 2004, of which 1.9 billion tonnes were wheat, rice and maize as shown in Fig. 4.6 [41]. Sugarcane was the largest single crop and provided the major food sweetener, sucrose, as did sugarbeet. The production of potatoes, soybeans, cassava, barley, oil palm fruit, sweet potatoes and tomatoes each amounted to more than one hundred million tonnes. Globally, crop production increased throughout the twentieth century, but the rate of increase became particularly impressive from the early 1960s to the mid 1980s as shown for the major crops, maize, wheat and rice in Fig. 4.7. This was a consequence of the "Green Revolution" that saw the introduction of high yielding crop varieties.

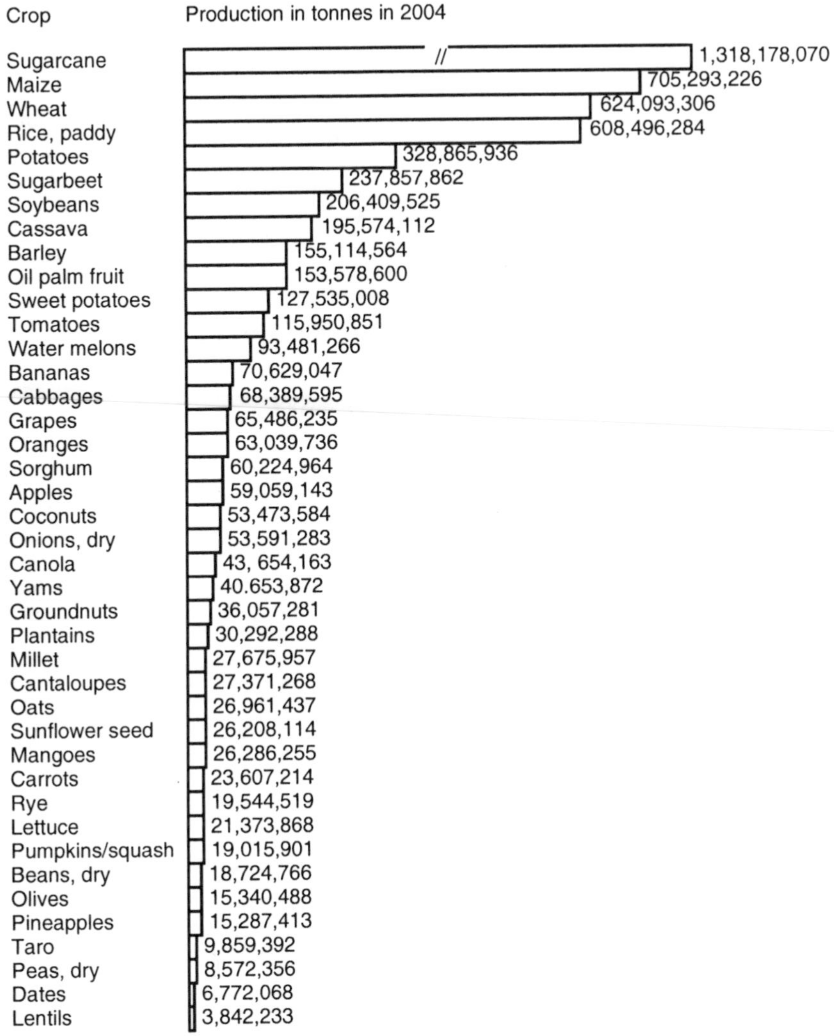

Crop	Production in tonnes in 2004
Sugarcane	1,318,178,070
Maize	705,293,226
Wheat	624,093,306
Rice, paddy	608,496,284
Potatoes	328,865,936
Sugarbeet	237,857,862
Soybeans	206,409,525
Cassava	195,574,112
Barley	155,114,564
Oil palm fruit	153,578,600
Sweet potatoes	127,535,008
Tomatoes	115,950,851
Water melons	93,481,266
Bananas	70,629,047
Cabbages	68,389,595
Grapes	65,486,235
Oranges	63,039,736
Sorghum	60,224,964
Apples	59,059,143
Coconuts	53,473,584
Onions, dry	53,591,283
Canola	43, 654,163
Yams	40.653,872
Groundnuts	36,057,281
Plantains	30,292,288
Millet	27,675,957
Cantaloupes	27,371,268
Oats	26,961,437
Sunflower seed	26,208,114
Mangoes	26,286,255
Carrots	23,607,214
Rye	19,544,519
Lettuce	21,373,868
Pumpkins/squash	19,015,901
Beans, dry	18,724,766
Olives	15,340,488
Pineapples	15,287,413
Taro	9,859,392
Peas, dry	8,572,356
Dates	6,772,068
Lentils	3,842,233

Fig. 4.6. Global production of some major plant foods in 2004. Data from [41].

Inevitably, there is a great emphasis on the wide variety of edible seed, fruit, leaves, and tubers of the crops produced that amounts to 29.6 percent by weight in the dry state. However, the inedible parts of the crop such as stalks, leaves, husks and roots amount to 70.4 percent in the dry state so that the ratio of the edible to inedible part of the crops, the global harvest index, is 0.42 [42]. These inedible crop residues contain substantial amounts of the major crop nutrients, nitrogen, phosphorus, and potassium, and lesser amounts of essential micronutrients required for healthy crop growth. They also represent a valuable source of nutrition for soil bacteria and insects and add to the soil humus level

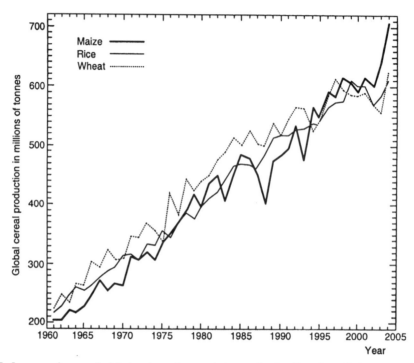

Fig. 4.7. Increases in annual global maize, wheat and rice production Data from [41].

that, in combination, help maintain soil fertility and structure and protect against erosion [42]. While substantial amounts of crop residues are used as household fuel, bedding and building material in the developing nations, and more generally as animal feed, in mushroom cultivation and as pulp for paper making, such uses are diminishing. As a consequence, 1-1.4 billion tonnes of crop residue are burnt annually *in situ* adding two billion tonnes of carbon dioxide to the atmosphere, and its value as humus and as a source of soil carbon and nutrients is lost. In addition, the baking effect of the burning damages soil structure and kills the microorganisms essential to its fertility. It is inevitable that a major reassessment on the use of crop residues will occur as humanity seeks to secure its food supply, moderate the effects of global warming and minimize environmental damage.

4.6. The Green Revolution

In 1968, Paul Erlich forecast in his book, "The Population Bomb", that, on the basis of contemporary global food production, an increasingly hungry humanity would experience an unprecedented level of starvation within a decade [43]. This did not happen and despite many people in developing nations being undernourished, global food supplies are more plentiful now than at any time in the twentieth century. The contrast between the 1968 prediction of increasing hunger and the more secure food supply at the beginning of the twenty-first century is a result of technological advances in agriculture, particularly in the greatly increased productivity of maize, rice and wheat crops. These were accompanied by a 6.87 and 3.48 fold increase in the use of nitrogenous and phosphorus fertilizers, respectively, a 3.48 fold increase in irrigation and a 1.1 fold increase in land under cultivation over thirty-five years [44]. Collectively, these advances have become known as the "Green Revolution" without which some 1.5 billion people would probably not be alive today [45].

Since the beginning of agriculture plant breeders have improved the yield of crops and the Green Revolution was a continuation of this process. A sophisticated focusing of a combined knowledge of plant genetics, irrigation, fertilizers, pesticides and herbicides on the productivity of maize, rice and wheat in the early 1960s increased global cereal production at an unprecedented rate. The origins of the Green Revolution can be traced back to 1943 when the American Nobel Laureate Norman Borlaug led a project to breed new high yielding wheat strains that were resistant to stem rust fungus that greatly reduced the size of the wheat crop in South America and parts of North America. After several years of research in Mexico, funded by the Mexican government and the Rockefeller Foundation, Borlaug's team bred short stemmed, disease resistant, high yielding hybrid wheat plants. In contrast to the 0.75 tonnes of grain per hectare commonly produced in Mexico at that time, the new hybrid yielded up to a staggering eight tonnes per hectare. Introduction of the new hybrid and increased fertilizer use saw a rapid increase in the Mexican wheat harvest from

Challenged Earth

0.75 tonnes per hectare in 1948 to 1.7 tonnes per hectare in 1961, and 3.0 tonnes per hectare in 1970 before levelling off at about 4 tonnes per hectare in 1980.

This spectacular success lead to the establishment of the Mexican research centre as the International Maize and Wheat Centre, or CIMMYT (Centro Internacional de Mejoramiento de Maiz y Trigo), funded by the Mexican government and the Rockefeller and Ford Foundations in 1963. Introduction of high yielding wheat hybrids to India and Pakistan followed with similar success from the early 1960s. Thus, Pakistan produced 3.8 million tonnes of wheat in 1961 at 0.82 tonnes per hectare that compared with 18.7 million tonnes of wheat produced at 2.2 tonnes per hectare in 1998. India produced 11 million tonnes of wheat in 1961 at 0.85 tonnes per hectare that compared with 66 million tonnes of wheat produced at 2.6 tonnes per hectare in 1998.

The Green Revolution rapidly became international and crop yields of maize, rice and wheat increased dramatically from the early 1960s. The increased yield of the new wheat and rice varieties was achieved through increasing the ratio of the weight of the grain to the weight of the leaves and stems to fifty percent or greater. This resulted in the new varieties generally having shorter and sturdier stems to support the greater weight of the grain. In contrast, the increase in maize yields was achieved through new varieties that tolerated being planted closer together than the varieties that they replaced.

The global production of cereals has more than doubled since 1961 as shown in Fig. 4.8 [41]. And, despite the rapid increase in population, the amount of cereals available per person increased until the 1980s when a plateau was reached. This levelling off was due to a variety of reasons including a deliberate decrease of cereal production in the European Community nations and loss of production following the demise of the Soviet Union. The effectiveness of the increase in crop yield is evident from the average global cereal yield increasing from 1.35 tonnes per hectare on 648 million hectares of cropland in 1961 to 2.97 tonnes per hectare on 708 million hectares of cropland in 1997. Without the advances of the Green Revolution traditional methods of cereal production would have required an expansion of cropland to 1,517 million hectares with a massively deleterious effect on wildlife and the environment. Largely on the basis of this record, the FAO has estimated that cereal production can keep up with population growth until 2020 at least.

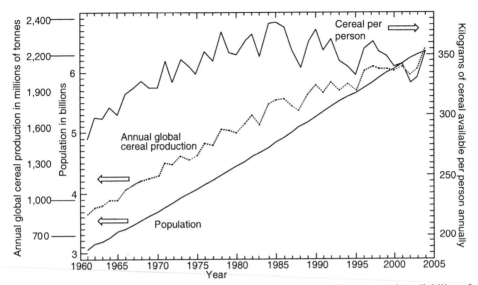

Fig. 4.8. Changes in cereal production and global population and the averaged availability of cereal based on FAO statistics [41].

The productivity of cereal farmers is not globally uniform as is shown for the major wheat producing countries in Fig. 4.9 [41]. Thus, while China is by far the

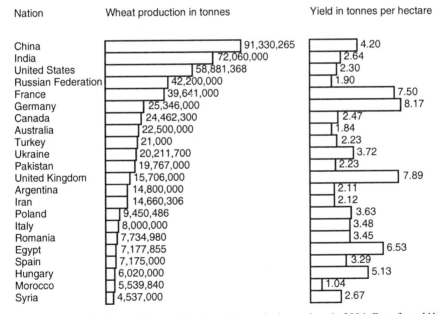

Fig. 4.9. Wheat production and its variation in major producing nations in 2004. Data from [41].

greatest wheat producer, its yield per hectare is well below those of France, Germany and the United Kingdom whose very fertile soils, temperate climates and good rainfalls are particularly conducive to wheat production. The use of pesticides and especially fertilizers also greatly influences wheat and other crop productivity. Consequently, their affordability impacts heavily on crop yields.

Paradoxically, as the Green Revolution has increased the food supply for the people of the developing nations, their diet has become deficient in iron, zinc, vitamin A and other micronutrients [46]. This is largely because the high yielding rice, wheat and maize varieties, that are often low in these nutrients, largely displaced the local fruit, vegetables and legumes that are high in them. The consequence has been an upsurge of anaemia and other illnesses resulting from these nutritional deficiencies. While food supplements can alleviate these problems in principle, the race is now on to produce high yielding rice and other cereals genetically engineered to contain much higher levels of vitamin A, iron and zinc [47-49]. However, the cropland soil itself becomes depleted in iron and zinc and other micronutrients if they are not replenished by micronutrient fertilizer application, as is becoming apparent in parts of India, and the full nutritional benefit of these new high yielding grain varieties will not be realized unless the soil micronutrients are continuously replenished [50].

4.7. Fertilizers

The great increase in cereal production since 1961 could not have occurred without the accompanying increase in fertilizer application to croplands [51]. The traditional sources of replenishment of soil fertility, animal, crop and human wastes were insufficient to sustain the growth in cereal production from 1961 to the beginning of the twenty-first century. As crops grow they remove nutrients from the soil and fertility is much decreased if they are not replaced. Three nutrients, nitrogen, phosphorus and potassium are particularly important, and this is reflected in the growth of fertilizer use as seen in Fig. 4.10 [52]. Nitrogenous fertilizer use has grown much more rapidly than has that of phosphate and potassium fertilizer. This is because plants need more nitrogen than phosphorus and potassium and also because nitrogenous fertilizers are usually more readily available. As a consequence the balance between the applications of the three fertilizers is often not optimal for maximizing crop yields. In some developing nations nitrogenous fertilizer has been applied almost exclusively, and the resulting increase in crop yield initially gained has depleted the soil of phosphate and potassium so that subsequent crop yields have decreased. This is largely due to phosphate and potassium fertilizers not being universally available because they are mined from mineral deposits that are being depleted. Such restrictions do not apply to nitrogenous fertilizers that are predominantly produced from atmospheric nitrogen through the Haber-Bosch process. This process is one of the most important technological innovations of the twentieth century for without it many millions of people now living would not exist because the fertilizer dependent Green Revolution would have been much less effective. This is reflected in a great increase of nitrogenous fertilizer use in Asia where high population growth has generated rapidly increasing food demands. However, in Africa lack of resources has restricted the increase in fertilizer use.

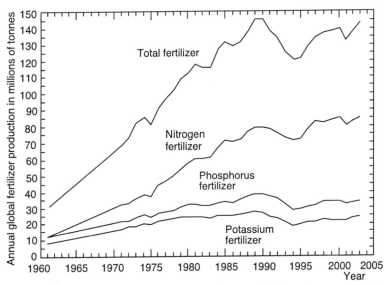

Fig. 4.10. Increases in fertilizer use since 1961. Nitrogen fertilizer is measured in tonnes of nitrogen, phosphorus fertilizer as tonnes of potash (P_2O_5) and potassium fertilizer as tonnes of potassium oxide (K_2O). Data from [52].

The collapse of the Soviet Union coincided with a much decreased use of fertilizer, as was also the case in Eastern Europe. In North America and Western Europe the use of fertilizers has levelled out as the optimum level beyond which increased fertilizer application causes little increase in yield has been reached. Typically, in the United States it is found that on unfertilized land maize yields of four tonnes per hectare are obtained and that this increases to seven tonnes per hectare when one hundred kilograms of nitrogen fertilizer per hectare is applied. However, the application of twice this amount only increases the maize yield to eight tonnes per hectare [15,53]. Clearly, there is an optimal amount of fertilizer necessary to gain a major increase in yield but the benefit gained diminishes rapidly above this amount. It is also known that adding fertilizer at specific stages in a crop's growing cycle can have a much greater effect on yield than does fertilizer addition at other times. In North America and Western Europe this knowledge is leading to a decreased and more effective use of fertilizers.

By comparison with carbon, oxygen and hydrogen, nitrogen is a minor component of living matter. Only four percent of the one hundred billion tonnes of nitrogen in terrestrial organic material is in living organisms, the rest is in dead organic matter. Of this, only fifteen percent is available as plant nutrient. The amount of nitrogen in the oceans is some ten times greater than on land and half exists in readily available nutrient forms with the remainder residing in dead organic matter [54]. While carbon, oxygen and hydrogen are readily available to plants through photosynthesis and water, and therefore to humans and animals through plant foods and water, nitrogen is far less readily available. Yet nitrogen is an indispensable component of amino acids, proteins and deoxyribonucleic acid (DNA) found in all living species. The difficulty is that most of Earth's 3.9 thousand trillion tonnes of nitrogen exists as dinitrogen (N_2) in which two nitrogen atoms are very strongly bound together and which represents 78.09 percent of the atmosphere by volume. Plants can only use nitrogen as a nutrient after the dinitrogen molecule has been split to form simple molecules in which nitrogen atoms are not bound directly to each other. This splitting process is called "nitrogen fixation". About one trillion tonnes of nitrogen are present in this "fixed" form as organic matter from dead animals and plants, in inorganic salts in sediments on the ocean floor and in rock strata through which water percolates. While this contact with water dissolves some of this fixed nitrogen that thereby becomes available as a plant nutrient, this source is much too small to supply Earth's terrestrial and aquatic plants. As a consequence plants are dependent on fixed nitrogen gained from the atmosphere. (Eighteen trillion tonnes of fixed nitrogen exists in sediments and rocks buried too deeply to be available to plants.)

Lightning can supply enough energy to break the dinitrogen bond and fix nitrogen in the presence of atmospheric oxygen and water to give nitric acid (HNO_3) that is carried to the surface in rain. There are some three billion lightning flashes a year each producing three to five kilograms of nitrogen in nitric acid [55]. Impressive as this is, it provides an insufficient amount of fixed nutrient nitrogen to sustain Earth's wild vegetation let alone intensive agriculture. The second natural nitrogen fixing process occurs when dinitrogen permeates soil either in air or dissolved in water. In soil, a range of nitrogen fixing bacteria exist that can break the dinitrogen bond that leads to the formation of nitrate (NO_3^-). Of these, the most important is *Rhizobium* that resides in nodules on the roots of leguminous plants such as peas, beans, clover and ferns. Thus, in addition to nitrate arising from lightning, leguminous plants and a few other plants gain nitrates from bacterial action. In addition, fixed nitrogen is also available as ammonia (NH_3) formed from nitrate by other bacteria in the soil. Both ammonia and nitrate are released from decaying animal and plant matter. In rivers, lakes and oceans cyanobacteria, often referred to as blue green algae, also produce ammonia and nitrate from dinitrogen. It is from these forms of fixed nitrogen that plants build amino acids, proteins and DNA. Thus, lightening and nitrogen fixing bacteria determine how much nitrogen is naturally available to plants and place a limit on the human population that can be supported in the absence of nitrogenous fertilizers.

To supplement natural nitrogen fixing processes it is obviously tempting to either imitate the chemistry of the nitrogen fixing bacteria or to use electrical discharges to simulate lightening. Although much research has been done into the first possibility, no economically viable process has yet been derived from it [56]. While electrical discharges fix nitrogen, production of fertilizer in this way is uneconomic unless cheap electricity is available. As a consequence most nitrogenous fertilizer is produced by a method developed in the early years of the twentieth century with a view to producing explosives as well as fertilizer. In 1899, a young German, Carl Bosch, began a study of chemical nitrogen fixation, but the major breakthrough came in 1909 when Fritz Haber, of Karlsruhe University, produced ammonia by reacting atmospheric nitrogen and hydrogen over a catalyst at high temperature and pressure [51,56]. Development of this process by Bosch resulted in the first industrial production of ammonia in 1913, just in time for the synthetic production of explosives for the First World War through the conversion of ammonia to nitrates. (Haber and Bosch received Nobel Prizes for Chemistry in 1919 and 1934, respectively.) The eponymous Haber-Bosch process is now used globally to produce nitrogenous fertilizer from dinitrogen and either hydrogen or natural gas that is available to farmers as either liquid ammonia, a highly volatile liquid that requires low temperature handling facilities, or ammonium nitrate or urea, both of which are white powders that are more easily handled. The amount of nitrogen fixed in this way is at least equal to the amount fixed naturally by lightening and bacteria.

Despite the now essential role of fertilizers in food production, there are penalties to be paid for their injudicious use. Modern intensive agriculture has cause massive changes to the natural recycling of phosphorus and nitrogen and inevitably there have been major consequential effects on the environment [57]. Phosphorus and nitrogen containing fertilizers dissolve readily in water and in this way enter rivers and lakes and the water supply mainly as phosphates and nitrates where they have deleterious effects as is discussed in Chapter 3. The amount of fertilizer in water runoff from agricultural land is directly proportional to the amount of fertilizer applied and it is clear that more judicious application of fertilizer should both reduce this considerably and save money. Thus, the amount required for optimum plant growth and soil type is increasingly carefully assessed and in some cases the variations of soil on farms are precisely mapped and the application of fertilizer is adjusted from area to area with the aid of satellite global positioning systems [58]. Inevitably such "precision farming" will spread throughout the developed nations as it not only lessens environmental stress but also decreases the amount of fertilizer used and therefore the cost of crop production. However, the price of this technology is likely to make its adoption less rapid in developing nations where fertilizer use is well below the optimal amount. It can be anticipated that as fertilizer use increases in the developing nations both fresh water and marine environments will suffer as fertilizer runoff increases unless great care is taken.

4.8. Pesticides and Pest Control

In addition to the introduction of higher yielding crops and increased fertilizer use, the Green Revolution also brought with it a strong emphasis on pest control and a substantial increase in chemical pesticide usage. The use of insecticides, in particular, is a two edged sword for, as well as killing the targeted insects, it may kill other insects that prey on them and other non pest insects. This has the compounding effect of destroying the food source of birds that feed on both the pest insect, its insect predator and other insects, thereby seriously disrupting a delicate ecological balance. In 1962, at the very beginning of the Green Revolution, Rachel Carson's book, Silent Spring, drew attention to this deleterious effect of the use of DDT (dichlorodiphenyltrichloroethane) and other organochlorine insecticides in the United States and, largely as a consequence, the use of DDT and similar insecticides are now banned in developed nations [59].

Even so, a wide range of chemical pesticides remains in use and their effect on all forms of life remains a matter of concern. Globally, herbicides account for fifty percent of pesticide sales, followed by insecticides at thirty percent and fungicides at twenty percent [60]. The United States is the largest user of pesticides in monetary value as shown by the 1997 sales of US$5.9 billion. Thus, it is not surprising that a 1999 study of the United States found that pesticides were widespread in streams and ground water and followed a seasonal and geographical variation that reflected their use [61]. In late spring and summer herbicides are detected in rain throughout the midwestern and northeastern United States with particularly high levels in the Corn Belt. Globally, hundreds of millions of agricultural workers and consumers are exposed to excessive use of pesticides and acute pesticide poisoning occurs frequently in developing nations. Of considerable concern is the potential for pesticide exposure to cause changes in the human immune system so that resistance to infectious diseases and cancer is lowered. Epidemiological studies indicate a link between the prevalence of Parkinson's disease and either working in the agricultural, chemical or pharmaceutical industries, or living in the countryside. It is known that rotenone, the active agent in hundreds of products ranging from tomato sprays to tick and flea powders, causes Parkinson-like symptoms in rats. While this does not prove that a similar effect is inevitable in humans, it adds emphasis to the importance of minimizing exposure to pesticides.

An alternative to the use of chemical pesticides is the use of natural pesticides. These are exemplified by twenty-five or so related crystalline proteins produced by the bacterium *Bacillus thuringiensis*, *Bt*, commonly found in soil and insect rich environments [62]. They are toxic to a range of insect pests, but have a lesser effect on other insects that are beneficial to agriculture. Either *Bt*, the proteins they produce, or combinations of both have been in use as insecticides for several decades but represent only a minor portion of the insecticides in use. While *Bt* insecticides appear to pose little if any threat to the environment, farmers and consumers, they tend to have short shelf lives, mainly act at the larval stage, kill pest insects slowly and have little residual activity. They are not systemic insecticides that pervade plant tissue and kill sap feeding insects and are generally not very effective when applied to a heavily infected crop. As a consequence, *Bt* based insecticides must be applied prior to an insect infestation and it is here that their potential for a novel form of application arises. It is possible to engineer the *Bt* gene that produces the insecticidal protein into genetically modified crops that thereby continually produce their own insecticide as is discussed in section 4.16.

The battle between humanity and the pests that either prey on, introduce disease to, or compete for space with food crops is likely to be a continually evolving one for as pesticides kill susceptible pests so pesticide resistant strains become dominant until once potent pesticides become of much decreased effectiveness. To a large extent this susceptibility of food crops to disease and attacks by pests is accentuated by the vast monoculture systems that now dominate agriculture [63,64]. While such systems have greatly increased the food supply, the disadvantages that they have brought through increasing pesticide use and narrowing of the gene pool on which the major food crops are based has increased the vulnerability to plant disease. A large scale experiment in the Chinese province of Yunnan on rice crops in 1998 and 1999 showed that intermixing of different varieties of rice greatly reduced the incidence of rice blast, a major rice disease caused by the fungus *Magnaporthe grisea*, to the point where it was no longer necessary to spray with fungicides [65]. This was the largest scale on which such an experiment had been conducted to that time and showed the effect of increasing the diversity of a single species on the spread of disease. It also increased the diversity of the pest fungus and thereby reduced its ability to cause disease, a principle that may have wider application in crop protection [66]. The possibility of further increasing crop disease resistance through growing different species, as opposed to different varieties, next to each other also exists [67]. A different approach to biological pest control is the introduction of natural selective insect and fungal predators for introduced weeds such as the prickly pear cactus and the water hyacinth [68].

4.9. Food from Animals

Land animals provide a lesser, but important, source of food in the form of meat, eggs and milk as shown in Fig. 4.11 [41]. Generally the diet of the developed nations contains a much higher proportion of food from animals than is the case for the developing nations. However, as affluence increases in the latter nations

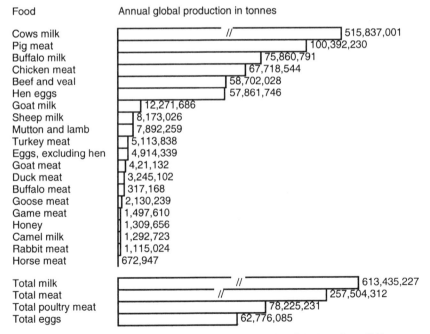

Fig. 4.11. Global animal food production in 2004, excluding fish. Data from [41].

so does the demand for animal sourced foods. This potentially imposes a limit on the availability of food if this demand continues to grow as it can only be met by using maize and other cereals as animal feed [33]. Such use of grain, that could otherwise feed humans, is an inefficient conversion of plant food to animal food as shown in Fig. 4.5, and may become a major problem in those nations with a rapidly growing population and a limited availability of agricultural land. In many nations the availability of water will become the limiting factor, and it is here that the relatively high water requirements of meat production may become prohibitive as seen from Fig. 4.12 [69,70]. While potatoes produce one kilogram of food for 500 kilograms, or half a cubic metre, of water, one kilogram of beef requires at least one 100,000 kilograms, or 100 cubic metres, of water to produce when the amount of water required to produce the forage and grain to feed the cattle is taken into consideration.

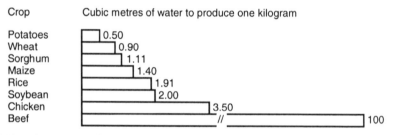

Fig. 4.12. The amount of water required to produce one kilogram of various plant and animal foods. Data from [69].

4.10. Soil: The Foundation of Agriculture

Fertile soil is usually quite dark in colour and moist to the touch. Closer inspection reveals the presence of pebbles and earthworms of varying sizes. Still closer inspection reveals the presence of decaying animal and plant fragments, usually called humus, and a variety of insects, nematodes and fungi of varying size. Smaller than the unaided eye can see, are myriads of bacteria and viruses. Thus, fertile soil teems with life that feeds on finely fragmented rock mixed with minerals formed by slow natural physical and chemical processes and also on decaying animal and plant tissue. As much as half of the volume of soil is made of small spaces, or pores, in between the soil particles that may be either partially or completely filled by either air or water depending on season and geographical location. It is this complex mixture of inanimate material and living organisms that is the foundation of agriculture and the major part of humanity's food supply. At the same time, soil is the interface through which animals and plants and their nutrients recycle between the atmosphere, biosphere, geosphere and hydrosphere as shown in Fig. 4.13 [71].

Soil is formed by the physical weathering of rocks that occurs when water percolates into rock fissures and expands when it freezes and breaks flakes from the rock. Alternatively, repeated expansion and contraction in the daytime heat of the sun and the cool of night also fragments rock. Chemical weathering occurs when water leaches soluble minerals and salts from rocks and weakens their

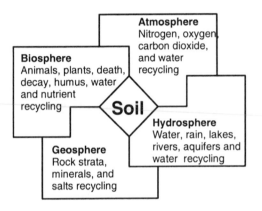

Fig. 4.13. Soil is at the interface of the atmosphere, biosphere, geosphere and hydrosphere.

structure particularly if the water has either become acid because of dissolution of atmospheric nitric and sulfuric acid, or has been acidified as it percolates through decaying plant material. Nitrogen fixing bacteria in the soil convert nitrogen into ammonia and nitrate while others reverse this process. Yet other bacteria, methanogens, convert dead animal and plant material into methane and leave behind mineral salts while others convert methane into carbon dioxide. These few examples of the processes occurring in soil are representative of the multitude that produces fertile soil. The slowest of these processes is the breaking down of rock that probably takes a hundred years to produce a centimetre depth of soil. The interactions with the atmosphere, biosphere and hydrosphere occur over much shorter timescales but they can hardly occur at all without the fragmented rock from which the nutrient minerals are extracted. Because of this there is increasing concern about the loss of soil as it is carried away by water and wind erosion. Soil fertility is reduced through chemical degradation, water logging that leaches out soil nutrients and excessive irrigation that causes salinization. Soil is also effectively lost through contamination with pesticides and heavy metals, the leaching of essential metals, such as calcium and zinc, and mobilization of toxic metals, such as aluminium, by acid rain and physical degradation through the compaction caused by heavy equipment and the disruption of soil structure by excessive cultivation.

4.11. Soil Degradation: Agriculture's Nemesis

Soil erosion by wind, water and chemical and physical action is a natural phenomenon that has occurred over billions of years. Probably the most spectacular example of this is the Grand Canyon in the United States where the Colorado River has cut a huge and scenic canyon in Earth's surface. Such natural erosion has both produced fertile soil and placed it in those areas where agriculture now thrives. The concern about human caused degradation is that fertile soil may be lost and the fertility of cropland decreased through intensive agriculture. There is no doubt that human caused soil loss and degradation has occurred and is occurring, on a large scale. Two of the more prominent early estimates of global soil lost through erosion were twenty-five and seventy-five billion tonnes annually [10,72]. Clearly, the effect of such losses on food production would be substantial, however, more recent and estimates suggest that global soil loss rate is probably closer to eleven billion tonnes annually in addition to the eleven billion tonnes lost through natural erosion [12]. This is still very serious and local soil loss can be very high as exemplified by Haiti where some 900,000 hectares, or one third of Haiti's arable land, has lost so much topsoil that it is barely cultivable.

Estimates of global soil degradation are shown in Fig. 4.14 together with the consequent loss of productivity [9,73,74]. This 1990 assessment shows that twenty-two percent of land used for permanent and annual crops, as permanent pasture and as forest and woodland had been degraded over the previous forty-five years with an accompanying 4.6 percent loss in productivity. This represents 0.5 percent of this land becoming degraded to some extent annually and an accompanying loss of 0.1 percent in productivity if it is assumed that this loss has occurred at a steady rate.

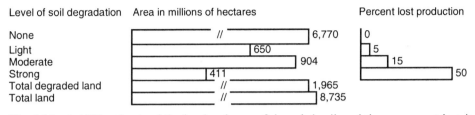

Fig. 4.14. A 1990 estimate of the level and area of degraded soil, and the consequent loss in productivity of crops, permanent pasture and forest. Based on [9].

The estimated extent of regional moderate and excessive degradation of soil that occurred in the forty-five years up to 1990 is shown in Fig. 4.15 [2,73]. Of this loss, water erosion accounted for 748 million hectares, wind erosion, 280 million hectares, chemical degradation, 147 million hectares and physical degradation, 39 million hectares. This resulted from deforestation, 552 million hectares, overgrazing, 352 million hectares, poor farming practice, 291 million hectares and a variety of other causes, 49 million hectares.

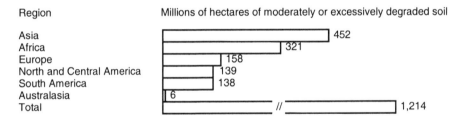

Fig. 4.15. A 1990 estimate of the regional degradation of soil. Based on [2, 73].

Such human caused degradation and losses of soil through erosion sounds disastrous but an analysis of the consequences is complex. Much of the "lost" soil may simply be transferred from one place to another. Thus, short surface flows of water following rain showers may simply carry soil form one part of a field to another while heavy storms may carry soil much further. Combined surface and sub surface water flows into rivers are part of the natural process that creates major accumulations of fertile soil. Prime examples of this are the fertile deltas and flood plains built up by silt from the Brahmaputra, Euphrates, Ganges, Mississippi, Nile, Tigris, Yangtze and other rivers that carry high silt loads. Similarly, wind blown soil may either be transported from one side of a field to another, or deposited hundreds of kilometres away and add to the depth of fertile soil already in place. Water or wind borne soil carried far out into the ocean is lost. The effect of soil loss on food production may be small in regions where the soil is many metres deep but may expose rock in shallow soil areas and greatly decrease the area's ability to support crops.

Globally, with the exception of some nine million hectares of severely eroded land, the degraded areas appear reclaimable with adoption of more enlightened agricultural practice [9,12,72-75]. Often, the combined effects of soil degradation produce deserts as in north central China where inappropriate agricultural practices, tree felling, and wind blown encroaching dunes caused the 14.9 million hectare desert areas comprising fifteen percent of the Chinese land area to expand at a rate of 15,600 hectares a year [76]. For this most populous of nations, with twenty-five percent of the Earth's population and only seven percent of the cropland, such desertification poses a serious threat to the food supply and is being actively combated by conservation and reclamation projects. At present it is estimated that at least fifteen percent of global soil erosion caused by water occurs in China with a loss of 5.5 billion tonnes of soil annually. This is accompanied by the loss of 27.5 million tonnes of organic matter, 5.5 million tonnes of nitrogen, 0.5 million tonnes of potassium and 60 thousand tonnes of phosphorus. The latter three losses accounted for 46.2, 63 and 2 percent of the total nitrogen, potassium and phosphorus fertilizer applied annually to Chinese cropland in the 1990s. Such losses increases agricultural dependence on fertilizers and thereby the cost of food.

Arid regions are particularly prone to desertification and for much of the twentieth century there was an impression that the deserts were growing [77]. The advent of satellite imaging facilitated a more quantitative assessment of Earth's deserts and a study of the largest desert, the Sahara, is particularly interesting [78]. The Sahara is bounded by the Atlantic Ocean to the west, by the Atlas Mountains and the Mediterranean Sea to the north, the Red Sea to the east and a semiarid grassland boundary region, the Sahel, to the south. While the western, northern and eastern Saharan borders are fixed, the southern border is not and shows considerable northward and southward movement depending on light rainfall variations. Satellite and ground mapping of Sahelian vegetation density and extent showed that over the eighteen years 1980 to 1997 the area of the Sahara reached a maximum area of 9,980,000 square kilometres in 1984 and a minimum area of 8,600,000 square kilometres in 1994. The average area over the same period was 9,150,000 square kilometres and no systematically increasing or decreasing trend in area was detected. Thus, it seems that the intuitive idea that the extent of largely natural vegetation will wax and wane with annual rainfall in arid regions applies to the Sahel with a corresponding variation of the size of the Sahara. However, clearing of native vegetation and inappropriate farming methods remove the capacity for regeneration and desertification often follows.

4.12. Fish: The Last Wild Harvest

Fish provided about fifteen percent of all animal protein in the human diet in the last decade of the twentieth century and at the beginning of the twenty-first century [79-81]. But, unlike other food production that is through cultivation, the dominant part of the ninety or so millions of tonnes of fish caught each year are wild fish from the oceans despite the growth of aquaculture. This ocean catch has levelled off since the 1980s and the global annual supply of fish would not be growing if it were not for China's greatly increased aquaculture production. The capture of wild fish from marine and inland waters represents a major source of food and the last vestiges of the hunter-gatherer for the bulk of humanity. However, the technology used in most wild fish capture is now so effective that, combined with an unwillingness to acknowledge the finite regenerative capacity of fish stocks, it has led to the depletion and collapse of major fisheries [82]. Simultaneously, aquaculture has grown rapidly as shown by the doubling of production from 1987 to 1998, and continues to do so. The 2002, production of fish and aquatic plants as a source of food amounted to 145,918,548 tonnes and was gained from the sources shown in Fig. 4.16. The value of fish from the wild and aquaculture was US$78 billion and US$53.8 billion, respectively, in 2002.

While the growth of aquaculture increases the supply of fish, it is substantially dependent on wild fish for its existence and has major environmental impacts [83,84]. These interactions were explored in a 1997 analysis and are shown in Fig. 4.17. It is seen that capture fisheries and aquaculture provided ninety-four million tonnes of fish and nine million tonnes of aquatic plants for human consumption. An additional thirty-two million tonnes were used to produce fishmeal and other food for farm animals, pets and aquaculture. This pattern has remained much the same since then, except that aquaculture production has increased.

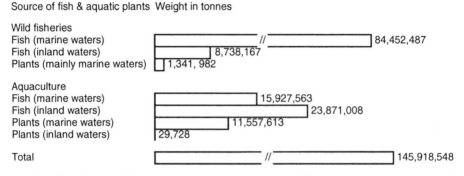

Fig. 4.16. The origins of fish and plants from marine and inland fisheries and aquaculture in 2002. Data from [80,81].

The annual global catch from wild fisheries has fluctuated over a range of eighty to ninety-five million tonnes in the last two decades and there has been a gradual change in the nature of this catch. For a fishery to remain viable only a moderate proportion of mature fish should be taken so that breeding may continue at a sustainable level. Commensurate with this, the proportionate populations of other fish and organisms in the food chain should not be diminished. Unfortunately, quite the opposite has happened in all of the major fisheries where the catches of the generally preferred mature carnivorous fish have greatly decreased [85]. Those that are caught are frequently immature and to compensate other fish much lower down the food change are being caught. In addition, there is evidence that trawling is damaging the seafloor habitat of many species that are part of the food chain [86]. Of the wild fish caught, some twenty percent are discarded as either being too small or of the wrong type in the so-called bycatch thereby decreasing the breeding stock and further diminishing the food chain. Unsurprisingly, most fisheries are now either being fished at their maximum capacity, are declining or are depleted. This has occurred despite the establishment of various overseeing bodies commencing with the establishment of the International Council for the Exploration of the Sea in Copenhagen in 1902. Some of the over fishing has been attributed to the unwillingness of scientists to make definitive recommendations on sustainable catch levels and to political considerations [87]. Fortunately, it appears that setting aside reserves, or "no take" zones, where fish can breed undisturbed may allow depleted fisheries to regenerate [88].

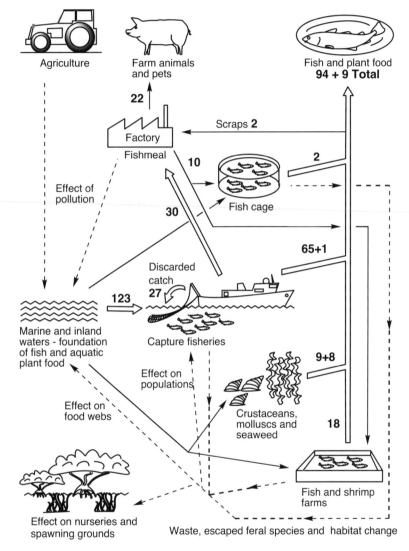

Fig. 4.17. The origins of fish and plants from marine and inland waters in 1997. The broad lines show the route of fish and plants through capture fisheries and aquaculture to human consumption and fishmeal production in millions of tonnes. Where two numbers appear they refer to amounts of fish and plants, respectively. The thin lines from marine and inland waters show other inputs that include filter feeding by fish and the capture of immature fish, fry and seed for aquaculture. The broken lines show the detrimental effects of over exploitation of wild fisheries and aquaculture. Adapted from [84].

4.13. The Wild Fisheries

Most wild fish are caught in the nutrient rich waters of the continental shelves and oceanic shallows. In 2002, the Northwest, Central, and Southeast Pacific, and Northeast Atlantic fisheries provided by far the largest catches as shown in Fig. 4.18 [80,81]. The Southeast Pacific fishery is very susceptible to the El Ninõ climatic phenomenon that greatly decreases the upwelling of nutrient rich waters off the South American coast and the anchoveta, or Peruvian anchovy, catch along with it. This is illustrated by the halving of the 1998 Peruvian and Chilean

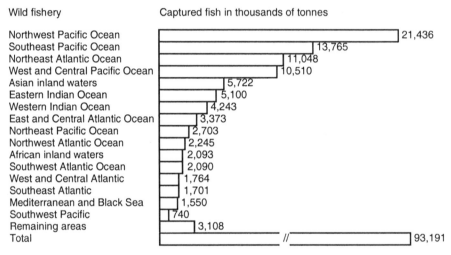

Fig. 4.18. Catches in the major wild fisheries in 2002 excluding aquatic plants [80,81].

catches of 4,338,437 and 3,265,293 tonnes, respectively, by comparison with their 1996 catches of 9,515,048 and 6,690,665 tonnes because of the effects of an El Niño event. Fortunately, the return of the more usual oceanic weather pattern, La Niña, and an increase in upwelling nutrients off the Chilean and Peruvian coasts usually sees the anchoveta population and catch rebuild. The variations of the catches of the other major fishing nations arise from a variety of causes including fishery depletion, changes in fishing fleet size and international agreements on catch size. A combination of these influences can cause major variations as shown by the global fish catch falling from 93,619,100 tonnes in 1996 to 86,299,400 tonnes in 1998 with El Ninõ having the greatest impact [80]. In 2002, China accounted for almost eighteen percent of the wild fish caught, almost twice the weight of wild fish caught by the next major fishing nation, Peru, as shown in Fig. 4.19. Of the fish caught in 2002, the El Niño sensitive anchoveta provided the largest global catch that was almost four times the size of the next largest catch, the Alaska pollock, as seen in Fig. 4.20.

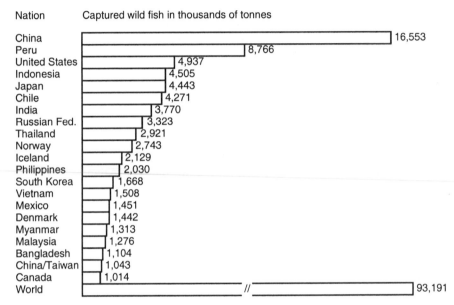

Fig. 4.19. The nations catching more than one million tonnes of fish in 2002. Data from [80].

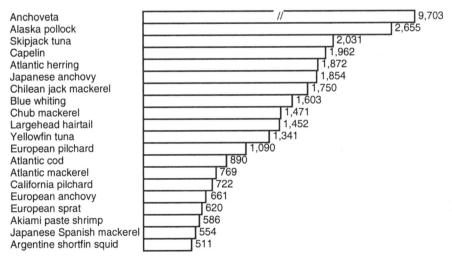

Fig. 4.20. The twenty wild fish caught in the largest amounts in 2002. Data from [81].

As nearly all of the traditional fisheries in water of less than two hundred metres depth have been depleted interest has grown in deep water fishing [89]. Deep water fish are usually rich in oil that renders them attractive as food fish because it helps retain their flavour and texture after freezing. The drawback is that deep water fish are slow growing and late maturing. The result is that fishing fleets take more fish in a short time than can be replaced so that after a few years of large catches fish populations suffer a rapid decline. Typical of this were catches of the New Zealand orange roughy that fell by ninety percent from 1972 to 1982. These fish only reach maturity at thirty years with forty centimetres long fish being between seventy and 150 years old. In the 1990s, the Patagonian toothfish catches off Chile, Argentina and South Africa successively suffered major declines as fishing fleets moved on from each fishery as it became depleted. Similarly, the catch of giant redfish on the Mid Atlantic Ridge was a thousand tonnes in 1996 but fell to a few hundred tonnes in 1997. Such depletion of these slower growing fish threatens to permanently change the deep water fish populations in favour of faster growing species as has already happened in shallow waters with the slow growing Atlantic halibut that has vanished from fisheries where it was once plentiful.

4.14. Aquaculture

In 2002, the global aquaculture production in marine and inland waters amounted to 39.8 million tonnes of fish and 11.6 million tonnes of aquatic plants (mainly seaweed) that together represented 35.2 percent of global total fish and aquatic plant production. By far the largest aquaculture producer was China with 27.8 million tonnes of fish amounting to 69.8 percent of global aquaculture production and 8.8 million tonnes of aquatic plants amounting to 76.0 percent of global aquaculture production. Although more than two hundred species of fish and shellfish are farmed, they are dominated by a relatively small number of species whose production is shown in Fig. 4.21 [81]. In monetary value, the Pacific cupped oyster and the silver carp were the most valuable fish produced by aquaculture in 2002, as shown in Fig. 4.22, while in third place was the tiger prawn whose production of 514,887 tonnes was relatively small.

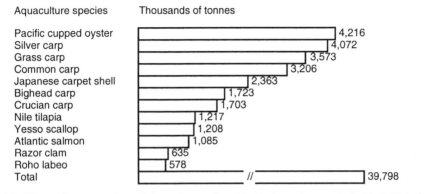

Fig. 4.21. The twelve aquaculture fish species produced in the largest amounts in 2002. Data from [81].

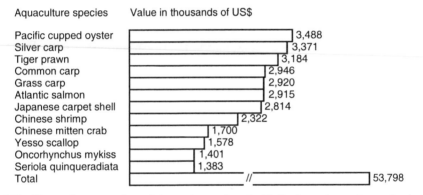

Fig. 4.22. The twelve aquaculture fish species produced with the largest monetary value in 2002 [81].

Important as aquaculture is likely to be to the future food supply, it is very dependent on wild fish for supply of seed, fry and immature fish and food for the cultured fish, as seen from Fig. 4.17, and is contributing to the declining viability of the wild fisheries. It is having particularly deleterious effects in Asia where most commercial shrimp farming is carried out [90-92]. At the beginning of the farming process, the taking of wild shrimp seed has brought the wild shrimp population close to extinction in several fisheries. The bulk of the farmed shrimp is grown on farms formed from either mangrove swamps or rice paddies, thereby removing coast protection and wild fish nurseries in the first case and using valuable rice growing land in the second. Shrimp require food containing amino acids in similar proportions to those found in their own proteins that are generally only available from fishmeal produced from wild fish. As a consequence, trawling for wild fish to produce fishmeal for shrimp farms has increased to the point where the viability of the wild fisheries is threatened. Prior to being harvested, shrimp consume three times their body weight of feed and only half of a shrimp is edible flesh. On this basis alone, this is an unsustainable method of shrimp farming. However, there is a further impost on the environment. To rear shrimps, a continuous change of water is necessary to the extent that up to 50,000 tonnes of water are used to produce one tonne of shrimps. Half of the water used in intensive shrimp farming must be fresh. Much of this is pumped from aquifers with a consequent lowering of water tables that causes seawater to percolate into the aquifers. The shrimp farm water is rapidly polluted and it is estimated that on the east coast of India 2.5 million tonnes of such water were discharged from shrimp farms daily in 1997; an amount that Thailand exceeded with 3.5 million tonnes.

Shrimp farming continues to expand in many parts of Asia with deleterious environmental effects. This is exemplified by estimates that of Thailand's original 180,000 hectares of mangrove swamps none will be left by 2015 if expansion of shrimp farms continues at its present rate. In contrast to mangrove swamps that sustain fishing indefinitely, within ten years and often in five years, shrimp farms become so polluted with dead shrimp and faeces that disease kills the shrimps. This is exemplified by the epidemic that saw Taiwanese shrimp production plunge from 80,000 tonnes in 1987 to almost nothing in 1991 [91]. As a consequence polluted farms are abandoned frequently and either more mangroves or agricultural land is taken over. The deserted farms take twenty to thirty years to regenerate during which time they produce little if any food. In 1997, India became so concerned about the effect of such practices that it decided to close virtually all commercial shrimp farms in five coastal states. Increasingly, restrictions are being placed on shrimp farming globally.

Fortunately, not all fish farming has developed such undesirable characteristics as shrimp farming. Even so, much of it is still very dependent on catching wild fish such as tuna to raise to maturity. In addition to shrimp in Asia and parts of South America, other examples where immature fish, or fry, are taken from the wild are milkfish in the Philippines and Indonesia, tuna in South Australia and eels in Europe and Japan [84]. While these examples of fish farming decrease the mortality of the captured wild fish and raise them to maturity, they do not represent real alternatives to wild fisheries. They not only decrease the populations of the selected fish in the wild, but also decrease the populations of other fish that are caught as both immature and mature fish along with the desired fry and are then discarded. When taking milkfish fry from the wild, eighty-five percent of the catch is usually fry and mature fish of other species that are discarded. It is estimated that the 1.7 billion milkfish fry caught for farming results in the discarding of about ten billion fry of other fish. In India and Bangladesh some 160 fry of other fish and shrimp fry are discarded for every fry of the giant tiger shrimp caught. The fry discarded annually at three collecting centres in West Bengal are thought to be between sixty million and three billion. In contrast, as much as forty percent of Atlantic salmon caught in the North Atlantic appear to have escaped from salmon farms. Over a quarter of a million Atlantic salmon have escaped into the Pacific since the 1980s and are caught by fishing boats from Alaska to Washington State and add to the wild salmon catch. This increased catch is welcome but it carries with it the possibility that diseases transmitted among intensively farmed salmon may be carried into the wild by escapees.

While aquaculture is producing vast amounts of fish, deleterious effects on wild fisheries offset much of this gain. A continuation of ecologically unviable practices is likely to lead to a decline in the global fish supply unless more sustainable aquaculture practices are adopted. Fortunately, considerable attention is now being given to the disposal of effluent and the other environmental aspects of aquaculture by both environmental protection agencies and aquafarmers, particularly in North America where aquaculture is a major food industry [92].

4.15. Fertilizing the Oceans

The major part of the global wild fish catch is made in shallow coastal waters where river flow often provides much of the nutrients for plants and fish. In contrast, vast areas of the oceans support much less life because low nutrient nitrogen and phosphorus levels limit the growth of the single celled plants, collectively called phytoplankton, that are the foundation of the oceanic food chain [93]. Phytoplankton, of which cyanobacteria are a major part, are the simplest of organisms able to carry out photosynthesis but, unlike most terrestrial plants that have an average life of about twenty years, they live for about a week and consequently only represent 0.2 percent of Earth's vegetation by mass despite the immensity of the oceans.

In phytoplankton the ratio of the number of carbon to nitrogen to phosphorus atoms is 106:16:1. While carbon is in plentiful supply from atmospheric carbon dioxide that dissolves in water, it is often considered that because of the 16:1 nitrogen to phosphorus ratio, the Redfield ratio, the supply of nutrient nitrogen is more critical than that of nutrient phosphorus. Despite the oceanic deficiency of these nutrients, about 178 billion tonnes of carbon dioxide from the atmosphere, equivalent to 48.5 billion tonnes of carbon, cycle through phytoplankton annually. (This estimate is based on satellite observation of phytoplankton growth and is about twice that of other estimates [93].) While the major part of this carbon recycles quite rapidly between the first 500 metres layer of the ocean and the atmosphere, it appears that about sixteen billion tonnes of carbon in dead phytoplankton and other dead marine species dependent on it in the food chain sink into the depths annually. This is equivalent to the removal of fifty-nine billion tonnes of carbon dioxide from the atmosphere annually for hundreds of thousands of years past. Thus, the ocean is a great carbon dioxide absorber, or sink. As carbon dioxide is removed from the ocean surface in this way, more can dissolve and this process is often referred to as a "biological pump" for carbon dioxide that is attracting increasing interest in the debate over global warming [94]. (The removal of carbon dioxide from the atmosphere is offset to some extent by carbon dioxide entering the atmosphere through volcanic action and the weathering of carbonate rocks and sediment as is discussed in more detail in Chapter 8.) There is also an oceanic deficiency of iron that is an essential micronutrient for nitrogen fixing phytoplankton, the major natural source of nutrient nitrogen in the oceans [95]. The oceanic surface waters inhabited by phytoplankton are dependent on wind blown dust for iron, aeolian iron, and on iron in water upwelling from the depths. However, only low oxidation state ferrous ion is soluble and available as a nutrient. This is rapidly oxidized to insoluble ferric iron in the wave aerated oceanic surface waters that are deficient in nutrient ferrous iron as a consequence.

In view of this deficiency of ferrous iron, it is not surprising that thought is being given to fertilizing the oceans in much the same way as agricultural land is fertilized to increase productivity [96]. The principle is simple. If the growth of phytoplankton is increased its effect should progress up the food chain to provide an increase in wild fish. Already, a series of experiments has shown that adding ferrous iron to the ocean surface produces huge blooms of phytoplankton lasting for several days. However, it has been found that the lifetime of the algal bloom is limited by the availability of silicic acid that indicates the complexity of the required nutrient supply [97]. Nevertheless, there are proposals to fertilize the oceans with iron, nitrogen, and phosphorus in the same way as fertilizers are used on land [98]. One such proposal envisaged greatly increasing the fish catch in the two million square kilometres of ocean surrounding the Marshall Islands in the Pacific. In a Norwegian venture called Maricult, it was proposed to study the effects of fertilizers on the composition of marine communities and ocean productivity. Another proposal envisaged pumping nitrogenous fertilizer into Chilean coastal waters to soak up more atmospheric carbon dioxide as a consequence of increased plankton growth that was also expected to feed more fish and boost catches. However, the understanding of the effect of fertilizers on the oceans is in its infancy and it is likely to be some time before increases in wild fish catches arise from this source. As with the use of fertilizers on land, any resulting increase in food production is likely to be accompanied by undesirable effects, in this case changes in ocean biogeochemistry and other deleterious ecological changes unless considerable caution is exercised.

4.16. Genetically Modified Crops and Food

Agriculture began about ten thousand years ago bringing with it the cultivation of crops and storage of food that slowly released most of humanity from the vagaries of the hunter-gatherer existence to achieve the sophistication of civilization [99]. The origin of this change was the selection of plants on the basis of the amount and type of edible seed, fruit and tubers that they produced both for food and seed. Later, crossbreeding of similar plants to produce higher yielding varieties represented the earliest human induced changes of plant genetics that may be viewed as the original form of genetic engineering. Thus, early crops of einkorn wheat, emmer wheat, barley, lentil, pea, chickpea and bitter vetch have their origins in the Fertile Crescent of the Middle East, rice, soybean, banana and plantain in Asia, sorghum and millet in Africa, and maize, beans, potato, sweet potato, cassava and groundnut in the Americas. Over millennia the traditional breeding of these and many other plants has produced new varieties that yield much greater amounts of food than do their wild ancestors. In the simplest case, the seeds collected from the highest yielding plants of an otherwise identical crop usually result in significant increases in the overall yield for the next generation grown. At a more sophisticated level induced cross-pollination between closely related plants produces new plant strains, or hybrids, that exhibit new characteristics such as changed leaf shape and size, height, fruit flavour and improved yields of fruit and seed. In this way agriculture has produced the high yield crops that currently produce most of humanity's food supply. The hallmark of this traditional breeding is that there is a transfer of genetic material between two similar species involving thousands of genes from each species to produce new strains [100]. In principle, such new strains could occur in the wild but they tend not to because the deliberate selectivity used by the plant breeder is absent.

Modern genetic engineering, or biotechnology, came to prominence in the last decade of the twentieth century [20]. It made possible the transfer of genetic material, that might be a single gene, between very dissimilar species that would be most unlikely to occur in the wild. This is exemplified by the transfer of a gene from the flounder, *Pseudopleuronectes americanus*, to potatoes to produce a new variety of potato that is much less susceptible to frost damage because the transferred gene produces an antifreeze protein that allows the flounder to live in icy waters and the engineered potato to avoid frost damage [101]. Another example is the transfer of the gene from *Bacillus thuringiensis* responsible for producing insect toxic proteins to maize and other plants to render them pest resistant [102]. In another example, the transfer of a gene from the small plant, *Arabidopsis thaliana*, into the tomato, *Lycopersicon esculentum*, resulted in a new tomato that grows in saline conditions, under which unmodified tomatoes grow poorly, to produce fruit of marginally changed salt content [103]. Salt tolerant plants such as these are increasingly attracting attention as large areas of irrigated cropland are becoming salinized [104]. The creation of such new plants is based on the increasing understanding of the genetic code contained in their genomes that distinguishes one lifeform from another, as is discussed in more detail in Chapter 5.

A plant is characterised by the sequence in which the billions of amino acid, or nucleotide base, pairs are linked to form the double helical molecule deoxyribonucleic acid, DNA. About three percent of this DNA appears to dominantly control the nature of the plant. Within this three percent, sequences of linked nucleotide base pairs, called genes, control the particular aspects of the chemistry that gives the plant most of its distinguishing features. In turn, genes are linked together in chromosomes that constitute the plant genome (*gene-*chromos*ome*). The first completely characterized plant genome was that of *Arabidopsis thaliana*, a member of the brassica plant family that includes broccoli and cauliflower [105]. Although of no value as a food plant, *Arabidopsis thaliana* was chosen for study because its genome is relatively small compared with that of many economically important plants such as wheat whose genome is 160 times larger. *Arabidopsis thaliana* contains five chromosomes where 1 to 5 contain 6,543, 4,036, 5,220, 3,825, and 5,874 genes, respectively, each of which produces proteins that determine particular plant characteristics. Such knowledge of a plant genome gives a powerful understanding of plant biology and insight into the possibilities for genetic modification to produce new plants.

Genetic modification has opened a vast range of possibilities for new food plant species [106]. Simultaneously, it has engendered an extensive debate about the safety of such new foods, the effect of genetically modified crops on the environment, equity in sharing this new technology between developed and developing nations, and the moral and ethical aspects of genetic engineering that significantly rest on the "unnatural" character of the transfer of genes between widely different species [107-109]. This debate was first heard predominantly in Europe but quickly spread to other developed nations, and concern is now being expressed in some developing nations also. While many of the concerns raised have substantial validity, there is a danger that the realization of the potential that genetic engineering holds for improving crop yields, nutrient levels and pest resistance generally could be diminished by this debate unless the differences between the major participants, the biotechnology companies, governments and the public can be safely resolved. This would be a pity, as it appears that

Challenged Earth

genetically modified plants hold considerable potential for helping the nutrition poor developing nations in particular.

4.17. Genetically Engineered Plants

If genes giving desirable characteristics in a plant or animal can be transferred to another plant a new genetically engineered, or transgenic, plant is obtained that may also show these desirable properties, although some gene transfers may also introduce undesirable traits. Such gene transfer, or gene splicing, is now carried out routinely and has resulted in a large range of new plants collectively named either as genetically modified plants, GM plants, or genetically modified organisms, GMOs. Examples of such modifications are those that seek to increase the nutritional value of rice that is the major food of half of humanity [47-49]. Many dominantly rice eaters suffer from vitamin A deficiency and anaemia because rice is low in vitamin A and most of the iron in rice is chemically bound by phytate that renders it largely unavailable for human digestion, in contrast to other cereal crops. To improve this situation, a gene from the daffodil that produces β-carotene, the precursor of vitamin A in the body, has been introduced into rice to give a new strain of saffron coloured "golden rice". Another strain of rice has been engineered through the introduction of a soybean gene that causes it to contain much more iron in a form readily digestible by humans.

Gene transfers conferring disease resistance on major food plants, such as rice, are particularly attractive as they offer increased crop yields. One of the more virulent pests affecting rice is bacterial blight caused by the bacterium *Xanthomonas oryzae* pv. *oryzae*, known as *Xoo*, that spreads swiftly and can reduce a rice crop yield by half. To combat this, a gene responsible for bacterial disease resistance isolated from a wild rice strain has been inserted into the genome of high yielding domesticated rice to produce blight resistant strains [110]. Another example of pest resistance gained through gene transfer is a new strain of sorghum resistant to the parasitic plant *Striga* [111]. Usually known as witchweed, *Striga hermonthica* and *Striga asiatica* attack cereal crops while *Striga gesneroides* attacks cowpeas and tobacco, and all three readily adapt to new hosts. Crop losses due to *Striga* infestation have ranged from fifteen to forty percent of Africa's cereal crop annually. Since 1995, the introduction of a gene from a resistant sorghum strain that limits *Striga*'s ability to germinate into several new cultivated strains of sorghum has conferred *Striga* resistance on them. This has allowed farmers to grow sorghum in large areas of sub-Saharan Africa that had been abandoned because of *Striga* infestation. The importance of this achievement is hard to underestimate, as, with maize and millet, sorghum is one of the major African cereal crops.

High aluminium levels render thirty to forty percent of arable land largely infertile for food production. The possibility of growing food crops on such land has emerged through the insertion of a gene from the bacterium *Pseudomonas aeruginosa* into the genome of papaya [112,113]. The inserted gene causes the plant roots to release five to six times more citrate into the soil than does the unmodified plant. Citrate immobilizes aluminium in the soil and prevents root damage and thereby allows GM papaya to grow well in soils containing tenfold higher aluminium levels than those tolerated by unmodified papaya. The potential of this development is shown by the possibility of growing GM papaya on three million hectares of Mexican land with high aluminium levels on which unmodified papaya will not grow. This compares with the total Mexican papaya crop presently being grown on 20,000 hectares. Extension of this genetic modification to other food crops holds considerable promise. These examples illustrate the potential for genetic engineering to improve humanity's food supply through a variety of crops [114]. However, the major plantings of GM crops have brought to light a complex variety of potential environmental problems that have to be addressed as is discussed below [115].

4.18. The Major Genetically Modified Crops

The first widely planted GM crops are engineered to be either herbicide tolerant so that competing weeds can be sprayed with herbicide without damage to the crop, or to produce the *Bacillus thuringiensis*, *Bt*, insecticide and become pest resistant, or to combine both of these characteristics [116,117]. Four crops, soybean, maize, cotton and canola presently dominate GM crop plantings as seen from Fig. 4.23. The area of GM crop planting has grown rapidly and most extensively in the United States, with Argentina and Canada in second and third places, as seen in Fig. 4.24. Other nations were at first slow to follow, but by

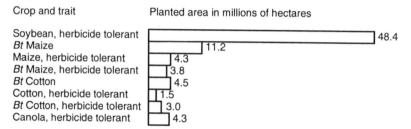

Fig. 4.23. The major GM crops planted in 2004 covered eighty-one million hectares. Data from [114].

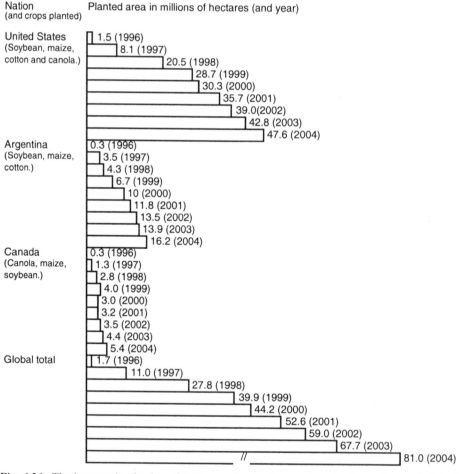

Fig. 4.24. The increase in planting of GM crops for the three major planting nations and globally from 1996 to 2004. Data from [114].

2004 fourteen other nations had reported the significant planting of GM crops: Brazil (5.0), China (3.7), Paraguay (1.2), India (0.50), South Africa (0.50), Uruguay (0.30), Australia (0.25), Romania (0.10), Mexico (0.075), Spain (0.058), the Philippines (0.052), and Colombia (0.010), where the numbers in brackets are plantings in millions of hectares.

The herbicide tolerant crops in Fig. 4.23 have inserted into their genome a gene that renders them impervious to a widely used non selective glyphosate herbicide. Typical of such crops is the Monsanto GM soybean, RoundupReady, that is unaffected by the glyphosate herbicide, Roundup, used to control the most persistent weeds. Thus, a crop of RoundupReady soybeans can be sprayed with Roundup to destroy weeds without harming the soybean crop. It appears that these soybeans are also tolerant of sulfosate herbicide [118]. While this is a great advantage to a soybean farmer who may expect increased crop yields, it is unlikely to be environmentally beneficial unless the overall incidence of spraying is reduced. Similar considerations apply to several varieties of GM canola that are resistant to glyphosate, phosphinothricin and oxynil herbicides, and also to other herbicide resistant crops likely to be developed [119].

Concerns about the effects herbicide spraying of herbicide resistant GM crops resulted in farm scale evaluations of herbicide resistant GM canola, sugarbeet, fodder beet and maize in the United Kingdom between 2000 and 2002 [120]. It was found that the populations and diversity of seeds, weeds, and insects in fields of GM crops were generally lower than in fields of conventional crops and thereby lessened the food supply for small animals and birds that feed on these populations. This appeared not to be caused by the GM crops themselves but by the herbicide spraying regimes to which they were subjected. Thus, the planting of such GM crops is likely to accelerate the decline of farmland wildlife that has coincided with the intensification of agriculture in the UK over the last fifty years unless herbicide spraying becomes much less damaging to wildlife. Given that there is substantial resistance to the planting of GM crops in the UK and Europe, these findings are unlikely to enhance the cause of GM crop adoption.

Insect pest resistance has been achieved by inserting into several crop plants a gene from *Bacillus thuringiensis* that produces an insecticidal protein that is toxic to leaf eating insects and caterpillars. (This natural pesticide has been used by organic farmers for many years and appears to be harmless to humans when used in a spray as discussed in section 4.8.) This should reduce the need to apply insecticides, be more environmentally friendly and lead to lower costs for widely planted *Bt* cotton (1990) and *Bt* maize (1992), where the dates in brackets are those when the gene transfer was first achieved. The same gene has also been inserted into *Bt* tomato (1987), *Bt* tobacco, *Bt* potato (1989) and *Bt* rice (1992). However, *Bt* crops cannot be regarded as a complete answer to the threat of insect pests that have a long history of becoming resistant to insecticides and have already shown some resistance to the insecticide produced by *Bt* plants [121]. The likelihood of *Bt* insect resistance occurring with *Bt* crops arises because the plant produces the *Bt* toxin continuously while the earlier method of using *Bt* sprays resulted in a short exposure to the biodegradable *Bt* toxin. Even without resistance, heavy infestation may sometimes overwhelm *Bt* crops as occurred over thousands of hectares of *Bt* cotton in the United States in 1996 when chemical insecticidal sprays had to be used [122]. In 2001, it was reported that a defective recessive gene in the tobacco budworm, that commonly attacks cotton and other crops in the United States, endows it with resistance wherein lies the path to *Bt* resistance [123].

The insecticidal effect of *Bt* plants is unlikely to be confined to the immediate vicinity of a crop. This concern was raised when it was found that larvae of the harmless Monarch butterfly die when they feed on milkweed dusted with pollen from *Bt* maize [124]. In the United States pollen blows for eight to ten days from the maize fields onto surrounding vegetation that often includes milkweed, the major food of the Monarch larvae. As the planting of *Bt* maize is expected to increase, this knowledge raises concerns for the future of the Monarch, but it may be that the alternative to spray traditional maize with insecticides is even more damaging to this butterfly. Nevertheless, American *Bt* crop farmers are required to plant borders of non *Bt* varieties of the same crop comprising twenty percent of the planted area to provide simultaneously a buffer zone for wind borne pollen to protect butterflies and other insects, and a refuge for *Bt* susceptible insect pests to slow the evolution of *Bt* resistance. Such provisions illustrate the increased sophistication of crop management that is necessary to reap the potential benefits of *Bt* crops. Unfortunately, nineteen percent of all maize growing farmers in Iowa, Minnesota and Nebraska failed to plant the required refuges in 2003 [125]. Meanwhile, concerns have been expressed over the release of *Bt* toxins from GM plant roots into the soil where it might engender *Bt* toxin resistance in rootworm pests [126].

There have been a number of incidents that have shaken public confidence in GM crop management and GM foods. In one case a Canadian company accidentally sent GM canola seed instead of traditional canola seed to Sweden and the United Kingdom [127]. A second and more damaging case occurred in the United States in September 2000 where the insecticidal protein, Cry9C, was detected in taco pastry [128]. This protein is produced by a gene from the *Bacillus thuringiensis* subspecies *tolworthi* engineered into StarLink *Bt* maize that produces ten to a hundred times as much Cry9C as is produced in other strains of *Bt* maize, and accordingly was only licensed for use as cattle feed. The finding of Cry9C in taco pastry resulted in a cascade of consequences: the recall of 800 different grocery products in the United States; Aventis, the seed company, being severely criticized by governmental agencies; calls by major food companies for more stringent governmental regulation of GM crops and mandatory reviews of all new GM crops before seed is licensed for sale; withdrawal from sale of StarLink seed and a lawsuit being brought against Aventis by farmers. Contamination of food products with Cry9C was also found in Taiwan and Japan. The basis of the concern was twofold. First, the unexplained widespread appearance of Cry9C in the taco shells and second, the fact that Cry9C only breaks down slowly when heated and resists digestion, two characteristics of food allergens.

The possibility exists that the gene engineered into Starlink, that constituted less than one percent of the 2000 United States GM maize crop, may have been accidentally transferred to another maize strain. A similar transfer may have occurred between another GM maize and unmodified maize in Mexico, although the evidence for this transfer has been disputed [129]. Such transfers occurred when sugarbeet engineered for herbicide resistance pollinated another sugarbeet strain, and herbicide resistant canola pollinated traditional canola in fields up to three kilometres distant [130]. Inevitably, this reinforces fears that crossbreeding between GM crops and wild plants, where the latter may be genetically close relatives, could produce hard to eradicate GM weeds [131]. (In principle, such crossbreeding may be avoided by engineering the desired gene into the DNA of the plant chloroplast rather than into that of the nucleus as has been achieved with tobacco [132]. Because pollen only transmits DNA from the nucleus, accidental cross-pollination from such GM plants should not transfer the genetic modification.) This, together with the fear that displacement of natural crops by genetically modified crops may have a deleterious effect on wildlife and the environment generally, and also make it difficult to keep organically grown crops GM free, has added significantly to public resistance to the introduction of GM crops and GM food [133].

4.19. Public Attitudes to Genetically Modified Crops and Food

Over many millennia humans have learned that plants range from being very nutritious to highly toxic. Thus, wild plants such as digitalis, hemlock, and oleander are known to be toxic and their ingestion is avoided. Many common food plants contain low levels of toxins that have little effect when eaten while others such as cassava and kidney beans have to be carefully prepared to reduce toxins to a safe level before eating. Every so often traditionally bred new strains of food plants have shown increased levels of toxins to the point where they are harmful. An example was the Lenape potato, introduced in the United States and Canada in the 1960s, that contained increased levels of the glycoalkaloid toxin, solanine [134]. Another example was a new strain of celery highly resistant to insect pests introduced in the United States in the mid 1980s that produced a high level of psoralen that induced severe skin rashes in people who came in contact with it.

On this basis, it is not surprising that a significant section of the population in developed nations, particularly in Europe, should think the safety of GM foods a matter of concern as usually the crops from which they are produced involve greater differences in the species between which genes are transferred than is the case with traditionally bred new strains. (Far more traditionally bred new crop varieties than GM varieties appear each year, as exemplified by the United Kingdom where 950 of the former and eighteen of the latter were tested in various ways in 1998.) Early examples of GM crops to have undesirable characteristics were canola and soybeans incorporating a Brazil nut gene to improve their nutritional value that were found to be highly allergenic and were not marketed [135]. The GM plants now being grown are the first wave where usually genetic modifications to change only one or two characteristics have been introduced. However, the possibilities for changing many genes to introduce a wide range of new characteristics are seemingly endless and it appears inevitable that a plethora of such GM plants will become available in due course. If such GM plants replace traditional plants to a major extent, subtle and widespread changes in diet may occur and, because diet is closely linked to health, this could lead either to an improvement, little change or a decline in the health of humanity. Should the latter be the case it may be difficult to unravel the cause of declining health if the general diet has changed to include a wide range of GM foods. (That such a possibility exists is shown by the diet change in Asia arising from the Green Revolution resulting in vitamin A, iron and zinc health diminishing diet deficiencies that are only now being addressed as discussed in section 4.6.)

The need to weigh the potential for benefit against that for harm leaves little option but to consider each new GM crop and food on a case by case basis, as now applies on either a voluntary or mandatory basis in developed nations. To this point most governmental food agencies have assessed GM modified foods on the basis of their equivalence in general composition to similar foods produced from traditional crops [136]. Critics argue that this is too simplistic a method of screening as it is seldom the major component of a food that leads to allergenicity or toxicity. Increasingly, calls for public involvement in controlling GM food regulation are heard. And by November 2000, fifteen European Community nations, Australia, Japan, New Zealand, Norway, Poland, South Korea and Switzerland had made labelled identification of GM food content mandatory and fifteen more including Brazil, India and the United States had either introduced labelling legislation or were considering it [137].

The public criticism that biotechnology companies have suffered is largely a consequence of ignoring public perceptions. There can be little argument that most of the ideas behind genetic modification of plants are potentially beneficial to consumers and the environment alike provided that due care in cultivation of GM crops and careful monitoring of GM foods occurs. However, surprising naivety with respect to public reaction on the part of scientists and biotechnology companies has damaged the future for GM foods in the short term at least. This is exemplified by the use of antibiotic marker genes inserted along with the modifying gene into a range of GM plants. It is not a simple matter to tell whether the plants grown from cell cultures subjected to modifying procedures have been genetically modified. Accordingly, a gene imparting antibiotic resistance is often used as a marker to detect genetic modification through treatment of the young plants with antibiotics. Those that die do not contain the marker gene and those that live contain both the marker gene and the modifying gene introduced simultaneously. This is a clever technique for use in the research laboratory but less so when retained in the seeds and fruit of GM food crops. There is a possibility that "horizontal" gene transfer from such GM food to bacteria in the human digestive system could yield new strains of antibiotic resistant bacteria to add to those already presenting major health problems [138].

The potential magnitude of this problem is apparent from the variety of antibiotic marker genes in use that confer resistance to β-lactam antibiotics such as ampicillin and amoxycillin used against chest infections, genamycin B used against meningitis and central nervous system infections, neomycin used in veterinary medicine, amikacin reserved for use against gentamycin resistant infections, streptomycin used against tuberculosis and spectinomycin used against gonorrhoea. Early tests failed to detect any transfer of the *bla* gene, that confers antibiotic resistance to ampicillin, from genetically modified maize to bacteria [139]. However, fear of the possibility of such transfers is likely to remain given the current concern about increasing antibiotic resistant bacteria generally and the knowledge that antibiotic resistant bacterial strains have resulted from the routine use of antibiotics in rearing poultry and other animals [140]. An alternative marker gene that causes the modified plant to germinate much earlier than the unmodified plant may provide a way out. This is a timely development as in 2001 the European Community passed legislation that from 2005 will not permit the licensing of GM plants incorporating antibiotic marker genes for agricultural production [141].

European public opposition to GM foods hardened over several years according to the responses of 16,082 people in an European Community survey conducted in November 1999 [133]. Fifty-three percent of respondents opposed the introduction of GM foods, many on the grounds that they "threaten the natural order", were "fundamentally unnatural", had "risks that are not acceptable" and "pose dangers for future generations". Such public reaction has manifest itself in the illegal destruction of GM crops matched by their illegal planting and European governments banning some GM crops [142]. The growth in consumer resistance towards GM food combined with the advent of stringent food labelling requirements became a major worry for American farmers who had earlier eagerly embraced GM maize, soybean and canola crops, much of which had previously been exported [143]. Substantially because of concerns about a hardening of foreign attitudes against GM foods and also because of growing concerns about food labelling in the United States, American export agencies and major food processors decided to give preference to traditional crops and not to accept mixtures of GM and traditional crops. Previously, American farmers generally allowed some mixing of their GM and traditional crops but the new edict heightened their fears about being able to sell their crops and about wind assisted cross pollination between GM and traditional crops.

Increased rigour in the regulation of GM plants and other organisms was adopted in January 2000 through the Cartagena Protocol on Biosafety that is intended to protect biodiversity [144]. This was one of the first legally binding international agreements governing the transfer of GMOs (living modified organisms, LMOs, in the language of the protocol) between nations. It requires a GMO exporting nation to share information with, and to seek the consent of, an importing nation before a GMO is imported and gives nations the right of refusal to import. There is also an obligation to assist the importing nation to assess the risk to its environment of release of GMOs and to register each exported GMO with a United Nations database.

4.20. Genetically Modified Crops and the Developing Nations

Unlike the freely accessible research that led to the Green Revolution that was dominantly governmentally and philanthropically funded, the research behind GM crops is largely commercially funded and is protected by many patents. This has raised concerns that the future of humanity's food supply could eventually rest in the hands of a few biotechnology companies in developed nations, and that the bulk of the population that lives in the developing nations will not be able to afford any benefits that GM crops might offer [113,145]. While the desire of biotechnology companies to recover their research costs and to make a living is readily understandable some of their actions are less so. These are exemplified by the insertion of a terminator gene into plants that renders their seed sterile. A United States patent for this biotechnology was defended by the owners on the basis that it protected their intellectual property and that farmers wishing to grow the crops containing the terminator gene would have to buy seed each year [146]. However, the situation is not quite this simple. While many farmers in developed countries buy seed afresh each year, this is too expensive an option for many farmers in developing nations where it is estimated that some 1.4 billion people are dependent on saved seed for farming [109]. In addition, pollen from GM plants is carried by wind and insects for kilometres so that genetically modified plants could cross pollinate neighbouring traditional crops and cause their seed to become sterile. Such was the disquiet caused by the terminator technology that the major biotechnology company, Monsanto, decided not to proceed with its introduction. Even so, there is a wide network of patents that protect biotechnology companies' intellectual property such that developing nations may be unable to afford the cost of GM crops and thereby be prevented from embarking on possibly a second Green Revolution. Fortunately, the European Community and the Rockefeller Foundation funded the research behind the nutrient enriched "golden rice" discussed in section 4.17. The biotechnology companies holding the many patents for the genetic engineering techniques involved have given royalty free licenses for their use. Thus, the GM rice strains will be passed on to the developing nations, for which they hold great promise, largely unfettered by commercial considerations. It is to be hoped that these expressions of goodwill are early indications that the potential that GM crops hold for improving humanity's food supply can be realized both equitably and safely.

4.21. Food Prospects for the Twenty-First Century

Long term predictions of human destiny have usually proved to be at the best inaccurate, and frequently wrong. With this in mind, the predictions for humanity's food supply are here largely restrained to the first twenty-five years of the twenty-first century. By 2025 the population of the developing nations will represent an even larger proportion of the population than they do now. Accordingly, it is in these nations that the greatest demand for food will arise, and it is probable that rice will be continue to their dominant food. At present, rice is the staple diet of Asia where it provides fifty to eighty percent of daily energy intake depending upon the nation, and its production is the single largest source of income and employment for rural people. It is projected that the rapidly growing Asian population will increase the 2010 total requirement for cereals by thirty-five percent and for rice by twenty-five percent over the 1993 requirements [147]. To achieve this the 250 million farmers of Asia will have to increase their current average yield from five tonnes per hectare to eight tonnes per hectare that is close to the present maximum attainable yield. This assumes that the proportion of food lost to pests before and after harvest remains at the present high levels. Reducing these losses through increased pesticide use, increased biological control, better storage and distribution could substantially reduce the pressure to grow more food.

Should China, the most populous nation, be unable to feed itself as its population levels off at around 1.5 million, the impact on the global food supply could be considerable depending on the size of the shortfall. There is little doubt that sufficient food could be imported from the food exporting nations. However, large food imports by China could price grain beyond the reach of poorer nations, particularly in Africa [148]. As a consequence China's ability to feed itself has attracted considerable attention. China's agriculture came under increasingly centralised control from the early 1950s. By 1958, huge agricultural communes comprising four thousand or so families farmed the land for the state with little individual stake in the success of the enterprise. The consequence was an inefficient production system that produced insufficient food so that an estimated thirty million Chinese died of starvation between 1959 and 1962. This resulted in a gradual dismantling of the communes and a return of an element of land tenure to farmers to provide incentive for increased agricultural productivity and innovation. The result was that food production increased substantially, but it is unlikely to reach its full potential until decision making is completely devolved to farmers, as is the case in more agriculturally productive nations. Nevertheless, it appears that China is likely to be able to feed itself to a major extent and import only a modest component of its food requirements.

While there is reason to be optimistic about China's future self-sufficiency in food, it is clear that many nations will find it beyond their ability to produce sufficient food and will need to import ten percent or more of their cereal needs in 2025 as shown in Fig. 4.25 [149]. Apart from a few developed nations such as Italy and Japan who will be well able to afford such imports, African nations, in particular, may find this beyond their ability and have to rely on food aid unless the terms of global trade change sufficiently for them to generate adequate export earnings.

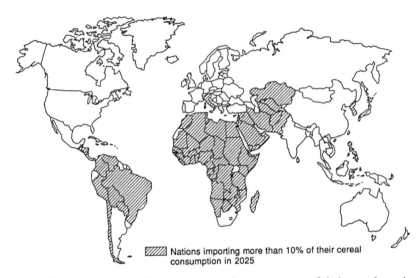

Fig. 4.25. The nations expected to import more than ten percent of their cereal requirements by 2025. Based on estimates in [149].

Despite many concerns, there is cause for optimism about the future of humanity's food supply, the safeguarding and increasing of which is likely to become a growing global enterprise [150]. There are many aspects of the drive to achieve this to which the sustainability of food production is pivotal. Thus, soil and fresh water supplies must be safeguarded, the genetic diversity of crops and potential crop breeding sources must be maintained, and adequate but judicious use of fertilizers and pesticides must be encouraged along with improved pest control. The distribution of food should become more equitable and efficient and human wastage of food must be greatly reduced. However, great prizes may be in store. Among these are GM crops and GM food, both of which have suffered a poor introduction, due to a rush to marketing, possible detrimental environmental and health effects, a lack of foresight and the problem of patenting restricting their cultivation in the developing nations that might most benefit from them.

Given the ingenuity of the research that led to the first GM crops, there is good reason to believe that later generations of GM crops will be environmentally benign and nutritious without risk to health. It should not beyond the wit of humanity to deal with the patenting problems equitably.

Finally, the impact of climate change on food production, while uncertain, appears likely to be manageable, although the developing nations may be disadvantaged by comparison with the developed nations [151]. Increasing carbon dioxide levels may change plant crop growth rates and while some regions may become drier others may become wetter so that the pattern of global crop cultivation may change. As climate change becomes better understood it should become possible to more accurately predict its impact humanity's food supply.

References

1. *The State of Food and Agriculture 1998* and *The State of Food and Agriculture 2003-2004*, Food and Agricultural Organisation of the United Nations, Rome, 1998 and 2004. http://www.fao.org/

2. D. Norse, A new strategy for feeding a crowded planet, *Environment*, **1992**, *34*, June, 6.

3. a) N. Alexandratos, ed., *World Agriculture: Towards 2010: An FAO Study*, Wiley, Chichester, UK, 1995. b) F. Pearce, Crop gurus sow some seeds of hope, *New Sci.*, **1996**, 9 Nov., 6. c) N. Alexandratos, World food and agriculture: Outlook for the medium and longer term, *Proc. Natl. Acad. Sci. USA*, **1999**, *96*, 5908. d) T. Dyson, World food trends and prospects to 2025, *ibid.*, 5929. e) V. Smil, *Feeding the World: A Challenge for the Twenty-First Century*, MIT Press, Cambridge, USA, 2000. f) M. W. Rosegrant and S. A. Cline, Global food security: Challenges and Policies, *Science*, **2003**, *302*, 1917.

4. G. Conway, Food for all in the 21st century, *Environment*, **2000**, *42*, Jan./Feb., 8.

5. Rome Declaration on World Food Security and World Food Summit Plan of Action, Rome, 13 -17 Nov. 1996, *Popn. Dev. Rev.*, **1996**, *22*, 807. b) World Food Summit, Rome 13-17 Nov., 1996, http://www.fao.org/wfs/ final/rd-e.htm

6. R. Rabbinge, The ecological background of food production, in crop protection and sustainable agriculture, *Ciba Foundation Symposium 177*, Wiley, Chichester, UK, 1993, p. 2.

7. a) P. R. Ehrlich, A. H. Erlich and G. C. Daily, Food security, population, and environment, *Popn. Dev. Rev.*, **1993**, *19*, 1. b) J. Bongaarts, Can the growing world population feed itself? *Sci. Am.*, **1994**, *270*, March, 18. c) H. W. Kendall and D. Pimentel, Constraints on the expansion of the global food supply, *Ambio*, **1994**, *23*, 198. d) J. Bongaarts, Population pressure and the food supply system in the developing world, *Popn. Dev. Rev.*, **1996**, *22*, 483. e) G. K. Heilig, How many people can be fed on Earth? In, *The Future Population of the World: What Can We Assume Today?*, W. Lutz, ed., Earthscan, London, UK, 1996, p. 106. f) T. Dyson, *Population and Food – Global Trends and Prospects*, Routledge, New York, USA, 1996. g) R. Naylor, W. Falcon, and E. Zavaleta, Variability and growth in grain yields, 1950-94: Does the record point to greater instability, *Popn. Dev. Rev.*, **1997**, *23*, 41. h) F. W. T. Penning de Vries, R. Rabbinge and J. J. R. Groot, Potential and attainable food production and food security in different regions, *Phil. Trans. Roy. Soc. B*,

1997, *352*, 917. i) B. R. Döös and R. Shaw, Can we predict the future food production? A sensitivity analysis, *Glob. Environ. Change*, **1999**, *9*, 261. j) N. V. Fedoroff and J. E. Cohen, Plants and population: Is there time? *Proc. Natl. Acad. Sci. USA*, **1999**, *96*, 5903. k) N. Ramankutty, J. A Foley and N. J. Olejniczak, People on the land: changes in global population and croplands during the 20th century, *Ambio*, **2002**, *31*, 251.

8.   a) D. Pimental, *et al.*, Impact of population growth on food supplies and environment, *Popn. Environ.*, **1997**, *19*, 9. b) A. Trewavas, Malthus foiled again and again, *Nature*, **2002**, *418*, 668.

9.   D. G. Johnson, The growth of demand will limit output growth for food over the next quarter century, *Proc. Natl. Acad. Sci. USA*, **1999**, *96*, 5915.

10.   D. Pimentel *et al.*, Environmental and economic costs of soil erosion and conservation benefits, *Science*, **1995**, *267*, 1117.

11.   a) P. Crosson, Will erosion threaten agricultural productivity? *Environment*, **1997**, *39*, Oct., 4. b) R. Lal, Degradation and resilience of soils, *Phil. Trans. Roy. Soc. B*, **1997**, *352*, 997. c) J. K. Syers, Managing soils for long-term productivity, *ibid.*, 1011. d) f) K. G. Cassman, Ecological intensification of cereal production systems: Yield potential, soil quality, and precision agriculture, *ibid.*, 5952. e) S. W. Trimble and P. Crosson, U.S. soil erosion rates - myth and reality, *Science*, **2000**, *289*, 248.

12.   J. Kaiser, Wounding Earth's fragile skin, *Science*, **2004**, *304*, 1616.

13.   a) C. Safina, The world's imperiled fish, *Sci. Am.*, **1995**, *273*, Nov., 30. b) R. M. Cook, A. Sinclair, and G. Stefánsson, Potential collapse of North Sea cod stocks, *Nature*, **1997**, *385*, 521. c) C. Safina, The world's imperiled fish, *Sci. Am. Quart., The Oceans*, **1998**, *9*, Fall, 58. d) D. Pauly *et al.*, Fishing down aquatic food webs, *Am. Sci.*, **2000**, *88*, Jan.-Feb., 46. e) D. MacKenzie, Cod's last gasp, *New Sci.*, **2001**, 27 Jan., 16. f) J. J. Magnuson, C. Safina and M. P. Sissenewine, Whose fish are they anyway? *Science*, **2001**, *293*, 1267. g) K. Kleiner, All fished out, *New Sci.*, **2002**, 23 Feb., 11. h) Q. Schiermeier, How many more fish in the sea? *Nature*, **2002**, *419*, 662. i) D. Malakoff and R. Stone, Scientists recommend ban on North Sea cod, *Science*, **2002**, *298*, 938. j) R. Hilborn *et al.*, State of the world's fisheries, *Annu. Rev. Energy Environ.*, **2003**, *26*, 359.

14.   a) B. Moss, A land awash with nutrients - the problem of eutrophication, *Chem. Indust.*, **1996**, 3 June, 407. b) A. Hendriksen, D. O. Hessen, and E. Kessler, Nitrogen: A present and a future threat to the environment, *Ambio*, **1997**, *26*, 253, and following papers in a dedicated issue of *Ambio*. c) S. W. Nixon, Enriching the sea to death, *Sci. Am. Quart., The Oceans*, **1998**, *9 (3)*, Fall, 48. d) D. Malakoff, Death by suffocation in the Gulf of Mexico, *Science*, **1998**, *281*, 190. e) J. Kaiser, The other global pollutant: nitrogen proves tough to curb, *Science*, **2001**, *294*, 1268. f) R. W. Howarth *et al.*, Nitrogen use in the United States from 1961-2000 and potential future trends, *Ambio*, **2002**, *31*, 88.

15.   R. H. Socolow, Nitrogen management and the future of food: Lessons from the management of energy and carbon, *Proc. Natl. Acad. Sci. USA*, **1999**, *96*, 6001.

16.   a) A. K. Sen, *Poverty and Famines. An Essay on Entitlement and Deprivation*, Oxford University Press, Oxford, UK, 1983. b) R. W. Kates and V. Haarmann, Where the poor live: Are the assumptions correct? *Environment*, **1992**, *34*, May, 4. c) J. R. Latham, There's enough food for everyone, but the poor can't afford to buy it, *Nature*, **2000**, *404*, 222.

17.   a) M. Gantz, *Currents of Change: El Niño's Impact on Climate and Society*, Cambridge University Press, Cambridge, UK, 1996. b) P. J. Webster and T. N. Palmer, The past and the future of El Niño, *Nature*, **1997**, *390*, 562.

18.   K. Broad and S. Agrewala, The Ethiopia food crisis - uses and limits of climate forecasts, *Science*, **2000**, 289, 1693.

19.   FAO news release, Bumper harvest in Ethiopia, but more than 2 million people need food aid, http://www.fao.org/newsroom/en/news/2005/89438/index.html.

20.   a) M. J. Reiss and R. Straughan, *Improving Nature? The Science and Ethics of Genetic Engineering*, Cambridge University Press, Cambridge, UK, 1996. b) M. G. Paoletti and D. Pimentel, Genetic engineering in agriculture and the environment, *BioScience*, **1996**, *46*, Oct., 665. c) A. A. Snow and P. M. Palma, Commercialization of transgenic plants:

potential ecological risks, *BioScience*, **1997**, *47*, Feb., 87. d) G. Conway and G. Toenniessen, Feeding the world in the twenty first century, *Nature*, **1999**, *402*, Impacts, C55. e) A. McHughen, *Pandora's Picnic Basket: The Potential and Hazards of Genetically Modified Foods*, Oxford University Press, Oxford, UK, **2000**. f) M. Marvier, Ecology of transgenic crops, *Am. Sci.*, **2001**, *89*, 160.

21.   a) G. R. Fleming and R. van Grandelle, The primary steps of photosynthesis, *Physics Today*, **1994**, *47*, 49. b) V. A. Szalai and G. W. Brudvig, How plants produce dioxygen, *Am. Sci.*, **1998**, *86*, 542. c) R. G. Jensen, Activation of rubisco regulates photosynthesis at high temperature and CO_2, *Proc. Natl. Acad. Sci. USA*, **2000**, *97*, 12937. d) D. J. des Marais, When did photosynthesis emerge on Earth? *Science*, **2000**, *289*, 1703. e) P. Hunter, Flower power, *New Sci.*, **2004**, 1 May, 28.

22.   These calculations are based on the photosynthetic reaction: $CO_2 + H_2O \rightarrow CH_2O + O_2$. Estimates of the amount of carbon annually incorporated into green plants may be found in: G. L. Ajtay, P. Ketner and P. Duvigneaud, Terrestrial primary production and phytomass, in *The Global Carbon Cycle*, B. Bolin *et al.*, eds., Wiley, Chichester, UK, 1979, p. 129; J. S. Olson *et al.*, Atmospheric Carbon Dioxide and the Global Carbon Cycle, J. R. Trabalka, ed., U.S. Department of Energy, Washington, DC, USA, 1985, p. 175; and J. H. Martin *et al.*, *Deep Sea Res.*, **1987**, *34*, 267. The amount of oxygen in the atmosphere is about 1,200,00 billion tonnes.

23.   B. Bugbee and O. Monje, The limits of crop productivity, *BioScience*, **1992**, *42*, July/Aug., 494.

24.   I. A. Shiklomanov, World fresh water resources, in *Water in Crisis: A Guide to the World's Fresh Water Resources*, P. H. Gleick, ed., Oxford University Press, Oxford, UK, 1993, pp. 13.

25.   a) H. D. Holland, *The Chemical Evolution of the Atmosphere and Oceans*, Princeton University Press, Princeton, USA, 1984. b) J. Copley, The story of O, *Nature*, **2001**, *410*, 862. c) G. C. Dismukes, Splitting water, *Science*, **2001**, *292*, 447. d) J. F. Kasting, The rise of atmospheric oxygen, *Science*, **2001**, *293*, 819.

26.   a) L. T. Evans, Adapting and improving crops: The endless task, *Phil. Trans. Roy. Soc. B*, **1997**, *352*, 901. b) C. C. Mann, Crop scientists seek a new revolution, *Science*, **1999**, *283*, 310. c) C. C. Mann, Genetic engineers aim to soup up crop photosynthesis, *Science*, **1999**, *283*, 314.

27.   a) V. Smil, *Cycles of Life: Civilisation and the Biosphere*, Chapter 2, Scientific American Library, New York, USA, 1997. b) J. A. Raven, Evolutionary options, *Nature*, **2002**, *415*, 375. c) C. Surridge, The rice squad, *Nature*, **2002**, *416*, 576. d) J. M. Hibberd and W. P. Quick, Characteristics of C_4 photosynthesis in stems and petioles of C_3 flowering plants, *ibid.*, 451.

28.   a) K. O'Dea, Traditional diet and food preferences of Australian aboriginal hunter-gatherers, *Phil. Trans. Roy. Soc. B*, **1991**, *354*, 233. b) K. Hawkes, J. F. O'Connell and N. G. Blurton Jones, Hunting income patterns among the Hadza: big game, common goods, foraging goals and the evolution of the human diet, *ibid.*, 243. c) K. Milton, Comparative aspects of diet in Amazonian forests dwellers, *ibid.*, 253. d) S. J. Ulijaszeh, Human diet change, *ibid.*, 271. e) D. A. Southgate, Nature and variability of human food consumption, *ibid.*, 281. E. M. Widdowson, Contemporary human diets and their relation to health and growth: overview and conclusions, *ibid.*, 289.

29.   G. Vines, Food, glorious food, *New Sci.*, Inside Science 104, **1997**, 18 Oct., 1.

30.   Dimensions of need: an atlas of food and agriculture, FAO, Rome, Italy, 1995, http://www.fao.org/docrep/u8480e/U8480E00.htm.

31.   Food, agriculture and food security: developments since the World Food Conference and prospects, Technical background document, World Food Summit, FAO, Rome, 1996. http://www.fao.org/wfs/.

32.   W. H. Bender, How much food will we need in the 21st century? *Environment*, **1997**, *39*, March, 6.

33.   D. MacKenzie, Protein at a price, *New Sci.*, **2000**, 18 March, 32.

34.   a) W. Bender, An end use analysis of global food requirements, *Food Policy*, **1994**, *19*, 381.
      b) M. B. Thomas, Ecological approaches and the development of "truly integrated" pest
      management, *Proc. Natl. Acad. Sci. USA*, **1999**, *96*, 5944.
35.   a) W. E. Fry, Resurgence of the Irish potato famine fungus, *BioScience*, **1997**, *47*, June,
      363. b) R. Nelson, The blight is back, *Sci. Am.*, **1998**, *278*, June, 18. c) CGIAR scientists
      develop new potato clones to counter late blight, world's worst agricultural disease,
      Consultative Group on International Agricultural Research press release, Washington, DC,
      USA, May 24, 1998, http:/www.worldbank.org/html/cgiar/press/potrell.html. d) Q.
      Schiermeier, Russia needs help to fend off potato famine, researchers warn, *Nature*, **2001**,
      410, 1011. e) J. B. Ristaino, C. T. Groves and G. R. Parra, PCR amplification of the Irish
      potato famine pathogen from historic specimens, *ibid.*, 695. f) G. Garielik, Taking the bite
      out of potato blight, *Science*, **2002**, *298*, 1702.
36.   a) J. Giles, Delays allowed foot-and-mouth epidemic to sweep across Britain, *Nature*, **2001**,
      410, 501. b) M. Woolhouse and A. Donaldson, Managing foot-and mouth, *ibid.*, 515. c) M.
      Enserink, Intensified battle against foot and mouth appears to pay off, *Science*, **2001**, *292*,
      410. d) N. M. Ferguson, C. A. Donnelly and R. M. Anderson, The foot and mouth epidemic
      in Great Britain: pattern of spread and impact on interventions, *Science*, **2001**, *292*, 1155. e)
      M. J. Keeling *et al.*, Dynamics of the 2001 UK foot and mouth epidemic: stochastic
      dispersal in a heterogeneous landscape, *Science*, **2001**, *294*, 813. f) M. Woolhouse *et al.*,
      Foot and mouth disease under control in the UK, *Nature* **2001**, *411*, 258. g) J. Giles, The
      killing fields, *ibid.*, 839. h) D. MacKenzie, Running wild, *New Sci.*, **2001**, 5 May, 5.
37.   S. H. Henry, *et al.*, Reducing liver cancer - global control of aflotoxin, *Science*, **1999**, *286*,
      2453.
38.   a) D. Taylor, BSE: our future in the balance, *Chem. Indust.*, **1998**, 1 June, 444. b) A.
      Coghlan, How it went so horribly wrong, *New Sci.*, **2000**, 4 Nov., 4. c) C. Ainsworth, A
      killer is born, *New Sci.*, **2000**, 4 Nov., 7. d) D. Mackenzie, Mad meat, *ibid.*, 8. e) D.
      MacKenzie, The human tragedy may be just beginning, *ibid.*, 9. f) D. MacKenzie, Burn,
      bovine, burn, *New Sci.*, **2000**, 9 Dec., 7. g) P. Aldhous, Inquiry blames missed warnings for
      scale of Britain's BSE crisis, *Nature*, **2000**, *408*, 3. h) P. Brown, Mad-cow disease in cattle
      and human beings, *Am. Sci.*, **2004**, *92*, 334.
39.   a) D. MacKenzie, Mad to deny it, *New Sci.*, **2000**, 2 Dec., 7. b) C. A. Donnelly, Likely size
      of the French BSE epidemic, *Nature*, **2000**, *408*, 787. c) A. Abbott, BSE fallout sends shock
      waves through Germany, *Nature*, **2001**, *409*, 275. d) D. MacKenzie, North America needs
      tougher BSE measures, *New Sci.*, **2003**, 5 July, 5. e) B. Hileman, Guarding against mad
      cow disease, *Chem. Eng. News*, **2003**, *81*, 4 Aug., 22. f) D. Normile, First U.S. case of mad
      cow sharpens debate over testing, *Science*, **2004**, *303*, 156.
40.   A. C. Ghani, N. M. Ferguson, C. A. Donnelly and R. M. Anderson, Predicting vCJD
      mortality in Great Britain, *Nature*, **2000**, *406*, 583.
41.   FAO Statistical Data Base. http://apps.fao.org.
42.   a) V. Smil, Crop residues: agriculture's largest harvest, *BioScience*, **1999**, *49*, April, 299. b)
      R. Lal, Soil carbon sequestration impacts on global climate change and food security,
      Science, **2004**, *304*, 1623.
43.   P. R. Ehrlich, *The Population Bomb*, Ballantine Books, New York, USA, 1968.
44.   D. Tilman, Global environmental impacts of agricultural expansion: the need for
      sustainable and efficient practices, *Proc. Natl. Acad. Sci. USA*, **1999**, *96*, 5995.
45.   a) N. E. Borlaug, The Green Revolution: Peace and Humanity, A speech on the occasion of
      the awarding of the 1970 Nobel Peace Prize in Oslo, Norway on 11 Dec., 1970.
      http://www.theatlantic.com/atlantic/issues/97Jan/borlaugh/speech.htm. b) G. Easterbrook,
      Forgotten benefactor of humanity, *The Atlantic Monthly*, **1997**, *279*, Jan., 75. c) M. Smale,
      The Green Revolution and wheat genetic diversity: some unfounded assumptions, *World
      Dev.*, **1997**, *25*, 1257. d) C. Mann, Reseeding the Green Revolution, *Science*, **1997**, *277*,
      1038.
46.   J. Seymour, Hungry for a new revolution, *New Sci.*, **1996**, 30 March, 2.

47.   a) B. Holmes, Grains of gold, *New Sci.*, **1999**, 14 Aug., 12. b) T. Gura, New genes boost rice nutrients, *Science*, **1999**, *285*, 994. c) Editorial, A golden bowl of rice, *Nature Biotechnol.*, **1999**, *17*, 831. d) M. L. Guerinot, The Green Revolution strikes gold, *Science*, **2000**, *287*, 241. e) X. Ye *et al.*, Engineering the provitamin A (β-carotene) biosynthetic pathway into (carotenoid-free) rice endosperm, *Science*, **2000**, *287*, 303. f) A. J. Bouchie, Golden handouts on the way, *Nature Biotechnol.*, **2000**, *18*, 911.

48.   a) B. Mazur, E. Krebbers and S. Tingey, Gene discovery and product development for grain quality traits, *Science*, **1999**, *285*, 372. b) D. DellaPenna, Nutritional genomics: manipulating plant micronutrients to improve human health, *Science*, **1999**, *285*, 375.

49.   a) F. Goto *et al.*, Iron fortification of rice seed by the soybean ferritin gene, *Nature Biotechnol*, **1999**, *17*, 282. b) M. L. Guerinot, Improving rice yields - ironing out the details, *Nature Biotechnol.*, **2001**, *19*, 417. c) M. Takahashi *et al.*, Enhanced tolerance of rice to low iron availability in alkaline soils using barley nicotianamine aminotransferase genes, *Nature Biotechnol.*, **2001**, *19*, 466.

50.   a) S. J. Jayaraj and R. J. Rabindra, The local view of the role of plant protection in sustainable agriculture in India, in *Crop Protection and Sustainable Agriculture, Ciba Foundation Symposium 177*, Wiley, Chichester, UK, 1993, p. 168. b) D. Sharma, The Green revolution turns sour, *New Sci.*, **2000**, 8 July, 44.

51.   a) V. Smil, Population growth and nitrogen, an exploration of a critical existential link, *Popn. Dev. Rev.*, **1991**, *17*, 569. b) V. Smil, Global population and the nitrogen cycle, *Sci. Am.*, **1997**, *277*, July, 58. c) V. Smil, Detonator of the population explosion, *Nature*, **1999**, *400*, 415. d) V. Smil, *Transforming the World: Synthesis of Ammonia and Its Consequences*, MIT Press, Cambridge, USA, 2000. e) K. F. Isherwood, *Mineral Fertilizer Use and the Environment, International Fertilizer Association*, Paris, **2000**. f) V. Smil, Nitrogen and food production: proteins for human diets, *Ambio*, **2002**, *31*, 126. g) P. E. Fixen and F. B. West, Nitrogen fertilizers: meeting contemporary challenges, *ibid.*, 169. h) G. J. Leigh, *The World's Greatest Fix: A History of Nitrogen in Agriculture*, Oxford University Press, Oxford, UK, 2004.

52.   International Fertilizer Industry Association, Oct, 2004, http://www.fertilizer.org/

53.   a) National Research Council, *Soil and Water Quality, An Agenda for Agriculture*, National Academy Press, Washington, DC, USA, 1992. b) P. A. Matson, R. Naylor and I. Ortiz-Monasterio, Integration of environmental, agronomic, and economic aspects of fertilizer management, *Science*, **1998**, *280*, 112. c) K. G. Cassman, A. Dobermann and D. T. Walters, Agrosystems, nitrogen-use efficiency, and nitrogen management, *Ambio*, **2002**, *31*, 132. d) R. N. Roy, R. V. Misra and A. Montanez, Decreasing reliance on mineral nitrogen – yet more food, *ibid.*, 177.

54.   a) A. Kinzig and R. Socolow, Human impacts on the nitrogen cycle, *Phys. Today*, **1994**, *47*, 24. b).W. H. Schlesinger, *Biogeochemistry; An Analysis of Global Change*, 2nd. edn., Cambridge University Press, Cambridge, UK, 1997.

55.   a) F. Pearce, Lightening sparks pollution rethink, *New Sci.*, **1997**, 25 Jan., 15. b) C. Price, J. Penner and M. Prather, NO_x from lightening 1. Global distribution based on lightening physics, *J. Geophys. Res.*, **1997**, *102*, 5929. c) C. Price and J. Penner, NO_x from lightening 2. Constraints from the global atmospheric electric circuit, *ibid.*, 5943.

56.   G. J. Leigh, Fixing nitrogen any which way, *Science*, **1998**, *279*, 506.

57.   a) P. M. Vitousek, *et al.*, Human alteration of the global nitrogen cycle: causes and consequences, *Ecol. Appns.*, **1997**, *7*, 737. b) D. O. Hessen *et al.*, Human impacts on the nitrogen cycle: a global problem judged from a local perspective, *Ambio*, **1997**, *26*, 321. c) A. S. Moffat, Global nitrogen overload problem grows critical, *Science*, **1998**, *279*, 988. d) E. M. Bennett, S. R. Carpenter and N. F. Caraco, Human impact on erodable phosphorus and eutrophication: a global perspective, *BioScience*, **2001**, *51*, March, 227. e) N. Nosengo, Fertilized to death, *Nature*, **2003**, *425*, 894.

58.   a) C. R. Frink, P. E. Waggoner and J. H. Ausubel, Nitrogen fertilizer: retrospect and prospect, *Proc. Natl. Acad. Sci. USA*, **1999**, *96*, 1175. b) J. V. Stafford, Improving agriculture's aim, *Chem. Indust.*, 2000, 7 Feb., 98.

59.   R. Carlson, *Silent Spring*, Houghton-Mifflin, Boston, USA, 1962.
60.   A. M. Thayer, Transforming agriculture, *Chem. Eng. News*, **1999**, *77*, 19 April, 21.
61.   a) R. Repetto and S. S. Baliga, *Pesticides and the Immune System: Public Health Risks*, World Resources Institute, Washington DC, USA, 1996. b) D. A. Goolsby *et al.*, Herbicides and their metabolites in rainfall: origin, transport, and deposition patterns across the Midwestern and Northeastern United States, 1990-1991, *Environ. Sci. Technol.*, **1997**, *31*, 1325. c) R. J. Gilliom *et al.*, Testing water quality for pesticide pollution, *Environ. Sci. Technol.*, **1999**, *33*, 164 A. d) L. Helmuth, Pesticide causes Parkinson's in rats, *Science*, **2000**, *290*, 1068.
62.   B. Lambert and M. Peferoen, Insecticidal promise of *Bacillus thuringiensis*, *BioScience*, **1992**, *42*, Feb., 112.
63.   a) J. A. Browning and K. Frey, Multiline cultivars as a means of disease control, *Annu. Rev. Phytopath.*, **1969**, *7*, 355. b) M. S. Wolfe, The current status and prospects of multiline cultivars and variety mixtures for disease resistance, *Annu. Rev. Phytopath.*, **1985**, *23*, 251.
64.   R. Edwards, Tomorrow's bitter harvest, *New Sci.*, **1996**, 17 Aug., 14.
65.   a) M. S. Wolfe, Crop strength through diversity, *Nature*, **2000**, *406*, 681. b) Y. Zhu *et al.*, Genetic diversity and disease control in rice, *ibid.*, 718.
66.   a) C. Lammou and C. C. Mundt, Evolution of a pathogen population in host mixtures: simple race-complex race competition, *Plant Pathol.*, **1996**, *45*, 440. b) K. A. Garrett and C. C. Mundt, Epidemiology in mixed host populations, *Phytopathol.*, **1999**, *89*, 984.
67.   a) S. L. Pimm, In search of perennial solutions, *Nature*, **1997**, *389*, 126. b) D. Tilman, C. L. Lehman and K. T. Thomson, Plant diversity and ecosystem productivity: theoretical considerations, *Proc. Natl. Acad. Sci. USA*, **1997**, *94*, 1857.
68.   G. A. Strobel, Biological control of weeds, *Sci. Am.*, **1991**, *265*, July, 50.
69.   D. Pimentel *et al.*, Water resources: agriculture, the environment, and society, *BioScience*, **1997**, *47*, Feb., 97.
70.   S. L. Postel, Water for food production: will there be enough in 2025? *BioScience*, **1998**, *48*, Aug., 629.
71.   a) D. Rimmer, Ultimate interface, *New Sci.*, *Inside Science, 115*, **1998**, 14 Nov., 1. b) D. H. Yaalon, Down to earth, *Nature*, **2000**, *407*, 301. c) E. Pennisi, The secret life of fungi, *Science*, **2004**, *304*, 1620. d) D. Wardle *et al.*, Ecological linkages between aboveground and belowground biota, *ibid.*, 1629. e) I. M. Young and J. W. Crawford, Interactions and self-organization in the soil-microbe complex, *ibid.*, 1634.
72.   L. Brown and E. Wolf, Soil erosion: quiet crisis in the world economy, Worldwatch paper 60, Worldwatch Institute, Washington DC, USA, **1984**.
73.   a) L. R. Oldeman, R. T. A. Hakkeling and W. G. Sombroek, *Global Assessment of Soil Degradation*, International Soil Information and Reference Center and United Nations Environmental Programme, Waneningen, The Netherlands, 1990. b) *World Map of the Status of Human-Induced Soil Degradation, Global Assessment of Soil Degradation*, United Nations Environmental Programme, Nairobi, Kenya, 1991.
74.   H. Dregne and N.-T. Chou, Global desertification and dimensions and costs, in *Degradation and Restoration of Arid Lands*, H. Dregne, ed., Texas Technical University, Lubbock, USA, 1992.
75.   G. C. Daily, Restoring the value of the world's degraded lands, *Science*, **1995**, *269*, 350.
76.   a) M. A. Fullen and D. J. Mitchell, Desertification and reclamation in North-Central China, *Ambio*, **1994**, 23, 131. b) M. A. Fullen *et al.*, Soil erosion and conservation in Yunnan province, China, *Ambio*, **1999**, *28*, 125.
77.   M. Mainguet, *Desertification: Natural Background and Human Mismanagement*, 2nd. edn., Springer-Verlag, Berlin, Germany, 1994.
78.   C. J. Tucker and S. E. Nicholson, Variations in the size of the Sahara desert from 1980 to 1997, *Ambio*, **1999**, *28*, 587.
79.   R. Grainger, Recent trends in global fishery production, FAO, Rome, Italy, 2000, http://www.fao.org/fi/trends/catch/catch.asp.

80. The state of world fisheries and aquaculture 2002, FAO, Rome, Italy, 2002. http://www.fao.org/docrep/w9900e02.htm.
81. Yearbooks of Fishery Statistics Summary tables-2002, FAO, Rome, Italy, 2000. http://www.fao.org/FI/STAT /summary/default.asp.
82. a) D. Malakoff, Extinction on the high seas, *Science*, **1997**, *277*, 486. b) L. W. Botsford, J. C. Castilla and C. H. Peterson, The management of fisheries and marine ecosystems, *ibid.*, 509. c) C. D. Stone, The crisis in global fisheries: can trade laws provide a cure? *Environ. Conservn.*, **1997**, *24*, 97. d) J. M. Casey and R. A Myers, Near extinction of a large, widely distributed fish, *Science*, **1998**, *281*, 690. e) M. Moghim and J. D. Neilson, Imminent collapse of the Caspian Sea stellate sturgeon (*Acipenser stellatus*): evidence from the Iranian fishery, *Ambio*, **1999**, *28*, 372. f) K. Christen, Sustaining global fish stocks, *Environ. Sci. Technol.*, **1999**, *33*, 452 A. g) D. Pauly *et al.*, Towards sustainability in world fisheries, *Nature*, **2002**, *418*, 689. h) R. W. Zabel *et al.*, Ecologically sustainable yield, *Am. Sci.*, **2003**, *91*, 150.
83. a) R. L. Naylor *et al.*, Nature's subsidies to shrimp and salmon farming, *Science*, **1998**, *282*, 883. b) J. Forster, Shrimp and salmon farming, *Science*, **1999**, *283*, 639.
84. R. L. Naylor *et al.*, Effect of aquaculture on world fish supplies, *Nature*, **2000**, *405*, 1017.
85. a) M. Hammer, A. Jansson and B.-O. Jansson, Diversity change and sustainability: Implications for fisheries, *Ambio*, **1993**, *22*, 97. b) N. Williams, Overfishing disrupts entire ecosystems, *Science*, **1998**, *279*, 809. c) D. Pauly *et al.*, Fishing down marine food webs, *Science*, **1998**, *278*, 860. d) B. Holmes, The rape of the sea, *New Sci.*, **1998**, 14 Feb., 4.
86. a) P. K. Dayton, Reversal of the burden of proof in fisheries management, *Science*, **1998**, *279*, 821. b) D. Malakoff, Papers posit grave impact of trawling, *Science*, **1999**, *284*, 2168.
87. a) E. Masood, Scientific caution 'blunts efforts' to conserve fish stocks, *Nature*, **1996**, *379*, 481. b) F. Pearce, Only stern words can save the world's fish, *New Sci.*, **1996**, 10 Feb., 4. c) F. Pearce, North Sea cod on the brink of collapse, *New Sci.*, **1997**, 8 Feb., 6. d) E. Masood, Fisheries science: all at sea when it comes to politics? *Nature*, **1997**, *386*, 105. e) D. Spurgeon, Canada's cod leaves science in hot water, *Nature*, **1997**, *386*, 107. f) E. Masood, Fishing by the numbers reveals its limits, *Nature*, **1997**, *386*, 110. g) Battle on the high seas for remaining catches, *The European*, 17-23 April, **1997**. h) C. M. O'Brien *et al*, Climate variability and North Sea cod, *Nature*, **2000**, *404*, 142.
88. a) J. N. Butler *et al.*, The Bermuda fisheries: a tragedy of the commons averted, *Environment*, **1993**, *35*, Jan./Feb., 6. b) K. F. Schmidt, 'No take' zones spark fisheries debate, *Science*, **1997**, *277*, 489. c) A. Hastings and L. W. Botsford, Equivalence in yield from marine reserves and traditional fisheries management, *Science*, **1999**, *284*, 1537. d) C. M. Roberts *et al.*, Effects of marine reserves on adjacent fisheries, *Science*, **2001**, *294*, 1920. e) H. Nicholls, Sink or swim, *Nature*, **2004**, *432*, 12.
89. a) S. R. Dovers and J. W. Handmer, Ignorance, the precautionary principle, and sustainability, *Ambio*, **1995**, *24*, 92. b) D. Mackenzie, In deep waters, *New Sci.*, **1997**, 8 Nov., 12.
90. a) B. Gujja and A. Finger-Stich, What price prawn? *Environment*, **1996**, *38*, Sept., 12. b) E. Masood, Aquaculture: a solution, or source of new problems? *Nature*, **1997**, *386*, 109. c) K. McKinsey, The promise and perils of aquaculture, *Sci. Am. Quart., The Oceans*, **1998**, *9*, Fall, 64. d) U. Barg *et al.*, Towards shrimp culture development: implementing the FAO Code of Conduct for Responsible Fisheries, FAO, Rome, 1999. http://www.fao.org /fi/faocons/shrimp/honduras.asp.
91. B. Holmes, Blue revolutionaries, *New Sci.*, **1996**, 7 Dec., 32.
92. K. Kreeger, Down on the fish farm: developing effluent standards for aquaculture, *BioScience*, **2000**, *50*, Nov., 949.
93. a) P. G. Falkowski, R. T. Barber and V. Smetacek, Biogeochemical controls and feedbacks on ocean primary production, *Science*, **1998**, *281*, 200. b) C. Field *et al.*, Primary production of the biosphere: integrating terrestrial and oceanic components, *ibid.*, 237.

94.   a) G. R. Bigg, *The Oceans and Climate*, Cambridge University Press, Cambridge, UK, 1996. b) J. Houghton, *Global Warming: The Complete Briefing*, 2nd, edn., Cambridge University Press, Cambridge, UK, 1997.

95.   a) P. G. Falkowski, Evolution of the nitrogen cycle and its influence on the biological sequestration of CO_2 in the ocean, *Nature*, **1997**, *387*, 272. b) J. A. Fuhrman and D. G. Capon, Nifty nanoplankton, *Nature*, **2001**, *412*, 593. c) J. P. Zehr *et al.*, Unicellular cyanobacteria fix N_2 in the subtropical North Pacific Ocean, *ibid.*, 635. d) K. Christen, Linking iron with carbon sequestration, *Environ. Sci. Technol.*, **2001**, *35*, 98 A.

96.   a) J. H. Martin, S. E. Fitzwater and R. M. Gordon, Iron deficiency limits phytoplankton growth in Antarctic waters, *Global Biogeochem. Cycles*, **1990**, *4*, 5. b) J. H. Martin *et al.*, Testing the iron hypothesis in ecosystem of the equatorial Pacific Ocean, *Nature*, **1994**, *371*, 123. c) Z. S. Kolber *et al.*, Iron limitation of phytoplankton photosynthesis in the equatorial Pacific Ocean, *ibid.*, 145. d) K. H. Coale *et al.*, A massive phytoplankton bloom induced by an ecosystem-scale iron fertilization experiment in the equatorial, Pacific Ocean, *Nature*, **1996**, *383*, 495. e) R. Maranger, D. F. Bird and N. M. Price, Iron acquisition by photosynthetic marine phytoplankton from ingested bacteria, *Nature*, **1998**, *396*, 248. f) S. Nadis, Fertilizing the sea, *Sci. Am.*, **1998**, *278*, April, 25. g) C. W. Mullineaux, The plankton and the planet, *Science*, **1999**, *283*, 801. h) M. J. Behrenfeld and Z. S. Kolber, Widespread iron limitation of phytoplankton in the South Pacific Ocean, *ibid.*, 840. i) T. Tyrrell, The relative influences of nitrogen and phosphorus on oceanic primary production, *Nature*, **1999**, *400*, 525. j) J. J. Cullen, Iron, nitrogen and phosphorous in the ocean, *Nature*, **1999**, *402*, 372. k) T. Tyrrell, Iron, nitrogen and phosphorous in the ocean, response, *Nature*, **1999**, *402*, 372. l) S. W. Chisholm, Stirring times in the Southern Ocean, *Nature*, **2000**, *407*, 685. m) P. W. Boyd *et al.*, A mesoscale phytoplankton bloom in the polar Southern Ocean stimulated by iron fertilization, *ibid.*, 695. n) E. R. Abraham *et al.*, Importance of stirring in the development of an iron-fertilized phytoplankton bloom, *ibid.*, 727. o) A. J. Watson *et al.*, Effect of iron supply on Southern Ocean CO_2 uptake and implications for glacial atmospheric CO_2, *ibid.*, 730. p) F. Pearce, Age of iron and ice, *New Sci.*, **2000**, 14 Oct., 11. q) P. G. Falkowski, The ocean's invisible forest, *Sci. Am.*, **2002**, *287*, Aug., 38. r) K. O. Buesseler and P. W. Boyd, Will ocean fertilization work? *Science*, **2003**, *300*, 67. s) Q. Schiermeier, *Nature*, The oresman, **2003**, *421*, 109.

97.   P. W. Boyd *et al.*, The decline and fate of an iron-induced subArctic phytoplankton bloom, *Nature*, **2004**, *428*, 549.

98.   a) R. Sorensen, Maricult - planting the seeds of a blue revolution, Tell'us Research Council of Norway, **1995**, *2*, 4. b) D. MacKenzie, Norway's fish plan 'a recipe for disaster', *New Sci.*, **1996**, 13 Jan., 4. c) I. S. F. Jones and H. E. Young, Engineering a large sustainable world fishery, *Environmental Consn.*, **1996**, *24*, 99. d) G. Schueller, Testing the waters, *New Sci.*, **1999**, 2 Oct., 34. e) R. Rawls, Sequestering CO_2, *Chem. Eng. News*, **2000**, *78*, 18 Sept., 66. f) F. Pearce, A cool trick, *New Sci.*, **2000**, 8 April, 18. g) S. W. Chisholm, P. G. Falkowski and J. J. Cullen, Discredited ocean fertilisation, *Science*, **2001**, *294*, 309.

99.   a) B. D. Smith, *The Emergence of Agriculture*, Scientific American Library, New York, USA, 1995. b) B. D. Smith, Between foraging and farming, *Science*, **1998**, *279*, 1651. c) H. Pringle, The slow birth of agriculture, *Science*, **1998**, 282, 1446. d) S. Pääbo, Neolithic genetic engineering, *Nature*, **1999**, *398*, 194. e) R.-L. Wang, *et al.*, The limits of selection during maize domestication, *ibid.*, 236. f) S. Lev-Yadun, A. Gopher and S. Abbo, The cradle of agriculture, *Science*, **2000**, *288*, 1602. g) D. R. Piperno *et al.*, Starch grains reveal early root crop horticulture in the Panamanian tropical forest, *Nature*, **2000**, *407*, 894. h) J. Diamond, Evolution, consequences and future of plant and animal domestication, *Nature*, **2002**, *481*, 700. i) J. R. McNeill and V. Winiwarter, Breaking the sod, humankind, history, and soil, *Science*, **2004**, *304*, 1627.

100. D. Barnes, Biotechnology: where now? *Chem. Indust.*, **2000**, 22 May, 338.

101. W. Jaffé and M Rojas, Transgenic potato tolerant to freezing, *Biotechnol. Dev. Monitor*, **1994**, *18*, 10.

102. D. L. Plucknett and D. L. Winkelman, Technology for sustainable agriculture, *Sci. Am.*, **1995**, *273*, Sept., 148.
103. H.-X. Zhang and E. Blumwald, Transgenic salt-tolerant tomato plants accumulate salt in foliage but not in fruit, *Nature Biotechnol.*, **2001**, 19, 765.
104. A. Coghlan, Supercrop thrives on saline soil, *New Sci.*, **2002**, 28 Sept., 17.
105. a) E. M. Meyerowitz, Today we have the naming of parts, *Nature*, **1999**, *402*, 731. b) X. Lin *et al.*, Sequence and analysis of chromosome 2 of the plant *Arabidopsis thaliana*, *ibid.* 761. c) K. Mayer *et al.*, Sequence and analysis of chromosome 4 of the plant *Arabidopsis thaliana*, *ibid.*, 769. d) The Arabidopsis genome initiative, Analysis of the genome sequence of the flowering plant *Arabidopsis thaliana*, *ibid.*, 796. e) A Theologis *et al.*, Sequence and analysis of chromosome 1 of the plant *Arabidopsis thaliana*, *Nature*, **2000**, 408, 816. f) M. Salanoubat *et al.*, Sequence and analysis of chromosome 3 of the plant *Arabidopsis thaliana*, *ibid.*, 820. g) S. Tabata *et al.*, Sequence and analysis of chromosome 5 of the plant *Arabidopsis thaliana*, *ibid.*, 823.
106. a) C. S. Gasser and R. T. Fraley, Transgenic crops, *Sci. Am.*, **1992**, *266*, June, 34. b) D. Concar, Brave new rose, *New Sci.*, **1998**, 31 Oct., 30. c) S. H. Strauss, Genomics, genetic engineering, and domestication of crops, *Science*, **2003**, *300*, 61.
107. a) J. Rifkin, Apocalypse when? *New Sci.*, **1998**, 31 Oct., 34. b) N. Williams, Agricultural biotech faces backlash in Europe, *Science*, **1998**, *281*, 768. c) G. Gaskell, *et al.*, Worlds apart? The reception of genetically modified foods in Europe and the US, *Science*, **1999**, *285*, 384. d) M. Enserink, GM crops in the cross hairs, *Science*, **1999**, *286*, 1662. e) M. Heylin, Agbiotech's promise clouded by consumer fear, *Chem. Eng. News*, **1999**, 77, 6 Dec., 73. f) T. Gura, The battlefields of Britain, *Nature*, **2001**, *412*, 760. g) K. Brown, Seeds of concern, *Sci. Am.*, **2001**, *284*, April, 40.
108. a) D. Mack, Food for all, *New Sci.*, **1998**, 31 Oct., 50. b) C. Macilwain, Access issues may determine whether agri-biotech will help the world's poor, *Nature*, **1999**, *402*, 341. c) Editorial, Collaborations essential for food in the developing world, *ibid.*, 829.
109. R. Paarlberg, Genetically modified crops in developing countries. Promise or peril, *Environment*, **2000**, *42*, Jan./Feb., 19.
110. P.C. Ronald, Making rice disease resistant, *Sci. Am.*, **1997**, *277*, Nov., 68.
111. C. Mann, Saving sorghum by foiling the wicked witchweed, *Science*, **1997**, *277*, 1040.
112. a) M. Barinaga, Making plants aluminum tolerant, *Science*, **1997**, *276*, 1497. b) M. de la Fuente *et al.*, Aluminum tolerance in transgenic plants by alteration of citrate synthesis, *ibid.*, 1566.
113. L. Herrera-Estrella, Transgenic crops for tropical regions: some considerations about their development and transfer to the small farmer, *Proc. Natl. Acad. Sci. USA*, **1999**, *96*, 5978.
114. a) G. M. Kishore and C. Shewmaker, Biotechnology: enhancing human nutrition in developing and developed worlds, *Proc. Natl. Acad. Sci. USA*, **1999**, *96*, 5968. b) A. S. Moffat, Crop engineering goes south, *Science*, **1999**, *285*, 370.
115. J. Rissler and M. Mellon, *The Ecological Risks of Engineered Crops*, MIT Press, Cambridge, USA, 1996.
116. C. James, International Service for the Aquisition of AgriBiotech Application Briefs 5 (1997), 8 (1998), 17 (1999), 21 (2000), 24 (2001), 27 (2002), 30 (2003), 32 (2004). http://www.isaaa.org.
117. A. M. Thayer, Agbiotech, *Chem. Eng. News*, **1999**, *77*, Oct., 21. c) L. Fletcher, GM crops are no panacea for poverty, *Nature Biotechnol.*, **2001**, *19*, 797.
118. K. Kleiner, Clash of the titans. Biotech's giants go to war over engineered crops, *New Sci.*, 12 Sept., **1998**.
119. a) J. Heritage, Will GM rapeseed cut the mustard? *Science*, **2003**, *302*, *401*. b) M. J. Wilkinson *et al.*, Hybridization between *Brassica napus* and *B. rapa* on a national scale in the United Kingdom, *ibid.*, 457.
120. a) The farm scale evaluations of spring-sown genetically modified crops, papers in a theme issue of *Phil. Trans. Roy. Soc. B*, **2003**, *358*, 1775-1899. b) J. Giles, Biosafety trials darken outlook for transgenic crops in Europe, *Nature*, **2003**, *425*, 751.

121. a) B. Holmes, resistance is useless, *New Sci.*, **1998**, 31 Oct., 48. b) M. Brookes and A. Coghlan, Live and let live, *New Sci.*, **1998**, 31 Oct., 46. c) R. Renner, Will *Bt*-based pest resistance management plans work? *Environ. Sci. Technol.*, **1999**, *33*, 410 A. d) F. Huang *et al.*, Inheritance of resistance to *Bacillus thuringiensis* toxin (Dipel ES) in the European corn borer, *Science*, **1999**, *284*, 965. e) F. Gould, Testing *Bt* refuge strategies in the field, *Nature Biotechnol.*, **2000**, *18*, 266. f) E. Stokstad, First light on genetic roots of Bt resistance, *Science*, **2001**, *293*, 778. g) B. E. Tabashnik, Breaking the code of resistance, *Nature Biotechnol.*, **2001**, *19*, 922. h) I. Denholm, G. J. Devine and M. S. Williamson, Insect resistance on the move, *Science*, **2002**, *297*, 2222. i) J. Randerson, Resistance to pesticides goes global in a flash, *New Sci.*, **2002**, 5 Oct., 15.

122. a) J. Kaiser, Pests overwhelm Bt cotton crop, *Science*, **1996**, *273*, 423. b) K. Kleiner, Monsanto's cotton gets the Mississippi blues, *New Sci.*, **1997**, 1 Nov., 4.

123. a) L. J. Gahan, F. Gould, and D. G. Heckel, Identification of a gene associated with Bt resistance in *Heliothis virescens*, *Science*, **2001**, *293*, 857. b) R. Renner, Bt breakthrough reveals good and bad news for pest resistance, *Environ. Sci. Technol.*, **2001**, *35*, 295 A.

124. a) J. E. Losey, L. S. Raynor and M. E. Carter, Transgenic pollen harms monarch larvae, *Nature*, **1999**, *399*, 214. b) J. Hodgson, Critics slam new Monarch *Bt*-corn data criticized, *Nature Biotechnol.*, **2000**, *18*, 1030. c) B. Hileman, *Bt* threat to monarch caterpillars affirmed, *Chem. Eng. News*, **2000**, *78*, 28 Aug., 7. d) B. Hileman, Engineered corn poses small risk, *Chem. Eng. News*, **2001**, *79*, 17 Sept., 11. e) A. R, Zangerl *et al.*, Effects of exposure to event 176 *Bacillus thuringiensis* corn pollen on monarch and black swallowtail caterpillars under field conditions, *Proc. Natl. Acad. Sci. USA*, **2001** *98*, 11908. f) K. S. Oberhauser *et al.*, Temporal and spatial overlap between monarch larvae and corn pollen, *ibid.*, 11913. g) J. M. Pleasants *et al.*, Corn pollen deposition on milkweeds in and near cornfields, *ibid.*, 11919. h) R. L. Hellmich *et al.*, Monarch larvae sensitivity to *Bacillus thuringiensis*-purified proteins and pollen, *ibid.*, 11925. i) D. E. Stanley-Horn *et al.*, Assessing the impact of Cry1Ab-expressing corn pollen on monarch butterfly larvae in field studies, *ibid.*, 11931. j) M. K. Sears *et al.*, Impact of *Bt* corn pollen on monarch butterfly populations: a risk assessment, *ibid.*, 11937.

125. T. Clarke, Pest resistance feared as farmers flout rules, *Nature*, **2003**, *424*, 116.

126. D. Ferber, New corn plant draws fire from GM food opponents, *Science*, **2000**, *287*, 1390.

127. Editorial, What hopes for GM food? *Nature*, **2000**, *407*, 431,

128. a) B. Hileman, Bt Corn strain found in wrong place, *Chem. Eng. News*, **2000**, *78*, 25 Sept., 13. b) B. Hileman, Aventis suspends Bt corn seed sales, *Chem. Eng. News*, **2000**, *78*, 2 Oct., 13. c) B. Hileman, Transgenic corn contamination saga continues, *Chem. Eng. News*, **2000**, *78*, 9 Oct., 10. d) E. Dorey, Taco dispute underscores need for standardized tests, *Nature Biotechnol.*, **2000**, *18*, 1136. e) J. Netting, Aventis gets short shrift over release of modified corn, *Nature*, **2000**, *407*, 395. f) K. Kleiner, Unfit for humans, *New Sci.*, **2000**, 2 Dec., 11. g) A. Thayer, Farmers sue Aventis over StarLink corn, *Chem. Eng. News*, **2000**, *78*, 11 Dec., 10. h) B. Hileman, Transgenic corn contamination saga continues, *Chem. Eng. News*, **2000**, *78*, 11 Dec., 10. i) M. Heylin, Biotechnology steps in it again down on the farm, *Chem. Eng. News*, **2000**, *78*, 11 Dec., 10. j) J. Kaiser, Panel urges further study of biotech corn, *Science*, **2000**, *290*, 1867. k) A. Thayer, Starlink corn derails ag chain, *Chem. Eng. News*, **2001**, *79*, 22 Jan., 23

129. a) D. Quist and I. H. Chapela, Transgenic DNA introgressed into traditional maize landraces in Oaxaca, Mexico, *Nature*, **2001**, *414*, 541. b) J. Pelley, Engineered genes contaminate corn's birthplace, *Environ. Sci. Technol.*, **2001**, *35*, 472 A. c) C. C. Mann, Transgene data deemed unconvincing, *Science*, **2002**, *296*, 236. d) J. Pelley, Complexity behind biotech corn not addressed, *Environ. Sci. Technol.*, **2002**, *36*, 227 e) A. F. Pearce, The great Mexican maize scandal, *New Sci.*, **2002**, 15 June, 14. f) *Anon.*, Is it worth worrying about? *Ibid.*, 17. g) J. Kaiser, Calming fears, no foreign genes found in Mexico's maize, *Science*, **2005**, *309*, 1000.

130. a) D. MacKenzie, Stray genes highlight superweed danger, *New Sci.*, **2000**, 21 Oct., 6. b) E. Stokstad, A little pollen goes a long way, *Science*, **2002**, *296*, 2314. c) M. A. Rieger *et al.*,

Pollen-mediated movement of herbicide resistance between commercial canola fields, *ibid.*, 2386.
131. a) M. Brookes, Running wild, *New Sci.*, **1998**, 31 Oct., 38. b) J. P. R. Martínez-Soriano and D. S. Leal-Klevezas, Transgenic maize in Mexico: no need for concern, *Science*, **2000**, *287*, 1399. c) R. Nigh *et al.*, Transgenic crops: a cautionary tale, *ibid.*, 1927. d) S. Abbo and B. Rubin, Transgenic crops: a cautionary tale, *ibid.*, 1927.
132. a) K. Kleiner, Let us spray. Weed killer-resistant tobacco plant keeps altered genes to itself. *New Sci.*, **1998**, 4 April, 16. b) A. J. Gray and A. F. Reynold, Reducing transgene escape routes, *Nature*, **1998**, *392*, 653. c) H. Daniell, *et al.*, Containment of herbicide resistance through genetic engineering of the chloroplast genome, *Nature Biotechnol.*, **1998**, *16*, 347.
133. G. Gaskell *et al.*, Biotechnology and the European public, *Nature Biotechnol.*, **2000**, *18*, 935.
134. a) P. Cohen, Strange fruit, *New Sci.*, **1998**, 31 Oct., 42. b) A. Coghlan, How safe is safe? *New Sci.*, **1999**, 16 Oct., 7.
135. J. A. Nordlee *et al.*, Identification of a Brazil-nut allergen in transgenic soybeans, *New Engl. J. Med.*, **1996**, *344*, 688.
136. a) P. Wymer, Genetically modified food: ambrosia or anathema? *Chem. Indust.*, **1998**, 1 June, 422. b) Report of the OECD Workshop on the Toxicological and Nutritional Testing of Novel Foods. SG/ICGB(98)1, Sept., 1998, OECD, Paris, France. c) D. Butler *et al.*, Long-term effect of GM crops serves up food for thought, *Nature*, **1999**, *398*, 651. d) E. Millstone, E. Brunner and S. Mayer, Beyond 'substantial equivalence', *Nature*, **1999**, *401*, 525. e) A. Trewavas and C. J. Leavert, Conventional crops are the test of GM prejudice, *Nature*, **1999**, *401*, 640. f) P. Kearns and P. Mayers, Substantial equivalence is a substantial tool, *ibid.*, 640. g) H. I. Miller, Substantial equivalence: its uses and abuses, *Nature Biotechnol.*, **1999**, *17*, 1042. h) Y. Uozum, Japan to bring in mandatory tests for GM foods, *Nature* **1999**, *402*, 846.
137. a) C. Macilwain, US food safety body hears protests over genetically modified food, *Nature*, **1999**, *402*, 571. b) C. Robinson, European GM foods labelling legislation: yet another straw on the camel's back? *Trends Food Sci. Technol.*, **1999**, *10*, 375. c) H. L. Miller, A rational approach to labeling biotech-derived foods, *Science*, **1999**, *284*, 1999. d) D. Dickson, GM debate must go global, says meeting ... and calls for openness and transparency, *Nature*, **2000**, *404*, 112. e) A. Coghlan, Judging gene foods, *New Sci.*, **2000**, 15 April, 4. f) K. S. Betts, GMO testing hurdles, *Environ. Sci. Technol.*, **2000**, *34*, 472 A. g) A. J. Bouchie, Australia/NZ label GM foods, *Nature Biotechnol.*, **2000**, *18*, 911. h) A. G. Haslberger, Monitoring and labeling for genetically modified products, *Science*, **2000**, *284*, 1471. i) A. McHughen, Uninformation and the choice paradox, *Nature Biotechnol.*, **2000**, *18*, 1018. j) J. Knight, Trade war looms as US launches challenge over transgenic crops, *Nature*, **2003**, *423*, 369. k) B. Hileman, Clashes over agbitech, *Chem. Eng. News*, **2003**, 9 *81*, June, 25. l) R. Winder, Bush pushes pro GMO stance, *Chem. Indust.*, **2003**, 7 July, 9. m) M. Louise, The widening GM divide, *ibid.*, 21 July, 9. n) P. Aldous, Time to choose, *Nature* **2003**, *425*, 655, 658,
138. a) C. F. Amábile-Cuevas and M. E. Chicurel, Horizontal gene transfer, *Am. Sci.*, **1993**, *81*, 332. b) S. Mayer, Let's keep the genie in its bottle, *New Sci.*, **1996**, 30, Nov., 51. c) M. Syvanen, In search of horizontal gene transfer, *Nature Biotechnol.*, **1999**, *17*, 833. d) K. Kleiner, Blowing in the wind, *New Sci.*, **1999**, 14, Aug., 18.
139. A. Coghlan, So far so good, *New Sci.*, **2000**, 25 March, 4.
140. C. Mlot, Antidotes for antibiotic use on the farm, *BioScience*, **2000**, *50*, Nov., 955.
141. a) A. Coghlan, Playing safe. *New Sci.*, **1999**, 4 Sept., 22. b) B. Hileman, Europe leaves biotech door ajar, *Chem. Eng. News*, **2001**, *79*, 19 Feb., 16.
142. a) A. Coghlan, Named and shamed. Genetic engineers who flouted the rules are brought to book, *New Sci.*, **1998**, 4. April, 4. b) H. Wallace and J. Beringer, Natural Justice? *New Sci.*, 14 Aug., **1999**. c) A. Coghlan, Playing safe. Crops modified using a new method can't spread resistance, *New Sci.*, **1999**, 4 Sept., 22. d) A. Coghlan, Coming a cropper, *New Sci.*, **1999**, 25 Sept., 5. e) A. Abbott, Swiss reject GM trial to protect organics, *Nature*, **1999**,

398, 736. f) M. Walker, Hold the radicchio. Bioengineered salads are off the menu in Europe, *New Sci.*, **1999**, 30 Oct., 12. g) A. Abbot and D. Dickson, Germany holds up cultivation of GM maize ... amid calls for international biotech panel, *Nature*, **2000**, *403*, 821. h) J. Hodgson, National politicians block GM progress, *Nature Biotechnol.*, **2000**, *18*, 918. i) A. Medola, Italian GMO ban could spread, *ibid.*, 1137.

143. a) S. Lehrman, GM backlash leaves US farmers wondering how to sell their crops, *Nature*, **1999**, *401*, 107. b) K. Kleiner, Farmers in the firing line, *New Sci.*, **1999**, 25 Sept., 18. c) M. Wadham, US processor rejects maize that EU won't take, *Nature*, **1999**, *398*, 736.

144. a) A. Gupta, Governing trade in genetically modified organisms. The Cartagena Protocol on Biosafety, *Environment*, **2000**, *42*, May, 22. b) L. Helmuth, Both sides claim victory in trade pact, *Science*, **2000**, *287*, 782. c) C. Macilwain, Rules agreed over GM food exports, *Nature*, **2000**, *403*, 473. d) B. Jank and H. Gaugitsch, Decision making under the Cartagena Protocol on Biosafety, *Trends Biotechnol.*, **2001**, *19*, 194.

145. a) I. Serageldin, Biotechnology and food security in the 21st century, *Science*, **1999**, *285*, 387. b) R. Gilmore, Agbiotech and world food security - threat or boon? *Nature Biotechnol.*, **2000**, *18*, 361. c) K. S. Fischer *et al.*, Collaborations in rice, *Science*, **2000**, *290*, 279. d) F. Pledge, Patenting, piracy and the global commons, *BioScience*, **2001**, *51*, April, 273. e) I. Serageldin, World poverty and hunger – the challenge for science, *Science*, **2002**, *296*, 54.

146. a) R. F. Service, Seed-sterilizing 'terminator technology' sows discord, *Science*, **1998**, *282*, 850. b) C. Macilwain and S. Lehrman, Developing countries look for guidance in the GM crops debate.....as Rockefeller head warns of backlash, *Nature*, **1999**, *401*, 831. c) J. Fox, USDA's terminator talks, *Nature Biotechnol.*, **2000**, *18*, 911.

147. a) *Sustaining Food Security Beyond the Year 2000: A Global Partnership for Rice Research, Rolling Medium Term Plan, 1999-2001*, International Rice Research Institute, Manila, The Philippines, 1998. b) K. Cassman, Ecological intensification of cereal production systems: yield potential, soil quality, and precision agriculture, *Proc. Natl. Acad. Sci. USA*, **1999**, *96*, 5952.

148. a) L. R. Brown, *Who Will Feed China? Wake-Up Call for a Small Planet*, Norton, New York, USA, 1995. b) V. Smil, Who will feed China? *The China Quarterly*, **1995**, *143*, 801. c) R. L. Paarlberg, Rice bowls and dust bowls: Africa, not China, faces a food crisis, *Foreign Affairs*, **1996**, *75*, 127. d) R. L. Prosterman, T. Hanstad and L. Ping, Can China feed itself? *Sci. Am.*, **1996**, *274*, Nov., 70. e) G. K. Heilig, Anthropogenic factors affecting land use change in China, *Popn. Dev. Rev.*, **1997**, *23*, 139.

149. International Water Management Institute, http://www.cgiar.org/iwmi/home/wsmap.htm

150. a) P. Crosson and N. J. Rosenberg, Strategies for agriculture, *Sci. Am.*, **1989**, *261*, Sept., 78. b) J. P. Reganold, R. I. Papendick and J. F. Parr, Sustainable Agriculture, *Sci. Am.*, **1990**, *262*, June, 72. c) M. J. Jones, Sustainable agriculture: an explanation of a concept, *Ciba Foundation Symposium 177*, Wiley, Chichester, UK, 1993, p. 30. d) F. Bray, Agriculture for developing nations, *Sci. Am.*, **1994**, *271*, July, 18. e) S. D. Tanksley and S. R. McCouch, Seed banks and molecular maps: unlocking genetic potential from the wild, *Science*, **1997**, *277*, 1063. f) P. E. Rasmussen *et al.*, Long-term agrosystem experiments: assessing agricultural sustainability and global change, *Science*, **1998**, *282*, 893. g) D. E. Ervin *et al.*, A new strategic vision, *Environment*, **1998**, *40*, July/Aug., 8. h) D. Hoisington *et al.*, Plant genetic resources: what can they contribute toward increased crop productivity? *ibid.*, 5937. i) V. W. Ruttan, The transition to agricultural sustainability, *Proc. Natl. Acad. Sci. USA*, **1999**, *96*, 5960. j) I. Serageldin and G. J. Persley, Promethean Science. *Agricultural Biotechnology, the Environment, and the Poor*, Consultative Group on International Agricultural Research, Washington, DC, USA, 2000. http://cgiar.org. k) *Transgenic plants and world agriculture*, The Royal Society, London, UK, 2000.

151. a) C. Rosenzweig and M. L. Parry, Potential impact of climate change on world food supply, *Nature*, **1994**, *367*, 133. b) M. Hulme *et al.*, Relative impacts of human-induced climate change and natural climate variability, *Nature*, **1999**, *397*, 688. c) R. E. Evenson,

Global and local implications of biotechnology and climate change for future food supplies, *Proc. Natl. Acad. Sci. USA*, **1999**, *96*, 5921.

Chapter 5

A New Biology

"We shall not cease from exploration
And the end of all our exploring
Will be to arrive where we started
And know the place for the first time."

T. S. Eliot, Little Gidding, 1942.

Quoted by the International Genome Consortium on the publication
of their first draft of the human genome in 2001.

5.1. A New Understanding: Genetics

Humans have never experienced such a rapid growth in the understanding of their own biology and that of other living things as at the junction of the twentieth and twenty-first centuries. Its genesis can be traced back to the nineteenth century and an unsettled time in Europe, particularly in Vienna, capital of the Austro-Hungarian Empire. There, amidst riots in 1848, Chancellor Klemens von Metternich resigned and went into exile and Francis I abdicated in favour of his son, Franz Joseph, destined to be the last Austrian emperor. In the midst of this turmoil a young monk, Gregor Mendel, arrived at the University of Vienna to train as a teacher [1,2]. However, after several years he was deemed unsuited to teaching and returned to his Augustinian monastery at Brno in the now Czech Republic in 1853. Some time later he began the experiments in the crossbreeding of garden peas to produce hybrids that were destined to revolutionize the understanding of biology. He found that, contrary to the then general expectation, the hybrid peas did not show averaged height and colour and other characteristics of the peas from which they were crossbred. Instead, there was a dominance of characteristics that derived from one or other of the crossbred peas. Subsequently, self-fertilization of the hybrids produced a second generation of peas in which the characteristics of one of the original crossbred peas dominated that of the other by a factor of three to one.

From these observations Mendel deduced that for each of its inherited characteristics a hybrid pea carries two factors, now called genes, one of which is dominant while the other is recessive. During fertilization, each of the pollen and ovule, or egg, of the hybrid peas contribute one gene for each characteristic so that the possible combinations of genes in the new plants are fourfold:

dominant gene-dominant gene

dominant gene-recessive gene

recessive gene-dominant gene

recessive gene-recessive gene

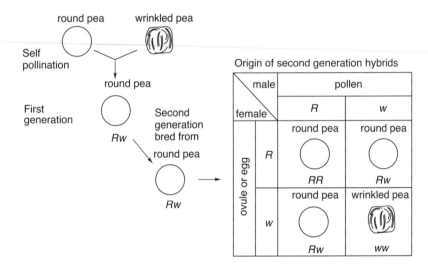

Fig. 5.1. Mendel's experiment in crossbreeding round and wrinkled garden peas where *R* is the dominant gene for roundness and *w* is the recessive gene for wrinkledness.

Thus, the dominant gene controls the characteristics of the peas from the first three combinations while the recessive gene controls that of the fourth. The effect of this is shown in Fig. 5.1 for the crossbreeding of peas that produce round seed with those that produce wrinkled seed where the gene for roundness, *R*, is dominant and that for wrinkledness, *w*, is recessive. When the round peas and the wrinkled peas cross-pollinate the first generation from this crossbreeding contains a gene from each parent, *R* and *w*. This first generation always produces round peas because the gene that determines roundness, *R*, is dominant. In the second generation both the pollen and the ovule carry the *R* and *w* genes.

Fertilization of the ovule by the pollen results in one quarter of the pea plants produced having two *R* genes and round peas, half having *R* and *w* genes and round peas and one quarter having two *w* genes and wrinkled peas. Mendel published his work in the Journal of the Brno Natural History Society in 1866 and it is now widely recognized as the beginning of the science of genetics although it attracted little attention at the time and languished for some thirty years before rediscovery.

5.2. Genetics at the Cellular Level: Prokaryotes and Viruses

A deeper understanding of genetics is gained from a brief look at the biological cell of which as few as one and as many as trillions and thousands of trillions compose living organisms from the smallest to the largest. While conventionally many living organisms are classified as animals, bacteria, fungi, plants and so on, biologists also use a more fundamental classification, prokaryotes and eukaryotes, according to the type of cell of which the organism is composed [3,4]. Bacteria are prokaryotes and were also the first form of cellular life to appear on Earth some 3.5 billion years ago. They consist of a single cell in which a strand of deoxyribonucleic acid, DNA, whose ends are usually linked together to form a circle. This DNA contains the bacterial chromosome, segments of which are the genes that control most aspects of bacterial life. Bacteria also contain shorter pieces of cyclic DNA containing genes, called plasmids, that are able to incorporate DNA segments from other bacteria and elsewhere [5]. This is the route through which bacteria incorporate genes that sometimes make them antibiotic resistant. It is also an ability that is often exploited in genetic engineering as is discussed in sections 5.13 and 5.14. Finally, bacteria also contain numerous smaller molecular assemblies, called ribosomes that are important participants in producing the proteins that control cell chemistry. The chromosome, plasmids and ribosomes are suspended in the watery cytoplasm that fills the bacterium within a containing membrane and cell wall as shown in simplified form in Fig. 5.2 where the elegant double helix of DNA is also shown in simplified form [6,7]. It transpires that two of the three domains of life possess the prokaryote cell [8,9]. These are the Bacteria and the monocellular Archaea that also do not have nuclei but, in addition to genes unique to themselves, contain genes also found in the Eukarya. The Eukarya are the third domain of life based on the eukaryote cells, also shown in Fig. 5.2, and from which animals and plants emerged as distinct entities hundreds of million years ago.

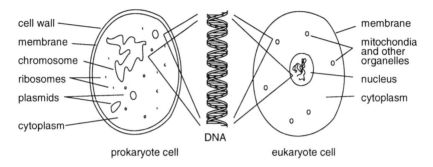

cell wall

membrane

chromosome

ribosomes

plasmids

cytoplasm

DNA

prokaryote cell

membrane

mitochondia and other organelles

nucleus

cytoplasm

eukaryote cell

Fig. 5.2. Prokaryote cells, typified by bacteria, are both different in structure and much smaller than the eukaryote cells of animals and plants. The cytoskeleton, a stiffening internal structure composed of proteins, of the eukaryote cell is not shown. The DNA in the chromosome, plasmid, nucleus and mitochondrion possesses a double helical structure as shown at the centre of the figure.

It is convenient at this point to introduce viruses that are not usually considered to be living organisms but are another DNA containing part of the biosphere. Viruses are usually much smaller than bacteria and are little more than DNA segments encased in a protective protein sheath. There are a great variety of viruses and all are dependent on cells of other organisms to reproduce themselves and are able to lie dormant for long periods [3,10,11]. They represent the boundary between the animate and inanimate. Nevertheless, they were probably in existence when the earliest forms of life appeared on Earth. They are now prolific as shown by estimates that viruses in the oceans contain 270 million tonnes of carbon, more than twenty times the amount of carbon in all of the whales [12].

One particularly well understood type of virus is the bacteriophage shown in Fig. 5.3a [11]. Its DNA is contained in a large head envelope composed of proteins. It reproduces by injecting its DNA into a bacterium through its protein sheath after attaching itself to the bacterial cell wall using its tail fibres. A viral enzyme destroys the bacterial DNA while the virus reproduces its own DNA so that many new copies of the virus form until the bacterium bursts, releasing them to repeat their reproductive cycle. An alternative mode of viral reproduction occurs when the viral DNA is incorporated into the bacterial DNA and reproduced along with it. Subsequently, the viral DNA cuts itself out of the bacterial DNA and takes over the bacterium to reproduce many copies of the virus until they are released when the bacterium bursts.

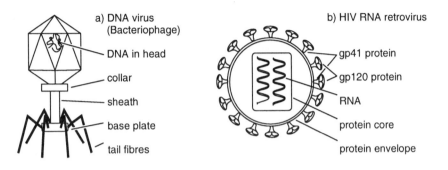

a) DNA virus (Bacteriophage)
DNA in head
collar
sheath
base plate
tail fibres

b) HIV RNA retrovirus
gp41 protein
gp120 protein
RNA
protein core
protein envelope

Fig. 5.3. a) A type of DNA virus called a bacteriophage that attacks bacteria. b) The HIV retrovirus that contains RNA within a protein core and envelope.

Instead of DNA, some viruses contain ribonucleic acid, or RNA, that resembles a single strand of the DNA double helix as is discussed in section 5.6. Such viruses are of two basic types: RNA viruses that invade cells to reproduce their RNA and protein coats, and retroviruses that use their RNA to generate DNA that reproduces in the invaded cell before reverting to viral RNA and forming more viruses. Probably the most feared virus is a retrovirus, the human immune deficiency virus, or HIV, shown in Fig. 5.3b [13]. The HIV attaches mainly to a lymphocyte, or T-cell, surface through its gp41 and gp120 proteins and injects its RNA into the T-cell together with an enzyme called reverse transcriptase. This enzyme allows the HIV RNA to produce a new viral DNA segment that inserts itself into the T-cell's DNA where it is reproduced over and over. Eventually, the viral DNA regenerates HIV RNA that produces many more HIVs that break out of the T-cell to invade further T-cells and eventually overwhelm the immune system of the victim who then becomes very vulnerable to diseases that a healthy immune system would usually overcome. (The HIV and autoimmune deficiency syndrome, AIDS, are discussed in more detail in Chapter 6.)

5.3. Genetics at the Cellular Level: Eukaryotes

The eukaryotes cells, that first appeared about 1.5 billion years ago, and of which animals and plants are constructed, are much larger than prokaryote cells [14]. In place of a single chromosome immersed in the cytoplasm, as in the prokaryote cells, eukaryote cells usually contain several chromosomes that are more complex than those of the prokaryotes in a membrane encapsulated nucleus. Mammalian eukaryote cells lack the stiffening cell wall of the prokaryote cell and rely on a flexible internal cytoskeleton made up of proteins for much of their rigidity. Dispersed throughout the cytoplasm are small structures called mitochondria that enclose circular DNA in a membrane as shown in Fig. 5.2. The mitochondria contain only a fraction of a percent of the amount of DNA contained in the nucleus and act as the cell's chemical power sources [15]. The mitochondrial DNA, mtDNA, also contains genes that for humans number thirteen. It appears that about two billion years ago some cells absorbed smaller ones that eventually became the mitochondria of eukaryote cells.

The number of chromosomes in the eukaryote nucleus varies greatly among animals and plants. Probably the most widely known are the forty-six human chromosomes contained in human DNA as twenty-three pairs. One half of each chromosome is inherited from the father and the other from the mother. For females, both chromosomes of each pair are identical in all twenty-three pairs with the twenty-third pair consisting of two of the sex-determining X chromosomes [16]. For males, twenty-two of the chromosomes contain identical pairs but the twenty-third pair comprises the different sex-determining X and Y chromosomes. Thus, a profound effect arises from a chromosomal difference that has provided humanity with an enduring passion, the attraction between the sexes and the procreation of children as is entertainingly discussed in Richard Dawkins' book "The Selfish Gene" [17].

Chromosomes contain many genes, each controlling a function of a human cell depending on the diversely specialized nature of the cells such as those found in brain, muscle, and skin. Ideally, the genes and chromosomes carry out their functions perfectly, but a defect or mutation in a gene or chromosome can result in genetically based diseases and disabilities. The occurrence of varying degrees of colour blindness in eight percent of males is due to a mutation in the single X chromosome of the male. For a female to suffer the same colour blindness both of her X chromosomes must carry the same mutation and so the incidence of this disability in females is eight percent of that in males, or 0.64 percent. While colour blindness is inconvenient, it is not life threatening as is haemophilia that predominantly afflicts males and is passed on through females carrying the chromosomal fault to approximately half of their sons, but rarely to their daughters as this would require both X chromosomes to carry the same defect. Thus, the two X chromosomes offer females a buffer against some genetic diseases whereas males do not have this protection. However, both sexes are affected by a wide range of genetic diseases, many of which are life threatening. As a consequence, the rapidly growing knowledge of genetics holds the promise of either prevention, amelioration or cure of presently intractable genetic diseases such as acute lymphatic leukaemia, cystic fibrosis and Alzheimer's disease that are associated with defects in chromosomes four, seven, and twenty-one, respectively. By early 2001, at least one disease related mutation had been discovered in 1,100 genes and more are discovered at frequent intervals [18]. Chromosome pairs are duplicated during cell division, or mitosis, to produce two new cells of the same type. Each human cell experiences mitosis fifty to ninety times, and in this way any faults in the chromosomes are passed to succeeding generations of cells. A fault in a chromosome may also be passed from parent to child.

5.4. A Molecular View of Genetics: Deoxyribonucleic Acid (DNA)

Although Walter Fleming discovered chromosomes as thread-like entities in cell nuclei in 1882, the critical breakthrough leading to an understanding of their genetic significance and that of DNA did not occur until 1953. In that year, James Watson and Francis Crick published two papers in the journal Nature in which they described the double helical structure of DNA and the very specific pairing of nucleotide bases within it [19]. In one of the greatest scientific understatements to be made, they wrote:

"It has not escaped our notice that the specific pairings we have postulated immediately suggest a possible copying mechanism for the genetic material."

They had brought the molecular basis for an understanding of genetics within reach for the first time and, together with Maurice Wilkins, were awarded the Nobel Prize for Medicine in 1962. James Watson gave a fascinating account of their discovery of the structure of DNA in his book "The Double Helix" [20].

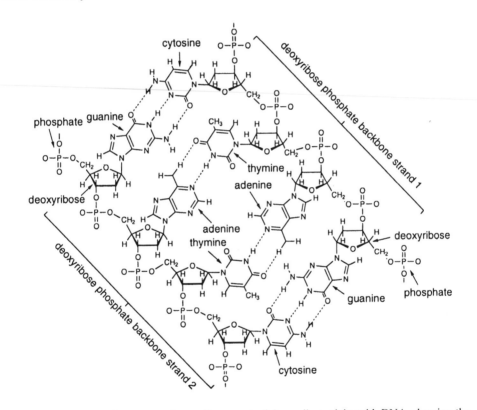

Fig. 5.4. The chemical structure of a small segment of deoxyribonucleic acid, DNA, showing the linking of the two deoxyribose phosphate strands by either the adenine and thymine or cytosine and guanine nucleotide base pairs. Hydrogen (H), oxygen (O), nitrogen (N) and phosphorus (P) atoms are represented by letters and a carbon atom exists where the solid lines representing chemical bonds intersect and no letter is shown. The broken lines between the paired nucleotide bases represent weak hydrogen bonds that break to allow the two strands to separate when either replication or ribonucleic acid, RNA, production occurs.

The giant DNA molecule consists of two parallel strands of negatively charged deoxyribose phosphate strands cross-linked through pairs of nucleotide bases, adenine and thymine or cytosine and guanine [7,21]. Each nucleotide base pair links the two strands like the steps in a spiral staircase as seen in Figs. 5.4 and 5.5. In an analogy to the twenty-six letter alphabet, the four nucleotide bases, adenine, thymine, guanine and cytosine, indicated by the letters, A, T, G and C in Fig. 5.5, may be thought of as the four letters of the DNA alphabet. Three

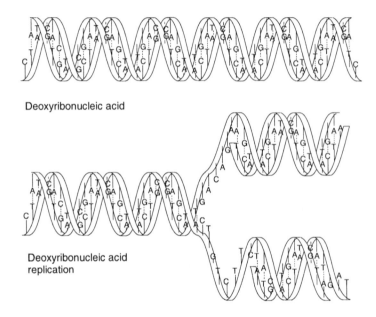

Deoxyribonucleic acid

Deoxyribonucleic acid
replication

Fig. 5.5. The double stranded deoxyribonucleic acid helical structure showing the two deoxyribose phosphate strands linked by the nucleotide base pairs where adenine, thymine, cytosine and guanine are represented by the letters A, T, C, and G, respectively. The lower part of the figure shows a deoxyribonucleic acid double helix replicating itself as two identical new double helices.

adjacent nucleotide bases attached to a single deoxyribosephosphate DNA strand are called a codon that may be likened to a word in the DNA language. Because there are four possible choices for each of the first, second, and third nucleotide bases in each codon, there are sixty-four possible codons, or sixty-four words, in the DNA language. Groups of linked codons make up genes that are analogous to sentences. Thus, the genes in the chromosome are equivalent to a book and each chromosome represents a book in the DNA library that is the genetic code, or genome, that contains the building and operating instructions for each living species. The number of chromosomes and genes varies greatly among species and some genes are common to a wide range of species. (The name "genome" is derived from the truncation of *"gene"* and "chromos*ome*".)

Given its complexity and large size, it is not surprising to find that DNA can suffer significant damage from interacting with chemicals and ultraviolet radiation, in particular. Such damage disrupts the DNA coding that, in turn, has a potential to disrupt the normal functioning of the cell containing the damaged DNA. Fortunately, DNA has efficient mechanisms for either tolerating or repairing this damage so that more often than not DNA damage is not critical to the cell [22]. However, if the damage is extensive, or the repair is ineffective, the DNA may malfunction with the consequence that the cell may either die, which most often is the case, or may function abnormally to form a viable mutant gene and lead to disease.

DNA fits into the eukaryote cell nucleus by being wound around proteins to make a thread that is then folded and refolded to form a very compact package. The great efficiency of this packaging is shown by a human cell that is only a hundredth of a millimetre in diameter and yet the DNA contained in its nucleus is two metres long. The complete sequence in which the genes are linked in the chromosomes represents the genome of a species and the sequence of the nucleotide bases in their DNA represents the genome of an individual. Nevertheless, the genomes of a human and a cabbage are more remarkable for their similarities than their differences. It is the differences that determine that a cabbage has leaves, carries out photosynthesis and gains its nutrition from the air and soil, while humans have complete freedom of motion and eat both animals and plants. Within the genome, it may be only a few instances where a single nucleotide base is substituted for another in a gene, or single nucleotide polymorphisms, or SNPs, that produce differences between individuals as in humans with blue or brown eyes, or fair or dark skin, and as in cabbages with green or red leaves. Comparisons of human genomes from different individuals show that as many as two million SNPs, or one difference in every 1,250 or so nucleotide bases, occur on average [23].

As animals and plants grow and maintain themselves they initially increase the number of most types of cells and then replace them through cell division when DNA reproduces itself [24]. In a chromosome the two strands of DNA separate, rather like a spiral zip fastener, and each then builds a matching new strand according to the nucleotide base pairing sequence to produce two perfect copies of the original DNA, one each for the two new cells, as shown in Fig. 5.5. This process is called replication. Every billion or so replications, the sequence of nucleotide bases is imperfectly reproduced and this represents a mutation, as does the similarly infrequent mistake in reproducing the sequence in which genes are linked in a chromosome. Such mutant DNA and the cells containing them seldom survive, but those that do are part of the evolutionary process that leads either to new species or genetic disease. The process of cell division, or mitosis, is choreographed to ensure that the two new cells produced contain exactly the same complement of chromosomes. This complex process is controlled by a centromere in each chromosome that at first holds each pair of newly produced chromosomes together and then controls their separation so that one of each type of chromosome is positioned in each half of the dividing cell that finally separates into two new cells with the same complement of chromosomes [25].

5.5. The Molecular Machinery

Despite the growing sophistication of the understanding of the molecular basis of genetics, much remains to be discovered. Only about three percent of human DNA composes the genes that determine most of the differences between and within species. Within each gene are exons, the segments of DNA whose nucleotide base sequences carry the code for the polypeptides and proteins that the gene produces and the other roles it performs. The exons are separated by segments of DNA called introns, as shown in Fig. 5.6. The role of introns is only partially understood but there is mounting evidence that nucleotide base sequences from introns insert themselves into exon sequences as part of the evolutionary process [26,27]. The introns and the longer intergenic DNA sequences were at one time called "junk DNA" because they do not produce proteins and were thought to be largely parts of the genome rendered redundant by evolution. However, while as yet less well understood than the DNA in genes, this DNA is now often referred to as "non coding" and has retained long and

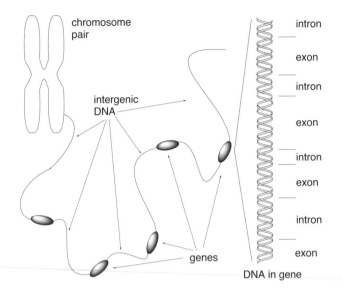

Fig. 5.6. A simplified illustration of the intergene, exon and intron segments of DNA present in a human chromosome and its genes.

frequently repeating sequences of nucleotide bases common to all mammals, including humans, over many millions of years of evolution. While this may to some extent arise from nucleotide base sequences made redundant by evolution, sequences of non coding DNA, or pseudogenes, can produce RNA that does not go on to produce proteins but instead interacts with DNA, possibly to control cell, and thereby embryo, development and differentiation. An example of a pseudogene is *makorin1p1* that is a greatly shortened version of the *makorin1* gene in the genome of mice that is dependent on *makorin1p1* to function [27]. It is thought that human DNA contains about 20,000 pseudogenes.

The number of human genes is 20,000 to 25,000 and the function of most is unknown, but it is thought that their activities in the cell are apportioned approximately as shown in Fig. 5.7 [28]. Thus, some genes produce RNA that controls the production of a plethora of proteins such as haemoglobin that binds oxygen in the blood of humans and chlorophyll that is an integral part of the photosynthetic chain in green plants. Other genes control the specific roles of different cells such as lymphocytes that are part of the immune system and pancreatic islet cells that produce insulin to metabolise sugars. Some genes control the replication of DNA and cell division while others signal when these and other events, including natural cell death, or apoptosis, are to occur. (Apoptosis plays an integral part in the maintenance of healthy tissue [29].) The building of cells into tissues is controlled by another set of genes and yet other genes control the immune system that organizes the body's defence against infection.

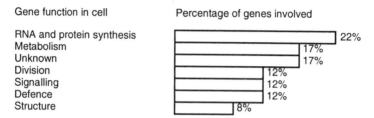

Fig. 5.7. The probable percentage of genes responsible for various functions in a typical human cell.

Each cell of a particular animal carries the same genome in its nucleus, and yet a muscle cell is quite different from a nerve cell as a consequence of different sets of genes activating the cellular chemistry. Similarly, a plant carries the same genome in the nucleus of each of its cells but a root cell is quite different from a leaf cell. It is clear from this that while the genome carries the complete design, or building code, for each species, it also carries within it the ability to activate different sets of genes in particular cells so that each group of cells carries out its unique function, an ability that is presently incompletely understood.

5.6. Ribonucleic Acid (RNA): The Genomic Enabler

While it is the DNA of the genome that largely contains the building and operating instructions of a species, the chemistry that controls the metabolism of cells requires the participation of another nucleic acid, ribonucleic acid, or RNA. Apart from having ribose in its backbone instead of the deoxyribose of DNA, where DNA has a thymine nucleotide base RNA has a uracil nucleotide base that can only pair with adenine and is indicated by U in Fig. 5.8. RNA is produced when a segment of the DNA helix unravels to act as an RNA template in a process called transcription. The DNA sequence of nucleotide bases built into, or transcribed into, the RNA strand is tightly controlled because of the strict limitations placed on nucleotide base pairing. In the presence of an enzyme for making RNA, one strand of the unravelled segment of DNA controls the

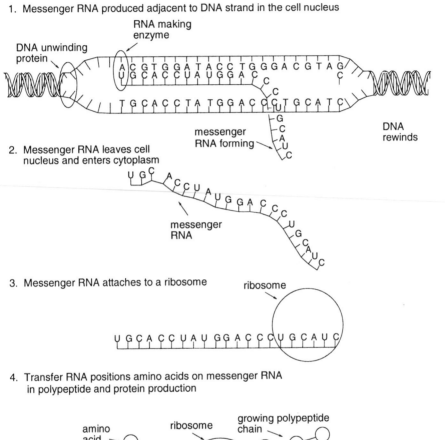

1. Messenger RNA produced adjacent to DNA strand in the cell nucleus

2. Messenger RNA leaves cell nucleus and enters cytoplasm

3. Messenger RNA attaches to a ribosome

4. Transfer RNA positions amino acids on messenger RNA in polypeptide and protein production

Fig. 5.8. A simplified illustration of the sequence of events producing messenger RNA that, in concert with transfer RNA, produces polypeptides and proteins. Unlike prokaryote messenger RNA, that of eukaryotes initially carries the exons and introns characterizing the DNA, but the latter are cut out before the messenger RNA participates in producing polypeptides and proteins.

sequence of addition of the nucleotide bases, derived from amino acids, to produce the growing RNA strand until it is complete [3,30]. It then leaves its DNA template that reforms its double helix as shown in Fig. 5.8.

The newly produced strand of RNA carries a message from the DNA in the sequence of the nucleotide bases attached to its ribose phosphate backbone and is called messenger RNA. The message is in the sequence of the three letter codons inserted into the messenger RNA by the DNA. This represents the sequence in which the amino acids are to be joined in either the polypeptide or protein and the length of the joined sequence to be formed. There are only twenty amino acids found in human polypeptides and proteins while there are sixty-four codons, each of which may contain the code for a particular amino acid. In practice, several codons code for the same amino acid, as is the case for UUA, CUA and CUG that code for leucine while AUG and UGG only code for methionine and tryptophan, respectively. Other codons such as UGA do not code for an amino acid but contain the code to stop increasing the length of the sequence of linked amino acids instead [31]. (Strictly speaking, amino acids only exist as single molecules, but it has become the convention to use their names for both the nucleotide bases formed from them in DNA and RNA, and also for the segments of polypeptides and proteins formed from them. By convention polypeptides contain up to one hundred amino acids linked together and those containing more are called proteins)

In the next step the messenger RNA becomes attached to a ribosome where it awaits the arrival of short segments of another type of RNA, transfer RNA, possessing three nucleotide bases and an additional amino acid [32]. The transfer RNA now pairs its nucleotide bases with those of the messenger RNA according to the allowed pairings, and the ribosome moves along the messenger RNA strand linking the amino acids into either a polypeptide or a protein that then discards the transfer RNA. This process is called translation because it translates the code of nucleotide base sequences originally imposed on the messenger RNA by the DNA of the gene to produce a specific sequence of linked amino acids in the polypeptide or protein.

The polypeptides and proteins produced in this way have specific roles in controlling the chemistry of each type of cell within an organism and it is on this that genetic engineering is based. Thus, by introducing either one or several new genes into a plant, proteins may be produced that in turn produces molecules that alter the colour, vitamin content, drought and pest tolerance and other characteristics of the plant [33,34]. Similarly, genes may be inserted that cause the engineered plant to produce a range of molecules for use in the food, pharmaceutical and plastics industries. Unsurprisingly, the introduction of a new gene into an animal can also alter the characteristics of the animal. This has been shown by the engineering of a growth hormone gene into coho salmon and trout that resulted in much greater growth rates in the genetically modified fish by comparison with those of their native relatives [35]. Sometimes the transferred gene is inactive as was the case at birth with ANDi, a genetically modified rhesus monkey and the first genetically modified primate, whose genome had inserted into it the gene that produces the green fluorescent protein of the glowing jellyfish [36]. ANDi was not fluorescent although the toenails and hair follicles of two aborted monkey foetuses resulting from the same study did glow green. However, Alba, a genetically modified rabbit, showed a green fluorescence as do viruses, bacteria, plants, fruitflies, zebrafish and mice that have had the jellyfish gene engineered into them.

5.7. Genetically Engineered Children

In May 2001, the revelation that the first genetically engineered children had been inadvertently produced as a consequence of an infertility treatment program stirred the unease that much of humanity feels about intervention in human genetics [37]. Some of the children produced through the program differ from all other humans in that they carry DNA from two women and one man. This came about through the treatment of a type of infertility where the composition of the cytoplasm of some eggs renders them infertile. When cytoplasm taken from an egg donated by a fertile woman is injected into the infertile egg, it too can become fertile. However, the transferred cytoplasm contains mitochondria and their genes so that the now fertile egg contains genetic material from both the infertile and the fertile woman. All of the thirty children born from the fertilized eggs treated in this way carried DNA from the father and infertile mother in their cell nuclei as expected. In addition, two of the twelve children examined carried the mitochondria of the infertile and the fertile women that can only be passed on to offspring through the female. The consequences of carrying mitochondria from two women are unknown but the children resulting from this infertility program appeared to be healthy. However, when the foetuses that did not proceed to live birth were examined they were found to have a higher than usual incidence of Turner's Syndrome, a chromosomal fault. Other infertility treatments also increase the incidence of genetic defects. This is a trend that reinforces the concerns about any attempt to clone humans through the much more radical transfer of nuclear DNA. This technique leads to a very high incidence of abnormalities in cloned animal foetuses, as is discussed in section 5.16.

5.8. The Grand Design: The Genome

Essential to the understanding of how genes function is a knowledge of the sequence in which the nucleotide bases line up as they link the two deoxyribose strands of DNA together. The determination of such DNA sequences began in 1977 with the 5,386-nucleotide base genome of the bacteriophage ϕX174 almost coincident with the isolation of the first human gene [38]. By 1986, the automated sequencing of DNA was developed and with it the rapid and complete sequencing of the DNA of the genome characterizing any organism became a possibility [39]. In 1995, the first genome from the Bacteria domain of life, that of *Haemophilus influenzae*, was sequenced. The sequencing of other genomes followed swiftly, including the first from the Archaea domain, that of *Methanococcus jannaschii*, in 1996, and the first from the Eukarya domain, that of *Saccharomyces cerevisiae*, or yeast, in 1998. The first genome of a multicellular organism, the nematode, or roundworm *Caenorhabditis elegans*, was also determined in 1998. Each newly sequenced genome attracted great attention and enhanced the growing knowledge of the genetic make up of life at the molecular level [40]. The sequencing of the first genome of a flowering plant in 2000, thale cress, *Arabidopsis thaliana*, attracted particular attention as most of humanity's food is based on flowering plants [41]. The greatest genomic prize of all, the sequence of the human genome, was announced in two working draft forms in early 2000 [42]. However, before exploring the human genome, it is informative to briefly review the picture of life that is emerging at the molecular level of the genome. It is based on the genomes of 599 viruses, 205 plasmids and 185 organelles from cells, thirty-one bacteria, seven archaea, one fungus, two animals and one plant sequenced before the human genome.

It might be expected that the greater the sophistication of a living organism, the greater should be the size of its genome. However, as knowledge of increasing numbers of genomes is gained it has become apparent that there is no simple link between genome size and the apparent sophistication of the organism. This is illustrated by a selection of the sequenced genomes shown in Fig. 5.9 [43,44]. As expected, the bacterial and archaeal genomes, of which one of the larger so far sequenced is that of *Pseudomonas aeruginosa* with 6,300,000 nucleotide base pairs, are smaller than those of animals and plants. Rather unexpectedly, the 16,000,000,000 nucleotide base pairs in the wheat genome exceed the number in the human genome sevenfold. The pea and barley genomes are also larger than the human genome but the reasons for this are as yet unclear. It is a little surprising that the 20,000-25,000 genes of the human genome are similar in number to the 25,496 genes of thale cress, and not much greater than the 13,600 of the fruitfly, *Drosophila melanogaster*, and the 19,099 of the roundworm. However, human genes appear to be able to produce a greater range of proteins and consequently may be more versatile [45].

While Fig. 5.9 shows mainly sequenced genomes, the sizes of many more are known and the variation of the number of base pairs in them is great [46]. The size of prokaryote genomes varies from 500,000 to 5,000,000 nucleotide base pairs, a tenfold variation. In contrast, the sizes of eukaryote genomes vary by greater than 80,000 fold with those of plants varying by more than 6,000 fold and those of animals by more than 3,000 fold. The largest eukaryote genome with 670 billion nucleotide base pairs is that of the amoeba, *Amoeba dubia*, a monocellular protozoan. This genome is 235 times larger than the human genome. That of its cousin, *Amoeba proteus*, is half the size at 290 billion nucleotide base pairs. The genome of the frog, *Bufo bufo*, has 6,900,000,000 nucleotide base pairs and those of the muntjak deer and the boa constrictor have 2,521,500,000 and 2,100,000,000 each. Initially, such huge variation in genome size puzzled biologists and was named the "C-value paradox" because of the seeming lack of relationship between the organism's complexity and the number of base pairs in its genome. However, this puzzlement is decreasing as understanding of the evolution of genomes and the differences in genes grows. Even so, it is likely that it will be some time before a comprehensive understanding of the causes of the size variations in genomes is gained.

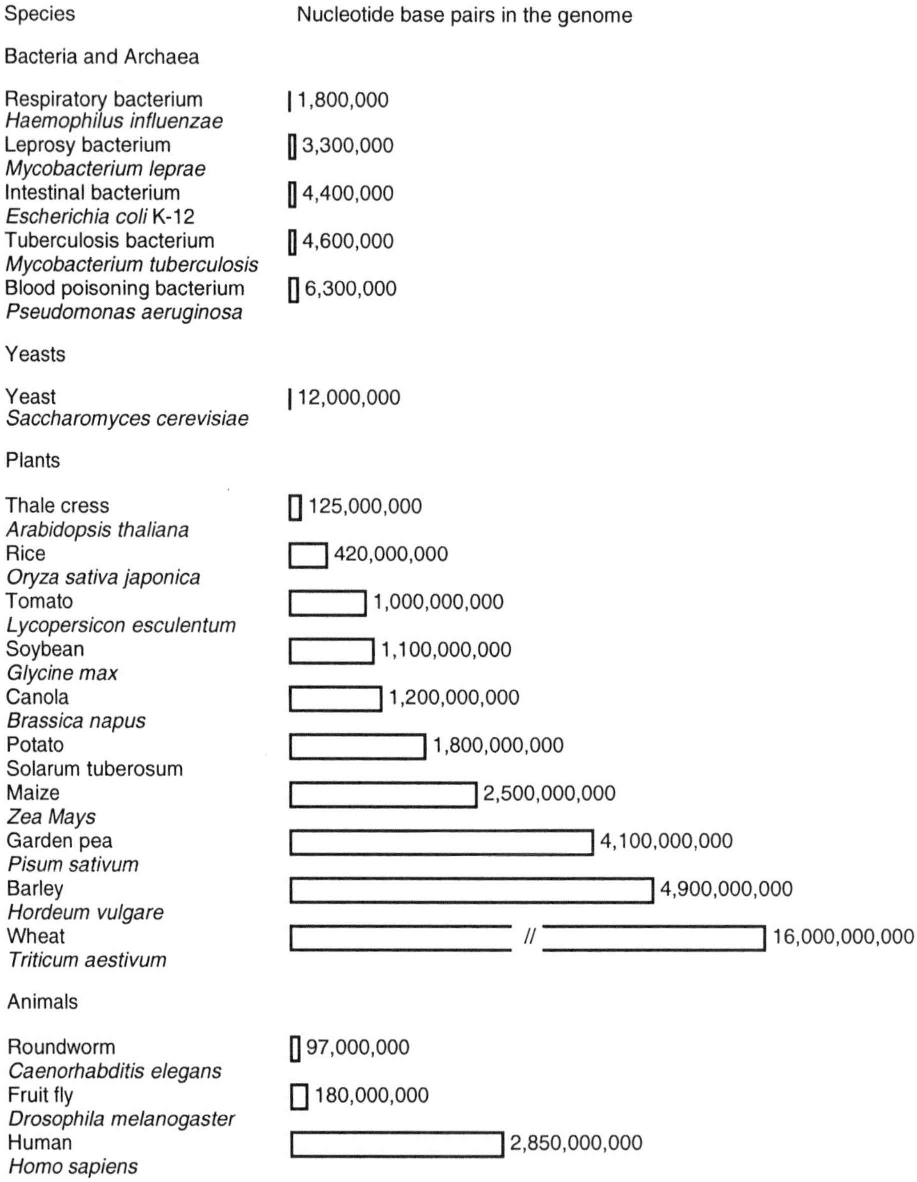

Species	Nucleotide base pairs in the genome

Bacteria and Archaea

Respiratory bacterium
Haemophilus influenzae ▌1,800,000

Leprosy bacterium
Mycobacterium leprae ▐ 3,300,000

Intestinal bacterium
Escherichia coli K-12 ▐ 4,400,000

Tuberculosis bacterium
Mycobacterium tuberculosis ▐ 4,600,000

Blood poisoning bacterium
Pseudomonas aeruginosa ▐ 6,300,000

Yeasts

Yeast
Saccharomyces cerevisiae ▌12,000,000

Plants

Thale cress
Arabidopsis thaliana ▯ 125,000,000

Rice
Oryza sativa japonica ☐ 420,000,000

Tomato
Lycopersicon esculentum ☐ 1,000,000,000

Soybean
Glycine max ☐ 1,100,000,000

Canola
Brassica napus ☐ 1,200,000,000

Potato
Solarum tuberosum ☐ 1,800,000,000

Maize
Zea Mays ☐ 2,500,000,000

Garden pea
Pisum sativum ☐ 4,100,000,000

Barley
Hordeum vulgare ☐ 4,900,000,000

Wheat
Triticum aestivum ☐ // ☐ 16,000,000,000

Animals

Roundworm
Caenorhabditis elegans ▯ 97,000,000

Fruit fly
Drosophila melanogaster ▯ 180,000,000

Human
Homo sapiens ☐ 2,850,000,000

Fig. 5.9. The numbers of nucleotide base pairs in some bacterial, plant and animal genomes. The Bacteria and Archaea genomes are plotted on a larger scale [46].

5.9. Life Looks at Life: The Human Genome

The commencement of the US$3 billion, fifteen year, National Institute of Health project to sequence the human genome, under the leadership of James Watson, was formally announced in 1990 in the United States [21,47]. It involved wide international collaboration through the International Human Genome Consortium. The aim of the project was to identify the continuous sequence of the nucleotide bases of each of the twenty-three chromosome pairs and to determine the exact position of every gene by 2003. This was an immense undertaking and intense debates sometimes arose as to the best approach to the project as the sequencing methods and technology rapidly developed. One outcome of this was that by 1998 Carl Venter had established a private company, Celera Genomics, that competed with the publicly funded consortium to publish the first draft of the human genome. Celera planned to make its sequencing data available on a commercial basis, whereas the Genome Consortium data were to be freely available. Large numbers of scientists and technologists undertook the sequencing of the human genome through either the Genome Consortium or Celera Genomics. The competition between the two was intense not only to be the first to decipher a working draft of the human genome but also because the two different methods used were subject to claims and counter claims of one method being superior to the other [48].

The Genome Consortium used a "clone by clone" recombinant DNA method where DNA was cut into fragments containing 150,000 nucleotide base pairs, and each fragment was then introduced into a bacterium that incorporated it into its DNA and exactly reproduced, or cloned, the fragment as it multiplied. After harvesting the rapidly multiplying bacteria, a plentiful supply of the particular fragment was gained that was then cut into smaller and smaller pieces until its sequence of nucleotide bases was completely determined. In this way the nucleotide base sequence of each gene was found that, together with a knowledge of the order in which the genes were linked, gave the nucleotide base sequence of the human genome. In contrast, Celera used a random or "shotgun" method where the entire genome was simultaneously cut into small fragments that were then reassembled to produce the nucleotide base sequence of the human genome. Because human DNA has many nucleotide base sequences that are repeated throughout the genome, the correct reassembly of all of the fragments produced by the shotgun method to give the correct sequence of the human genome presented a very major challenge. However, this was largely met by employing massive computing power to achieve this reassembly.

Despite the intense rivalry between the Genome Consortium and Celera, they agreed to publicly reveal their first working drafts of the human genome simultaneously at the White House on 26 June 2000. Unsurprisingly, the most intimate look ever at the underpinning of humanity produced statements that went far beyond the science of the achievement. President Bill Clinton said [49]:

> "With this profound new knowledge, mankind is on the verge of
> gaining immense new power."

The elation at reaching a major milestone in the sequencing of the human genome was reflected by Francis Collins, Director of the National Human Genome Research Institute in Bethesda, Maryland, USA, who said:

> "It is humbling for me and awe-inspiring to realize that we have
> caught the first glimpse of our own instruction book, previously
> known only to God."

Complementary sentiments were expressed by Craig Venter, President of Celera Genomics, who said:

> "The complexities and wonder of how the inanimate chemicals that
> are our genetic code give rise to the imponderables of the human
> spirit should keep poets and philosophers inspired for the
> millenniums."

However, a cautionary note had been sounded some months earlier by Eric Lander and Robert Weinberg who were heavily involved in the human genome project and who wrote [2]:

> "At the same time, we must confront this new world soberly and
> with some trepidation. The genetic diagnostics that can empower
> patients to seek personalized medical attention may also fuel
> genetic discrimination. The understanding of the human genetic
> circuitry that will provide cures for countless diseases may also

lead some to conclude that humans are but machines designed to play out DNA cassettes supplied at birth - that the human spirit and human potential are shackled by double-helical chains. So the most serious impact of genomics may well be on how we choose to view ourselves and each other. Meeting these challenges, some quite insidious, will require our constant vigilance, lest we lose sight of why we are here, who we are, and what we wish to become."

These statements reflect the realization that the long evolutionary process has produced sufficient intelligence in humanity to understand the most intimate details of life at a molecular level and perhaps to alter the course of evolution. They also reflect very clearly an appreciation either directly, or in the manner of their expression, that humans are much more than the sum of their molecular parts [50].

However, there remained much work to be done to complete the sequencing of the human genome as the Genome Consortium's first draft contained the sequences of eighty-four percent of the nucleotide bases, and that of Celera, eighty-three percent, with both indicating the presence of 35,000 or so genes [51]. By early 2003 more than ninety-eight percent of the human genome had been sequenced with 99.99 percent accuracy and the estimate of the number of genes had decreased to less than 30,000 [52]. In October 2004, the International Human Genome Consortium reported that the almost completely sequenced human genome contains 2.85 billion nucleotide bases and 20,000 to 25,000 genes [53]. Of these genes, 1,183 appear to have evolved recently by comparison with the majority of genes, while thirty-seven appear to have undergone recent mutations that cause them to be inoperative.

Percentage of human genes similar to those in other species

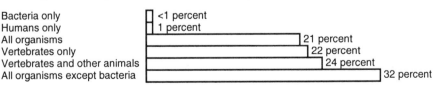

Bacteria only — <1 percent
Humans only — 1 percent
All organisms — 21 percent
Vertebrates only — 22 percent
Vertebrates and other animals — 24 percent
All organisms except bacteria — 32 percent

Fig. 5.10. The shared genetic heritage of humans and other species.

The growing knowledge of the human genome makes possible a comparison with the genomes of other species. As humanity evolved from single cellular species along with other multicellular organisms, it is unsurprising that the human genome shares many genes with other species as shown in Fig. 5.10 [53].

5.10. The Genome and Disease

The sequencing of the human genome brought a new challenge; the gaining of an understanding of the variations of the nucleotide base sequence in the genome that results in genetic disease. As indicated in section 5.3, specific genes have been implicated in a wide range of diseases. However, there remain a great variety of genetic diseases attributed the action of several genes but whose exact genomic origin is unclear. As a first step towards gaining knowledge of such origins, the International HapMap project was set up in October 2002 as a collaboration between Canada, China, Japan, Nigeria, the United Kingdom and the United States to develop a haplotype map, or HapMap, of the human genome [23,55,56]. Although about ninety-nine percent of the sequences of nucleotide bases in the genomes of unrelated people are identical, the remaining one percent differs and appears to be the origin of different susceptibilities to genetic disease and environmental conditions. Thus, one individual may have one nucleotide base different from another in a particular chromosome, or a single site polymorphism, or SNP. Each form of the chromosome is called an allele and all of the alleles characterizing an individual make up the genotype. About ten million SNPs exist in the genome of the human population and the alleles arising from them in the region of a chromosome are called haplotypes. To identify all of the SNPs existing in a person's chromosomes would be extremely time consuming. Fortunately, in the region of most chromosomes there are only a few common haplotypes that occur with a frequency of at least five percent and this is the cause of most of the variations between individuals within a population. Thus, in a particular chromosome region there may be many SNPs but the judicious selection of a few "tag" SNPs that identify most of the character of the nucleotide base variation in the chromosome region, as shown in Fig. 5.11, greatly simplifies determination of the frequency with which particular haplotypes occur.

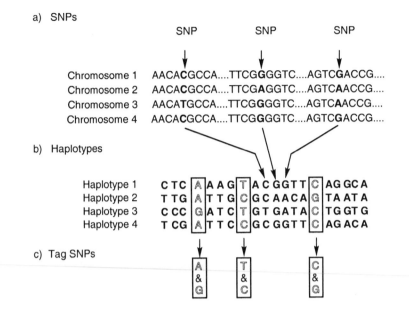

Fig. 5.11. a) A short length of DNA from four versions of the same chromosome region from four different people. The sequence of nucleotide bases is identical in these chromosomes except in three places where SNPs occur. The alternative nucleotide bases are C (cytosine) and T (thymine), A (adenine) and G (guanine), and A and G in the first, second, and third SNPs, respectively. b) Four haplotypes each made up of a combination of 20 SNPs across 6,000 nucleotide bases of DNA (only the SNPs are shown, including the three in a). For this region, most of the chromosomes in the population surveyed possessed haplotypes 1-4. c) The three tag SNPs selected from the twenty SNPs are sufficient to uniquely identify the four haplotypes. Thus, if chromosomes have the patterns A-T-C, A-C-G, G-T-C, and A-C-C at the three tag SNPs this matches the patterns for haplotypes, 1-4, respectively. Adapted from [56].

While most of the common haplotypes occur in all human populations, the frequency of occurrence differs such that one closely related group is characterised by a relatively small range of haplotypes in a particular chromosome region while another group is characterised by a different small range of haplotypes. Accordingly, it is necessary to study several different populations to choose tag SNPs. Consequently, the HapMap is being constructed initially from 270 DNA samples provided one each by forty-five people from China, forty-five from Japan, ninety Yorubans from Nigeria and ninety people from Utah whose ancestors were from north and western Europe. These groups were selected on the basis that they show significant differences in the frequency of the occurrence of a range of haplotypes in their genomes.

Challenged Earth

Except in the cases of sperm and ova cells, the chromosomes in human cells occur in pairs. While half of each chromosome pair is inherited from the father and half from the mother, they do not pass from one generation to the next intact. Instead, on fertilization of the ova, or egg, by a spermatazoan the chromosomes exchange segments in a recombination process. Thus, in the course of breeding in an isolated population group the ancestral chromosomes pass through repeated recombinations. However, some segments, or haplotypes, of the ancestral chromosomes survive intact. While these haplotypes are separated by chromosomal segments where recombination has occurred, they reflect humanity's distant past. Humans are a young species on the evolutionary timescale and are descended from African ancestors of some 200,000 years ago. Consequently, haplotypes of populations outside Africa are usually subsets of African haplotypes and are longer. This is because the African population has been larger than others through much of humanity's existence and has had more time for recombination to disrupt haplotypes. The greater length of non African haplotypes reflects the probability that groups leaving Africa carried a smaller selection of haplotypes and were often isolated from other groups so that recombination through breeding was less disruptive. As a result, the isolated non African groups of humanity inherited different proportions of haplotypes that persist into the twenty-first century. The identification of these haplotypes in the HapMap and the subsequent search for particular haplotypes associated with population groups showing a prevalence of either a particular genetic disease or an environmental susceptibility promises new approaches to diagnosis, prevention and treatment of disease.

In parallel with the HapMap project another endeavour targeting genetic disease, and only made possible with the completion of the sequencing of the human genome, is in progress. The genome determines the function of each cell through the chemistry of the proteins it produces. Each type of cell contains a different selection of proteins that changes as it passes through its lifespan, and the plethora of human cells produces vast numbers of different proteins that are collectively called the human proteome [57]. These proteins are integral to the proper functioning of healthy cells but when defective genes produce aberrant proteins they cause the chemistry of the cell to change detrimentally and thereby become the agents of genetic disease. This new endeavour, proteomics, seeks to determine the structure of the proteins in the proteome and to find those whose aberrant actions can be blocked by specifically designed drugs and thereby ameliorate genetic diseases. The immensity of this task is readily appreciated when it is realized that at the extreme end of gene capability, one fruitfly gene is capable of encoding for 38,000 different proteins and human genes may exceed this capability.

5.11. The Genomic Pandora's Box

Each advance in knowledge usually represents a Pandora's box from which, once opened, the potential for good and evil is released. Knowledge of the human genome and other genomes is no different. While understanding of the function of genes identifies the predisposition of humanity to disease and through that the opportunity for preventative care and therapy [23,58], the likelihood that employers and life insurance companies may seek to use such genetic information to screen out individuals on the basis that they possess genes predisposing them to disease is considerable [59]. Thus, there arises the possibility that while knowledge of the human genome may lead to an increase in health, for many it could strengthen humanity's all too strong tendency to discriminate between its members simply on the basis of genetic characteristics of which no individual has a choice.

It is an irony that well before the announcement of the first drafts of the human genome, the first widespread use of the DNA sequence was concerned with one of humanity's less endearing characteristics, criminality, although it also serves the cause of one of humanity's better ideas, justice. The genetic fingerprinting technique that makes use of the uniqueness of the DNA of each individual to help either convict or exonerate suspects through comparison of their DNA with that left behind at crime scenes is now widespread [60]. However, by 2001, the avidity with which DNA was being routinely collected from people taken into custody and their DNA profiles stored was causing disquiet about the weakening of the right to privacy.

An intriguing use of the genetic fingerprinting technique was in identifying the remains of Tsar Nicholas II and other members of the executed Romanov family after seventy years in the ground [61]. While most human DNA is contained in the cell nucleus, a much smaller and circular DNA, mtDNA, containing 16,569 nucleotide base pairs in its genome, is found in the mitochondria of the cell [62]. The much larger nuclear DNA contains nucleotide base sequences from both parents while those of mtDNA are inherited from the mother alone. Because mtDNA survives longer in dead tissue than does nuclear DNA, the Romanovs were identified from mtDNA taken from their bones. That of Tsar Nicholas II was identified through a comparison of his mtDNA with that taken from the bones of his brother, Georgij Romanov, and with mtDNA samples provided by the great-great-great-granddaughter and great-great-grandson of the Tsar's grandmother, Louise of Hesse Cassel. The bones of the Tsarina and those of three of her daughters were identified through a comparison of their mtDNA with mtDNA samples provide by Prince Philip of the United Kingdom whose maternal grandmother was the Tsarina's sister. By the same method, the claim of Anna Anderson Manahan to be Anastasia, the Tsar's daughter, was shown to be false [63]. A massive example of identification through DNA analysis arose from the destruction of the World Trade Centre twin towers and the attack on the Pentagon by terrorists on 11 September 2001 that killed almost three thousand people, many of whom were burnt beyond recognition [64].

5.12. Commercialising the Genome

The commercialisation of access to genetic information comes to the fore in some areas of genomic research. This arises from the substantial expense and effort involved in sequencing DNA and identifying genes. When companies are involved in such research, the necessity of recovering costs and the desire to make a profit may result in the release of the information gathered on a commercial basis only [65]. In addition, it usually results in the patenting of sequencing and related technology and of DNA segments and complete genes, including human genes. In an effort to accelerate the understanding of the occurrence of and transmission of genetic disease, isolated communities with well established health records and a minimum of mixing with other communities are particularly attractive for genomic study [66]. This is the case for Estonia, Iceland and Latvia that have been the subjects of such studies. In return for carrying out the research, the companies concerned have been granted almost sole right of access to the full range of genomic information gathered. As a consequence much of this knowledge of the human genome is only commercially available, a situation that has the potential to place many of the benefits flowing from this knowledge beyond the financial reach of public health systems and much of humanity.

The substantial expense incurred in plant genome sequencing raises the question as to who will fund such sequencing. The first draft sequencing of the genome of a crop plant, that of the rice variety, *Oryza savita japonica*, was announced by Monsanto in 2000 who made it publicly available by donated it to the International Rice Genome Sequence Project (IRGSP) that was publicly funded by Japan with US$100 million and by China, India, and the United States with US$10 million each [67]. The rice genome is composed of some four hundred million nucleotide bases and contains twelve chromosomes and sixty thousand genes. Subsequently, Syngenta published a draft sequence of the genome of the same rice variety in April 2002 simultaneously with the publication of a draft sequence of the major rice variety planted in China, *Oryza sativa* L. ssp. *indica*, by the Beijing Genomics Institute and other Chinese academies [68]. The Syngenta draft is proprietary, which reflects common practice where companies investing in research patent their discoveries to gain commercial advantage. However, in this case the new knowledge pertains to a basic food that feeds half of humanity. Any restriction of knowledge of the rice genome raises concerns that the improvements in rice that could result from it may not be available to developing nations dependent on rice as a major food. Fortunately, Syngenta promised that information and technology for rice crop improvement would be made available to subsistence rice farming research without royalty or technology fees and offered to share genomic data to speed up the finalizing of the completion of the sequence of the rice genome. Even so, it seems inevitable that commercial restrictions placed on access to genomic information will cause considerable debate and antagonism in the future.

So great is commercial interest in genomics itself and its integration with other established technologies that there seems little doubt that the knowledge gained and the control of that knowledge will have a major impact on the global economy [69]. In a defensive move, the publicly funded Maize and Wheat

Improvement Centre, CIMMYT, decided to patent its genetic information to ensure that it remains available to scientists working for sustainable agriculture in developing nations.

5.13. Genetic Engineering: Bacteria

Genetic engineering has become broadly understood to mean the modification of bacteria, animals and plants through the direct introduction of changes in their genomes. This involves transfer of selected genes from one species to another to produce genetically modified (GM), or transgenic, species. While this innovative science is alarming to many, the conventional crossbreeding of animals and plants involves the transfer of many more genes and is a less direct form of genetic engineering. Within nature, bacteria readily exchange DNA and crossbreeding among animals and plants is part of the evolutionary process. The new evolutionary aspect is that genetic engineers have the gained the ability to transfer genes between very different species, an occurrence not encountered in nature. This has attracted widespread interest and concern. In the latter case, the issues of appropriateness, morality and the possibilities of inadvertently releasing harmful genetically modified organisms into the environment became the subject of intense debate at the junction of the twentieth and twenty-first centuries [70]. This debate has already been discussed to some extent in Chapter 4 and is further alluded to below as genetic engineering is explored together with the benefits and hazards that may flow from it.

Some of the earliest benefits gained from genetic engineering involved the hormone insulin, a protein produced in the pancreas that controls the metabolism of sugar in the body. When the pancreas fails to produce insulin, diabetes mellitus results and is fatal if untreated. The first treatment of diabetes sufferers with insulin from pigs in the 1930s represented a major breakthrough in combating diabetes. Unfortunately, the medical use of material from either animals or humans carries with it the possibility of contamination, particularly that of pathogenic viruses. The risk of such contamination can be virtually eliminated and the availability of therapeutic agents such as insulin can be greatly increased through the use of genetically engineered bacteria. To achieve this it is necessary to insert the gene producing the therapeutic agent into the plasmid of a bacterium by adapting naturally occurring processes. The required gene is excised from the selected organism by an enzyme, restriction endonuclease, that cuts the DNA of the genome between selected nucleotide base sequences. Then, the DNA of the bacterial plasmid is similarly cut between the same set of nucleotide base sequences. This cutting leaves one of the strands at each end of both the plasmid DNA and the excised gene DNA longer than the other strand so that several nucleotide bases are left unpaired on these so-called "sticky ends". However, the nucleotide bases in the end of the longer DNA strand of the cut plasmid match those in the longer end of the excised gene DNA to form nucleotide base pairs of double stranded DNA. Under the influence of another enzyme, ligase, the matching sticky ends join to form a circular plasmid now incorporating the excised gene. Such DNA is usually referred to as recombinant DNA. The human insulin gene is found in chromosome 11 and is quite small. In 1978, the two sequences of DNA corresponding to those in the human insulin gene were chemically synthesized. A year later they were inserted into the *Escherichia coli* plasmid to give genetically modified bacteria that both multiplied rapidly and produced, or expressed, the two protein components of human insulin. These were harvested and joined to give insulin that first became available as a pure and reliable supply in 1982 and is now the dominant source of insulin used by diabetics [71].

A similar genetic engineering of *Escherichia coli* resulted in a reliable supply of human growth hormone used to treat children with slow growth rates. It also eliminated the risk of contamination with the prion causing Creutzfeld Jakob disease, CJD, associated with the hormone extracted from the pituitary glands of cadavers that was previously used, sometimes with tragic consequences [72]. Human interferons that are produced by cells exposed to viruses and give them some protection against viral attack have also been produced by similar methods [73]. Genetically engineered *Escherichia coli* have become a major vehicle for producing a wide range of amino acids, polypeptides and proteins that is a technique that is under continuous development [74]. While such genetic engineering usually involves a single gene, whole chromosomes, including human chromosomes, have also been engineered into *Escherichia coli* [75].

Potentially, a wide range of bacteria may be genetically engineered to produce desirable proteins and other molecules. However, there is also the potential to increase the virulence of pathogenic bacteria and viruses for use in biological warfare. Such warfare is so feared that an international treaty banning the manufacture and use of biological weapons, the 1972 Biological and Toxin Weapons Convention, was agreed to by 140 nations [76]. With good reason, most of the expertise gained in this type of weaponry is classified and closely guarded. However, every so often the potential for genetic engineering to make biological warfare even more fearsome is revealed inadvertently as was the case in early 2001 [77]. Australia suffers from huge and economically damaging periodic infestations of mice that are an introduced species with no natural predator. Accordingly, genetic engineers took the mousepox virus that causes relatively mild illness in mice and sought to modify it to cause infertility so that release of infected mice into the mouse population and subsequent cross infection would decrease fertility generally. Initially the virus was modified to carry a mouse egg protein to trigger an antibody response and rejection of mouse eggs to cause infertility. While this worked with one strain of mice, it did not with a second. Consequently, the interleukin-4 gene was also engineered into the virus to produce interleukin-4 to further boost the immune system. The startling result was that the virus now suppressed the immune response against it, was little affected by a mousepox vaccine, multiplied rapidly and killed the mice. This gave rise to the fear that a similar modification of the smallpox virus could produce a particularly virulent strain of smallpox that would be even more deadly towards humans than the natural strain. Because of this and other developments, including the chemical synthesis of a poliovirus [78], which could aid bioterrorists, the editors of a number of prestigious scientific journals agreed in 2003 to censor publication details that could afford such aid [79].

5.14. Genetic Engineering: Plants

The vast majority of plants grown in gardens, orchards and fields have been selectively crossbred to produce flowers of exquisite beauty, great quantities of flavoursome fruit and high yielding vegetable and grain crops [80]. Such crossbreeding has occurred with increasing rapidity and success over many millennia, and without it the human population could neither have reached six million nor have been sustained at that level and beyond. This represents the great success of genetic engineering by traditional methods whereby crossbreeding of closely related plants and the transfer of many genes simultaneously have produced new plant varieties of great benefit to humanity. As with bacteria, the selective transfer of genes into plant genomes using genetic engineering techniques has become commonplace bringing with it a wide range of possibilities for new plant varieties.

To directly introduce a new gene into a plant genome requires a sophisticated understanding of plant biology that was first practically achieved in the early 1980s by using a route that normally causes plant disease [81]. The bacterium, *Agrobacterium tumefaciens*, carries genes that it injects into plant cells that are incorporated into the plant genome with the result that the infected plant forms galls, or tumours, and prolific root masses that are ideal environments for the proliferation of the infecting bacteria. This infection is known as crown gall disease. It was realized that if the disease causing genes of *Agrobacterium tumefaciens* were replaced by others that could be incorporated into a plant genome to give desired characteristics, plant infection by the modified bacterium would be an effective route, or vector, through which this could be achieved. And so it proved to be. In addition to *Agrobacterium tumefaciens* and others of the same genus, *A. rhizogenes*, *A. vitia* and *A. rubi*, the bacteria *Sinorhizobium meliloti and Mesorhizobium loti* may also be used to insert genes into plant genomes [82].)

Plant leaves contain stem cells that can develop into all of the other specialized cells to produce a complete plant under favourable conditions. Such stem cells are described as totipotent and are fairly widespread in the leaf structure. As a consequence, infection of part of a leaf with the desired genes by *Agrobacterium tumefaciens* results in genetically modified plants growing from the infected leaf cells. However, not every leaf part is infected and it is difficult to distinguish unmodified from genetically modified, or GM, plants until an advanced stage of growth. Hence, it was necessary to introduce a further modification to distinguish between plants at an early stage. This was achieved by modifying *Agrobacterium tumefaciens* to carry not only the desired gene in its plasmid but also an antibiotic resistant gene as a marker gene. Thus, when treated with an antibiotic, the unmodified plants die leaving the modified antibiotic resistant ones to grow to maturity from which they may either be propagated or their seed collected. This sequence of events is shown in Fig. 5.12. The presence of the marker gene presents a difficulty as it may induce antibiotic resistance in bacteria, and regulatory authorities increasingly require that the introduction of DNA not directly needed to induce the genetic modification should be kept to a minimum. As a consequence, techniques for elimination of the marker gene are under development that should go some way to eliminating the possibility of inducing bacterial antibiotic resistance [83].

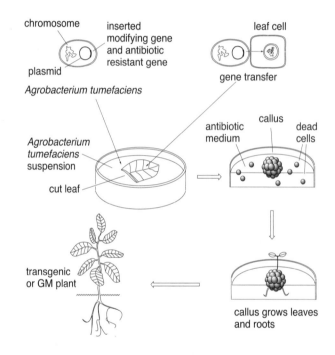

Fig. 5.12. The engineering of a transgenic, or GM, plant by infection with *Agrobacterium tumefaciens* genetically modified to carry the gene to be transferred together with an antibiotic gene. The uninfected plant cells die when treated with an antibiotic, leaving the infected cells to form calluses that, after growing leaves and roots, are transplanted to grow into mature GM plants.

While the use of *Agrobacterium tumefaciens* as a vector for transferring genes to plants is widespread, mechanical methods have also been developed. These methods involve the firing of either tiny gold or tungsten particles coated with the required genetic material from a "gene gun" into a cell culture, or by simply shaking a suspension of the cells and the foreign DNA with small sharp crystals of hard silicon carbide. These methods punch holes in the cell membranes through which the foreign DNA enters with a fair expectation that it will incorporate itself into the cell genome. Alternatively, the application of either an electric field or a chemical may be used to open pores in the cellular membrane through which foreign DNA may enter. If the cells are large enough, direct injection of the foreign DNA into the cell may be used. These are known as vectorless methods since no biological carrier, or vector, is used [33].

By mid 1992, more than forty new species of plant had been genetically engineered including apple, cauliflower, cranberry, kiwi fruit, papaya, rice, sugarbeet and tomato. In addition to inserting genes to manufacture the insect toxin of *Bacillus thuringiensis* and generate herbicide resistance, other genes have been inserted that delay the ripening of harvested fruit and thereby their shelf life, and to give a blue colour to petunias [33,81]. By the mid 1990s, genetic engineers had become so confident in the understanding of their technology that they were considering plants to be production systems, or bioreactors, that could be genetically manipulated to produce a wide range of sophisticated chemicals. These include carbohydrates, lipids, pharmaceutical proteins, industrial enzymes and biodegradable plastics. Substantial effort has been devoted to genetically engineering plants to produce antigens to a range of diseases including cholera, hepatitis B, influenza, malaria and rabies. Such antigens cause an immune response to a particular disease agent in the same way as traditionally produced vaccines. It appears that it may be possible to genetically engineer bananas, potatoes, tomatoes and other foods to contain such antigens, which on eating would act as orally administered vaccines; potentially a great advantage in developing countries where the storage, distribution, and administration of vaccines are often difficult [84]. However, the possibility that unwitting consumption of such pharmaceutical and vaccine carrying foods could have harmful consequences if they became mixed with normal foods has changed the choice of such genetic engineering to non food plants [85].

5.15. Genetic Engineering: Animals

In February 1997, the announcement of the birth of a lamb by a team led by Ian Wilmut of the Roslin Institute in Edinburgh attracted widespread attention [86,87]. This Finn Dorset lamb, named Dolly, was remarkable in possessing exactly the same genome as that of a Finn Dorset ewe and none of the genetic material of the Scottish Blackface ewe that gave birth to her. In other words, Dolly was an exact genetic copy, or clone, of the Finn Dorset ewe from which her genome had been taken quite contrary to expectation had she been conceived naturally. This came about because Dolly grew in her mother's uterus from the enucleated egg, or oocyte, of a Scottish Blackface ewe in which the nucleus had been replaced by that taken from a mature, or somatic, cell of the udder of a six year old Finn Dorset ewe as shown in Fig. 5.13. This represented a major

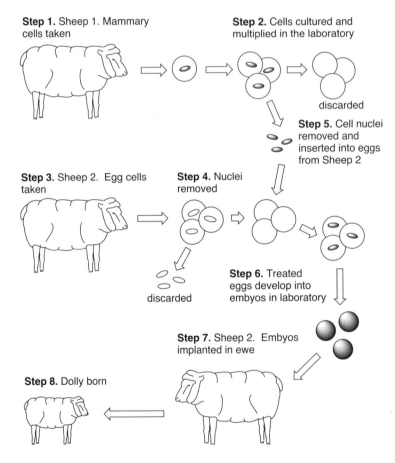

Fig. 5.13. The events leading to the birth of the cloned Finn Dorset lamb, Dolly.

breakthrough as somatic cells such as udder, muscle or skin cells are highly specialized and normally only reproduce themselves whereas a foetus grows from the totipotent cells of a fertilized egg that produces every type of cell necessary to build the complete animal. Thus, the genetic material in the transferred nucleus from the Finn Dorset udder cell had regained its ability to produce other types of cells in the Scottish Blackface enucleated egg cell. While Dolly was the only lamb born in this way from 277 enucleated Scottish Blackface eggs in which the nuclei had been replaced by those from Finn Dorset udder cells, her birth raised the possibility of cloning mature animals with desirable characteristics, such as durable wool quality, high milk production and good meat texture.

Dolly was followed in quick succession by cattle, goats, mice, mules, cats and horses cloned from somatic cells and it seems that there are likely to be few limits to the animals that might be cloned [88]. A novel offshoot from this technology is the cloning of extinct or rare animals [89]. To achieve this, somatic cells are taken from either fresh or preserved tissue of the animal to be cloned and their nuclei are excised and inserted into enucleated eggs of a surrogate mother from a closely related species. This has been attempted with the African wildcat, the bongo antelope, the gaur, the moflin sheep, the eland, and a rare red deer with mixed success. The cheetah, Sumatran tiger and giant panda are also possible candidates for such cloning. The first such clone was Noah, a gaur calf born on 8 January 2001. Unfortunately, Noah died two days later from a bacterial infection common in young calves.

Prior to the birth of Dolly, calves were quite commonly produced from embryonic stem cells [90]. However, the number of cells in an embryo is obviously less than the millions of specialized somatic cells of a chosen type that may be taken from an adult donor. Clearly if a greater success rate can be achieved in producing clones from somatic cells this is likely to become the cloning method of choice. However, a note of caution was sounded when an examination of Dolly's DNA showed that her cells were older than she was. This appeared to indicate that that Dolly's cells retained the age of the adult sheep from which the somatic cells and the egg were taken, and that Dolly might age prematurely. This prediction was seemingly borne out by her development of premature arthritis in late 2001 and an incurable lung cancer in February 2003 as a consequence of which she had to be put down [91]. In contrast, cells of mice and calves cloned from somatic cells in the same way as Dolly appeared to have been rejuvenated, suggesting that these clones may have a normal life span, and possibly an extended one [92]. The origins of these differences between the cloned animals are unclear, and it will be sometime before the issue of their longevity is settled. However, cloned mice were found to have multiple gene defects, to be more susceptible to disease than naturally conceived mice of the same strain and to die earlier [93]. Genetic defects have also been found in cloned cattle and sheep embryos.

To place the cloning of Dolly in perspective, it should be remembered that frog clones were produced from nuclei taken from the totipotent cells of embryos and from somatic cells of tadpoles using similar methods to those that produced Dolly, although when somatic cells were used the success rate was much less than when embryonic cells were used [94]. This development and the accompanying pictures of thirty cloned albino frogs raised great interest in 1977. In the plant world it has been common practice for thousands of years to propagated, or clone, selected plants by taking small portions of the plant, or cuttings, and either planting them directly in soil or grafting them onto a mature plant to produce exact copies of the plant providing the cuttings. Today, plant nurseries produce huge numbers of plant clones in culture media. However, this is much less complex than producing animal clones as the totipotent cells in plant fragments are much more robust toward handling and proliferation into specialized cells than are animal totipotent cells.

Cloning from somatic cells offers great advantages in the production of genetically modified animals that have had human genes inserted into their genome so that they produce sorely needed therapeutic proteins in their milk [95]. This was exemplified by the genetically modified pig Genie that generated the human protein C that stops bleeding in haemophiliacs. Prior to Dolly's birth, genetically modified animals were produced by injecting embryos with the required human gene and then breeding those few that incorporated the gene into their genome. Shortly after Dolly's arrival, Polly and two other genetically modified lambs cloned in a similar way to Dolly, but carrying a human gene to produce a blood clotting protein, were also born at the Roslin Institute [96]. Such cloning promises to build up numbers of genetically modified animals to produce large amounts of pharmaceuticals in substantial quantities that may be unavailable in significant amounts through more conventional sources. However, scientific success in this area does not guarantee commercial success as shown in

2003 by a major downsizing of PPL Therapeutics set up so exploit this technology in cloned animals [97]. A further problem is that this appears set to become a litigious area of endeavour. Given the low cloning success rates achieved by current technology and the high cost involved, it is not surprising that most animal cloning techniques have either been patented or patents have been applied for. As a consequence legal challenges to patent rights have appeared [98].

A more revolutionary potential use for cloned animals is in xenotransplantation, the transplanting of animal organs into humans when an organ has failed. While transplantation of organs such as hearts, kidneys and lungs from human donors to recipients works reasonably well, the recipients must thereafter take immunosuppressive drugs that prevent their immune systems rejecting the transplanted organ. There is also a limited supply of human organs for transplantation and the surgery and associated treatment is extremely expensive. Although the transplantation of animal organs, particularly those of pigs whose organs are of similar size and nature to those of humans, potentially provides an alternative source of organs, the problem of genetic incompatibility remains. However, it has been proposed that by genetically engineering pigs it may prove possible to gain a much larger supply of replacement organs. Nevertheless, such xenotransplantation still has to overcome the barrier posed by rejection of pig organs by human recipients' immune systems and it is hoped that genetic engineering might lower this barrier. A major cause of rejection of pig organs by the human immune system appears to be the presence of a sugar, galactose-α-1,3-galactose [99]. It appears that the cloning of pigs from which the gene producing this sugar has been removed might be a major step towards overcoming rejection of xenotransplants. However, the possibility that xenotransplantation might also transfer porcine viruses to the human recipient and possibly generate new viral diseases is a matter of concern [100]. The transfer of a porcine retrovirus occurred when pig pancreatic cells were transplanted into mice gave credence to such fears [101]. However, such viral transfers appear not to have happened in humans who have been treated with pancreatic islets and other cells from pigs [102].

5.16. The Cloning of Humans

With the cloning of Dolly, it was immediately recognized that humans might be
cloned in the same way, a prospect that quickly became a controversial issue
[103]. Human clones are not new, however, as the millions of naturally
conceived identical twins possess exactly the same genome as they come from a
fertilized egg that spontaneously split into two in their mother's uterus.
Nevertheless, the objections to cloning humans as Dolly was cloned range from
it being considered an unnatural interference with nature to concerns about the
possibility of producing superhumans, objections given some impetus by an
announced intention to clone humans in Italy and the United States that swiftly
attracted the attention of legislators. However, a much more real danger exists
should attempts to clone humans occur. This is that in replacing the nucleus of an
egg with that from a somatic cell, damage may occur that results in abnormalities
in the implanted eggs and resultant embryos such that few pregnancies would go
to full term, and the possibilities of abnormalities and low survival rates in the
full term children would be considerable. Evidence for the occurrence of such
damage comes from the high abortion rate and high death rate soon after birth for
cloned calves and lambs, and abnormalities in those that survive [104]. Some
cloned animals become huge and often sickly during gestation, traits that have
been linked to them possessing lower than normal levels of the IGF2R protein
that controls the growth of the foetus [105]. This is attributed to the loss of
methyl groups ($-CH_3$) from the DNA of the gene producing the protein during
the cloning procedure. In March 2001, Ian Wilmut, the leader of the team that
cloned Dolly, coauthored a cogent argument against cloning humans on these
grounds, quoting the 1997 conclusion of the American National Bioethics
Committee [106]:

> "At present, the use of this technique to create a child would expose
> the foetus and the developing child to unacceptable risks."

Nevertheless, the first human clones were reported in October 2001 from the
laboratories of a private company [107]. They were produced using a procedure

similar to that used to produce animal clones but consisted of only a few cells and were not intended for implantation into women to continue to full term babies. Rather, the intention was to test the possibility of producing human embryonic stem cells that could be used therapeutically. In February 2004, a South Korean group announced that it had cloned human embryos, some of which reached the blastocyst stage and permitted the extraction of embryonic stem cells, also with a view to therapeutic use [108]. The ideas for therapeutic use of embryonic stem cells are based on the knowledge that a fully formed child develops from the small number of cells composing the embryo, the embryonic stem cells, and that they develop into the multitude of specialized cells in the child. Such differentiation of embryonic stem cells into specialized cells also occurs in culture media [109] and it appears possible that replacement tissue and organs for transplantation might be grown from embryonic stem cells as is discussed in more detail in section 5.17.

Because the harvesting of embryonic stem cells destroys embryos from which children might grow, therapeutic cloning raises ethical questions and is subject to legal restriction in some nations [110]. Despite these restrictions, an American company, Geron Corporation, was mistakenly granted two British patents in January 2000 that appeared to give it commercial rights to human embryos created through cloning [111]. This raised substantial concern and the patent phrasing was quickly corrected to exclude human embryos.

It is both ironic and perplexing that as knowledge of the possibility of improving health and prolonging life through therapeutic cloning using embryonic stem cells increases it is inevitable that it will result in the loss of embryos, the beginnings of life [112]. However, the number of patients who might benefit from therapies that arise from stem cell research is large as is shown by an estimate for the United States in Fig. 5.14 [113]. It is likely that the arguments between those who see the sanctity of early human life at the embryo stage as pre-eminent and those who see the alleviation of suffering as being of greater importance will continue while much of humanity will experience mixed feelings. However, there are early signs that there may be a way out of this perplexing situation [114]. Most tissue contains somatic, or adult, stem cells that both reproduce themselves for most of the human lifetime and repair the tissue by turning into the cells composing the tissue. It had been thought that such stem cells could only transform into the type of cell of the particular tissue concerned, but it now appears that somatic stem cells are able to transform into a variety of other cells under appropriate conditions. Haematopoietic stem cells taken from adult bone marrow have been found to develop into blood cells and possibly brain cells in culture media [115]. Mesenchyal stem cells from the same source develop into muscle, bone, cartilage, tendon and airway epithelial cells. It has also been shown that neuronal cells taken from foetal brain develop into neuronal cells, glia cells that support neurons, and perhaps blood cells. Such transformations indicate that somatic stem cells may be usable both in research and therapeutic use.

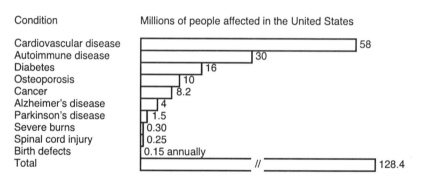

Condition	Millions of people affected in the United States
Cardiovascular disease	58
Autoimmune disease	30
Diabetes	16
Osteoporosis	10
Cancer	8.2
Alzheimer's disease	4
Parkinson's disease	1.5
Severe burns	0.30
Spinal cord injury	0.25
Birth defects	0.15 annually
Total	128.4

Fig. 5.14. An estimate of the numbers people whose conditions may be helped by stem cell research and gene therapy. Data from [113].

5.17. New Cells for Old: Stem Cells

A development, closely related to the cloning of complete animals, is the cloning of human tissue and organs for use in repair and transplantation surgery [116]. The possibility of growing either complete organs, tissue or healthy cells from stem cells taken from the patient that might be used to either replace or regenerate a diseased organ provide attractive alternatives to present transplantation techniques as genetic incompatibility and the possibility of viral transfer would be eliminated [117,118]. (A variety of types of skin cultured from skin cells from both patients and donors are commonly used in transplants for burns victims [119].)

Every human developed from one totipotent stem cell capable of differentiating into every cell necessary to take the fertilized egg through to the newly born child. The fertilized egg grows into the multicell embryo and its later stage, the blastocyst, consisting of several hundred cells. The cells of the outer envelope of the blastocyst form into the placenta and the amniotic sac membrane when the blastocyst attaches to the wall of the womb, while the inner cell mass consists of pluripotent, and possibly totipotent, stem cells that develop into all of the cells making up the foetus and the child. (The differentiation between the descriptions of cells as pluripotent and totipotent that can develop into either some or all of the cells of the organism, respectively, has become a little blurred in current usage.) Most adult tissues contain somatic stem cells that normally replace only the limited selection of cells of the tissue lost through either apoptosis or injury and may be described as marginally pluripotent. The embryonic stem cells of the inner cell mass of the blastocyst are more versatile than this and are particularly amenable to having their development directed to produce a chosen tissue in the laboratory. It is as yet incompletely understood what signals cause stem cells to develop into different types of cells and then into a particular tissue, but it appears that the levels of a "gatekeeper protein", Oct-3/4, in a stem cell is one controlling factor in its differentiation into specialized cells [120].

Once the factors controlling the development of stem cells become better known, the possibility of using cells and tissue grown in the laboratory to replace diseased organs in humans may become feasible [118,121]. One way in which this might be achieved is shown in Fig. 5.15 where embryonic stem cells are generated from the nucleus of a patient's excised cell, transformed into specialised cells and transplanted back as either somatic cells or tissue to replace diseased tissue [122]. Alternatively, it may prove possible to avoid utilising the blastocyst stage by taking somatic stem cells from the patient and directly transforming them into the required cells for transplantation. Alternatively, either healthy cells to replace diseased cells or viruses engineered to carry a normal replacement for a defective gene might be injected into either the bloodstream or the defective tissue to regenerate it. Already, it appears that a cure for diabetes may be within reach through the injection of healthy pancreatic islets, or β-cells, into the inactive diabetic pancreases. It was reported in 2000 that diabetics treated in this way with β-cells from cadavers began to produce insulin so that they no longer required insulin injections, although they did have to take mild

immunosurpressing drugs to prevent rejection of the foreign β-cells [123]. As
this source of β-cells is limited, the discovery that β-cells can either be induced to
reproduce themselves or that they may be produced from somatic stem cells,
suggests that an alternative source of material for this therapy may become
available [124].

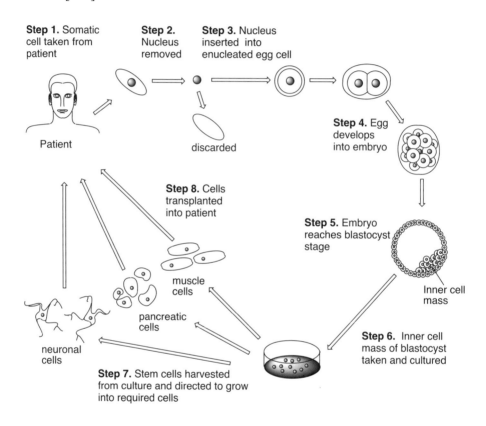

Fig. 5.15. A possible therapeutic cloning cycle for the generation of replacement cells and tissue
for transplantation into humans.

While, the ability of stem cells to develop into healthy specialised cells
appears to have few limits, their ability to develop into mutant, and possibly
cancerous cells, cannot be overlooked. Indeed, there is evidence that the
mutation of stem cells may be an initiator of cancer [125]. In 2005, it was shown
that adult stem cells grown outside the body through several generations can
become cancerous, a cautionary note for therapies requiring such stem cell
culturing [126].

It is salutary to reflect that the perfection of some of the medical abilities discussed above may ultimately give humans some of the capacity for self-repair possessed by other animals. Thus, an earthworm cut in two, repairs the severed ends to become two identical clones, and some lizards, upon being caught by the tail, drop their tails to escape and grow a replacement quite rapidly. A variety of ways in which humans might develop such a self repair capacity with medical assistance after birth, and perhaps in the womb, is sometimes collectively called cell therapy but more often gene therapy [127].

5.18. Gene Therapy

One in every hundred children suffers from a serious genetic defect that greatly decreases their quality of life and often shortens it [128]. Therefore, it is little wonder that treatments for such illness are vigorously sought. On 14 September 1990, Ashanti DeSilva became the first patient to undergo government approved gene therapy in the United States [129]. Ashanti suffered from severe combined immunodeficiency, SCID, where the gene needed to produce an enzyme essential to the proper functioning of the immune system was defective leaving her susceptible to a plethora of infections that a healthy immune system combats with ease. The gene therapy Ashanti received involved insertion of functioning copies of the defective gene into her T-cells, a type of white blood cell that is part of the immune system, using a retrovirus to carry the gene. The result was little short of miraculous, for it transformed Ashanti from a frail child who seldom left her house for fear of infection into a healthy child who had grown into a thriving thirteen year old by April 2000.

Gene therapy is subject to considerable risks as shown by the initial elation at an apparently successful treatment being followed by a tragic setback that occurred when two of ten French children treated for SCID-X1 caused by a defect in a gene in the X chromosome developed a rare form of leukaemia as a result of the treatment [130,131]. The treatment involved the insertion of the corrective gene into a retrovirus stripped of most of its viral genes to prevent it causing dangerous infection prior to administration to the children. Nine of the children were cured of SCID-X1 including two three-year-old boys who later developed leukaemia. The therapy works through the retrovirus inserting itself and the corrective gene into the patient's DNA. However, there is presently no control over the point at which the retrovirus inserts itself into the DNA, and the possibility exists that another gene may be disrupted as a consequence and cause disease. In the cases of the French children it appears that the retrovirus inserted itself near the *LMO2* gene that is associated with leukaemia. As a consequence of this emergence of the high risk of the treatment and fears that the other treated children might also develop leukaemia, this type of gene therapy was stopped globally in January 2003 for reassessment. Upon reflection, it was considered that the benefits of the therapy outweighed the risks for desperately ill patients and the therapy recommenced in 2004. Unfortunately, a third child developed leukaemia in 2005 and the treatment was stopped again in France and the United States [132,133].

Many people have received a range of gene therapies; sometimes with good results but in other cases the results have ranged from minimal to life threatening [134]. As a consequence, this revolutionary therapy that is in its infancy is presently under a cloud. There have been some very unfortunate experiences in the rush to develop gene therapy in the face of incomplete knowledge. A particularly widely discussed case was that of eighteen year old Jesse Gelsinger who died after undergoing an experimental adenovirus gene therapy treatment in September 1999 [135]. Patients undergoing gene therapy for cancer have also died, possibly as a consequence of the treatment [136]. Gene therapy has also produced adverse effects in patients as exemplified by Parkinson's disease suffers whose condition was greatly worsened in some cases [137]. Despite these discouragements, the hope for the eventual perfection of gene therapy was strengthened by the generation of mouse inner ear, or cochlear, hair cells from embryonic stem cells and the restoration of hearing to deaf guinea pigs by the insertion of the of the *Atoh1* gene into their inner ear tissue [138]. Hearing loss afflicts a significant portion of humanity and is very often caused by the destruction of the sound sensing hair in the inner ear.

An alternative approach to introducing a single gene to correct the activity of a malfunctioning gene is to introduce the complete chromosome containing the replacement gene. The chromosome is continually reproduced during cell division and the therapeutic benefits of the replacement gene are retained indefinitely. To this end the construction of the desired chromosomes is being studied and the first artificial human microchromosomes have been produced [139].

5.19. Genetic Engineering and Humanity's Future

Where the extraordinary molecular insights into biology gained in the last decades of the twentieth century will take humanity in the twenty-first century is a matter of speculation, but with judicious application the potential for benefit in diverse areas is great. It promises new strains of crops to grow on unproductive salinized land and higher producing and more nutritious crops to better feed humanity as population peaks towards the end of the twenty-first century. However, this requires wider screening of the nutritional and environmental effects of such plants than has so far occurred. It also requires a more thoughtful approach to the release of genetically modified plants to farmers than was the case at the end of the twentieth century where sometimes inappropriate commercial practices greatly threatened the future of the genetic modification of plants. Similarly, the cloning of domesticated animals may lead to more efficient production of better quality meat, milk, eggs and wool and perhaps to more sophisticated drugs and vaccines. However, these probabilities pale into insignificance by comparison with the myriad outcomes possible from engineering the human genome. Gene therapy and related techniques promise a substantial improvement in the treatment of hereditary diseases and cancer and could bring about revolutionary changes in organ transplant and renewal practice. They could even lead to such a retardation of the ageing process that the human lifespan might be extended to the point where a kind of immortality, but not invulnerability, becomes feasible [140]. At the beginnings of life, it is probable that many women will choose to have their fertilized eggs screened for genetic characteristics outside the womb so that only those free of genes predisposing them to disease will be implanted into the womb to go to full term. Each of these human interventions into the course of biology will change the evolutionary process while at the same time raising substantial ethical issues. It is to be hoped that humanity will show its greatest and eponymous attribute in overseeing these exciting and challenging prospects.

256 *Challenged Earth*

References

1. V. Orel, Gregor Mendel: *The First Geneticist*, Oxford University Press, Oxford, UK, 1996.
2. E. S. Lander and R. A. Weinberg, Genomics: journey to the center of biology, *Science*, **2000**, *287*, 1777.
3. M. J. Reiss and R. Straughan, *Improving Nature? The Science and Ethics of Genetic Engineering*, Cambridge University Press, Cambridge, UK, 1996.
4. P. Nurse, The incredible life and times of biological cells, *Science*, **2000**, *289*, 1711.
5. R. V. Miller, Bacterial gene swapping in nature, *Sci. Am.*, **1998**, *278*, Jan., 67.
6. J. Kingsland, Border control, *New Sci.*, *Inside Science 132*, **2000**, 15 July, 1.
7. a) R. E. Dickerson, The DNA helix and how it is read, *Sci. Am.*, **1983**, *249*, Dec., 86. b) L. Hood and D. Galas, The digital code of DNA, *Nature*, **2003**, *421*, 444. c) The double helix – 50 years. A series of articles in *Nature*, **2003**, *421*, 395. d) DNA: 50 years of the double helix. A series of articles in *New Sci.*, **2003**, 15 March, 35.
8. a) C. R. Woese, O. Kandler and M. L. Wheelis, Towards a natural system of organisms: proposals for the domains Archaea, Bacteria and Eucarya, *Proc. Natl. Acad. Sci. USA*, **1990**, *87*, 4576. b) P. J. Keeling and W. F. Doolittle, Archea: Narrowing the gap between prokaryotes and eukaryotes, *Proc. Natl. Acad. Sci. USA*, **1995**, *92*, 5761. c) J. L. Stein and M. I. Simon, Archeal ubiquity, *Proc. Natl. Acad. Sci. USA*, **1996**, *93*, 6228. d) M. T. Madigan and B. L. Manx, Extremophiles, *Sci. Am.*, **1997**, *276*, April, 67. e) J. A. Lake, R. Jain and M. C. Rivera, Mix and match in the tree of life, *Science*, **1999**, *283*, 2027. f) E. Pennisi, Is it time to uproot the tree of life? *Science*, **1999**, *284*, 1305.
9. M. Blackwell, Terrestrial life - fungal from the start? *Science*, **2000**, *289*, 1884.
10. a) J. H. Strauss and E. G. Strauss, With a little help from the host, *Science*, **1999**, *283*, 802. b) J. Randerson, Its vast, for a virus, *New Sci.*, **2003**, 15 April, 18. c) B. La Scola, A giant virus in amoebae, *Science*, **2003**, *299*, 2033. d) A. E. Smith and A. Helenius, How viruses enter animal cells, *Science*, **2004**, *304*, 237. e) L. P, Villarreal, Are viruses alive? **2004**, *291*, 29 Dec., 76. f) A. Zlotnick, Viruses and the physics of soft condensed matter, *Proc. Natl. Acad. Sci. USA*, **2004**, *101*, 15549. g) R. Zandi *et al.*, Origin of icosahedral symmetry in viruses, *ibid.*, 15556.
11. P. Davies, The living dead, *New Sci.*, *Inside Science 144*, **2001**, 13 Oct., 1.
12. M. Balter, Evolution on life's fringes, *Science*, **2000**, *289*, 1866.
13. a) J. N. Weber, HIV infection: the cellular picture, *Sci. Am.*, **1988**, *259*, Oct., 81. b) W. C. Greene, AIDS and the immune system, *Sci. Am.*, **1993**, *269*, Sep., 67. c) M. A. Nowak and A. J. McMichael, How HIV defeats the immune system, *Sci. Am.*, **1995**, *273*, Aug., 42. d) J. M. Coffin, S. F. Hughes, and H. E. Varmus, eds., *Retrovirus*, Cold Spring Harbor Laboratory Press., USA, 1998. e) J. G. Bartlett and R. D. Moore, Improving HIV therapy, *Sci. Am.*, **1998**, *279*, July, 65. f) R. A. Weiss, Gulliver's travels in HIV land, *Nature*, **2001**, *410*, 963.
14. a) P. Davis, Control centre, *New Sci.*, *Inside Science 122*, **1999**, 17 July, 1. b) Y. Nakaseko and M. Yanagida, Cytoskeleton in the cell cycle, *Nature*, **2001**, *412*, 291.
15. L. A. Grivell, Mitochondrial DNA, *Sci. Am.*, **1983**, *248*, March, 60.
16. a) S. S. Hall, James Watson and the search for biology's "Holy Grail", *Smithsonian*, **1990**, *20*, Feb., 41. b) J. Rennie, Grading the gene tests, *Sci. Am.*, **1994**, *270*, June, 66. c) The genome has landed, *New Sci.*, **2000**, 20 May, 14. A series of articles by several authors.
17. R. Dawkins, *The Selfish Gene*, Oxford University Press, Oxford, UK, 1989.
18. a) D. Malkin *et al.*, Germ line p53 mutations in a familial syndrome of breast cancer, sarcomas, and other neoplasms, *Science*, **1990**, *250*, 1233. b) B. J. Culliton, Hubert Humphrey's bladder cancer, *Nature*, **1994**, *369*, 13. c) R. H. Reeves, Recounting a genetic story, *Nature*, **2000**, *405*, 283. d) G. Jiminez-Sanchez, B. Childs and D. Valle, Human disease genes, *Nature*, *ibid.*, 853. e) L. Peltonen and V. A. McKusick, Dissecting disease in the postgenomic era, *Science*, **2001**, *291*, 1224. f) A. W. Futreal *et al.*, Cancer and genomics, *Nature*, **2001**, *409*, 850. g) K. Bendall, Genes, the genome and disease, *New Sci.*, *Inside Science 138*, **2001**, 17 Feb., 1. h) A. Chakravarti and P. Little, Nature, nurture and

human disease, *Nature*, **2003**, *421*, 413. i) J. I. Bell, The double helix in clinical practice, *ibid.*, 414. j) C. Dennis, The rough guide to the genome, *Nature*, **2003**, *425*, 758. k) J. Couzin, Two new asthma genes uncovered, *Science*, **2004**, *304*, 185.

19. a) J. D. Watson and F. H. C. Crick, Molecular structure of nucleic acid, *Nature*, **1953**, *171*, 737. b) J. D. Watson and F. H. C. Crick, Genetical implication of the structure of deoxyribonucleic acid, *ibid.*, 964. c) R. Olby, Quiet debut of the double helix, *Nature*, **2003**, *421*, 402. d) B. Maddox, The double helix and the 'wronged heroine', *ibid.*, 407.

20. J. D. Watson, *The Double Helix*, Atheneum, New York, USA, 1968.

21. L. Roberts *et al.*, A history of the human genome, *Science*, **2001**, *291*, 1195.

22. E. C. Friedberg, DNA damage and repair, *Nature*, **2003**, *421*, 436.

23. a) M. Stoneking, From the evolutionary past..., *Nature*, **2001**, *409*, 821. b) A. Chakravarti, ...to a future of genetic medicine, *ibid.*, 822. c) The International SNP Map Working Group, A map of human genome sequence variation containing 1.42 million single nucleotide polymorphisms, *ibid.*, 928.

24. a) J. Marx, How DNA replication works, *Science*, **1995**, *270*, 1585. b) B. Alberts, DNA replication and recombination, *Nature*, **2003**, *421*, 431.

25. a) E. Pennisi, Closing in on the centromere, *Science*, **2001**, *294*, 30. b) M. G. Schuler *et al.*, Genomic and genetic definition of a functional centromere, *ibid.*, 109.

26. a) W. Makalowski, Not junk after all, *Science*, **2003**, *300*, 1246. b) G. Lev-Maor *et al.*, The birth of an alternatively spliced exon: 3' splice-site selection in *Alu* exons, *ibid.*, 2003. c) W. W. Gibbs, The unseen genome: gems among the junk, *Sci. Am.*, **2003**, *289*, Nov., 46.

27. S. Hirotsune *et al.*, An expressed pseudogene regulates the messenger-RNA stability of its homologous coding gene, *Nature*, **2003**, *423*, 91. d)

28. W. A. Haseltine, Discovering genes for new medicines, *Sci. Am.*, **1997**, *276*, March, 78.

29. a) R. C. Duke, D. M. Ojcius, and J. D.-E. Young, Cell suicide in health and disease, *Sci. Am.*, **1996**, *275*, Dec., 48. b) K. Weissman, Life and cell death, *Chem. Br.*, **2003**, *39*, Aug., 19.

30. a) P. Cohen, Lights camera action, *New Sci.*, **2000**, 20 May, 18. b) P. Cohen, Life, the sequel, *New Sci.*, **2000**, 30 Sep., 32.

31. The codons for the amino acids are: alanine; GCU, GCC, GCA, GCG; arginine; CGU, CGC, CGA, CGG; aspargine; GAU, GAC; cysteine; UGU, UGC; glutamine; CAA, CAG; glutamic acid; GAA, CAG; glycine; GGU, GGC, GGA, GGG; histidine, CAU, CAC; isoleucine, AUU, AUC, AUA; leucine, CUU, CUC, CUA, CUG; lysine, AAA, AAG; methionine, AUG; phenylalanine; UUU, UUC; proline, CCU, CCC, CCA, CCG; serine, UCU, UCC, UCA, UCG; threonine, ACU, ACC, ACA, ACG; tryptophan, UGG; tyrosine, UAU, UAC; and valine, GUU, GUC, GUA, GUG; and terminations are indicated by UAA, UAG, UGA.

32. M. Ibba, Discriminating right from wrong, *Science*, **2001**, *284*, 70.

33. a) T. A. Holton *et al.*, Cloning and expression of cytochrome P450 genes controlling flower colour, *Nature*, **1993**, *366*, 276. b) R. Walden and R. Wingender, Gene transfer and plant regeneration techniques, *Trends Biotechnol.*, **1995**, *13*, 324.

34. a) O. J. Goddijn and J. Pen, Plants as bioreactors, *ibid.*, 379. b) H. S. Mason and C. J. Arntzen, Transgenic plants as vaccine production systems, *ibid*, 388. c) H. J. Bohnert and R. G. Jensen, Strategies for engineering water stress tolerance in plants, *Trends Biotechnol.*, **1996**, *14*, 89. d) T. H. Schuler *et al.*, Insect-resistant transgenic plants, *Trends Biotechnol.*, **1998**, 16, 168. e) F. Varoquaux *et al.*, Less is better: new approaches for seedless fruit production, *Trends Biotechnol.*, **2000**, 18, 233. f) J. K.-C. Ma, Genes, greens, and vaccines, *Nature Biotechnol.*, **2000**, *18*, 1141. g) G. Giddings *et al.*, Transgenic plants as factories for biopharmaceuticals, *ibid.*, 1151. h) Q. Schiermeier, Designer rice to combat diet deficiencies makes its debut, *Nature*, **2001**, *409*, 551. i) P. Cohen, Fighting over pharming, *New Sci.*, **2003**, 1 March, 22.

35. a) R. H. Devlin, *et al.*, Extraordinary salmon growth, *Nature*, **1994**, *371*, 209. b) E. Niiler, FDA considers transgenic fish, *Nature Biotechnol*, **2000**, *18*, 143. c) R. H. Devlin *et al.*, Growth of domesticated transgenic fish, *Nature*, **2001**, *409*, 781.

36. a) G. Vogel, Infant monkey carries jellyfish gene, *Science*, **2001**, *291*, 226. b) A. W. S. Chan *et al.*, Transgenic monkeys produced by retroviral gene transfer into mature oocytes, *Science*, **2001**, 291, 309. c) A. Coghlan and E. Young, Too close for comfort. If its monkeys today, will it be humans tomorrow? *New Sci.*, **2001**, 20 Jan., 6. d) A. W. S. Chan *et al.*, Transgenic monkeys produced by retroviral gene transfer into oocytes, *Science*, **2001**, *291*, 309. e). G. Schueller, But is it art? *New Sci.*, **2001**, 6 Jan., 34.

37. a) E. Marshall, Moratorium urged on germ line gene therapy, *Science*, **2000**, *289*, 2023. b) J. A. Barritt *et al.*, Mitochondria in human offspring derived from ooplasmic transplantation, *Hum. Reprod.*, **2001**, *16*, 513. c) E. Parens and E. Juengst, Inadvertently crossing the germ line, *Science*, **2001**, *292*, 397. d) A. Coghlan and J. Marchant, My two mums, *New Sci.*, **2001**, 12 May, 7.

38. a) M. Szekely, φX174 sequenced, *Nature*, **1977**, *265*, 685. b) F. Sanger *et al.*, Nucleotide sequence of bacteriophage φX174 DNA, *Nature*, **1977**, *265*, 687. c) F. Sanger *et al.*, The nucleotide sequence of bacteriophage φX174, *J. Mol. Biol.*, **1978**, *125*, 225. d) F. Sanger *et al.*, Nucleotide sequence of bacteriophage λ DNA, *J. Mol. Biol*, **1982**, *162*, 729.

39. a) L. Smith *et al.*, Fluorescence detection in automated DNA sequence analysis, *Nature*, **1986**, *321*, 674. b) R. D. Fleischmann *et al.*, Whole-genome random sequencing and assembly of *Haemophilus influenzae* Rd, *Science*, **1995**, *269*, 496. c) C. J. Bult *et al.*, Complete genome sequence of the methanogenic Archaeon, *Methanococcus jannaschi*, *Science*, **1996**, *273*, 1058. d) R. A. Clayton *et al.*, The first genome from the third domain of life, *Nature*, **1997**, 387, 459. e) A. Goffeau *et al.*, The yeast genome directory, *Nature*, *Supplement*, **1997**, *387*, 5. f) The *C. elegans* Sequencing Consortium, Genome sequence of the nematode *C. elegans:* a platform for investigating biology, *Science*, **1998**, *282*, 2012.

40. a) O. Goga Vukmirovic and S. M. Tilghman, Exploring genome space, *Nature*, **2000**, *405*, 820. b) M Snyder and M Gerstein, Defining genes in the genomics era, *Science*, **2003**, *300*, 258.

41. a) E. M. Meyerowitz, Today we have the naming of parts, *Nature*, **1999**, *402*, 731. b) X. Lin *et al.*, Sequence and analysis of chromosome 2 of the plant *Arabidopsis thaliana*, *Nature*, *ibid.*, 761. c) K. Mayer *et al.*, Sequence and analysis of chromosome 4 of the plant *Arabidopsis thaliana*, *ibid.*, 769. d) V. Walbot, A green chapter in the book of life, *Nature*, **2000**, *408*, 794. e) The Arabidopsis Genome Initiative, Analysis of the genome sequence of the flowering plant *Arabidopsis thaliana*, *ibid*, 796. f) A Theologis *et al.*, Sequence and analysis of chromosome 1 of the plant *Arabidopsis thaliana*, *Nature*, *ibid.*, 816. g) M. Salanoubat *et al.*, Sequence and analysis of chromosome 3 of the plant *Arabidopsis thaliana*, *Nature*, *ibid.*, 820. h) S. Tabata *et al.*, Sequence and analysis of chromosome 5 of the plant *Arabidopsis thaliana*, *ibid*, 823.

42. a) T. G. Wolfsberg, J. McEntyre, and G. D. Schuler, Guide to the draft genome, *Nature*, **2001**, *409*, 824. b) D. Baltimore, Our genome unveiled, *ibid*, 815. c) The International Human Genome Mapping Consortium, Initial sequencing and analysis of the human genome, *ibid.*, 860. d) J. C. Venter *et al.*, The sequence of the human genome, *Science*, **2001**, *291*, 1304. e) L. Philips, A history of the human genome project, *Science*, **2001**, 291, 1195.

43. a) F. R. Blattner *et al.*, The complete genome sequence of *Escherichia coli* K-12, *Science*, **1997**, *277*, 1453. b) S. T. Cole *et al.*, Deciphering the biology of Mycobacterium tuberculosis from the complete genome sequence, *Nature*, **1998**, *393*, 537. c) D. Adam, Now for the hard ones, *Nature*, **2000**, *408*, 792. d) M. D. Adams *et al.*, The genome sequence of *Drosophila melanogaster*, *Science*, **2000**, *287*, 2185. e) E. P. Greenberg, Pump up the versatility, *Nature*, **2000**, *406*, 947. S. T. Cole *et al.*, Massive gene decay in the leprosy bacillus, *Nature*, **2000**, *409*, 1007. f) A. Coghlan, Shotgun wedding, *New Sci.*, **2001**, 19 May, 7. g) E. Pennisi, The human genome, *Science*, **2001**, *291*, 1177.

44. C. K. Stover *et al.*, Complete genome sequence of *Pseudomonas aeruginosa* PA01, an opportunistic pathogen, *Nature*, **2000**, *406*, 959.

45.   a) G. M. Rubin, Comparing species, *Nature*, **2001**, *409*, 820. b) R. Tupler, G. Perini and M. R. Green, Expressing the human genome, *ibid.*, 832. c) P. Cohen and A. Coghlan, Less is more: our genes are subtler than we thought, *New Sci.*, **2001**, 17 Feb., 6.

46.   a) E. R. Winstead, Sizing up genomes: *Amoeba* is king, 2001, 21 Feb., http//www. celera.com/genomics/news/articles/02_01/Sizing_genomes.cfm. b) T. R. Gregory and P. D. N. Herbert, The modulation of DNA content: proximate causes and ultimate consequences, *Genome Res.*, **1999**, *9*, 317. c) W. H. Li, *Molecular Evolution*, Sinauer Associates Inc., Sunderland, Massachusetts, USA, 1997. d) S. Kraiec and M. Riley, Organisation of the bacterial chromosome, *Microbiol. Rev.*, **1990**, *54*, 502. e) A listing of many genomes may be found at http//www.cbs.dtu.dk/databases/DOGS/abbr_table.bysize.txt.

47.   L. Roberts, Controversial from the start, *Science*, **2001**, *291*, 1182.

48.   a) J. C. Venter *et al.*, Shotgun sequencing of the human genome, *Science*, **1998**, *280*, 1540. b) P. Little, Two routes to the genome, *Nature*, **1999**, *402*, 468. c) M. V. Olson, Clone by clone by clone, *Nature*, **2001**, *409*, 816.

49.   C. Macilwain, World leaders heap praise on human genome landmark, *Nature*, **2000**, *405*, 983.

50.   a) A. Mauron, Is the genome the secular equivalent of the soul? *Science*, **2001**, *291*, 831. b) S. Pääbo, The human genome and our view of ourselves, *Science*, **2001**, *291*, 1219.

51.   P. Bork and R. Copley, Filling in the gaps, *Nature*, **2001**, *409*, 818.

52.   a) E. Pennisi, Reaching their goal early, sequencing labs celebrate. *Science*, **2003**, *300*, 409. b) F. S. Collins, M. Morgan and A. Patrinos, The human genome project: from large-scale biology, *ibid.*, 286. c) C. Henry, Human genome project finished, *Chem. Eng. News*, **2003**, *81*, 21 April, 12.

53.   a) L. D. Stein, End of the beginning, *Nature*, **2004**, *431*, 915. b) International Human Genome Sequencing Consortium, Finishing the euchromic sequence of the human genome, *ibid.*, 931.

54.   a) A. Coghlan, Beg, steal or borrow, *New Sci.*, **2000**, 20 May, 17. b) W.-H. Li *et al.*, Evolutionary analysis of the human genome, *Nature*, **2001**, *409*, 847. c) J. O. Andersson, W. F. Doolittle and C. L. Nesbe, Are there bugs in our genome? *Science*, **2001**, *292*, 1848. d) S. L. Salzberg *et al.*, Microbial genes in the human genome: lateral transfer or gene loss? *ibid.*, 1903. e) S. Pääbo, The mosaic that is our genome, *Science*, **2003**, *421*, 395.

55.   a) J. Couzin, New mapping project splits the community, S. B. Gabriel *et al.*, *Science*, **2002**, *296*, 1391. b) The structure of haplotype blocks in the human genome, *ibid.*, 2225. c) J. Couzin, Consensus emerges on HapMap strategy, *Science*, **2004**, *304*, 671. d) D. Altshuler and A. G. Clark, Harvesting medical information from the human family tree, *Science*, **2005**, *307*, 1052. e) D. A Hinds *et al.*, Whole-genome patterns of common DNA variations in three human populations, *ibid.*, 1072.

56.   The International Hapmap Consortium, The International HapMap Project, *Nature*, **2003**, *426*, 789.

57.   a) Editorial, The promise of proteomics, *Nature*, **1999**, *402*, 703. b) A. Abbott, A post-genomic challenge: learning to read patterns of protein synthesis, *ibid.*, 715. c) D. Eisenberg *et al.*, Protein function in the post-genomic era, *Nature*, **2000**, *405*, 823. d) A. Pandey and M. Mann, Proteomics to study genes and genomes, *ibid.*, 837. e) A. Coghlan, Land of opportunity, *New Sci.*, **2000**, 4 Nov., 30. f) P. Cohen, High protein, *ibid.*, 38. g) C. Ezzell, Beyond the human genome, *Sci. Am.*, **2000**, *283*, July, 52. h) S. Fields, Proteomics in genomeland, *Science*, **2001**, *291*, 1221. i) G. Grandi, Antibacterial vaccine design using genomics and proteomics, *Trends Biotechnol.*, **2001**, *19*, 181. j) K. H. Lee, Proteomics: a technology-driven and technology-limited discovery, *ibid.*, 19, 217.

58.   a) N. Risch and K. Merikangas, The future of genetic studies of complex human diseases, *Science*, **1996**, *273*, 1516. b) W. F. Bodmer, Genetic diversity and disease susceptibility, *Phil. Trans. Roy. Soc., B*, **1997**, *352*, 1043. c) J. Bell, Genetics of common disease: implications for therapy, screening and redefinition of disease, *ibid.*, 10. d) T. Friedmann, Overcoming the obstacles, *Sci. Am.*, **1997**, *276*, June, 80. e) N. Boyce, In sickness and in health, *New Sci.*, 25 Oct., **1997**, 20. f) N. J. Risch, Searching for genetic determinants in the

new millennium, *Nature*, **2000**, *405*, 847. g) A. D. Roses, Pharmacogenetics and the practice of medicine, *ibid.*, 857. h) C. Sander, Genomic medicine and the future of health care, *Science*, **2000**, *287*, 1977. i) J. Marchant and M. Day, Health and happiness.....or doom and gloom, *New Sci.*, **2000**, 20 May, 20. j) K. Brown, Close cousins, *New Sci.*, **2000**, 4 Nov., 42. k) J. Marchant, Know your enemy, *ibid.*, 46. l) A. Coghlan and J. Marchant, What's next, *New Sci.*, **2001**, 17 Feb., 5. m) A. Abbott, With your genes? Take one of these, three times a day, *Nature*, **2003**, *425*, 760.

59.   a) K. L. Hudson *et al.*, Genetic discrimination and health insurance: an urgent need for reform, *Science*, **1995**, *270*, 391. b) M. Wadman, New Jersey outlaws genetic discrimination, *Nature*, **1996**, *384*, 202. c) E. Masood, UK insurers oppose moratorium plea on use of genetic data, *Nature*, **1998**, *391*, 3. d) R. J. Pokorski, A test for the insurance industry, *ibid.*, 835. e) M. Murphy, Book of death? *Chem. Indust.*, **2000**, 10 July, 427. f) E. Young, Premium genes. Britain sanctions genetic tests for insurance purposes, *New Sci.*, **2000**, 21 Oct., 7. g) J. M. Jeffords and T. Daschle, Political issues in the genome era, *Science*, **2001**, *291*, 1249. h) W. Nowlan, A rational view of insurance and genetic discrimination, *Science*, **2002**, *297*, 195. i) K. H. Rothenberg and S. F. Terry, Before its too late – addressing fear of genetic information, *ibid.*, 196. j) T. Duster, Race and reification in science, *Science*, **2005**, *307*, 1050.

60.   a) P. G. Debenham, Probing identity: the changing face of DNA fingerprinting, *Trends Biotechnol.*, **1992**, *10*, 96. b) E. Inder, DNA fingerprinting: Science, law, and the ultimate identifier, in *The Code of Codes: Scientific and Social Issues, in The Human Genome Project*, D. J. Kevles and L. Hood, eds., Harvard University Press, Cambridge, USA, **1992**. c) C. Couet, J. McBride and S. Oehlschlager, PCR - a repeating performance, *Chem. Br.*, **1994**, *30*, Nov., 903. d) R. Hoyle, The FBI's national data base, *Nature Biotechnol.*, **1998**, *16*, 987. e) J. M. Butler and B. C. Levin, Forensic applications of mitochondrial DNA, *Trends Biotechnol.*, **1998**, 16, 158. f) A. Watson, A new breed of high-tech detectives, *Science*, **2000**, *289*, 850. g) Editorial, We're all suspects now, *New Sci.*, **2001**, 5 May, 3. h) D. Concar, What's in a fingerprint? *ibid.*, 9. i) D. Concar, Could it be you? *ibid.*, 10. j) R. Williamson and R. Duncan, DNA testing for all, *Nature*, **2002**, *418*, 585.

61.   a) P. Gill *et al.*, Identification of the remains of the Romanov family by DNA analysis, *Nature Genet.*, **1994**, *6*, 130. b) P. L. Ivanov *et al.*, Mitochondrial DNA sequence heteroplasmy in the Grand Duke of Russia Georgij Romanov establishes the authenticity of the remains of Tsar Nicholas II, *Nature Genet.*, **1996**, *12*, 417.

62.   a) S. Anderson *et al.*, Sequence and organisation of the human mitochondrial genome, *Nature*, **1981**, *290*, 457. b) D. Malakoff, The year of living dangerously, *Science*, **2001**, *294*, 2443. c) J. Paradise, L. Andrews and T. Holbrook, Patents on human genes: an analysis of scope and claims, *Science*, **2005**, *307*, 1566.

63.   P. Gill *et al.*, Establishing the identity of Anna Anderson Manahan, *Nature Genet.*, **1995**, *9*, 9. Erratum, *ibid*, 218.

64.   A. Lawler, Massive DNA identification effort gets under way, *Science*, **2001**, *294*, 278.

65.   a) N. Boyce and A. Coghlan, Your genes in their hands, *New Sci.*, **2000**, 20 May, 15. b) K. Brown, The human genome business today, *Sci. Am.*, **2000**, *282*, June, 40. c) M. Bobrow and S. Thomas, Patents in the genetic age, *Nature*, **2001**, *409*, 763. d) C. Curran, Who's right to patent our genes? *Chem. Indust.*, **2001**, 7 May, 268.

66.   a) E. Hagelberg *et al.*, DNA from ancient Easter Islanders, *Nature*, **1994**, *369*, 25. b) M. Enserink, Physicians wary of scheme to pool Icelander's genetic data, *Science*, **1998**, *281*, 890. c) C. Weijer and E. J. Emanuel, Protecting communities in biomedical research, *Science*, **2000**, *289*, 1142. d) L. Frank, Estonia prepares for national DNA database, *Science*, **2000**, *290*, 31. e) G. Pálsson and P. Rabinow, The Icelandic genome debate, *Trends Biotechnol.*, **2001**, *19*, 166. f) A. Abbot, Hopes of biotech interest spur Latvian population genetics, *Nature*, **2001**, *412*, 468. g) J. Kaiser, Population data bases boom, from Iceland to the U.S., *Science*, **2002**, *298*, 1158.

67. a) D. Butler and P. Pockley, ...as Monsanto makes rice genome public, *Nature*, **2000**, *404*, 534. b) E. Niiler, Monsanto releases rice data to academia, *Nature Biotechnol.*, **2000**, *18*, 484.
68. a) D. Dickson and D. Cyranoski, Commercial sector scores success with whole rice genome, *Nature*, **2001**, *409*, 551. b) R. J. Davenport, Syngenta finishes, consortium goes on, *Science*, **2001**, *291*, 807. c) M. Bevan, The first harvest of crop genes, *Nature*, **2002**, *416*, 590. d) D. Normile and E. Pennisis, Rice: Boiled down to bare essentials, *Science*, **2002**, *296*, 32. e) R. P. Cantrell and T. Reeves, The cereal of the world's poor takes center stage, *ibid.*, 53. f) P. Ronald and H. Leung, The most precious things are not jade and pearls.... *ibid.*, 58. g) J. Yu *et al.*, A draft sequence of the rice genome (*Oryza sativa* L. ssp. *indica*), *ibid.*, 79. h) S. A. Goff *et al.*, A draft sequence of the rice genome (*Oryza sativa* L. ssp. *japonica*), *ibid.*, 92. i) A. Coghlan, Complete rice sequence goes public at last, *New Sci.*, **2003**, 11 Jan., 5. j) The Rice Chromosome 10 Sequencing Consortium, In-depth view of structure, activity, and evolution of rice chromosome 10, *Science*, **2003**, *300*, 1566.
69. a) J. Enriquez, Genomics and the world's economy, *Science*, **1998**, *281*, 925. b) R. Dalton, Cereal gene bank accepts need for patents..., *Nature*, **2000**, *404*, 534.
70. a) R. S. Crespi, Biotechnology patents and morality, *Trends Biotechnol*, **1997**, *15*, 123. b) J. C. Polkinghorne, Ethical issues in biotechnology, *Trends Biotechnol.*, **2000**, *18*, 8.
71. a) R. Crea *et al.*, Chemical synthesis of genes for human insulin, *Proc. Natl. Acad. Sci. USA*, **1978**, *75*, 5765. b) D. V. Goeddel *et al.*, Expression in *Escherichia coli* of chemically synthesised genes for human insulin, *Proc. Natl. Acad. Sci. USA*, **1979**, *76*, 106. c) M. J. Gait, Synthetic genes for human insulin, *Nature*, **1979**, *277*, 429. d) G. I. Bell *et al.*, Sequence of the human insulin gene, *Nature*, **1980**, *284*, 26. e) D. Owerbach *et al.*, The insulin gene is located on chromosome 11 in humans, *Nature*, **1980**, *286*, 82. f) P. Newmark, Insulin on tap, *Nature*, **1982**, *299*, 293. g) D. Shapley, Human insulin, *Nature*, **1982**, *Proc. Natl. Acad. Sci. USA*, 100.
72. a) D. V. Goeddel *et al.*, Direct expression in *Escherichia coli* of a DNA sequence coding for human growth hormone, *Proc. Natl. Acad. Sci. USA.*, **1979**, *76*, 544. b) R. Walgate, Hormone growth, *Nature*, **1980**, *288*, 528.
73. a) P. Newmark, Engineered E. coli produce interferon, *Nature*, **1980**, *283*, 323. b) S. Pestka, The purification and manufacture of human interferons, *Sci. Am.*, **1983**, *249*, Aug., 28.
74. G. Hannig and S. C. Makrides, Strategies for optimising heterologous protein expression in *Escherichia coli*, *Trends Biotechnol.*, **1998**, *16*, 54.
75. H. Shizuya, *et al.*, Cloning and stable maintenance of 300-kilobase-pair fragments of human DNA in *Escherichia coli* using an F-factor-based vector, *Proc. Natl. Acad. Sci. USA*, **1992**, 89, 8794.
76. a) L. A. Cole, The specter of biological weapons, *Sci. Am.*, **1996**, *275*, Dec., 30. b) E. Marshall, Uncertainty on bioweapons treaty, *Science*, **2001**, *291*, 2288. c) M. Wheelis, Detering bioweapons development, *ibid.*, 2089. d) Editorial, A call to arms, *Nature*, **2001**, *411*, 223. e) S. M. Block, The growing threat of biological weapons, *Am. Sci*, **2001**, *89*, 28.
77. a) R. J. Jackson *et al.*, Expression of mouse interleukin-4 by a recombinant ectromelia virus suppresses cytolytic lymphocyte responses and overcomes genetic resistance to mousepox, *J. Virol.*, **2001**, *75*, 1205. b) R. Nowak, Disaster in the making, *New Sci.*, **2001**, 13 Jan., 4. c) E. Finkel, Engineered mouse virus spurs bioweapon fears, *Science*, **2001**, *291*, 585. d) L. Ember, Bioengineering work gone wrong, *Chem. Eng. News*, **2001**, *79*, 29 Jan., 13. e) C. Denis, The bugs of war, *Nature*, **2001**, *411*, 232. f) M. Enserink and R. Stone, Dead virus walking, *Science*, **2002**, *295*, 2001. g) D. MacKenzie and M. Le Page, "Act now" plea on bioterror threat, *New Sci.*, **2002**, 28 Sept., 4. h) P. Cohen, Recipes for bioterror, *New Sci.*, **2003**, 18 Jan., 10. i) A. S. Fauci, Biodefence on the research agenda, *ibid.*, 787.
78. J. Cello, A. V. Paul and E. Wimmer, Chemical synthesis of poliovirus cDNA: generation of infectious virus in the absence of natural template, *Science*, **2002**, *297*, 1016.
79. a) M. Le Page, Journal editors agree to censor research papers, *New Sci.*, **2003**, 22 Feb., 5. b) E. Check, US officials urge biologists to vet publications for bioterror risk, *Nature*, **2003**,

421, 197. c) Editorial, Statement on the consideration of biodefence and biosecurity, *ibid.*, 771.

80.   a) B. D. Smith, *The Emergence of Agriculture*, Scientific American Library, New York, USA, 1995. b) B. D. Smith, Between foraging and farming, *Science*, **1998**, *279*, 1651. c) H. Pringle, The slow birth of agriculture, *Science*, **1998**, *282*, 1446. d) R.-L. Wang *et al.*, The limits of selection during maize domestication, *Nature*, **1999**, *389*, 236. e) S. Pääbo, Neolithic genetic engineering, *Nature*, **1999**, *398*, 194. f) S. Lev-Yadun, A. Gopher, and S. Abbo, The cradle of agriculture, *Science*, **2000**, *288*, 1602. g) D. R. Piperno *et al.*, Starch grains reveal early root crop horticulture in the Panamanian tropical forest, *Nature*, **2000**, *407*, 894.

81.   a) M.-D. Chilton, A vector for introducing new genes into plants, *Sci. Am.*, **1983**, *248*, June, 36. b) C. S. Gasser and R. T. Frayley, Transgenic crops, *Sci. Am.*, **1992**, *266*, June, 34. c) E. Pennisi, New genome a boost to plant studies, *Science*, **2001**, *294*, 2266. d) D. W. Wood *et al.*, The genome of the natural genetic engineer *Agrobacteriun tumefaciens* C58, *ibid.*, 2317. e) B. Goodner, Genome sequence of the plant pathogen and biotechnology agent *Agrobacteriun tumefaciens* C58, *ibid.*, 2323.

82.   a) S. B. Gelvin, Agricultural biotechnology: Gene exchange by design, *Nature*, **2005**, *433*, 583. b) W. Broothaerts *et al.*, Gene transfer to plants by diverse species of bacteria, *ibid.*, 633.

83.   a) D. W. Ow, The right chemistry for marker gene removal? *Nature Biotechnol.*, **2001**, *19*, 115. b) J. Zuo *et al.*, Chemical-regulated, site specific DNA excision in transgenic plants, *ibid.*, 157. c) *Anon.*, Molecular shears provide safer GM crops, *Chem. Indust.*, **2001**, 5 Feb., 63.

84.   a) W. H. R. Langridge, Edible vaccines, *Sci. Am.*, **2000**, *283*, Sept., 48. b) Y. Thanavala *et al.*, Immunogenicity in humans of an edible vaccine for hepatitis B, *Proc. Nat. Acad. Sci. USA*, **2005**, *102*, 3378.

85.   a) Editorial, Too tempting: There is just one problem with edible vaccines, *New Sci.*, **2005**, 19 Feb., 3. b) *Anon.*, Why vaccination by potato got chopped, *ibid.*, 19.

86.   a) I. Wilmut *et al.*, Viable offspring derived from foetal and adult mammalian cells, *Nature*, **1997**, 385, 810. Erratum, *Nature*, **1997**, *386*, 200. b) E. Pennisi and N. Williams, Will Dolly send in the clones, *Science*, **1997**, *275*, 1415. c) C. Stewart, Anudder way of making lambs, *Nature*, **1997**, *385*, 769. d) T. Beardsley, The start of something big? *Sci. Am.*, **1997**, 276, May, 10. e) D. Butler, French clone provides support for Dolly, *Nature*, **1998**, *392*, 113. f) D. Solter, Dolly is a clone - no longer alone, *Nature* **1998**, *394*, 315. g) D. Ashworth *et al.*, DNA microsatellite analysis of Dolly, *ibid.*, 329. h) E. N. Signer *et al* DNA fingerprinting Dolly, *ibid.*, 329

87.   a) J. B. Gurdon and A. Colman, The future of cloning, *Nature*, **1999**, *402*, 743. b) A. McLaren, Cloning: pathways to a pluripotent future, *Science*, **2000**, *288*, 1775.

88.   a) Y. Kato *et al.*, Eight calves cloned from somatic cells of a single adult, *Science*, **1998**, *282*, 2095. b) T. Wakayama *et al.*, Full-term development of mice from enucleated oocytes injected with cumulus cell nuclei, *Nature*, **1998**, *394*, 369. c) N. Boyce, Go forth and multiply, *New Sci.*, **1998**, 25 July, 4. d) E. Pennisi and D. Normile, Perseverance leads to cloned pig in Japan, *Science*, **2000**, *289*, 1118. e) A. Onishi *et al.*, Pig cloning by microinjection of foetal fibroblast nuclei, *ibid.*, 1188. f) R. S. Prather, Pigs is pigs, *Science*, **2000**, *289*, 1886. g) I. A. Polejaeva *et al.*, Cloned pigs produced by nuclear transfer from adult somatic cells, *ibid.*, 86. h) M. Westhusin and J. Piedrahita, Three little pigs worth the huff and puff? *Nature Biotechnol.*, **2000**, *18*, 1144. i) T. Wakayama *et al.*, Cloning of mice to six generations, *Nature*, **2000**, *407*, 318. j) T. Shin *et al.*, A cat cloned by nuclear transplantation, *Nature*, **2002**, *415*, 859. k) C. Holden, First cloned mule races to finish line, *Science*, **2003**, *300*, 1354. l) C. Galli *et al.*, A cloned horse born to its dam twin, *Nature*, **2003**, *424*, 635.

89.   a) R. P. Lanza, B. L. Dresser, and P. Damiani, Cloning Noah's ark, *Sci. Am.*, **2000**, *283*, Nov., 66. b) G. Vogel, Cloned gaur a short-lived success, *Science*, **2001**, *291*, 409.

90.   a) E. Pennisi, After Dolly, a pharming frenzy, *Science*, **1998**, *279*, 646. b) J. B. Cibelli, *et al.*, Cloned transgenic calves produced from nonquiscent foetal fibroblasts, *Science*, **1998**, *280*, 1256.

91.   a) G. Vogel, In contrast to Dolly, cloning resets telomere clock in cattle, *Science*, **2000**, *288*, 586. b) D. Adam, Clone pioneer calls for health tests, *Nature*, **2002**, *415*, 103. c) O. Dyer, Dolly's arthritis dents faith in cloning, *Br. Med. J.*, **2002**, *324*, 67. d) N. Williams, Dolly clouds cloning hopes, *Curr. Biol.*, **2002**, *12*, R79. e) A. Coghlan, A sad farewell for Dolly the sheep, the world's first cloned mammal, *New Sci.*, **2003**, 22 Feb., 5.

92.   a) C. Tenove, Forever young, *New Sci.*, **2000**, 6 May, 4. b) R. P. Lanza *et al.*, Extension of cell life-span and teleomere length in animals cloned from senescent somatic cells, *Science*, **2000**, *288*, 665.

93.   a) N. Ogonuki *et al.*, Early death of cloned mice, *Nature Genet.*, **2000**, *30*, 253. b) N. Ogonuki *et al.*, Early death of mice cloned from somatic cells, *ibid.*, 253. c) P. Cohen, Live fast, die young, *New Sci.*, **2002**, 16 Feb., 14. d) C. Holden, Multiple gene defects found in clones, *Science*, **2002**, *297*, **1991**. e) F. Xue *et al.*, Aberrant patterns of X chromosome inactivation in bovine clones, *Nature Genet.*, **2002**, *31*, 216. f) I. Wilmut *et al.*, Somatic cell nuclear transfer, *Nature*, **2002**, *419*, 583.

94.   J. B. Gurdon, The Croonian Lecture, 1976. Egg cytoplasm and gene control in development, *Proc. Roy. Soc., B*, **1977**, *198*, 211.

95.   W. H, Velander, H. Lubon and W. N. Drohan, Transgenic livestock as drug factories, *Sci. Am.*, **1997**, *268*, Jan., 54.

96.   a) A. Schnieke, *et al.*, Human factor IX transgenic sheep produced by transfer of nuclei from transfected foetal fibroblasts, *Science*, **1997**, *278*, 2130. b) R. Rawls, Making drugs the transgenic way, *Chem. Eng. News*, **1997**, *75*, 6 Oct., 33. c) A. Mitchell, The science of the lambs, *Nature*, **1998**, *391*, 21. d) I. Wilmut, Cloning for medicine, *Sci. Am.*, **1998**, *279*, Dec., 30. e) M. Suraokar and A. Bradley, Targeting sheep, *Nature*, **2000**, *405*, 1004. f) K. J. McCreath *et al.*, Production of gene-targeted sheep by nuclear transfer from cultured somatic cells, *ibid.*, 1066.

97.   G. Vogel, Sheep fail to produce golden fleece, *Science*, **2003**, *300*, 2015.

98.   a) P. Aldhous, Cloning's owners go to war, *Nature*, **2000**, *405*, 610. b) A. Dove, Milking the genome for profit, *Nature Biotechnol.*, **2000**, *18*, 1045.

99.   a) J. L. Platt, New risks, new gains, *Nature*, **2000**, *407*, 27. b) D. Butler, Xenotransplant experts express caution over knockout piglets, *Nature*, **2002**, *415*, 103. j)

100. a) G. Vines, Pig transplants win ethical backing, *New Sci.*, **1996**, 9 March, 4. b) P. L. Tissier *et al.*, Two sets of human-tropic pig retrovirus, *Nature*, **1997**, *389*, 681. c) D. Butler *et al.*, Last chance to stop and think on risks of xenotransplants, *Nature*, **1998**, *391*, 320. d) J. L. Platt, New directions for organ transplantation, *Nature, Supplement*, **1998**, *392*, 11. e) R. A. Weiss, Transgenic pigs and virus adaptation, *ibid.*, 327. f) Editorial, The trials of xenotransplantation, *Nature*, **2000**, *406*, 661. g) R. S. Boneva, T. M. Folks and L. E. Chapman, Infectious disease issues in xenotransplantation, *Clin. Microbiol. Rev.*, **2001**, *14*, 1. h) J. Couzin, Wanted: Pig transplants that work, *Science*, **2002**, *295*, 1008.

101. L. J. W. van der Laan *et al.*, Infection by porcine endogenous retrovirus after islet xenotransplantation in SCID mice, *Nature*, **2000**, *406*, 90.

102. a) W. Heneine *et al.*, No evidence of infection with porcine endogenous retrovirus in recipients of porcine islet-cell xenografts, *Lancet*, **1998**, *352*, 695. b) K. Paradis *et al.*, Search for cross-species transmission of porcine endogenous retrovirus in patients treated with living pig tissue, *Science*, **1999**, *285*, 1236.

103. a) N. Williams, Cloning sparks calls for new laws, *Science*, **1997**, *275*, 1415. b) A. Kahn, Clone mammals ... clone man? *Nature*, **1997**, *386*, 119. c) D. Kestenbaum, Cloning plan spawns ethics debate, *Science*, **1998**, *279*, 315. d) G. Vogel, Cloning: could humans be next? *Science*, **2001**, *291*, 808. e) A. Abbot, Trepidation greets plan for cloning humans, *Nature*, **2001**, *410*, 293. f) J. Pickrell, Experts assail plan to help childless couples, *Science*, **2001**, *291*, 2061. g) D. W Brock, Human cloning and our sense of self, *Science*, **2002**, *296*, 314.

104. a) R. G. McKinnell and M. A. Berardino, The biology of cloning: history and rationale, *BioScience*, **1999**, *49*, Nov., 875. b) E. Pennisi and G. Vogel, Clones: a hard act to follow, *Science*, **2000**, *288*, 1722.

105. a) P. Cohen, Bad copies, *New Sci.*, **2001**, 3 Feb., 7. b) L. E. Young *et al.*, Epigenetic change in IGF2R is associated with fetal overgrowth after sheep embryo culture, *Nature Genet.*, **2001**, *27*, 153.

106. a) R. Jaenisch and I. Wilmut, Don't clone humans, *Science*, **2001**, *291*, 2552. b) P. Cohen and D. Concar, The awful truth, *New Sci.*, **2001**, 19 May, 14.

107. a) D. Adam, First human clones get a cool response, *Nature*, **2001**, *414*, 477. b) E. Marshall and G. Vogel, Cloning announcement sparks debate and skepticism, *Science*, **2001**, *294*, 1802. c) J. B. Cibelli *et al.*, The first human cloned, *Sci. Am.*, **2002**, *286*, Jan., 43.

108. a) G. Vogel, Scientists take step toward therapeutic cloning, *Science*, **2004**, *303*, 937. b) W. S. Hwang *et al.*, Evidence of a pluripotent human embryonic stem cell line derived from a cloned blastocyst, *ibid.*, 1669. c) H. Pearson, Cloning success marks Asian nations as scientific tigers, *Nature*, **2004**, *427*, 664.

109. a) M. J. Shamblott *et al.*, Derivation of pluripotent stem cells from cultured human primordial germ cells, *Proc. Natl. Acad. Sci. USA*, **1998**, *95*, 13726. b) J. A. Thomson *et al.*, Embryonic stem cell lines derived from human blastocysts, *Science*, **1998**, *282*, 1145. c) B. E. Reubinoff *et al.*, Embryonic stem cell lines from human blastocysts: somatic differentiation in vitro, *Nature Biotechnol.*, **2000**, *18*, 399.

110. a) E. Marshall, A versatile cell line raises scientific hopes, legal questions, *Science*, **1998**, *282*, 1014. b) P. Aldous, A world of difference, *Nature*, **2001**, *414*, 838.

111. a) G. Vogel, Company gets rights to cloned human embryos, *Science*, **2000**, *287*, 559. b) J. Grisham, EPO admits patent mistake, *Nature Biotechnol.*, **2000**, *18*, 366.

112. a) G. J. Annas, A. Caplan, and S. Elias, Stem cell politics, ethics and medical progress, *Nature Medicine*, **1999**, *5*, 1339. b) F. E. Young, A time for restraint, *Science*, **2000**, *287*, 1424. c) N. Lenoir, Europe confronts the embryonic stem cell research challenge, *ibid.*, 1425.

113. D. Perry, Patients' voices: The powerful sound in the stem cell debate, *Science*, **2000**, *287*, 1423.

114. a) C. R. Bjornson *et al.*, Turning brain into blood: a hematopoietic fate adopted by adult neural stem cells in vivo, *Science*, **1999**, *283*, 534. b) D. L Clarke *et al.*, Generalized potential of adult stem cells, *Science*, **1999**, *288*, 1660. c) G. Vogel, Can old cells learn new tricks? *Science*, **2000**, *287*, 1418. d) A. Coghlan, Back to the source, *New Sci.*, **2000**, 19 Aug., 15. e) D. Martindale, Old cells, new tricks, *New Sci.*, **2000**, 19 Aug., 16. f) A. Coghlan and E. Young, Reprogram your body. We won't need to clone embryos if we can recycle adult cells instead, *New Sci.*, 3 March, **2001**, 6. g) G. Vogel, Can adult stem cells suffice? *Science*, **2001**, *292*, 1820. h) C. Dennis, Take a cell, any cell.... *Nature*, **2003**, *426*, 490.

115. a) G. Vogel, Harnessing the power of stem cells, *Science*, **1999**, *283*, 1432. b) D. J. Mooney and A. G. Mikos, Growing new organs, *Sci. Am.*, **1999**, 280, April, 60. c) G. Wang *et al.*, Adult stem cells from bone marrow stroma differentiate into airway epithelial cells: Potential therapy for cystic fibrosis, *Proc. Natl. Acad. Sci. USA*, **2005**, *102*, 186.

116. a) R. A. Pedersen, Embryonic stem cells for medicine, *ibid.*, 68. b) R. S. Langer and J. P. Vacanti, Tissue engineering: the challenge ahead, *ibid.*, 86. c) I. L. Weissman, Translating stem and progenitor cell biology to the clinic: barriers and opportunities, *Science*, **2000**, *287*, 1442. d) A. Coghlan, Highly cultured, *New Sci.*, **2000**, 19 Aug., 15. e) D. Martindale, Making more of yourself, *New Sci.*, **2000**, 19 Aug., 16. f) S. Cohen and J. Leor, Rebuilding broken hearts, *Sci. Am.*, **2004**, *291*, Nov., 23. g) D. W. Hutmacher, M. Sittinger and M. V. Risbud, Scaffold-based tissue engineering: rationale for computer-aided design and solid free-form fabrication systems, *Trend. Biotechnol.*, **2004**, *22*, 354.

117. a) J. Gearhart, New potential for human embryonic stem cells, *Science*, **1998**, *282*, 1061. b) R. P. Lanza, J. B. Cibelli, and M. D. West, Human therapeutic cloning, *Nature Med.*, **1999**, *5*, 975.

118.	a) F. M. Watt and B. L. M. Hogan, Out of Eden: stem cells and their niches, *Science*, **2000**, *287*, 1427. b) D. van der Kooy and S. Weiss, Why stem cells? *ibid.*, 1439. c) K. D'Armour and F. H. Gage, New tools for human developmental biology, *Nature Biotechnol.*, **2000**, *18*, 381. d) A. Coghlan, Everything you ever wanted to know about stem cells, *New Sci.*, **2000**, 19 Aug., 14. e) M. Buckingham, A cellular cornucopia, *Nature*, **2000**, *408*, 773. f) R. Gardner *et al.*, *Stem cell research and therapeutic cloning: an update*, The Royal Society, London, UK, 2000.

119.	a) R. S. Kirsner, V. Falanga and W. H. Eaglstein, The development of bioengineered skin, *Trends Biotechnol.*, **1998**, *16*, 246. b) N. Parenteau, Skin: the first tissue-engineered products, *Sci. Am.*, **1999**, *280*, April, 83. c) L. G. Griffith and G. Naughton, Tissue-engineering - current challenges and expanding opportunities, *Science*, **2002**, *295*, 1009.

120.	a) H. Niwa, J. Miyazaki and A. G. Smith, Quantitative expression of Oct-3/4 defines differentiation, dedifferentiation or self-renewal of ES cells, *Nature Genet.*, **2000**, 24, 372. b) R. Nowak, Decision time, *New Sci.*, **2000**, 8 April, 4.

121.	a) J. Rossant and A. Nagy, *Nature Biotechnol.*, In search of the *tabula rasa* of human cells, **1999**, *17*, 23. b) J. A. Thomson and J. S. Odorico, Human embryonic stem cell and embryonic germ cell lines, *Trends Biotechnol.*, **2000**, *18*, 53.

122.	a) A. Smith, Cell therapy: in search of pluripotency, *Curr. Biol.*, **1998**, *8*, R802. b) D. Solter and J. Gearhart, Putting stem cells to work, *Science*, **1999**, *283*, 1468. c) A. Colman and A. Kind, Therapeutic cloning: concepts and practicalities, *Trends Biotechnol.*, **2000**, 18, 192.

123.	A. M. Shapiro *et al.*, Islet transplantation in seven patients with type 1 diabetes mellitus using glucocoticoid-free immunosuppressive regimen, *N. Engl. J. Med.*, **2000**, *343*, 230.

124.	a) N. Lumelsky *et al.*, Differentiation of embryonic stem cells to secreting structures similar to pancreatic islets, *Science*, **2001**, *292*, 1389. b) M. C. Gershengorn *et al.*, Epithelial-to-mesenchumal transition generates proliferative human islet cells, *Science*, **2004**, *306*, 2261. c) K Zaret, Self-help for insulin cells, *Nature*, **2004**, *429*, 30. d) Y. Dor *et al.*, Adult pancreatic β-cells are formed by self-duplication rather than stem-cell differentiation, *ibid.*, 41.

125.	a) M. F. Clarke, At the root of brain cancer, *Nature*, **2004**, *432*, 281. b) C. Sawyers, Targeted cancer therapy, *ibid.*, 294. c) P. A. Beachy, S. S. Karhadkar and D. M. Berman, Tissue repair and stem cell renewal in carcinogenesis, *ibid.*, 324. d) S. K. Singh *et al.*, Identification of human brain tumour initiating cells, *ibid.*, 396.

126.	a) A. Coghlan, Old stem cells can turn cancerous, *New Sci.*, **2005**, 21 April, 18. b) D. Rubio *et al.*, Spontaneous human adult stem cell transformation, *Cancer. Res.*, **2005**, *65*, 3035. c) J. S. Burns *et al.*, Tumorigenic heterogeneity in cancer stem cells evolved from long-term cultures of telomerase-immortalized human mesenchymal stem cells, *ibid.*, 3126.

127.	a) P. D. Robbins, H. Tahara, and S. C. Ghivizzani, Viral vectors for gene therapy, *Trends Biotechnol.*, **1998**, *16*, 35. b) F. H. Gage, Cell therapy, *Nature*, **1998**, 392, 18. c) W. F. Anderson, Human gene therapy, *ibid.*, 25. d) J. L Fox, Future tense for inutero gene therapy, *Nature Biotechnol.*, **1998**, *16*, 1002. e) P. R. Billings, *In utero* gene therapy - the case against, *Nature Med.*, **1999**, *5*, 255. f) H, Schneider and C. Coutelle, *In utero* gene therapy - the case for, *Nature Med.*, **1999**, *5*, 256. g) A. Mountain, Gene therapy: the first decade, *Trends Biotechnol.*, **2000**, 18, 199. h) J. Marchant, Generation game, *New Sci.*, **2000**, 2 Dec., 16. i) M. A. Kay, J. C. Glorioso and L. Naldini, Viral vectors for gene therapy: the art of turning infectious agents into vehicles of therapeutics, *Nature Med.*, **2001**, *7*, 33. j) J. M. Isner, Myocardial gene therapy, *Nature*, **2002**, *415*, 234. k) M. Murphy, Therapy in theory, *Chem. Indust.*, **2002**, 4 Feb., 9. l) J. Couzin and G. Vogel, Renovating the heart, *Science*, **2004**, *304*, 192.

128.	I. N. Verma, Gene therapy, *Sci. Am.*, **1990**, *263*, Nov., 34.

129.	a) W. F. Anderson, Gene therapy, *Sci. Am.*, **1995**, *273*, Sept., 96. b) R. M. Blaese *et al.*, T Lymphocyte-directed gene therapy for ADA-SCID: initial trial results after 4 years, *Science*, **1995**, *270*, 475. c) W. F. Anderson, The best of times, the worst of times, **2000**, *288*, 627.

130.   a) R. H. Buckley, Gene therapy for human SCID: dreams become reality, *Nature Med.*, **2000**, *6*, 623. b) M. Cavazzana-Calvo *et al.*, Gene therapy of human severe combined immunodeficiency (SCID)-X1 disease, *Science* **2000**, *288*, 669.

131.   a) E. Marshall, Gene therapy a suspect in leukemia-like disease, *Science*, **2002**, *289*, 34. b) E. Marshall, What to do when clear success comes with an unclear risk? *ibid.*, 510. c) E. Check, Regulators split on gene therapy as patient shows signs of cancer, *Nature*, **2002**, *419*, 545. d) E. Check, A tragic setback, *Nature*, **2002**, *420*, 116. e) E. Check, Safety panel backs principle of gene-therapy trials, *ibid.*, 595. f) H. Philips, Genes can come true, *New Sci.*, **2002**, 30 Nov., 29. g) E. Check, Second cancer case halts gene-therapy trials, *Nature*, **2003**, *421*, 305. h) J. Kaiser, Seeking the case of induced leukemias in X-SCID trial, *Science*, **2003**, *299*, 495. i) D. A. Williams and C. Baum, Gene therapy – new challenges ahead, *Science*, **2003**, *302*, 400. j) S. Hacein-Bey-Abina *et al.*, *LMO2*-Associated clonal T-cell proliferation in two patients after gene therapy for SCID-X1, *ibid.*, 415. k) M. Cavazzana-Calvo, A. Thrasher and F. Mavilio, The future of gene therapy, *Nature*, **2004**, *427*, 779.

132.   a) E. Check, Gene therapy put on hold as third child develops cancer, *Nature*, **2005**, *433*, 561. b) J. Kaiser, Panel urges limits on X-SCID trials, *Science*, **2005**, *307*, 1544.

133.   J. Couzin and J. Kaiser, As Gelsinger case ends, gene therapy suffers another blow, *Science*, **2005**, *307*, 1028.

134.   a) J. L. Fox., gene therapy issues come to the fore, *Nature Biotechnol.*, **1999**, *17*, 1153. b) J. L. Fox, Investigation of gene therapy begins, *Nature Biotechnol*, **2000**, *18*, 143. c) J. L. Fox, Scrutiny of gene therapy broadens, intensifies, *ibid.*, 377. d) T. Beardsley, Gene therapy setback, *Sci. Am.*, **2000**, *282*, Feb., 21. e) T. Friedmann, Principles for human gene therapy studies, *Science*, **2000**, *287*, 2163. f) P. Cohen, What went wrong? *New Sci.*, **2001**, 20 Jan., 8.

135.   a) J. L. Fox, Gene-therapy death prompts lawsuit, *Nature Biotechnol*, **2000**, *18*, 1136. b) E. Check, Sanctions agreed over teenager's gene-therapy death, *Nature*, **2005**, *433*, 674.

136.   a) Editorial, The increasing opacity of gene therapy, *Nature*, **1999**, *402*, 107. b) Editorial, Gene therapy's trials, *Nature Biotechnol.*, **2000**, 599. c) Editorial, Miracle cures and embryonic science, *Nature Biotechnol.*, **2001**, 19, 287. d) D. Concar, C. Ainsworth, and E. Young, Catastrophe or cure? *New Sci.*, **2001**, 24 March, 10.

137.   C. R. Freed *et al.*, Transplantation of embryonic dopamine neurons for severe Parkinson's disease, *N. Engl. J. Med.*, **2001**, *344*, 710.

138.   a) H. Li *et al.*, Generation of haircells by stepwise differentiation of embryonic stem cells, *Proc. Natl. Acad. Sci. USA*, **2003**, *100*, 13495. b) M. Izumikawa *et al.*, Auditory hair cell replacement and hearing improvement by *Atoh1* gene therapy in deaf mammals, *Nature Med.*, **2005**, *11*, 271.

139.   a) A. W. Murray and J. W. Szostak, Artificial chromosomes, *Sci. Am.*, **1987**, *257*, Nov., 60. b) P. E. Warburton and D. Kipling, Providing a little stability, *Nature*, **1997**, *386*, 553. c) J. J. Harrington *et al.*, Formation of *de novo* centromeres and construction of the first generation human artificial microchromosomes, *Nature Genet.*, **1997**, *15*, 345.

140.   J. Harris, Intimations of immortality, *Science*, **2000**, *288*, 59.

Chapter 6

Health and Disease: An Evolutionary Struggle

"And therefore never send to know for whom the bell tolls; it tolls for thee."

John Donne, Devotions on Emergent Occasions, 1624.

6.1. In Sickness and in Health

Sickness is a lonely experience, and most in their suffering are insufficiently contemplative to appreciate that they are in the midst of one of the innumerable skirmishes that are part of an evolutionary process. All humanity experiences sickness to some extent and each year about sixty million die mainly as a consequence of disease as shown in Fig. 6.1 [1]. Irrespective of age, most die of a disease that either represents a struggle between a pathogen and the immune

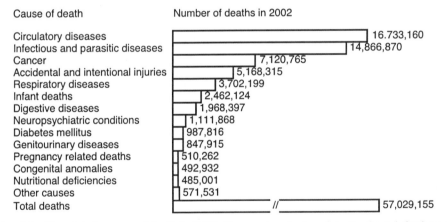

Fig. 6.1. The global causes of death in 2002. The great majority of deaths from infectious and parasitic diseases occurred in the developing nations. Data from [1].

system, a gene malfunction, or a disease of old age attributable to genetic limits to the human lifespan and to wear and tear [2]. In this chapter, disease is viewed largely in the context of the evolution of old pathogens, the appearance of new ones, the continual adaptation of the immune system to destroy pathogens and the evolution of humanity's response to disease in the form of increasingly sophisticated interventions that sometimes cause pathogens to evolve into new forms [3].

Some diseases that are major causes of death are either assisted by or caused by smoking, alcohol abuse, poor diet or poor exercise patterns [4,5]. This may also be viewed as evolutionary in origin as humans are the only species sufficiently sophisticated to develop life threatening habits that have little relation to the necessities of life. Most of the minority of deaths not due to disease arise from human inventiveness that is accompanied by new perils such as traffic accidents, firearms that only too easily lead to either intended or accidental deaths, and the use of mood altering drugs that can kill directly and indirectly through causing involvement in dangerous activities. This pattern is shown for the United States in 2000 in Fig. 6.2 where each death is assigned to a single cause to avoid double counting [4]. Thus, although there were 43,000 deaths due to traffic accidents, 16,653 of these were caused by alcohol consumption. The deaths attributed to microbial agents are those considered either avoidable through immunization or other treatment. Those attributed to sexual behaviour substantially arose through infection with the human immune deficiency virus, HIV. Accordingly, about half of the 2,403,351 deaths occurring in the United States in 2000 were probably preventable. Other developed nations show variations of this pattern of preventable deaths.

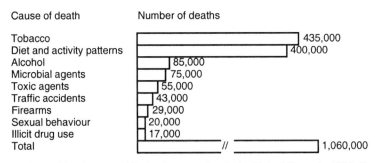

Cause of death	Number of deaths
Tobacco	435,000
Diet and activity patterns	400,000
Alcohol	85,000
Microbial agents	75,000
Toxic agents	55,000
Traffic accidents	43,000
Firearms	29,000
Sexual behaviour	20,000
Illicit drug use	17,000
Total	1,060,000

Fig. 6.2. The profile of preventable deaths occurring in the United States in 2000. Data from [4].

Inevitably, such statistics draw attention to the differences between nations. In the developed nations life expectancy is high and the possibility of dying young from disease is small. Such is not the case in developing nations where diseases such as malaria, tuberculosis, respiratory diseases, viral infections and typhus kill many millions annually, with cholera, dysentery and measles killing smaller, but still substantial, numbers. In earlier times the now developed nations similarly suffered from disease as is illustrated by a catastrophic pandemic that devastated Europe in the Middle Ages [6]. This was the Black Death that, apart from exacting a huge death toll, impacted deeply on the social and political fabric of Europe. Autoimmune deficiency syndrome (AIDS) is similarly affecting sub-Saharan Africa in the twenty-first century and could greatly disrupt other regions unless an affordable and effective control over the HIV is found very soon [7].

6.2. The Black Death: An European Catastrophe

One hundred million people lived in Europe, North Africa and nearby parts of the Middle East in 1346. Within a few years a quarter of them had died of a terrifying disease travelling westward along the Silk Road, the great trade route along which silk and other merchandise were brought from China to Europe [8-10]. Unfortunately, the traders often carried unwanted merchandise with them in the form of disease. In 1346, it was a combination of bubonic, septicaemic and pneumonic plague that burgeoned into a great pandemic that became known as the Black Death. The plague bacterium responsible was *Yersinia pestis* that is still carried by rodents, particularly black rats, in China, India and neighbouring nations and by chipmunks and woodchucks in the western United States. It is transmitted by fleas that ingest the bacterium when they bite an infected rodent. Subsequently, the bacterium multiplies in the flea's gut and is transmitted to a variety of hosts, including humans, that the flea next bites. The victim develops bubonic and septicaemic plague, where the latter can be passed between humans by fleas and lice, and pneumonic plague that is passed on through coughs and sneezes.

Rumours of great plagues in China, India and Mesopotamia reached Europe in 1346 at the same time as the first plague outbreaks occurred in Astrakhan and Saray, trading posts on the lower reaches of the Volga. It is likely that plague infected fleas were carried westward as far as Burma by the horse borne Mongol armies. From there, the plague moved further westward along trade routes as black rats from India became established in the settlements along the Silk Road and eventually reached Europe. It is thought that the skins of marmots, large rodents living in central Asia whose fur was then much prized, may have been infested with infected fleas that fed off black rats along the trade route and were a major factor in the westward spread of the plague. By 1347 the plague had reached Kaffa on the Black Sea. It was then carried by ships with infected crews and rats to Constantinople, modern Istanbul, that had lost forty percent of its population to a similar plague eight hundred years earlier. From Constantinople, it travelled onwards to southern Greece, the Egyptian coast, the eastern Adriatic coast, the west coast of Italy, Sicily, Sardinia and southern France as seen in Fig. 6.3. The spread continued onwards to encompass most of the coast of the Mediterranean, the Atlantic coast of Spain and Portugal and reached southern Ireland, England and the coast of northern France by 1348. The plague's advance now became inexorable as it enveloped all of central Europe, all of Ireland, much of England and Scotland, most of Scandinavia and the Baltic nations and surged deep into Russia by 1352, at the end of which year it subsided. The reason for this subsiding and subsequent quite rapid disappearance of the plague was a mystery for a long time but is now understood in terms of population dynamics based on black rats being the reservoir for *Yersinia pestis* [11].

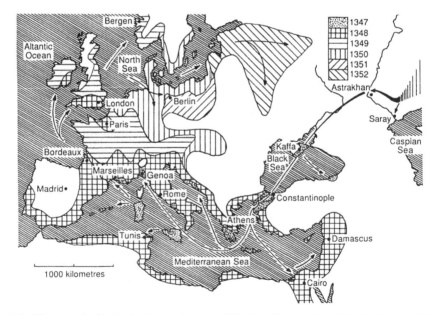

Fig. 6.3. The spread of bubonic plague along the Silk Road from central Asia to Europe. Adapted from [9].

The plague was extremely virulent with up to eighty percent of those infected dying. Death generally occurred within five days of infection and was preceded by painful swellings in either the lymph nodes of the armpit, neck or groin or in a combination of these locations. Fever followed three days later accompanied by black swellings, or buboes, arising from subcutaneous bleeding. Death was usually accompanied by some or all of several pneumonia-like symptoms in addition to agonizing bursting of swollen lymph glands and massive bleeding. The burial of the dead overwhelmed villages and towns where the dead often outnumbered the living and hasty burials in shallow graves frequently resulted in animals disinterring corpses for food. Often the stench of putrefaction became all pervasive.

Death on the massive scale of the Black Death destabilized the fabric of European society. An overwhelming feeling of powerlessness produced terror, hysteria, anger and a vengeful seeking of scapegoats. The medical faculty of the University of Paris unhelpfully advised the Pope and the French king that the plague was a result of a corruption of the atmosphere caused by a conjunction of Saturn, Mars and Jupiter [10]. Others had less exotic but far more dangerous ideas about the source of the plague. Many who did not readily fit into society such as strangers, cripples and beggars often fell under suspicion of being agents of the plague and were killed. Jews were singled out for massive persecution as exemplified by the killing of sixteen thousand in Strasbourg alone, but one example of many similar persecutions.

The plague reappeared with frequent regularity until the end of the fourteenth century. It is estimated that about twenty-five million died in the first onslaught starting in 1346 and that subsequent outbreaks carried off another twenty million by the end of the century so that a third of the European population was lost. This depopulation spread a mournful quiescence across much of Europe. Even today, the overgrown remains of villages whose populations vanished with the plague can still be found. It also had a profound and lasting effect in weakening the feudal system. The much decreased demand for food undermined prices at the same time that the cost of labour rose due to a decreased population. People became more self-assertive as a consequence of being hardened in surviving the plague, and the authority of the Catholic church was challenged by secular concerns, on one hand, and by religious reformers such as Huss in Bohemia and Wycliffe in England, on the other.

The last outbreak of plague in England, the Great Plague, beginning in 1665 occurred in London and ended amidst the Great Fire in 1666 [12]. The last epidemic occurred in France from 1720 to 1722. The displacement of the black rat by the Norwegian rat that does not carry a flea that both bites humans and carries *Yersinia pestis* eliminated the risk of plague in Europe in the nineteenth century. However, substantial outbreaks of plague occurred in China and India up to the early twentieth century, and isolated outbreaks have occurred there since, particularly in regions where the populations of black rats and fleas are high [13]. In total, there have been three great plague pandemics, the first being the Justinian plague of the fifth to seventh centuries, the Black Death and the third starting in southern China in 1894, during which Alexandre Yersin isolated the *Yersinia pestis* bacterium that bears his name. In total, 200 million have died from plague but now it is a readily treated disease and understanding of its nature has been greatly increased through the sequencing of the *Yersinia pestis* genome [14]. Under normal circumstances, a plague pandemic is unlikely to occur again. However, *Yersinia pestis* is potentially one of the most lethal biological warfare agents, as was realized by besieging Tartars who catapulting the bodies of plague victims into the Black Sea town of Kaffa in 1347, and which today could be sprayed in an aerosol to cause pneumonic plague.

While the Black Death still lingers in memories of Europeans, they are less familiar with their role in carrying infectious diseases such as measles, smallpox and epidemic typhus to the Americas and Australasia from the fifteenth century onwards [15]. As the European explorers and colonists spread, starting with Columbus in 1492, up to fifty-six million indigenous inhabitants of the Americas to the south of the present United States died with lesser, but still large numbers, dying in Australasia. Many were killed by the Europeans, but most died from diseases carried by the invaders against which their immune systems had developed little defence. Even in this century some of the last Amazonian tribes to be contacted by Europeans in the 1960s and 1970s lost up to three quarters of their number through contracting diseases against which they had little resistance. It is sometimes asserted that the indigenous Americans where partly revenged through infecting Europeans with syphilis, but the origin of this once deadly disease is uncertain [16]. Certainly tobacco came from the Americas and has killed millions of Europeans [17].

6.3. Knowing the Enemy: Pathogens

For most of humanity's existence the major threats to life have been starvation, physical injury, extreme weather and predatory large animals. These were readily recognized enemies but the causes of diseases that disabled and killed were unseen and unknown enemies. Both genetic and infectious diseases assailed humanity from the earliest times. However, while humans existed as nomadic hunter-gatherers in small and isolated groups infectious diseases were probably not so great a threat as they later became. With the advent of agriculture and the domestication of animals some ten to twenty thousand years ago, the spread of infectious diseases became more prevalent as settlements were established and the size of communities increased [18-21]. At the same time humans came into increasingly close contact with domesticated animals through consumption of milk and meat. It is also probable that water and food became increasingly contaminated by animal faecal matter. The result was that the transfer from animals to humans of large parasites, such as worms, and smaller ones such as protozoa, bacteria and viruses, collectively called microbes, or pathogens when agents of disease, often resulted in the transfer of infectious diseases from animals to humans in a process called zoonosis. So common is zoonosis, as is seen from Fig. 6.4, that it is thought that most infectious diseases were originally transferred to humans from animals [21].

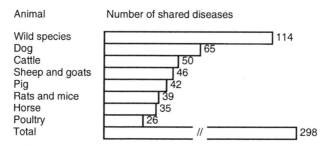

Fig. 6.4. Human diseases shared with animals through zoonosis. Some diseases are shared between many species. Data from [21].

The vast majority of microbes are invisible to the unaided eye and it was not until Antoni van Leeuwenhoek made his simple, but powerful, microscopes in the Netherlands in the seventeenth century that the teeming life of microbes was revealed [21]. Leeuwenhoek had an unpromising background for a person who was to become the founder of the science of microbiology. Born in 1643, he received little formal education and spent most of his life in Delft as a draper. He probably became familiar with optical lens through using them to examine cloth. Having realized his ability to make lenses with a two hundred fold magnification, he turned his attention to examining pond water that he found to contain prolific microscopic life. He then examined samples taken from himself and animals and therein discovered bacteria, protozoa, spermatozoa and also the circulation of red blood corpuscles in the capillaries of rabbits' ears. He wrote of his findings in many letters to the newly founded Royal Society of London of which he was made a Fellow in 1680. So impressive were his letters that more than a hundred were published in the prestigious Philosophical Transactions of the Royal Society and he remains the most prolific author in that journal.

Among the microbes that Leeuwenhoek discovered was at least one that produces disease, the protozoan *Giardia lamblia*, that caused him and many before and since to suffer from diarrhoea. Despite the newly gained ability to see bacteria and protozoa, pathogenic or otherwise, the generally much smaller viruses had to await the invention of the electron microscope before they could be seen. It is now well known that humanity not only shares the biosphere with a vast variety and population of microbes, but is also home to a large population of microbes of many types. These invaders use the human body as a source of nutrition and a basis for reproduction. Fortunately, most of them are benign and many are beneficial. However, some are pathogens that cause diseases varying from the mildly inconvenient, such as the *Rhinovirus* and the common cold, to the deadly, such as the bacterium *Salmonella typhi* and epidemic typhus that infected some sixteen million people annually in the developing nations, of whom 600,000 died each year, at the beginning of the twenty-first century [22]. Some pathogens cause a range of diseases such as the bacterium *Streptococcus pneumoniae* that causes bacteraemia, pneumonia and a form of meningitis that kills over three million children and even more elderly people annually [23].

In 1884, the Danish physician, Christian Gram, subdivided bacteria into two classes when he discovered that the outer membranes of some bacteria, such as *Escherichia coli*, previously stained with iodine and a dye called crystal violet, were decolourised when treated with acetone and alcohol. These he called Gram-negative, while those that were not decolourised such as *Staphylococcus aureus* and *Streptococcus pneumoniae*, he called Gram-positive; a classification still used today [24]. This classification is of more than academic interest for it indicates that the membranes of the two classes of bacteria differ and therefore respond differently to some antibiotics. It is now known that Gram-negative bacterial membranes contain lipopolysaccharides that cause the human immune system to secrete cytokines as part of its response in the fight against bacterial infection. In extreme cases, this secretion may produce lethal toxic shock resulting in fever, blood coagulation defects, lung dysfunction, kidney failure and circulatory collapse initiated by the response of the immune system to infection.

Largely because of their simplicity, compared with large species such as humans, many bacteria and viruses can multiply up to many billions of times in twenty-four hours and thereby mutate into new forms relatively rapidly [25]. They are able to change from innocuous organisms to pathogens and *vice versa*, and thereby generate a continuous evolution of disease to which the human immune system has to adapt. Thus, *Escherichia coli* that is usually a benign bacterial inhabitant of the human digestive tract is able to mutate into a pathogen. One such mutation transforms the benign *Escherichia coli* K-12 strain into the pathogenic *Escherichia coli* O157-H7 strain that causes often fatal food poisoning [26]. Similarly, the bacteria causing dysentery are eight separate mutants of *Escherichia coli*, possibly 35,000 to 270,000 years old, rather than a distinct bacterium, named *Shigella*, as was once thought. (Some 150 million years earlier *Escherichia coli* and *Salmonella enterica* mutated from the same bacterial ancestor.) Despite the rapidly increasing knowledge of the microbial domain, there remains considerable uncertainty about the diversity and identity of the microorganisms inhabiting the body, some of which are probably pathogens that cause unexplained illnesses and deaths [27].

6.4. Prions: The Unexpected Pathogens

The pathogens so far discussed all carry genetic material in the form of either deoxyribonucleic acid, DNA, or ribonucleic acid, RNA. It was therefore a great surprise when a new pathogen that contained no genetic material, the prion, was discovered by Stanley Prusiner who was awarded a Nobel Prize in 1997 [28]. Prions, whose name is derived from "proteinaceous infectious particle", are proteins that are present in most mammalian organs and most abundantly in the brain. Their function is unclear, other than that they can lead to fatal degenerative diseases of the brain collectively known as prion diseases, or transmissible spongiform encephalopathies (TSEs). The name "spongiform" comes from the appearance in the victim's brain of areas full of holes that resemble a sponge. This results from the formation of insoluble material, or plaques, that destroy brain cells with the result that the victim progressively loses mental capacity and bodily control.

One type of TSE, scrapie, has been known in sheep in the United Kingdom since the fifteenth century. It is so named because infected sheep continually scrape themselves against posts and trees as if suffering from skin irritation before the disease progresses to the point where control of limbs is lost. The appearance of this disease is infrequent and sporadic and caused little concern until a similar disease commonly known as "mad cow disease", and clinically as bovine spongiform encephalopathy (BSE), appeared with increasing frequency among cattle in the United Kingdom from 1984 onwards [29]. Its appearance was linked to a new practice of adding to cattle feed meal made from abattoir refuse that contained the remains of sheep that may have suffered from scrapie. This resulted in a huge cull of cattle in the United Kingdom in an attempt to eliminate the disease, and the addition of the meal to cattle feed was discontinued. However, BSE spread to parts of Europe and Japan and was detected in North America in 2003.

In 1996, the first human victim of a new disease similar to the degenerative brain disease that sometimes afflicts the elderly, Creutzfeld-Jacob disease (CJD), appeared and was named variant CJD, or vCJD [30]. A disturbing pattern emerged; the victims were much younger than expected for CJD, some being in their teens. A connection was made between vCJD and the consumption of beef from BSE infected cows thereby suggesting a transmission of a pathogen from sheep to cattle to humans. A prion, common in the plaques found in the brains of TSE affected sheep, cattle and humans, was identified as the cause of the disease.

In healthy brain cells prions exist in soluble forms that appear to be harmless and are collectively identified as PrP^C prions where C stands for cell. Prions, like other proteins, are composed of long helical strands of linked amino acids that form sheets and more complex folded structures [31]. It is thought that while in healthy brain cells PrP^C prions have one form of folded structure, they can, under as yet incompletely understood circumstances, fold into a different structure that is insoluble and forms the plaques leading to scrapie, BSE and vCJD. This form of prion is identified as PrP^{Sc} where Sc stands for scrapie. It is the PrP^{Sc} prion that appears to pass from traces of the central nervous system in beef to humans and there converts PrP^C to PrP^{Sc} to cause vCJD several years later.

It is feared that, because the extent of vCJD infection is unknown as a consequence of the time taken for the disease to incubate, an epidemic may be underway that could kill 136,000 people or more in the United Kingdom, particularly if vCJD can be contracted through consumption of scrapie infected mutton in addition to BSE infected beef [32]. The method of transmission of prions between species and the manner in which PrP^C prions are converted to PrP^{Sc} prions remains a matter of debate [33]. In a similar way, cannibalism in the Fore tribe of Papua New Guinea caused an epidemic of kuru, a vCJD-like disease [34]. A glimmer of hope for a cure for vCJD and other prion diseases appeared with experiments that showed that an antibody could stop mouse cells from producing PrP^{Sc} and that PrP^{Sc} already present was cleared from the cells [35].

While they do not appear to arise from infection with PrP^{SC}, a range of other diseases, among them CJD, Gerstmann-Sträussler-Scheinker disease and Alzheimer's disease, appear to be caused by insoluble proteins, such as β-amyloid protein produced by defective genes [36].

6.5. The Immune System and Human Resistance To Disease

Despite of, or because of, the myriad threats of disease humanity has evolved with a powerful resilience toward pathogens. All of the exposed human surfaces such as skin, the digestive tract, and the linings of orifices are densely populated by viruses, bacteria, protozoa and fungi. There, they fight among themselves for nutrients and living space and every so often invade the tissues, particularly in the case of viruses and bacteria [37]. A minority of these guests are distinctly unfriendly and together with worms, or helminths, represent at least 1,415 known pathogens as is seen from Fig. 6.5 [20]. They are the cause of a plethora of diseases. While humans come into contact with pathogens everyday through eating, breathing, skin abrasion, sex and contact with animals, plants, insects and soil they infrequently contract a disease. This is a reflection of the activity of the immune system that is able to recognize all of the constituents within a tissue of an individual as "self" and foreign entities, including pathogens, as "non-self" that it seeks to destroy with a sophisticated interlocking defensive system. The immune system has evolved through the continual struggle between diverse and rapidly changing pathogens and their higher animal and human hosts. This struggle is re-enacted throughout the life of each individual during which pathogens are prevented from building up to a level at which they can destroy tissue [20,38]. At the same time antipathogenic activity usually does not reach a level at which organs are damaged to the point where death might follow.

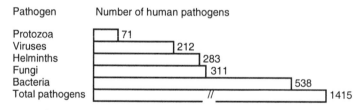

Fig. 6.5. The types and numbers of pathogens that attack humans. Data from [20].

Humans are born with an innate immune system that is able to recognize a range of pathogens and destroy them [39]. This is reinforced by an adaptive immune system that is built up by assembling gene segments into defensive systems to combat the great variety of pathogens encountered throughout life. The first stage of a pathogenic attack is the penetration of either mucusol linings or skin by pathogens that enter the blood stream and tissue. However, a significant infection may not occur as the pathogen may not pass through the mucosa and skin, that contain natural antipathogens, in either sufficiently good condition or sufficient numbers to survive. Those that do, encounter the first major line of defence, the innate immune system. Here they may either be destroyed by blood proteins, collectively called complement, or they may be coated with complement that attracts amoeba-like phagocytes that envelope the pathogen and digest it as they do abnormal and old body cells and cell fragments. However, some bacteria are coated with polysaccharides that render complement ineffective. Fortunately, larger phagocytes called macrophages can digest many such pathogens. Proteins, called toll-like receptors (TLRs [40,41]), attached to the phagocyte's surface possess molecular cavities, or receptor sites, that closely fit parts of the pathogen's polysaccharides and proteins and thereby "recognises" the pathogen. Such recognition of a pathogen by the TLRs, of which there are many types, triggers the innate immune system through the release of messenger molecules, or cytokines, and readies the second line of defence, the adaptive immune system.

Pathogens that survive the innate immune system and continue to multiply cause a noticeable infection and symptoms of disease. However, they are then faced with the adaptive immune system and the inflammatory, or humoral, response triggered by dendritic cells that also have TLRs attached to their surfaces that detect pathogens. This causes the release of cytokines such as interferon and interleukin-2 that direct white blood cells, or B-cells, to the infected site. The B-cells, also called B-lymphocytes, are produced in the bone marrow and can produce complex molecules called antibodies that attack the pathogen. They also produce memory cells that retain a memory of the pathogen and stimulate an attack on it upon reinfection that is the basis of humoral immunity.

Should the humoral response fail and the pathogen population continue to grow and invade tissue cells, the dendritic cells trigger a third and more powerful response. This is the cellular response that stimulates the production of T-lymphocytes, or T-cells, by the thymus that are of two main types, CD4$^+$ and CD8$^+$. Some of them act as helper T-cells through secretion of cytokines that stimulate the B-cells to greater activity, some destroy the pathogen and are known as killer T-cells. Others become memory T-cells that detect reinfection and trigger a fast response, the basis of cellular immunity. Inflammation, fever and pain usually accompany the cellular response. It is a two edged sword that destroys pathogens but while doing so may also destroy tissue to the point where either permanent tissue damage or death may occur. (Some 400,000 people in Europe and the United States die in this way annually [41].) This is exemplified by the immune response to several strains of the bacterium *Staphylococcus aureus*, or "golden staph", that can be so powerful that healthy tissue is destroyed [42]. Allergic reactions to food constituents, antibiotics and pollen have similar effects and, in the extreme case of autoimmune diseases that afflict up to five percent of humanity, the immune system largely loses the ability to distinguish "self" from "non-self" and attacks normal cells [43].

Human ingenuity has greatly aided the struggle against pathogens through the discovery and development of drugs, antibiotics and vaccines, collectively called pharmaceuticals, that ameliorate the symptoms of disease and help the immune system fight them. However, this ingenuity has also changed the evolution of pathogenic viruses and bacteria so that some once powerful drugs and antibiotics have been rendered largely ineffective and they have had to either be modified or replaced by new treatments to keep up with the evolving pathogens.

6.6. Strengthening the Defences

With the words:

> "There is a bark of an English tree, which I have found by
> experience to be a powerful astringent, and very efficacious
> in curing anguish and intermitting disorders."

written in a letter to the Royal Society of London in 1763, the Englishman, Edmund Stone, introduced the first systematic study of a folk medicine in Europe [44]. The tree was the willow, *Salix alba*, and the active agent in the bark was the glycoside of salicylic acid destined to become the basis of the well-known drug acetylsalicylic acid, or aspirin. Aspirin was first synthesized by the German, Felix Hofmann, in 1898 and remains the most widely used single pain reliever. Low doses of aspirin lessen the likelihood of heart attack and stroke largely through lessening blood clotting. Higher doses reduce pain and fever and still higher doses reduce the inflammation and swelling of joints caused by rheumatic fever, gout and rheumatoid arthritis. These effects are largely a result of aspirin decreasing the body's production of prostaglandin hormones that are involved in inflammation and pain producing processes. (This is quite unlike the pain relieving effect of morphine that acts on the pain sensing parts of the brain.) From the 1960's onwards, aspirin was supplemented by a range of similarly acting pain relieving drugs collectively called non-steroidal anti-inflammatory drugs, or NSAIDs [45]. Inevitably, the interaction with the body's metabolism of aspirin and NSAIDs has some adverse affects, the most common of which is stomach irritation and ultimately ulceration with long term use. Nevertheless, their beneficial effects generally outweigh their adverse effects and tens of thousands of tonnes of aspirin and NSAIDs are consumed globally each year.

Aspirin typifies the first phase of humanity's fightback against disease, that of palliation, or reducing the symptoms, from which the improvement of the quality of life has been considerable. It also typifies the way in which the pharmaceutical industry began by refining old remedies largely derived from the use of plants known to have therapeutic effects. Notable among these are the pain relievers, opium, morphine, codeine and heroin derived from the opium poppy. Digitalis and strophantin, derived from foxglove and African dogbane, respectively, are still used to treating heart disease. However, it was not until the end of the nineteenth century that the evolution of the manufacturing pharmaceutical industry began [46,47]. Initially, it was a hit and miss affair but a range of drugs of varying effectiveness appeared as understanding of the nature of disease and pharmaceutical science grew. Usually, these drugs were either taken orally or applied to the skin to relieve pain, fever, inflammation and soreness and treated the symptoms of disease rather than the disease itself. There were some notable exceptions to this such as Paul Erlich's arsenic compound 606 of 1909 that provided the first effective treatment for syphilis, a deadly disease until then. A huge breakthrough by the Canadians, Frederick Banting and Charles Best, came in 1921 with the isolation of the hormone, insulin, that made possible the treatment of diabetes mellitus initially with porcine insulin and subsequently with genetically engineered human insulin. Later, the antimalarial drug quinine extracted from the cinchona tree was replaced by the still widely used antimalarial drug, chloroquine, that attacks the malaria causing plasmodium and was first synthesized in Germany in 1934. Also in Germany, in 1935, Nobel Laureate Gerhard Domagk tested the forerunner of the sulfonamide, or sulfa, drugs that arrested the proliferation of bacteria so that the immune system could overcome infection. By the beginning of the twenty-first century, a plethora of drugs was available for the relief of a wide range of conditions and illnesses such as anxiety, pain, blood pressure and nausea. However, most such drugs continued to act by either suppressing or increasing metabolic processes and it is unsurprising that they sometimes had catastrophic side effects. Such a drug was thalidomide.

On Christmas Day, 1956, a baby girl was born without ears in the German town of Stolberg [48]. She was the first victim of a catastrophe that was to either kill or maim tens of thousands of children in the womb over the next decade in many nations. The drug involved was a mild sedative, thalidomide, that was also prescribed for morning sickness. Although it quickly became apparent that the drug often caused deformities in babies that where frequently manifest as hands and feet being directly connected to the torso, and sometimes also caused irreversible nerve damage in adults in Germany, Australia, Canada and the United Kingdom, its use continued and the toll mounted. The United States was saved from similar disasters by the vigilance of Frances Kelsey of the Food and Drug Administration although she suffered considerable harassment in doing so. An Australian obstetrician, William Macbride, played a major part in the exposure of the deforming effects of thalidomide through publication of his studies in 1961 [49]. Nevertheless, the pharmaceutical companies responsible sought to obscure the terrible side effects of thalidomide before it was finally taken off the market, and then fought for years against paying modest compensation to the victims. Ironically, thalidomide now shows promise in treating leprosy, some autoimmune diseases and other diseases [48,50].

A second example of the side effects of drugs taken by parents appearing in children arose with the synthetic oestrogen, diethylstilbestrol, or DES, prescribed to prevent miscarriages from the 1940s to 1971 [47]. Daughters were found to be prone to a rare form of vaginal cancer and infertility, and both daughters and sons were prone to reproductive tract abnormalities. The overall effects of such tragedies have been increasingly stringent laboratory and clinical testing and monitoring of prescribed drugs that have lessening the likelihood of side effects being either undetected or ignored. However, this stringency has greatly increased the cost of drug development.

6.7. Vaccines: Training the Immune System

A very different approach to counteracting disease, that of reinforcing the body's defences against infection, was initiated by the English physician, Edward Jenner [51]. He was fascinated by claims that dairymaids who became infected with a mild disease, cowpox, were immune to lethal smallpox that is caused by the variola virus. He reasoned that cowpox induced a change in the body's ability to fight smallpox infection and that this ability should be transferable. In 1796, he effected this transfer by taking the fluid from dairymaids' cowpox sores and rubbing it into cuts on the skin of people who had never been infected with either cowpox or smallpox. He called the fluid a *vaccinae*, after the Latin for cow, *vacca*, and ushered in the now universal practice of vaccination, or immunization, against a wide range of diseases that has saved many millions of lives. This led to the to the global elimination of smallpox in 1977, a disease thought to have killed, crippled, blinded and disfigured ten percent of all humans who ever lived. Much earlier, the Chinese had sought similar protection by breathing in small amounts of powdered scabs taken from the skin of smallpox victims [52].

More than two thousand years before Jenner, Thucydides, in his history of the Peloponnesian wars, described the plague of Athens and the immunity acquired by survivors of the plague [53]:

> "It was with those who had recovered from the disease that
> the sick and dying found most compassion. These knew what
> it was from experience, and had now no fear for themselves;
> for the same man was never attacked twice - never at least
> fatally."

This is probably the first description of acquired immunity to disease on which vaccines are based. Such immunity is the strongest guarantee of a long and disease free life and is best acquired in childhood. In recognition of this, the United Nations Children's Fund, UNICEF, and the World Health Organization, WHO, have embarked on child immunization programs that save three million or more lives annually and billions of dollars in medical costs [54]. In 1974, only five percent of children were immunized against diseases globally. By the mid 1990s, more than ninety percent had been immunized against tuberculosis and seventy-five to eighty-five percent against diphtheria, pertussis, whooping cough, tetanus, measles and mumps, a great achievement and a huge ongoing task to maintain this level of protection for the children born each year. Amidst these efforts to protect children, a particularly spectacular event occurred on 6-7 December 1996 when 121 million Indian children where given the oral poliomyelitis vaccine. By 1997, a new vaccine against the diarrhoea causing rotavirus that kills some 800,000 children annually was being field tested. Despite these achievements, it was estimated that in 1998 some twelve million children died in developing nations from preventable diseases such as measles, pneumonia, diarrhoea and malaria.

The basis of immunization is the training of the memory components of the immune system to recognize the pathogen, or fragments of it, and to trigger a rapid and powerful attack that destroys the pathogen before it can produce disease [55]. Because pathogens and the diseases they cause vary greatly, the response to immunization also varies with the consequence that a range of different types of vaccine has been developed. Some widely used vaccines, such as those for influenza and pertussis, use dead pathogens, as does the injectable Salk poliomyelitis vaccine. Fragments of pathogens are used in meningitis, pneumonia and hepatitis B vaccines. Such vaccines cannot invade tissue cells and so usually impart only a short protection, or humoral memory. They usually require booster shots and then provide protection over a period as short as six months in the case of influenza and several years for other diseases.

A stronger immune response is produced when vaccines based on weakened live, or attenuated, pathogens are used. The attenuation of the ability of the pathogen to cause significant disease is usually achieved by selectively growing successive generations of pathogens in culture media. The attenuated pathogen invades tissue cells to produce a mild form of the disease and provide a cellular immune response and a much longer lasting immunity, often for life. This is the basis of the oral Sabin poliomyelitis vaccine and for the combined vaccine against measles, mumps and rubella, MMR. However, such live vaccines carry the small risk that the attenuated pathogen might induce a serious infection in a person with a weak immune response. This is exemplified by the probability of this occurring being about one in 750,000 for the oral poliomyelitis vaccine. There is also the risk that a particular batch of vaccine might either contain insufficiently attenuated virus or that it might mutate into a more active form as has occurred in some isolated cases with the oral poliomyelitis vaccine [56].

As understanding of the action of pathogens and genomics has grown so has the ingenuity and variety of methods for producing vaccines increased [57]. These are exemplified by the genetic or "naked DNA" approach. Usually this is based on a bacterial plasmid, a small circle of DNA that is incapable of causing disease. One or more genes that produce the proteins that identify the pathogen to the immune system and sensitise it are inserted into the plasmid that remains incapable of causing the particular disease. Such modified plasmids are usually injected into muscle where they are taken up by tissue cells and incorporated into their nuclear DNA that then produces the pathogen protein. Inside the cell the pathogen protein induces a cellular memory of the pathogen while that which remains outside the cell produces a humoral memory. Clinical trials of such DNA vaccines against hepatitis B, herpes, HIV, influenza and malaria have commenced in addition to those against prostate and colon cancer. While cancers do not appear to be infectious diseases, they are composed of cells foreign to the body that have bypassed the immune system and the possibility exists that vaccines that sensitise the immune system may help combat cancer [58].

Another innovative approach to producing vaccines has arisen from genetic engineering of plants [59]. This is achieved by inserting into the genomes of food plants, such as bananas and potatoes, genes that produce pathogenic proteins and antibodies to respectively sensitise and assist the immune system against a range of diseases without inducing the disease itself. Such transgenic food promises the great advantage of oral delivery that is both more acceptable than are injections and is much cheaper. (However, reservations have been expressed about this method of immunization as the possibility of unwitting consumption of the transgenic food plants arises as discussed in section 5.14.) Oral immunization has the advantage of inducing a mucosal immune response to infection as almost invariably the pathogen has to pass through a mucosal layer before it enters the bloodstream. For the same reason vaccines delivered by nasal spray and vaginal spray have high potential in combating respiratory and sexually transmitted infections.

While vaccines are powerful weapons in the armoury against pathogens they seldom provide full protection from disease and many diseases exist for which there is no vaccine. Further, it should not be forgotten that the mutation of pathogens, as in the case of pertussis in Europe in 1997 [60] and as is frequently the case for influenza as discussed in section 6.12, may render a previously effective vaccine less so. The effect of immunization on large populations is complex as it appears that those vaccines that act by either reducing the growth of pathogens or lessening the effects of the toxins they produce may favour the rise of more virulent pathogens that cause more severe disease in those not immunized [61]. In contrast, vaccines that prevent infection do not have this effect and may favour the rise of less virulent pathogens.

6.8. Antibiotics: The Fallible Defence

In 1929, the British Nobel Laureate Alexander Fleming noticed that the mould, *Penicillium notatum*, the harbinger of the antibiotic age, was destroying a colony of staphylococcus bacteria growing on agar gel and thereby discovered penicillin [47]. This was not the first observation of the antibiotic action of penicillin for in 1896 the Frenchman, Ernest Duchesne, found that the mould *Penicillium glaucum* killed the bacterium *Escherichia coli* and the typhus bacterium *Rickettsia typhi* that causes scrub typhus. However, it was not until 1940 that research led by the Australian Nobel Laureate, Howard Florey, isolated penicillin, the antibiotic that saved the lives of hundreds of thousands of wounded in the Second World War and the lives of many millions of others suffering from bacterial infections since. In 1944, the tuberculosis defeating antibiotic, streptomycin, was isolated from *Streptomyces griseus*, one of a series of soil microorganisms that were to prove a rich source of antibiotics.

Penicillin and streptomycin are only effective against Gram-positive and Gram-negative bacteria, respectively. However, in 1948 the American, Benjamin Dugger, isolated the first of the tetracycline antibiotics, chlortetracycline, or aureomycin, from *Streptomyces aureofaciens*, the first broad spectrum antibiotic that is active against at least fifty pathogenic bacteria of both types. Thus, although the 1940s encapsulated the devastating Second World War, this was a decade in which the tide was turned in the battle against pathogenic bacteria as more and more antibiotics appeared in this and the following decade. These were exemplified by nystatin isolated from *Streptomyces noursei* in 1950, erythromycin isolated from *Streptomyces erythreus* in 1952, novobiocin isolated from *Streptomyces spheroides* in 1955, vancomycin isolated from *Streptomyces orientalis* in 1956 and kanamycin isolated from *Streptomyces kanamyceticus* in 1957.

Most antibiotics kill bacteria by either disrupting the structure of the bacterial cell wall and membrane, by disrupting the reproduction of DNA of the bacterial chromosome and plasmids or by disrupting the production of bacterial proteins. Within these three types of antibiotics are further subdivisions according to the antibiotic molecular structure as shown in Fig. 6.6. Human tissue is made up of

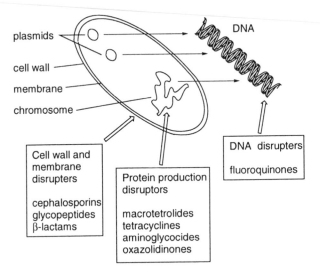

Fig. 6.6. A simplified picture of a bacterium showing the containing cell wall and membrane, the cyclic chromosome, that contains most of the bacterial DNA, and circular plasmids that contain the remaining DNA. The three major classes of antibiotics and their modes of action are shown in boxes.

cells that have some similarity to bacteria and it might be expected that antibiotics should also disrupt them. For some antibiotics this does happen and as a result they have never entered clinical use. For the antibiotics in clinical use such effects are minimal because they exploit the differences between bacteria and human cells. An example of this is vancomycin that disrupts the cell wall that surrounds the bacterial membrane and gives bacteria structural strength. However, human cells lack a cell wall and much of their rigidity arises from a flexible internal cytoskeleton [62]. As a result they are little affected by vancomycin and similar antibiotics.

6.9. Antibiotic Resistance: An Evolutionary Fight

Antibiotics have saved countless lives and continue to do so, but the ability of bacteria to mutate rapidly has the consequence that within a short time they develop resistance to newly introduced antibiotics. Thus, within months of the introduction of penicillin in 1941, *Staphylococcus aureus*, or "golden staph", that causes infections ranging from boils to pneumonia had developed strains that were resistant to penicillin [63,64]. At the time this did not cause widespread concern largely because an increase in dosage usually overcame the resistant strain and also because many other antibiotics were coming into use. In 1941, only one percent of *Staphylococcus aureus* showed penicillin resistance, but by 1946 it was fourteen percent and by 1997 it had risen to more than ninety percent [65]. Consequently, vancomycin became the effective antibiotic of choice for combating this bacterium. However, in mid 2002, a vancomycin resistant strain of *Staphylococcus aureus* emerged, but fortunately the newer antibiotics linezolid and quinupristin/dalfopristin proved effective against this new strain and reduced the fear of such resistance to some extent [66]. However, bacterial strains resistant to these antibiotics will inevitably emerge. By the early 1990's, it was clear that a major antibiotic resistance problem was developing with some bacteria being resistant to ten or more different antibiotics. Thus, despite more than one hundred and fifty antibiotics being available, a patient could still die from an antibiotic resistant strain of a bacterium that two decades earlier did not have resistance to so wide a range of antibiotics [67].

Antibiotic resistance arises because of the many different strains making up a bacterial population some are less susceptible to antibiotics than others. Thus, on treatment with antibiotics the less susceptible strains last longest and, if the treatment is not carried through to the point where the combined effect of the antibiotic and the immune system has eliminated the bacteria completely, they will form a new and more resistant population as shown in Fig. 6.7. While there is a necessary preoccupation with pathogenic bacteria, it should not be forgotten that benign bacteria are also affected by antibiotics and develop resistance. This leads to the possibility that the use of antibiotics could eventually produce a new pathogenic bacterial strain from previously benign bacteria.

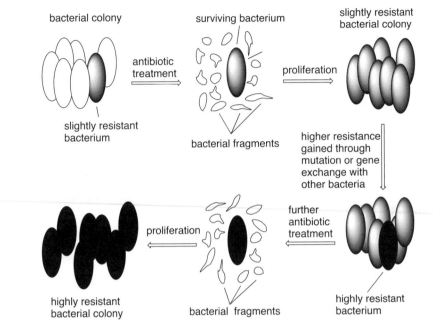

Fig. 6.7. A simplified picture of growth in population of highly antibiotic resistant bacteria.

Irrespective of the presence or otherwise of antibiotics, bacteria continually mutate into new strains by gaining genes from other bacteria and other species. Such is the rate of this mutation that, in evolutionary terms, a year for a bacterium is equivalent to a million years for a mammal. The use of antibiotics favours those bacteria that can resist them. This resistance results from the possession of a gene that either produces a protein that inactivates the antibiotic, differs from that which the antibiotic normally targets, renders the bacterial membrane impermeable to the antibiotic or pumps the antibiotic out of the bacterium [68]. Even so, all is not necessarily lost when an antibiotic resistant bacterial strain appears, for within a bacterial population each strain competes for nutrition and reproduces at different rates. It appears that for some bacteria the susceptible strains outgrow the resistant strains in the absence of antibiotics so that susceptibility to some antibiotics is re-established. However, continual mutation may render some resistant strains highly competitive with susceptible strains. Unfortunately, in the general community and in hospitals, bacteria are continually exposed to antibiotics with the consequence that antibiotic susceptibility is unlikely to be re-established unless the use of antibiotics is more strictly controlled [69].

While the risk of bacterial resistance is inevitable, even with judicious medical antibiotic use, the heavy use of antibiotics in agriculture has accelerated the rise of antibiotic resistance [70]. Many of the antibiotics used in agriculture and aquaculture are closely related to those used by humans. Penicillins are used to treat mastitis in cows and tetracyclines are used to treat respiratory infections in pigs, poultry, aquaculture and orchards. In addition, antibiotics are routinely added to animal feed as growth promoters. In 2001, it was reported that six out of the nineteen antibiotics approved as animal growth promoters were also used as antibiotics by humans [71]. Further, the 11.4 million kilograms of antibiotics fed to cattle, pigs and poultry as growth promoters globally compared with 1.4 million kilograms used by humans. Of particular concern is the use of avoparcin that is closely related to vancomycin, often the antibiotic of last resort in treating streptococcal and staphylococcal infections resistant to other antibiotics in humans. Worryingly, vancomycin resistant bacteria have become increasingly common [72]. Between 1992 and 1996, Australia used 582 kilograms of vancomycin in medicine annually and 62,642 kilograms of avoparcin in animal feed [73]. Similar proportions of these antibiotics were used in other nations, and the fear that vancomycin resistant bacteria would arise in farm animals and be passed on to humans in food and through faecal matter contamination has been realized. Sweden became particularly concerned about such possibilities and banned all use of antibiotics in agriculture in 1986 and no impairment of livestock appears to have occurred [74,75]. This was followed by the banning by the European Community in 1998 of four classes of antibiotics that include avoparcin. By 2001, the United States was considering the restriction of antibiotic use in agriculture [74,76]. While the agricultural industry is unhappy about being tagged as a major cause of the increase in antibiotic resistance in medicine, a similar antibiotic resistance is also occurring in agriculture itself.

The rise of antibiotic resistance in pathogenic bacteria could have been greatly slowed if the early emergence of resistance been more widely recognized as an inevitable evolutionary bacterial response, and that as a consequence a continual review of antibiotic use and the development of new antibiotics was necessary. However, amidst the euphoria generated by the almost miraculous cures of infections produced by the early antibiotics and the rapid discoveries of new ones, a complacency developed whereby it was largely assumed that the battle against bacterial diseases was won. As a result the search for new antibiotics slowed to a crawl by the 1970s.

By the 1990s a fear that humanity might be faced with a return to pre-antibiotic medicine arose because of increasing antibiotic resistance and a growing number of deaths from bacterial infections [77]. While the available one hundred and fifty or so antibiotics were capable of dealing with a wide range of pathogenic bacteria, there was little incentive for the pharmaceutical companies to embark on the expensive process of developing new antibiotics. Consequently, as the understanding of bacteria and the source of antibiotic resistance grew rapidly together with new strategies for developing antibiotics, newly available antibiotics were few and far between at the beginning of the twenty-first century. Among the paths to new antibiotics are chemical modification of presently used antibiotics, broadening the search for naturally occurring antibiotics, designing and synthesising new antibiotics that have no natural origin and targeting specific bacterial genes as the functioning of their genomes becomes better understood [78]. Nevertheless, while it is clear that humanity must maintain a continuous development of antibiotics to keep up with the evolution of antibiotic resistance, the high cost of antibiotic development and the attraction of more lucrative pharmaceutical products have resulted in a decrease of interest in antibiotics on the part of pharmaceutical companies in the developed nations [79]. As a consequence, it appears that their future may be in the hands of non profit agencies some of which are in the developing nations.

6.10. Changing Patterns of Disease

A remarkable change in the pattern of disease occurred in the developed nations during the twentieth century as shown for the United States in Fig. 6.8 [80,81]. Tuberculosis, pneumonia and diarrhoea, the top three killers of 1900, no longer pose major threats as a consequence of improved nutrition, housing and hygiene, better healthcare and the availability of vaccines and antibiotics, although the emergence of antibiotic resistant strains of tuberculosis is now causing concern.

They have been displaced by heart disease, cancer and diabetes, diseases much less amenable to treatment as yet and to some extent reflecting changing lifestyles. So dramatic has been the effect of immunization in the prevention of infectious diseases in the developed nations, and so effective were antibiotics in treating them when infection did occur, that the United States Surgeon General, W. H. Stewart, testified to Congress in 1969 that the time had come to "close the book on infectious diseases" [82]. However, this optimistic suggestion was made at a time when it was not widely apparent that evolutionary processes were generating strains of pathogenic bacteria that were not susceptible to the available antibiotics as discussed in section 6.9.

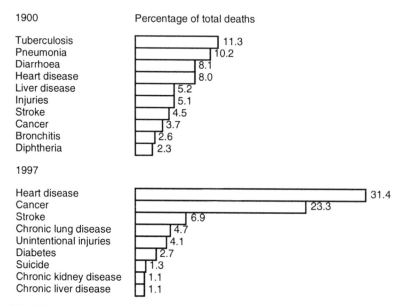

Fig. 6.8. The changing pattern of the ten major causes of death in the United States in the twentieth century. Data from [80].

The impact of vaccines in reducing the incidence of infectious diseases in the developed nations during the twentieth century has been remarkable as is seen from Fig. 6.9 that shows the decline of seven infectious diseases in the United States from their peak years to 1996 [83]. Unfortunately, this decline is not reflected by the developing nations where infectious diseases such as cholera, malaria, tuberculosis, epidemic typhoid and measles are rampant. This difference is largely a consequence of the greater wealth of the developed nations and their correspondingly much greater emphasis on clean water supply, sanitation and very much higher expenditures on healthcare through both public and private agencies. Even so, smallpox, a disease that once killed millions annually, has been eliminated and poliomyelitis, that once crippled hundreds of thousands annually, is close to elimination as a result of global immunization programs [52,56].

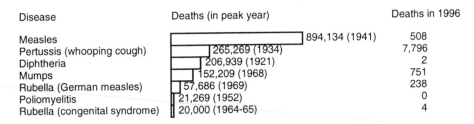

Disease	Deaths (in peak year)	Deaths in 1996
Measles	894,134 (1941)	508
Pertussis (whooping cough)	265,269 (1934)	7,796
Diphtheria	206,939 (1921)	2
Mumps	152,209 (1968)	751
Rubella (German measles)	57,686 (1969)	238
Poliomyelitis	21,269 (1952)	0
Rubella (congenital syndrome)	20,000 (1964-65)	4

Fig. 6.9. The impact of vaccines on the incidence of infectious diseases in the United States. Data from [83].

6.11. The Health and Wealth of Nations

It is generally found that the wealthier is a nation the healthier are its citizens, although within even the wealthiest nations substantial inequality in access to healthcare exists [84]. This is shown by World Bank figures where nations with averaged life expectancies of forty years in 1997 usually had an annual per capita income of US$700. By comparison, those with a life expectancy of seventy-five years or more enjoyed an annual per capita income of US$10,000 or greater. At first glance, this improvement in life expectancy and health with increase in income appears simply explained by greater purchasing power giving greater access to healthcare and associated services. However, it is possible that the correlation between good health and high income may partially apply in the opposite direction [85]. It has been argued that healthier populations are more productive because they are more energetic and mentally robust and lose fewer working days due to illness or the need for them to care for ill relatives. Generally, healthier people who live longer have greater incentives to invest in education with an expectation that they will benefit from it over a long period with the consequence that educational achievement rises. A healthy and educated workforce is likely to attract foreign investment with additional employment opportunities. Simultaneously, longer life expectancy increases the need to save for retirement. Such savings represent capital for investment and further improvements in employment and health.

Since 1950, life expectancy has increased globally by twenty years coincident with an increase in wealth. In the developing nations these increases have been uneven. This is shown by the nations of East Asia where the working population grew several times faster than the dependent population as life expectancy increased substantially as a consequence of the increased access to healthcare and improvements in the safety of drinking water and sanitation coincident with increasing prosperity. In contrast, sub-Saharan Africa is burdened with disease, low life expectancy and the impoverishment of the vast majority of families who spread their few resources among many children such that there is little accumulation of wealth for investment and employment generation. The overall result is a high birth rate-low life expectancy poverty trap, now worsened by the AIDS pandemic with increasing numbers of sub-Saharan Africans becoming HIV infected and a consequently decreasing life expectancy [86]. This tragedy has slowed economic development and the ability to combat other endemic diseases.

The Russian Federation provides another example of declining life expectancy seemingly linked to economic disruption [87]. There, the abrupt transition from a centrally controlled economy to a free market economy in the early 1990s produced a chaotic situation. Incomes plummeted such that diet deteriorated for many and health services were much diminished for most. Coincident with this was the fall of life expectancy for Russian males below those of many developing nations in the mid 1990s. This was reflected in 1.4 to 1.6 million premature deaths during 1990-1995, many of which were attributable to alcohol abuse, stress and the related rise in accident and injury rates.

A telling reflection of the correlation between wealth and healthcare is the great difference between expenditure on pharmaceuticals of the developed nations and the developing nations as shown in Fig. 6.10 [88]. Because most of the great cost of pharmaceutical design, development and clinical testing is borne by commercial companies, new pharmaceuticals are usually aimed at customers who can afford to buy them thereby enabling the manufacturers to recoup their research and development costs and to make profits. As a consequence, of the 1,233 new drugs that appeared in the period 1975-1997 only thirteen were designed specifically to combat tropical diseases; the diseases that are major afflictions of most of the developing nations [89]. The WHO estimates that one third of humanity does not have access to essential pharmaceuticals that are defined as "those that satisfy the healthcare needs of the majority of the population and therefore should be available at all times in adequate amounts and in appropriate dosage forms". Given the great differences in wealth between nations and the fact that the vast majority of pharmaceuticals are manufactured in the wealthy developed nations, this sad situation is not surprising. Somewhat belatedly, the developed nations' pharmaceutical industries have begun to address the developing nations' plight [90].

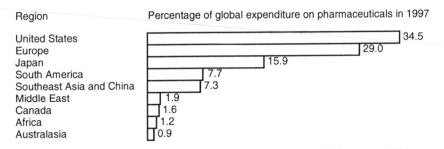

Region	Percentage of global expenditure on pharmaceuticals in 1997
United States	34.5
Europe	29.0
Japan	15.9
South America	7.7
Southeast Asia and China	7.3
Middle East	1.9
Canada	1.6
Africa	1.2
Australasia	0.9

Fig. 6.10. Percentage of global expenditure on pharmaceuticals in 1997. Data from [88].

It has taken the HIV pandemic that afflicts both developed and developing nations to induce a rethink of global healthcare. Thus, as a consequence of cruelly HIV afflicted South Africa's plea for assistance, developed nations and their pharmaceutical companies agreed in 2001 to make available the expensive drugs that slow the onset of AIDS after HIV infection and the transfer of the HIV from mother to foetus at a much lower price than they are available at in developed nations, albeit with some initial reluctance [91]. However, strict commercial conditions apply and, even at the reduced price, the pharmaceuticals remain very expensive for most South Africans and HIV/AIDS sufferers in other sub-Saharan nations. Prior to these agreements, some of the major pharmaceutical companies had taken on philanthropic roles and donated new drugs that treat major disease in poor countries and relaxed patent rights applying to them. Some private philanthropic foundations are now massively funding research and development into treatments for major diseases afflicting developing nations. However, it remains the case that few effective pharmaceuticals are available in the poorest developing nations and in many cases the facilities for their safe storage and wide distribution barely exist. Clearly, there is a long and hard struggle ahead before the developing nations gain access to the level of healthcare enjoyed by developed nations.

While the developed nations do not suffer most of the tribulations of the developing nations, they have little reason to be complacent, for devastating pandemics can arise quite suddenly from unexpected sources. This is illustrated by the great influenza or "Spanish Flu" pandemic of 1918 that took many millions of lives.

6.12. Influenza: The Perennial Threat

At 11 a.m. on 11 November 1918, an armistice came into effect on the Western Front in France and Belgium where the bulk of the vast armies of the First World War faced each other. The combatant nations breathed a sigh of relief at the cessation of a carnage that had claimed millions of lives and maimed millions more, and looked forward to the reconstruction of peace. However, a few months before the armistice, an unseen enemy, the Spanish influenza virus, launched an offensive. It was to claim at least twenty, probably forty, and possibly one hundred million lives over two years, far more than those lost in four and a half years of war [92]. While influenza was at that time a familiar winter illness, and still is, Spanish influenza was particularly virulent with two and a half percent of those contracting it dying, five times the usual influenza death rate. It was followed in 1957 by the Asian influenza pandemic that claimed a million lives, in 1968 by the Hong Kong influenza pandemic that claimed 700,000 lives and by the Russian influenza pandemic that took many thousands of lives in 1977. As a consequence great vigilance is exercised in the watch for early signs of new influenza pandemics. (While the Asian, Hong Kong, and Russian influenza pandemics were named after their perceived origins, the Spanish influenza pandemic appears to have been so named because its existence was first revealed in the uncensored Spanish press whilst the combatant nations' presses were censored.)

Although Spanish influenza may have originated in China, the pandemic appears to have started in army camps in Kansas early in 1918 and quickly spread across the United States. The first wave of influenza in the spring and summer of 1918 was very contagious but seldom killed its victims. However, in August 1918 a more virulent strain emerged and spread globally within six months with the main wave of the pandemic occurring in September and November. At its height, ten thousand people were dying every week in some American cities and almost a third of Americans became ill with some 675,000 dying out of a population of 103 million. The virus swiftly spread with similarly devastating consequences in Africa, Asia, Europe and the Pacific islands probably aided by the global movement of vast numbers of troops in the final stages of the First World War. However, it appears that a particularly virulent strain of influenza infected British and French troops in 1916 and 1917. This may have been either a first appearance of the 1918 virus or of a close relative, and its re-emergence in Europe in 1918 may have assisted in the remarkably rapid spread of the 1918 pandemic. And then the virus vanished leaving few clues to its nature and the origin of its virulence. It is possible that other strains of less virulent influenza viruses sensitised humanity's collective immune system to the virulent 1918 strain so that, being unable to reproduce itself, it died out. Nevertheless, because of the growing realization of the possibility of further influenza pandemics, and because of the extraordinary virulence of the 1918 strain, an understanding of the origin of its virulence became of major importance in preparing for new influenza pandemics.

6.13. The Influenza Virus

At first it was thought that influenza was caused by the bacterium *Bacillus influenzae*, and it was not until the 1930s that the responsible virus was isolated of which there are three types [93,94]. Type A is the most common and causes the most severe illness, infects humans, pigs, horses, seals, whales and birds and caused the pandemics of the twentieth century. Type B is similar to type A but is less common and only infects humans to cause regional epidemics rather than pandemics. Type C is even less common and causes illness akin to the common cold. Influenza attacks the lining, or epithelium, of the upper respiratory tract, and causes inflammation, much mucus, soreness and coughing and weakens the immune system so that secondary infections such as pneumonia occur. It is the secondary infection that usually leads to death in severe cases. Now, secondary bacterial infections can usually be cleared up with antibiotics, an unavailable option in 1918.

Fortunately, it is possible to prepare a vaccine to activate the immune system against each strain of the influenza virus from either an attenuated or a dead virus. Thus, immunization against influenza is now commonplace and effective provided that the vaccine is made from the virus currently posing the threat of infection [95]. It is the last qualification that is critical because the influenza virus can mutate very rapidly and catch public health authorities off-guard. As a consequence, there is always considerable anxiety when a new viral mutation appears. Usually, successful immunization is only effective for six months or so and repeat vaccinations are required every year. At present influenza infects some 120 million people in the United States, Europe and Japan annually, and causes twenty to forty thousand deaths in the United States alone.

The influenza virus mutates rapidly, first because its genome is composed of eight different strands of RNA, unlike many other viruses whose genome consists of a single strand of RNA, and second because the virus usually passes readily not only between humans but also between birds, pigs and humans. This provides a remarkable opportunity for different viral strains to exchange one or more strands of RNA to produce a new genome and a new strain of influenza. The eight RNA strands of the viral genome are enclosed inside a protein envelope on the surface of which there are two other proteins, hemagglutinin and neuraminidase as shown in Fig. 6.11. Hemagglutinin binds the virus to the surface of the cell to be infected and fuses the virus envelope with the cell membrane, whereupon the RNA of the virus is injected into the infected cell. There it is reproduced many times over before forming prolific new incipient influenza viruses in the membrane of the infected cell. Now the second surface protein, neuraminidase, comes into play. The incipient viruses are coated with sialic acid that causes them to stick together and to the cell membrane. Neuraminidase destroys the sialic acid and liberates the viruses to go on to infect other cells and multiply further.

This understanding of the influenza virus has led to some quite effective drugs that lessen the impact of infection [96]. The first such drugs, amantadine and rimantadine, were discovered some thirty years ago and interfere with a viral protein, M2, that controls passage of ions through the protein envelope of the

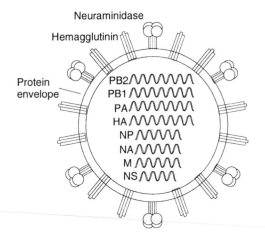

Fig. 6.11. Simplified cross-section of the influenza virus showing the eight RNA strands of the genome and the surface proteins hemagglutinin and neuraminidase. The protein envelope is 0.1 of a micrometre in diameter.

influenza A virus and thereby lessen its virulence. Unfortunately, these drugs have neurological side effects and are only used for high risk patients. As the influenza B virus does not possess the M2 protein, these drugs are ineffective against it. A newer type of drug, the neuraminidase inhibitor, stops the viral neuraminidase destroying sialic acid and thereby stops the virus from multiplying. Although neuraminidase mutates between viral strains, it retains a cleft into which sialic acid fits and is there destroyed. Recently several molecules that block this cleft have been designed so that neuraminidase cannot destroy sialic acid and the reproduction of the influenza virus is much reduced. These new drugs, two of which are zanamir and GS4101, reduce the intensity of influenza A and B infections by twenty-five to forty percent and their duration by about thirty percent.

Both hemagglutinin (H) and neuraminidase (N) proteins are made up of sequences of linked amino acids. These sequences mutate to produce different strains of influenza A viruses that are named according to the particular version of each of these proteins that they possess as exemplified by H1N1, H1N2, H2N2 and H7N7. It is known from the analysis of stored tissue samples and tissue from victims of the 1918 influenza pandemic buried in the permafrost in Alaska and Spitzbergen that the 1918 virus was a H1N1 strain [97]. The Asian,

Hong Kong and Russian pandemics arose from H2N2, H3N2 and H1N1 strains, respectively. Both hemagglutinin and neurominidase are produced by the two separate stands of RNA, HA and NS, in the viral genome. It is the reassorting of these strands between different strains of the virus within the lungs of carriers that produces mutations and the resulting new influenza viruses are usually referred to as "reassorted" viruses.

Influenza also afflicts a variety of mammals and birds. Waterbirds and ducks, in particular, act as reservoirs for influenza A viruses and it is thought that they are the primordial source of all influenza viruses in other animals including humans. Usually waterbirds that carry influenza viruses do not develop influenza themselves but the viruses are passed in their faeces into rivers, lakes and seawater where other animals may be infected. Among such animals is the pig from which humans are able to acquire the influenza virus. In southern China, in particular, ducks and pigs are often simultaneously reared in conjunction with fish farming so that faecal matter containing influenza viruses contaminates the fishponds from which both pigs, ducks and wild waterbirds drink. This provides an ideal situation for the reassortment of RNA strands between influenza viruses already inhabiting the pig and those picked up from waterbirds and humans. Thus, the stage is set for cross infection between pigs, humans and wild waterbirds where the latter carry the reassorted viruses over vast distances to infect other animals and ultimately more humans [98].

Every so often surprises occur such as the discovery that an avian influenza A virus, H5N1, in chickens was responsible for a 1997 influenza outbreak in Hong Kong in which eighteen or more humans were infected of whom six died [99]. It is both unusual for influenza viruses to transfer directly from birds to humans and for a H5 type virus to infect humans. The more common infections arise from either H1, H2 or H3 type viruses. Given the very high Hong Kong death rate, this virus had the potential to produce a deadly pandemic among humans but fortunately it did not appear to readily pass between humans. Nevertheless, thousands of chickens, the source of the virus, were destroyed. This was followed by the emergence of other human infecting avian influenza viruses, H9N2, in Hong Kong in 1999 and H7N7 in the Netherlands in 2003.

While influenza is an old enemy and humanity's defences against it are now quite good, this cannot be said of the newly emerged human immunodeficiency virus, HIV, that threatens to dwarf the toll taken by influenza and against which humanity is fighting a desperate battle.

6.14. The Acquired Immune Deficiency Syndrome (AIDS) Pandemic

In June 1981, the U.S. Centres for Disease Control became aware that five cases of a very rare form of pneumonia caused by the protozoan, *Pneumocystis carinii*, had been diagnosed in young homosexual men in Los Angeles [100]. At about the same time, twenty-six cases of the rare Kaposi's lymphoma, usually only seen in elderly men, were diagnosed in young homosexual men in California and New York City along with other diseases associated with much weakened immune systems. It was realized that a new condition had been detected, acquired immune deficiency syndrome, or AIDS. A hitherto undetected retrovirus, the human immunodeficiency virus, HIV, was found to be responsible by French and American researchers [101]. This was the first systematic detection of the beginning of the AIDS/HIV pandemic, that by 4 July 1988 resulted in 66,464 HIV infections and 37,535 deaths of adults and children in the United States as seen in Fig. 6.12.

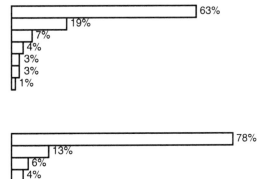

Adult cases by group
(Total number = 65,410)

Percentage of cases in each group

Homo- or bisexual men — 63%
Heterosexual intravenous drug users — 19%
Homo- or bisexual intravenous drug users — 7%
Heterosexual men and women — 4%
Recipients of blood or blood products — 3%
Other or unknown — 3%
Hemophilia or coagulation disorder sufferers — 1%

Children by group
(Total number = 1,054)

Born to HIV infected mothers — 78%
Recipients of blood or blood products — 13%
Hemophilia or coagulation disorder sufferers — 6%
Other or unknown — 4%

Fig. 6.12. Cases of HIV/AIDS among adults and children in the United States until 4 July 1988. Data from [100].

The major viral strain, HIV-1, is the most virulent and widespread, while HIV-0 and HIV-2 are slower acting and cause localized epidemics in parts of Africa. The HIV-1 enters the bloodstream either through exchange of bodily fluids during sexual intercourse or through injection of infected blood. The effects of this are clearly seen from the 1988 American infection patterns in Fig. 6.12. Those of other developed nations were similar and largely remain so. A particularly poignant aspect of these patterns was the infection with the HIV-1 of children born to HIV infected mothers and of those who had not engaged in practices rendering them vulnerable to the HIV-1 but became infected through blood transfusions [102]. This placed even more pressure on blood banks to screen their supplies for disease.

The number of Americans infected with the HIV by 1988 was small by comparison with the five million mainly sub-Saharan Africans infected in the same period [103], and is dwarfed by the tens of millions that carried the HIV globally in 2004 as is shown by Fig. 6.13 [104]. By the end of 2004, some sixty million people had been infected with the HIV of whom twenty million had died of a variety of diseases that healthy immune systems usually ward off but which AIDS suffers cannot. Of these deaths, about 17.4 million occurred in sub-Saharan Africa where 25.4 million HIV infected people lived. So vast is this pandemic that, by the end of 2004, seven in every one thousand adults were HIV infected on a global basis, with fifteen to thirty-six percent of the adult population of some sub-Saharan African nations being infected [103-105]. The growth of the pandemic shows little sign of abating with South and Southeast Asia and China threatening to match sub-Saharan Africa infection rates.

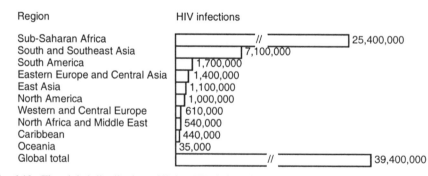

Fig. 6.13. The global distribution of living HIV infected persons in 2004. Data from [104].

By January 2000, the United Nations had become so concerned that it declared HIV/AIDS a threat to global security. The futures of nations are now under threat as the pandemic places an unbearable strain on health services and life expectancy plunges as exemplified by ten sub-Saharan African nations where the fall in life expectancy for children born in 2000 dropped from 50 down to 38 years for Mozambique and from 70 down to 39 years in Botswana. Controversially, the belief that poverty rather than HIV infection resulted in AIDS became widespread in South Africa at the very time when the immensity of the problem that AIDS was causing for that nation was evoking global concern [106]. This threatened to derail HIV/AIDS education programs and the persuasion of foreign pharmaceutical companies to lower prices for antiretroviral drugs and resulted in the "Durban Declaration" in which the evidence for the HIV being the sole cause of AIDS was unequivocally restated.

6.15. The Human Immunodeficiency Virus (HIV)

In the early stages of the pandemic, HIV infection and the subsequent development of AIDS usually resulted in death within seven to twelve years in the developed nations, as it still does in developing nations. Because of HIV's deadly nature and its resilience against treatment, the expenditure on research into HIV infection and its progression to AIDS has been unprecedented. The HIV mainly attacks the immune system's CD4$^+$ T-cells wherein it reproduces many times over before going on to spread the infection. This complex process is shown in Fig. 6.14 where the HIV is seen attaching itself through its gp41 and gp120 proteins to CD4$^+$ receptors and coreceptors on the CD4$^+$ T-cell surface [107]. This enables the HIV to merge with the cell membrane and inject its two RNA genome copies, each containing eight genes, into the cell. There, the HIV reverse transcriptase enzyme converts them into HIV DNA that is inserted into the DNA of the CD4$^+$ T-cell nucleus by another enzyme, HIV integrase. The metabolism of the T-cell is then taken over to produce more segments of the HIV DNA, while also producing HIV proteins. Subsequently, the HIV DNA segments are released from the CD4$^+$ T-cell DNA and revert to HIV RNA that emerges from the nucleus together with HIV proteins. HIV protease then modifies the HIV proteins before they join HIV RNA to form the protein coat and core proteins of proliferating nascent HIVs in the cell membrane prior to breaking free as new HIVs.

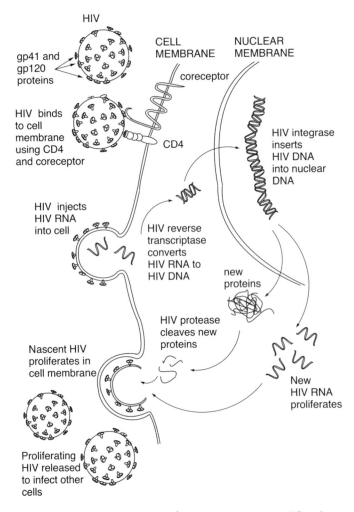

Fig. 6.14. The process of HIV infection of a CV4$^+$ T-cell followed by prolific release of new HIVs to infect more cells.

In the initial stage of HIV infection, the virus reproduces up to ten billion times a day and the CD4$^+$ T-cell blood level drops dramatically from the normal concentration of eight hundred thousand or more in every cubic centimetre of blood [107,108]. This is usually accompanied by influenza-like symptoms, enlarged lymph nodes and rashes that diminish as the immune system gains some control over the virus. Even so, the immune system seldom completely eliminates the HIV. Usually, after twenty-five weeks or so, the HIV blood level drops to a lower fairly stable level and CD4$^+$ T-cell levels remain high enough to ward of most infections. The length of this stage varies greatly from person to person during which time few health problems are experienced and there is little outward sign of HIV infection. However, the CD4$^+$ T-cell level gradually falls to an inadequate level to combat other infections and the onset of AIDS occurs. As the CD4$^+$ T-cell level drops below one hundred thousand in each cubic centimetre of blood the HIV level rises and the sufferer is at the mercy of a variety of infections that a healthy immune system would overcome, and death usually occurs within a year or two. There is an enormous variability in survival rates with some patients dying within a year of HIV infection while survival for more than twenty years after infection is known.

6.16. HIV Treatment and Control

The present understanding of HIV infection indicates several stages at which it might be possible to disrupt the HIV reproduction. These are the initial binding of the HIV to the receptors on the surface of the CD4$^+$ T-cell and the actions of HIV reverse transcriptase, HIV integrase and HIV protease. Such strategies seek to stop the HIV from reproducing itself through the use of reverse transcriptase and protease inhibitors [107,109]. These antiretroviral drugs became available in 1996 and by the end of 2004 many variants of them were available with sixteen being in major use in the developed nations. Typically, combinations of two or more antiretroviral drugs are more effective than any one of them alone. As a consequence, there are many combinations in use that depend on the sufferer's history and the inevitable side effects. The antiretroviral regimen must be strictly adhered to and may involve taking as few as three tablets twice a day and up to thirteen or more tablets twice a day. It must also accommodate the high mutation rate of the HIV arising from its fast and inaccurate reproduction, the appearance of antiretroviral resistant mutants and the possibility that the HIV level could soar out of control [110]. This is usually accomplished by adjusting the cocktail of antiretroviral drugs administered depending on the resistant HIV mutants particular patients may possess. This has become known as highly active antiretroviral therapy, or HAART.

   While it is possible to decrease the HIV to undetectable levels in blood during antiretroviral treatment, it has so far proved impossible to eliminate it from the central nervous system, the reproductive tract, the digestive tract and other tissues. Thus, although the onset of AIDS may be delayed for long periods, in some cases indefinitely, a cure for HIV infection has yet to be achieved. Deaths from AIDS declined between the first half of 1996 and that of 1997 by forty-four percent in the United States, the first decline since 1981, and was attributable to the new antiretroviral treatments. However, the cost of these treatments is enormous as indicated by the expenditure of US$1,600,100,000 for reverse transcriptase inhibitors and US$933,000,000 on protease inhibitors in the United States alone in the period February 1999 to February 2000 [111]. In the meantime improvement in antiretroviral treatments are sought as increasingly attention turns to making an HIV vaccine.

Although, the immune system generates antibodies and $CD4^+$ and $CD8^+$ T-cells in response to HIV infection, it is seldom able eliminate the virus [112]. This is a discouraging observation for those seeking to produce an HIV vaccine. Nevertheless, it is possible that an immune system presensitised toward the HIV by a vaccine might respond faster and more strongly to an HIV infection than the normal response and eliminate the HIV before it reaches a sustainable level. The classical production of a vaccine from either the attenuated or dead HIV has so far given disappointing results in the laboratory. Parallel tests on monkeys of a vaccine for the simian immunodeficiency virus, SIV, a virus closely related to the HIV, have also produced mixed results. Great disappointment arose in February 2003 when it was reported that the first vaccine to be tested extensively on humans showed little effect. Alternative approaches dispense with the HIV and instead seek to use HIV proteins to sensitise the immune system.

A more complex approach involves DNA engineered to carry the HIV genes that produce the HIV proteins. This "naked DNA" approach has the potential to produce HIV vaccines but once again the results have been mixed in the laboratory. A related approach involves insertion of HIV genes into weakened *Salmonella typhi* and *Listeria monocytogenes* bacteria that are able to live in the dendritic cells of the immune system [113]. There, they produce HIV proteins that are attacked by the immune system and sensitise it to the HIV that is attacked when infection occurs. Immunization in this way provides mice with immunity to HIV and may pave the way for development of a human HIV vaccine. Even so, an HIV vaccine appears to be some years away yet. Another glimmer of hope comes from small numbers of African prostitutes and homosexual men continuously exposed to the HIV but who remain uninfected for extended periods [114]. Understanding their immunity may eventually lead to the development of an HIV vaccine.

In the absence of either a cure or a vaccine for HIV, great emphasis has been placed on programs promoting "safe sex" through the use of condoms and clean needle exchange programs for drug users, programs that have greatly slowed the rate of HIV infection in developed nations, in particular, and in developing nations to a lesser extent [115]. However, five or so million people are newly infected with the HIV annually and a global total of one hundred million infections is possible within a decade.

6.17. The Origin of the HIV

Inevitably, the question arises as to where the HIV came from and why it only emerged in the latter part of the twentieth century [116]. There seems little doubt that the HIV-1 was derived from the simian immunodeficiency virus, SIV, transmitted to humans by the chimpanzee subspecies, *Pan troglodytes troglodytes*, found in Gabon, Equatorial Guinea and Cameroon. It appears that the common viral ancestor of the HIV-1 transferred to humans as early as the nineteenth or early twentieth century. It was most likely transmitted to hunters who may have either mixed butchered chimpanzee blood with their own through surface cuts or have been bitten. It seems that between 1915 and 1941 this common ancestor mutated into the first of the HIV-1s but this occurred in small localised communities who did not travel far and so the virus did not spread. The less common HIV-2 strain was probably transferred from other monkeys, the sooty mangabey, *Cercocebus atys*, and the macaques, *Macaca*.

As the twentieth century progressed, bringing with it the independence of former colonies, the opportunities for the spread of HIV-1 increased dramatically. Two probable major causes were increased mobility of population, much of it caused by wars and insurrections, and the reuse of unclean needles in extensive immunization programs that transferred HIV infected blood from patient to patient [117]. This may explain the difference in the pattern of adult HIV infection in sub-Saharan Africa that is dominated by heterosexual men and women who experience similar levels of infection, whereas in developed nations it is dominantly homosexual males and intravenous drug users who are HIV infected. The earliest sample of the HIV-1 came from a 1959 clinical sample from Kinshasa in the Democratic Republic of Congo and shows that the HIV-1 was in the African population in the 1950s [118]. The earliest definite finding of the HIV-1 in a developed nation came from tissue samples taken from a Norwegian father, mother and daughter infected in the 1960's who died of AIDS in the 1970s [119]. The father was a sailor who had visited Africa and is thought to have become infected there. Similarly, young Haitian men who participated in UNESCO educational programs in the Democratic Republic of Congo in 1960 to 1975 are thought to have carried the HIV to Haiti from where it was spread throughout the Caribbean and on to the United States [120]. In Haiti HIV/AIDS has lowered the life expectancy of a child born in 2000 to forty-nine years compared with the fifty-six years expected in the absence of the pandemic. A disturbing theory that an oral polio vaccine inadvertently contaminated with a chimpanzee virus in the 1950s led to the HIV pandemic appears to have been disproved [121].

A prime example of pathogens' continual attacks on humanity is the assistance that the HIV pandemic has given to a disease that has been with humanity for thousands of years, tuberculosis, that kills many of those who develop AIDS following the destruction of their immune system by the HIV.

6.18. Tuberculosis: "The Captain of All These Men of Death"

Tuberculosis kills three million people a year and two billion people carry the dormant form of *Mycobacterium tuberculosis*, the bacterium that causes the disease [80,81,122]. It probably transferred from cattle to humans ten to twenty thousand years ago when cattle were first domesticated. It resides in the deeper recesses of the human lung and only causes the disease called tuberculosis in about ten percent of those who carry the bacterium. The bacterium is transferred through sneezes and coughs and in most cases is either enveloped by macrophages and destroyed or it takes up residence in the lung and multiplies in lesions called tubercles, from which the disease gets its name. For ninety percent of those infected, the bacteria remain dormant within the tubercles throughout life without generating major disease symptoms. However, for ten percent of those infected the bacteria spread widely in the body either soon after initial infection or after lying dormant for several years. A common form of the disease is a progressive breakdown of the lung as the bacteria proliferate, albeit at a the slow rate of doubling every twenty-four hours, resulting in fever, weakness, emaciation, chest pain, coughing, bloody sputum and death in forty to sixty percent of sufferers if untreated. Other symptoms may include swollen glands, fusion of vertebrae and spinal deformation. While this variety of symptoms often caused tuberculosis to be thought of as several different diseases, it seems certain that in recorded history it has been the major cause of death in Europe and the United States.

In Hippocrates' time tuberculosis was known as "phthisis" after the Greek for emaciation. In medieval England and France it was commonly thought that the touch of a newly crowned king would cure the disease that was known as "King's Evil". Those showing the swollen neck glands sometimes symptomatic of the disease were said to have "scrofula". The overwhelming nature of tuberculosis was powerfully expressed by John Bunyan in 1680 [123]:

> "The captain of all these men of death that came against him to take
> him away, was the consumption, for it was that brought him down
> to the grave"

where "consumption" was another widespread name for tuberculosis.

Robert Koch identified *Mycobacterium tuberculosis* as the cause of tuberculosis in 1882. In 1908, Albert Calmette and Camille Guérin cultured *Mycobacterium bovis* that causes bovine tuberculosis, a less virulent form of tuberculosis that sometimes infects humans, to produce the Bacille Calmette Guérin attenuated vaccine, or BCG. It gives protection to more than two thirds of those immunized in developed nations and is still the most widely used vaccine against tuberculosis. However, all is not well with the BCG vaccine that exists as several variants and has been found almost completely ineffective in southern India, Malawi and some other developing nations [124]. It is thought that the children of these nations have either been exposed to a range of mycobacteria that interfere in BCG's ability to stimulate immunity to *Mycobacterium tuberculosis*, or that the BCG used in these nations may have lost fragments of its genome over the ninety years during which it has been in use and that this has decreased its effectiveness.

A major advance in the battle against tuberculosis occurred in 1947 with the use of the antibiotic, streptomycin, to destroy the bacterium. To this was soon added, ethionamide, isoniazid, pyrizinamide and rifampicin that, whilst ineffective against the dormant *Mycobacterium tuberculosis*, are very effective when the bacterium is actively multiplying. Even so, the bacterium is unusual in that it has a particularly complex cell wall and only reproduces once a day in its active form. This causes treatment to be a long process and it is probable that a cause of the present spread of tuberculosis is that treatment is not continued until a complete cure has been effected, particularly in developing nations. It is in these nations that tuberculosis is particularly prevalent as exemplified by Botswana, Zimbabwe and Cambodia where fifty-six in every ten thousand people suffer from the active form of the disease. Among Tibetan refugees in India the incidence is as high as two percent [125].

Inevitably and worryingly, a form of *Mycobacterium tuberculosis* resistant to the antibiotic izonazid has appeared largely as a consequence of patients not adhering to the strict six months long course of antibiotic treatment required to destroy the bacterium [126]. This antibiotic resistance first became apparent in the 1980s and 1990s in New York and London and by 2001 arose in two percent of cases globally, in fourteen percent of cases in Estonia and in twenty-five percent in the Russian Federation's prisons. The identification of the genetic mutation that renders isonazid ineffective shows the complexity of the fight against *Mycobacterium tuberculosis* [127]. Ethionamide, isoniazid and pyrazinamide are inactive until converted to their active forms by an enzyme produced by the bacterium that they then kill. The isoniazid resistant bacterium has deleted from its genome the gene producing this enzyme with the result that isoniazid remains inactive.

The *Mycobacterium tuberculosis* genome was completely sequenced in 1998 and found to contain four thousand genes among which are those that could generate all of the forms of antibiotic resistance discussed in section 6.9 in addition to resistance to isoniazid [128]. Fortunately, this knowledge also provides insight for future design of antibiotics to combat *Mycobacterium tuberculosis*. Such a new antibiotic is diarylquinoline R207910 that was showing promise in animal and clinical trials in 2005 [129].

6.19. Malaria: A Mosquito Borne Tropical Scourge

Many diseases are restricted to particular regions of Earth because the disease transmitting agent, or vector, can only survive in them. Such a disease is malaria, a fearsome disease of the tropics that was well known in the ancient world as indicated by its name derived from, *mala aria*, Latin for bad air, that the Romans believed to be the disease's origin. Malaria is caused by four types of protozoan parasites all transmitted to humans by female mosquitoes that feed on blood, in contrast to male mosquitoes that feed on nectar [130]. The four protozoa are: *Plasmodium falciparum*, *Plasmodium vivax*, *Plasmodium ovale*, and *Plasmodium malariae*, where the first of these causes up to two and a half million cases of the most deadly form of malaria annually of which at least one million, mainly children, die. Annually, 300 to 500 million suffer from malaria, ninety percent of them in sub-Saharan Africa, with two thirds of the remainder living in India, Brazil, Sri Lanka, Vietnam, Colombia and the Solomon Islands in decreasing order of prevalence of malaria. Overall, one third of humanity lives in areas where malaria is endemic and more people are infected at the beginning of the twenty-first century than ever before.

Of the 380 different species of mosquitoes, about fifty carry the protozoa that cause malaria in humans. The most deadly, combination is the female *Anopheles gambiae* mosquito, the vector, and the *Plasmodium falciparum* protozoan whose complex life cycle is shown in Fig. 6.15. Humans are infected with the sporozoite form of *Plasmodium falciparum* in the saliva of the female *Anopheles gambiae* mosquito as it bites and ingests blood. The sporozoites invade the liver within thirty minutes where they change into merozoites that multiply over five or so days before breaking out and invading red blood cells. There, they further multiply before emerging from the infected red blood cells to either re-invade more red blood cells to double their number in the blood every two days or to change into male and female gametocytes. The gametocytes are ingested with blood by the female mosquito where they sexually reproduce to form oocysts that release sporozoites into the salivary glands of the mosquito to begin the cycle again.

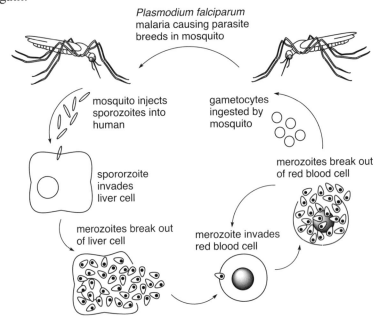

Fig. 6.15. The life cycle of the malaria causing parasitic protozoan, *Plasmodium falciparum.*

It is not until the merozoites enter the blood stream that the symptoms of malaria appear initially as a high fever and shivering that recurs every two days or so and causes a progressive exhaustion that can lead to death, particularly in undernourished sub-Saharan children. Late in this sequence anaemia develops, sometimes in company with cerebral malaria that arises when infected red blood cells stick together to block capillaries in the brain that if not treated usually leads to death [131]. Those who survive cerebral malaria are often left with speech impairment and neurological damage. The progress of malaria is dependent on age and the extent of prior exposure to the disease. Thus, for Americans and Europeans the first infection can be lethal. However, many early infections in African children are not severe and after several infections a partial immunity may be gained that limits parasite growth and disease intensity in subsequent infections. Adults living in areas where malaria is endemic seldom die from malaria. However, pregnant women are at much greater risk, particularly in the case of a first pregnancy. There is a major risk to the foetus, and children born to infected mothers are usually small and at great risk of dying.

Many, particularly new mothers, die from the anaemia associated with malaria that appears to be a consequence of depletion of red blood cells and a lessening of the bone marrow's ability to produce sufficient replacements. In the case of *Plasmodium vivax* malaria, the pathogen is often retained in a dormant form in body tissue and periodically causes further malarial attacks up to twenty years after the initial infection.

As for many diseases, some individuals are less susceptible to malaria than others and this has resulted in a genetic selection in humans. Among those living in much of Africa, the Mediterranean area, the Middle East, Central, South and Southeast Asia there exists a variety of mutations in the gene producing the oxygen carrying iron protein, haemoglobin. This causes it to have an abnormal structure that results in malfunctioning red blood cells, anaemia, bone marrow and bone abnormalities, and enlargement of the spleen; the symptoms variously found in people suffering from α- and β-thalassemia and sickle cell anaemia [132]. Generally, people possessing two genes predisposing them to either thalassemia or sickle cell anaemia develop the disease while those carrying one such gene do not and enjoy substantial immunity from malaria. This understanding of these genetic aspects may eventually yield a cure for sickle cell anaemia [133]. Another red blood cell genetic defect is also often found in the populations of these regions. In this case the result is a less effective version of the enzyme glucose-6-phosphate dehydrogenase that protects red blood cells from damage arising from their oxygen carrying function [134]. Despite these disadvantages, those carrying these genetic defects have a higher resistance to malaria and as a consequence there is a much higher incidence of such defects in those whose ancestry is in the regions where malaria has long been endemic.

6.20. Defences Against Malaria

For a long time the closely related alkaloids, quinine and cinchonine, extracted from the bark of the cinchona tree, were powerful suppressants for *Plasmodium vivax* and effective cures for *Plasmodium falciparum* malaria, but now resistance is building up against these drugs. However, in the mid twentieth century, effective synthetic antimalarial drugs such as atabrine, paludrine, and chloroquine became available. Chloroquine has become the antimalarial drug of last resort over the past twenty years although resistance to it is now common. Fortunately, a Chinese antimalarial drug, artemisinin, extracted from the wormwood shrub *Artemesia annua*, and its improved synthetic relatives, arteflene, artemether, and artesunate remain effective against malaria and are likely to become the drugs of choice as chloroquine resistance grows [135]. Even more effective synthetic drugs were on the horizon by 2004 [136].

The most successful attack on malaria was the widespread use of the insecticide dichlorodiphenyltrichloroethane, DDT, to destroy mosquitoes in the 1950s and 1960s. Removal of the mosquito as part of the malaria protozoan breeding cycle virtually eradicated malaria from large regions of Earth [137]. In less than twenty years spraying houses with DDT reduced Sri Lanka's annual malaria infection from 2.8 million and 7,300 deaths to seventeen infections and no deaths. In India, the toll dropped from seventy-five million infections and 800,000 deaths annually to 100,000 new malaria cases and no deaths twelve years after the start of the DDT campaign that eliminated malaria from ninety percent of the affected regions. However, with the decline of funding and rising concern about the effects of DDT on wildlife in the developed nations, the mosquito suppression programs diminished precipitately with the consequence that malaria began a rapid and disastrous re-emergence. Sri Lanka recorded 500,000 cases of malaria in 1969 and by 1999 India's toll had risen to 2.85 million infections and at least three thousand deaths. A re-emergence of malaria also occurred in South America. Thus, developed nations' concerns about DDT arising mainly from its use as an agricultural insecticide have substantially eroded its much smaller use in tropical areas where the *Anopheles gambiae* mosquito breeds. As a consequence, the incidence of malaria continues to grow in many developing nations and now stands at an all time high. Fortunately, it is possible that the use of insecticide impregnated bed nets designed to protect sleepers from malaria infected mosquitoes may help stem the rising tide of malaria infections although insecticide resistant mosquitoes are appearing with increasing frequency [138].

DDT resistant strains of *Anopheles gambiae* mosquitoes began to emerge simultaneously with *Plasmodium falciparum* developing resistance to chloroquine [139]. Thus, at the beginning of the twenty-first century, sub-Saharan Africa, in particular, faced the daunting prospect of a diminishing ability to control *Anopheles gambiae*, an increasingly drug resistant *Plasmodium falciparum*, a rapid rise in deaths from malaria and antimalarial research funded at about one thousandth the level of research into HIV/AIDS. The latter reflected the great divide in disease research funding whereby the developed nations quickly responded to any disease threat to themselves but were slow to respond to the diseases that were largely confined to the developing nations [140]. Somewhat belatedly in 1998, the World Health Organization (WHO) initiated a program in Africa called "Roll Back Malaria" whose aim was to reduce malaria induced deaths by fifty and seventy-five percent by 2010 and 2015, respectively, through a variety of means. The timing of this initiative may not have been entirely unrelated to the number of malaria infections in the European Community increasing from, 2,882 in 1981 to 12,328 in 1997, and the resurgence of malaria in Armenia, Azerbaijan, Tajikistan, and Turkmenistan. It also coincided with the curtailment of disease control programs with the demise of the Soviet Union and the appearance of a few cases of malaria in the United States, believed to be the result of mass air travel to and from areas where malaria is endemic [141]. Unfortunately, the "Roll Back Malaria" program had made little progress by 2005 largely due to insufficient funding.

No effective malaria vaccine is presently available despite much research. In principle vaccines could work against *Plasmodium falciparum* at either the sporozoite, merozoite or gametocyte stages of its life cycle and thereby decrease the chances of a person becoming infected, decrease the severity of the disease and the possibility of death, and decrease the spread of malaria, respectively. An experimental vaccine based on genetically weakened sporozoites that conferred immunity against malaria in mice appeared in 2005 [142] and may lead to a similar vaccine for human use [143].

It is probable that the complete sequencing of the *Plasmodium falciparum* genome that contains fourteen chromosomes and some 6,500 genes will give insights into new possibilities for antimalarial drugs and vaccines [144,145]. Similarly, the sequencing of the genome of *Anopheles gambiae* may lead to a better understanding of the opportunities for reducing the effectiveness of this malaria vector [146]. However, long and hard battles lie ahead before malaria is brought under control using genomic and other methods.

Any discussion of malaria highlights the crucial role of the mosquito as the vector through which the disease is transmitted to humans. In addition, mosquitoes also transmit filiariasis, West Nile fever and yellow fever, which emphasizes their role as a formidable foe in the struggle against disease [147]. The historical impact of mosquitoes is illustrated by the completion of the Panama Canal by the United States in 1914 only being made possible by effective mosquito control greatly reducing the occurrence of malaria and yellow fever, diseases that twenty years earlier were the major cause of the French abandoning the project.

6.21. Re-emerging and New Pathogens

Despite achieving much progress in combating disease, the evolutionary struggle between the immune system and pathogens will be ongoing. Much of this struggle will be against new strains of known pathogens and new pathogens as they appear. These will emerge as humanity's antipathogen activities grow and other activities expand with increasing travel, trade and technological abilities and the growth of population. An early example of this was the emergence of poliomyelitis as a major threat in the early and mid 1900s. The great improvements in hygiene that eliminated epidemic typhus as a major disease greatly diminished early childhood exposure to the poliomyelitis virus. Consequently, the heightened immune response was lost and paralytic poliomyelitis became a serious threat that was only defeated through the development of vaccines.

The appearance of a new pathogen can be both frightening and initially mysterious as exemplified by the Marburg and Ebola viruses [148]. In 1967, several research workers in a laboratory in Marburg, Germany, contracted a new and virulent disease, as did personnel from laboratories in Frankfurt and Belgrade. The disease that infected the laboratory staff rapidly transferred to others and a quarter of those infected died from a combination of fever, diarrhoea, vomiting, extensive bleeding, circulatory failure and toxic shock. The cause was a virus transmitted by monkeys from Uganda being used in the laboratories for poliomyelitis vaccine preparation and related research. This virus, named Marburg, is a hemorrhagic fever, or filovirus, that quickly disappeared following its European debut. It was not seen again until a single case of infection in South Africa in 1975. However, in 1976, two close viral relatives emerged almost simultaneously in the rainforests of Sudan and the north of the Democratic Republic of Congo, Ebola Sudan and Ebola Zaire, respectively. Hundreds were infected of whom eighty-five percent died. A few other isolated cases of Marburg infection occurred in Africa in the 1970s and 1980s followed by larger outbreaks in the Democratic Republic of Congo in between 1998 and 2005 and in Angola in 2005. These outbreaks, together with strains of Ebola, have killed hundreds in sub-Saharan Africa with death rates of fifty to ninety percent. Transmission between humans is through either direct physical contact such as a handshake or a cough as a consequence of high levels of the virus in the skin and bodily fluids.

The reservoir for the Marburg and Ebola viruses is thought to be either a rainforest monkey or rodent and they appear to be transmitted through either meat, urine or faeces. The appearance of the Marburg and Ebola viruses in humans is attributed to population growth causing encroachment of farmers into African rainforest areas as the pressure to produce more food increases and a coincidentally increased contact between humans and the animals carrying the viruses. By 2000, the development of a vaccine against the Marburg and Ebola viruses seemed close. However, because of the difficulty of predicting when and where the sporadic infections are likely to occur, mass immunization does not appear a reasonable response. Rather, it is probable that the medical staff seeking to control Marburg and Ebola viral infections and scientists working on the causative viruses are likely to be among the first beneficiaries of the vaccine.

Another new pathogen appeared in early 2003 in the southern Chinese province of Guangdong [149]. This was a variant of the coronavirus usually associated with the common cold. However, in its new mutation it caused an acute form of pneumonia named sudden acute respiratory syndrome, or SARS, with a death rate of about five percent of those infected across a wide age group that is a similar rate to that of the 1918 influenza pandemic. This mutated virus appears to have arisen from a transfer of the coronavirus between humans, animals and birds living in close proximity much as new forms of influenza viruses arise. It attracted global attention when a medical professor from Guangdong died in a Hong Kong hospital after passing on the infection to others in his hotel and hospital. The newness of the virus was not recognized before SARS was spread further in Hong Kong and by air travellers to Canada, Singapore, Vietnam and elsewhere, while it was also spreading beyond Guandong in China. This caused the WHO to issue global health warnings concerning SARS that, together with the tension generated by the coincident war in Iraq and more widespread threats of terrorism, caused a precipitate decline in air travel and tourism.

Cholera represents a re-emerging disease from which tens of thousands die annually as a result of the severe diarrhoea, dehydration and debilitation caused by infection with the bacterium, *Vibrio cholerae* [150]. However, it is not the bacterium that does the damage, but the CTXφ virus that invades the bacterium and inserts its DNA into that of the genome of *Vibrio cholerae* that then produces the toxic protein that causes water to pass much more readily through the intestinal wall and results in the cholera symptoms of infected humans. Diseases resembling cholera appear in Indian and Greek records of 2,000 to 2,500 years ago, and Vasco da Gama encountered an epidemic disease resembling cholera that killed 20,000 on the southwestern coast of India in 1498 [151]. The first modern cholera pandemic, in 1817 to 1832, was associated with the Oman War and the war between Persia and Turkey. The second pandemic of 1829 to 1851 probably started in Russia and was more widespread. It reached the United States in 1832 and emerged in London in 1849 where John Snow realized that the disease was spread by drinking water contaminated with sewerage. Four more cholera pandemics occurred with the last of these ending in 1923. Thereafter, cholera became a rare infection and it seemed that improvements in the cleanliness of drinking water might have eliminated it as a serious disease. However, from 1961 onwards, new pandemics started and now embrace substantial regions of Africa, Asia and South America. This is coincident with the 1961 appearance of the new El Tor strain of *Vibrio cholerae* 01 first found in Indonesia. It was followed by the appearance of a further new strain in Bengal, *Vibrio cholerae* 0139. Usually, outbreaks of cholera are associated with poor sanitation and crowded living conditions close to water. A strong association with the sea is thought to arise from the ability of *Vibro cholerae* to infect zooplankton, shellfish and finned fish, the movement of infected fishermen and mariners from cholera affected areas and the discharge of ballast water containing infected zooplankton. Fortunately, while the new *Vibrio cholerae* strains rendered earlier cholera vaccines ineffective, new oral vaccines prepared from both dead and attenuated bacteria now provide sixty to one hundred percent protection [152]

The transportation of a pathogen from a region where it is endemic to one where it is not, but where it can thrive, can have startling consequences. This was the case in New York where, in the summer of 1999, sixty-two citizens sickened and seven died from West Nile virus infection and, in 2002, more than 1,400 sickened and more than sixty died [153]. This was not a particularly high toll compared with those caused by more common diseases from which New Yorkers suffer. However, by 2002 West Nile viral infections had become widespread in the United States and Canada and were a matter of major concern. While as yet the disease afflicts a small number of people compared with those suffering from other diseases in North America, an unusual aspect was the sudden appearance of a disease endemic to Africa, Asia and parts of Europe. The virus resides in a wide variety of birds, some of which seem little affected while others die quite rapidly. Mosquitoes pass it to humans who, while suffering the consequences, do not transmit the virus to each other. It is possible that the virus reached New York in an infected imported bird from which it was spread by mosquito to other birds and people. The virus has now been detected in birds in many parts of the United States and could well be carried by migratory birds to South America. There is as yet no treatment for West Nile virus infection. Because the levels of infection in birds and mosquitoes change from month to month and year to year, along with their populations, the outbreaks of human infection are sporadic, vary in size and are difficult to predict. However, it is not a virus to be taken lightly as shown by a 1997 outbreak in Bucharest, Romania, where 527 people were infected and fifty died. At best, people infected with the West Nile virus experience lingering influenza-like symptoms, others suffer from neurological damaged, muscle weakness and poliomyelitis-like symptoms and, at worst, a minority die.

The increasing desire to restore the environment and to reintroduce wild animals can have unexpected consequences as represented by the re-emergence of Lyme disease named after a small town in Connecticut where it re-appeared in 1975 [154]. There were five hundred recorded cases of Lyme disease in the

United States in 1982 and by 1996 it had risen to 16,000. It is the most common vector borne disease in the United States, Europe and parts of Asia. The disease is caused by the bacterium *Borrelia bugdorferi*. It begins with influenza-like symptoms and a rash on the skin where the bacterium starts its attack. In seventy percent of infections, the bacterium invades one or more of the brain, eyes, joints, heart and nerves, resulting in central nervous system damage and chronic arthritis, but seldom death. In the forests of the north central, northeastern and western United States, the white footed mouse is the major reservoir for *Borrelia bugdorferi* that is ingested by the newly hatched larvae of blacklegged ticks, *Ixodes scapularis*, as they feed on blood sucked from this host in mid summer. The larvae then transform into nymphs and spend the winter on the forest floor. When spring and summer come they seek out animals, white tailed deer in particular, and also humans whom they may infect with *Borrelia bugdorferi* as they bite and ingest blood. The larvae develop into the adult tick on the white tailed deer where they mate before dropping to the ground in autumn and laying their eggs in the soil in spring and early summer to start the two year cycle over again.

The re-emergence of Lyme disease coincides with reforestation of much of the northeastern and north central United States as farmland and other areas are abandoned, and the encouragement of deer and other wild animals into areas from which they had been almost totally expelled by agriculture and suburbia. As a consequence, parts of the United States now more closely resemble America before white settlement when Lyme disease is thought to have been common. Thus, this environmental restoration carries with it the risks that have always existed, competition between species for space and resources of which Lyme disease is a manifestation.

While the emergence of new diseases and the re-emergence of old ones have been largely associated with the developing nations, it is the developed nations that are suffering from an increasing incidence of asthma [155]. Asthma was quite a rare disease until 1900 but its prevalence has doubled in the last twenty years. Its occurrence has reached epidemic proportions in developed nations such as the United States where fifteen million people suffer from asthma and five thousand die from it each year. Globally, 180,000 die from asthma annually. The disproportionately high incidence of asthma in the developed nations has been attributed to children being increasingly protected from bacteria and other pathogens so that their immune systems are not so strongly developed as in preceding generations. They and their parents spend more time indoors and so are more exposed to household allergens such as dust, mites, cockroaches and pets. Other possible causes are air pollution and a generally greater exposure to a wide range of chemicals, but from these possibilities no culprit has been unequivocally identified.

6.22. Cancer: The Global Nemesis

Cancer kills over seven million people annually, as seen in Fig. 6.1, and each year some ten million cancers are diagnosed globally as seen in Fig. 6.16 [156].

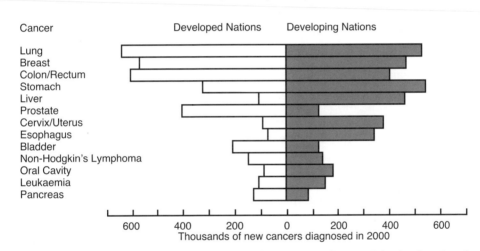

Fig. 6.16. The numbers and types of most prevalent cancers diagnosed in developed and developing nations in 2000. Adapted from [156].

Cancers of all types are composed of mutated cells that often contain from sixty to ninety chromosomes [157], an increase described as aneuploidy [158]. These cells multiply rapidly and overwhelm the particular tissue that they arise in. They are capable of invading every tissue of the body and appear in many different forms. At the first stage, a mutation in a single cell occurs that causes it to proliferate to produce a cluster of multiplying cells that still look normal in a particular tissue. This early stage is called hyperplasia. In time, another mutation occurs that gives the proliferating cells an abnormal appearance in a second stage called dysplasia. A further mutation may occur so that the proliferating cells become yet more abnormal in appearance but are still contained within a particular tissue and represent a localized cancer. Up to this point the cancer may often be effectively removed by surgery. The next stage, invasive cancer, is very dangerous. Now further mutations cause the cancer to invade other surrounding tissues and to shed cells into the blood circulatory system and the lymphatic system where they may circulate throughout the body to lodge elsewhere and initiate additional cancers in a process called metastasis.

Despite the abnormal nature of cancer cells, they are able evade the immune system. This is probably because the cancer cells are produced from healthy cells and do not differ as much from them as an invading pathogen does. While cancer strikes at all ages, the incidence of cancer increases with age to the point where among the elderly one in three women and one in two men die of cancer [159]. This is coincident with the general weakening of the immune system with age that may be part of the cause of the increased vulnerability of the aged to cancer. This idea gains support from the knowledge that persons whose immune systems are impaired either as a consequence of illness or medical therapy have an increased tendency to develop cancer. An additional possibility is that as tissue cells continuously reproduce themselves throughout life each reproduction has a finite possibility of making a mistake and reproducing DNA imperfectly to produce an abnormal cell. Usually such cells do not survive because of the complex cellular processes that lead to apoptosis, or natural cell death [160]. However, it may be that the accumulated number of cell reproductions that accompany ageing results in sufficient abnormal cells that can reproduce and result in a cancer. Such an explanation is very general, for unlike healthy tissue cells, or somatic cells, that can only reproduce themselves for a finite number of times and are therefore "mortal", cancer cells are able to reproduce themselves continuously and are "immortal".

A fundamental difference between healthy cells and cancer cells resides in the DNA of the cell. At either end of a strand of DNA are segments called telomeres that are essential to DNA reproduction [161]. Each reproduction shortens the telomere so that eventually it becomes too short for DNA to reproduce itself with the result that the cell undergoes apoptosis. Cancer cells would suffer the same fate if they did not contain an enzyme called telomerase that continually rebuilds the telomeres and bestows a kind of immortality on cancer cells. Telomerase is a recent discovery and is found in a very few specialist healthy tissue cells as well, notably in the germline cells, sperm or ova of adults, and stem cells of the developing embryo. Understanding of the action of telomerase and the process of apoptosis represent breakthroughs in the understanding of cancer and may prove to be a key weapon in the fight against it. The gaining of such understanding also raises the possibility of extending the human lifespan by treating ageing tissue with telomerase.

6.23. The Causes of Cancer

The occurrence of most cancers varies markedly between nations as seen in Figs. 6.16 and 6.17, and every type of cancer is rare somewhere. It is also known that

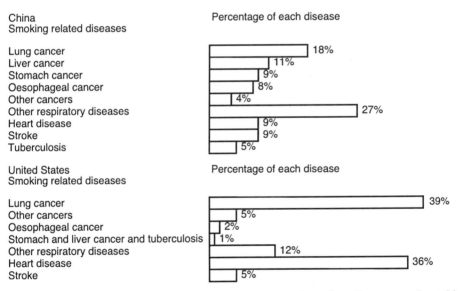

Fig. 6.17. Deaths occurring in persons below seventy years of age from diseases accelerated by smoking in China and the United States in 1990. Cancers of the mouth, pharynx, larynx, bladder and pancreas are collected under "Other cancers". Data from [156].

The pattern of cancer among immigrants changes to that of their new nation over a generation. Both of these observations suggest that the causes of cancer arise substantially from lifestyle and environment, and over the last twenty years many such causes have been identified [162]. An obvious difference between nations is the types of food consumed that is typified by the higher levels of meat, sweetened processed food and food additives consumed in the developed nations. Generally, the consumption of animal fats is thought to increase the risk of cancer while the consumption of fresh vegetables and fruit that contain antioxidants is thought to reduce the risk. On this basis it might be expected that the risk of cancer caused by diet would be lower in the developing nations. But this is not necessarily so as in many developing nations there is a great dependence on stored grain for food, and poor storage can result in moulds growing that contain aflotoxins that are implicated in the development of liver cancer. Isolating the effect of diet on the incidence of cancer from the plethora of other impacts that diet has on health is difficult but the general consensus appears to be that a varied diet low in animal fats and salt and high in antioxidants is likely to be beneficial.

Some of the major causes of cancer are becoming increasingly clear. The most important of these are smoking, excessive alcohol consumption, obesity and cancer causing viruses. However, the origins of the frequently occurring breast, prostate, and colon and rectum cancers remain unclear. Even so, there is widespread belief that cancer is largely a preventable disease [163]. The most effective cancer trigger is smoking where the tars inhaled in tobacco smoke cause cell abnormalities that develop into cancers, particularly those of the lung [164]. It is well established that the risk of developing lung cancer increases rapidly among those who continue to smoke so that the risk is greatest among those who begin to smoke when young and continue to do so throughout life. A graphic example of this comes from the United Kingdom where the great increase in smoking by men that occurred during and after the First World War resulted in an enormous increase in the prevalence of lung cancer in those born around 1900. Thus, by 1955 the lung cancer rate among males under fifty-five was the highest in the world. Recognition of the dangers of smoking led to a reduction of tar levels in cigarettes and smoking by males in the United Kingdom and a consequent decrease in the lung cancer rate below age fifty-five to the point where it is now amongst the lowest in the developed world and continues to decline. Twenty years later, the United States experienced a similar rise and fall in the incidence of lung cancers in males coincident with a great increase in smoking during the Second World War and a later recognition of its dangers. Women in the Western Europe and North America began smoking later than men and have shown a lesser tendency to stop. The consequence is that their lung cancer rates are either still increasing or are falling less rapidly. Meanwhile, in Eastern Europe and the developing nations, male lung cancer rates are still increasing coincident with a high and increasing level of smoking. The WHO estimates that smoking will become the largest global health problem by 2020, causing 8.4 million deaths annually.

Smoking also increases the incidence of cancer of the pancreas, bladder, kidney, larynx, mouth, pharynx, oesophagus, stomach, liver and cervix [164]. The incidence of different types of cancers and other diseases varies greatly between developed nations and developing nations as discussed earlier. As a consequence, the additional effect of smoking in increasing the incidence of these diseases similarly varies as shown in Fig. 6.17 in which the impacts of smoking on the occurrence of cancer and other diseases in China and the United States are compared.

A substantial number of viruses, bacteria and parasites have been identified as triggers for cancer although their mode of action is often unclear [165]. Thus, the many sexually transmitted papillomaviruses trigger cervical cancer and segments of their DNA are found in the cancerous cells. They are also responsible for other genital and anal cancers. Papillomaviruses may also be responsible for head, oesophagus, neck and skin cancers. Both the hepatitis viruses B and C predispose people to liver cancer. Each alters the level of calcium in liver cells that may result in mutation of DNA in the nucleus that in turn may cause the cell to become cancerous [166]. A bacterium that is carried by eighty-five percent of adults, *Helicobacter pylori*, sometimes causes stomach infections and gastric ulcers that in turn are major factors in the development of stomach cancer [167]. Elimination of these various infections would greatly decrease the incidence of cancer of the cervix, stomach and liver that account for five, six and nine percent, respectively, of cancers globally. On a less general scale, the Epstein-Barr virus is associated with B-cell malignancies and nasopharanyngeal cancer and, in association with malaria, Bukitt's lymphoma. The human T-cell lymphotropic virus type 1 is associated with T-cell leukaemia and lymphoma, the HIV with non-Hodgkin's lymphoma and the human herpes virus 8 with Kaposi's lymphoma. Schistosomiasis infection, or bilhazia, is associated with bladder and colon cancer and liver fluke with cholangiosarcoma. Indeed, there is a school of thought that attributes most cancers and other diseases to pathogens although such links are not proven in many cases [168].

Environmental factors are also responsible for some cancers. Probably the best known is the causing of skin cancer by ultraviolet-A and -B solar radiation [169]. This is particularly the case in fair skinned people where the closer they live to the equator the greater is their tendency to develop skin cancer. This is a consequence of the sun's ultraviolet radiation damaging DNA and also suppressing the immune system, particularly in childhood, and producing skin cancers later in life. A widespread environmental hazard in developed nations was the use of asbestos in building materials, as insulation and in automobile brake linings [170]. By 1960, it was established that inhalation of asbestos dust in any of its three forms, crocidolite, amosite and tremolite, predisposed workers to lung cancer and mesothelioma, a form of lung cancer, but it was not until the mid 1970's that the phasing out of asbestos based materials began. It is estimated that this exposure to asbestos prior to 1980 could result in a quarter of a million or more mesotheliomas in Western Europe alone. Because of their cancer producing potential, industrial exposure to radioactivity, ultraviolet radiation and chemicals are strictly controlled in developed nations, but such controls are much harder to effect in the general environment where atmospheric pollution in large cities contains carcinogens [171].

While living in the countryside greatly decreases exposure to industrial pollution, some of the more idyllic areas that contain granite hold a hidden, if small, threat. The continuous natural radioactive decay that occurs in Earth's crust, results in radioactive radon gas seeping from granite substrata and also from building material containing granite [172]. Granite is particularly high in uranium-238 that decays to radon-222, and thorium-232 that decays to radon-220. Both of these isotopes of radon are chemically harmless, but emit radiation that can result in DNA damage and cause a cancer to develop. Radon-222, that constitutes ninety-six percent of radon, emits alpha particles as it decays to solid polonium-218 and polonium-214 that also emit alpha particles. Alpha particles cause great damage to DNA that is the origin of their carcinogenic effect. In open countryside, radon is quickly dispersed into the atmosphere and poses no significant threat. However, radon seeping into poorly ventilated buildings can build up to a potentially dangerous level, although exactly what this level is debateable.

A predisposition to cancer may arise as a consequence of a mutation in a gene that might otherwise act to suppress cancer, such as the arcanely designated *VLH, mlh1, p16* and *p53* genes [173]. This transformation of "normal" genes into cancer causing genes, or "oncogenes", is aided by all causes of cancer. However, as the human is made up of some hundred trillion cells, each of which could become cancerous, it seems that there is a very effective damage limitation mechanism that prevents the multiplication of such mutated cells. This new and increasing knowledge of the molecular basis of cancer holds out the hope of more sophisticated treatments for cancer in the twenty-first century [174]. However, studies of the *p53* gene in mice suggest that there may be a price to pay for the cancer suppressing effect of this gene. Mice deficient in the *p53* gene are particularly susceptible to cancer probably because the availability of this gene to limit cell proliferation is low. Conversely, mice with enhanced *p53* activity are highly resistant to cancer but age prematurely. This infers that ageing may be a side effect of a natural protection from cancer. Nevertheless, the first *p53* gene therapy drug has been approved in China and drugs enhancing the function of the *p53* protein in cancerous cells are under study.

6.24. The Fight Against Cancer

In March 1998, it was announced that in the United States the incidence of cancer and death from it that had been steadily increasing since the 1930's had peaked in 1992 and was declining [175]. This was a consequence of changes in lifestyle, diet and therapy amongst other causes, and a reason for celebration. Most cancers are treated through removal of the cancer by surgery followed by radiation therapy and chemotherapy to clear up any remaining traces [176]. Even so, cancer is a hard disease to combat and the medical definition of a "cure" as five years without recurrence is a rather short period compared with the public expectation of a cure. Ideally, it is desirable to prevent cancer through immunization and to cure it through treatment analogous to the use of drugs and antibiotics in combating pathogens. There is now a range of anticancer drugs available that can at least temporarily cause a cancer to reduce in size and sometimes to disappear [177]. One such class of drugs is cisplatin and its derivatives, platinum based drugs that bind to DNA and disrupt the reproduction of cancer cells and often cure testicular, ovarian and neck cancer. Drugs such as 5-fluoracil and methotrexate block the action of enzymes that cause cells to become cancerous, while doxorubicin, cyclophosphamide and chlorambucil disrupt DNA in cancer cells. Other drugs restrict the growth of the capillaries that are essential for a cancer to grow. While the development of anticancer drugs is a major enterprise, the ability of cancers to develop a drug resistance is reminiscent of the development of resistance of other diseases to drugs as discussed earlier. To some extent this problem may be addressed by administration of a cocktail of drugs to outmanoeuvre the development of resistance. The alternative approach is to stimulate the immune system into attacking cancer cells and represents a growing area of development as understanding of molecular biology and genomics increases.

The most desirable therapies against cancer are vaccines that activate the immune system against cancer cells so that cancers do not develop at all [178,179]. However, such is the diversity of cancers and the differences between the genomes of cancer sufferers that the development of such vaccines presents a very major challenge even after a century of effort. Most such approaches are still being tested in the laboratory although a few are showing promise in clinical trials. This approach dates back to the 1890's when a New York cancer surgeon, William Coley, linked the sudden recovery of a bone cancer patient to the patient also suffering two attacks of the severe skin infection, erysipelas, caused by the bacterium *Streptococcus pyogenes*. Evidently, the patient's immune response to *Streptococcus pyogenes* also triggered an effective response to the bone cancer. Subsequently, Coley cured another bone cancer patient with an extract containing *Streptococcus pyogenes*. While Coley continued to develop this form of cancer immunology against a range of cancers with mixed results until his death in 1936, it was not widely adopted in the face of often disappointing results and adverse responses to the treatment. In addition to bacteria, viruses have been used with limited success in the treatment of cancer on the basis that their well-known ability to invade healthy cells and destroy them might be even more effective against cancer cells [180]. However, therapeutic infection with pathogenic bacteria and viruses presents the danger that the patient might develop a severe form of the associated disease and die of that instead. Fortunately, genetic engineering makes possible the systematic modification of bacteria and viruses to reduce the risk to healthy cells and make them potentially specific and destructive for cancer cells [181].

Another approach is to develop whole cell vaccines on the basis that they may stimulate the immune system to recognize the proteins, or antigens, produced on the surface of the cancer cell and attack the cancer cell. One method is to take a small piece of the cancer from the patient in a biopsy, grow its cells in a culture medium, kill them with ultraviolet or nuclear radiation and inject them back into the patient. The antigens of the killed cancer cells are unaffected by the radiation and cannot cause cancer, but are able to induce an immune response to attack the original cancer. This is called an autologous vaccine and contains the exact mix of the antigens of the patient's cancer at the time of the biopsy. However, the genetic similarity of the cancer cells to healthy cells may result in a weak immune response and the antigens may change as the cancer develops. Alternatively, an allogenic vaccine may be prepared from another patient's cancer and the genetic mismatch of the cells of the two patients usually generates a stronger immune response [179]. Vaccines consisting solely of the antigens that stimulate the immune system to react against the cancer provide another possible approach to vaccine development.

A related, but newer approach is to isolate dendritic cells from a cancer sufferer, culture them in the laboratory, expose them to cancer cell antigens and then injection them back into the patient in the hope that they will activate $CD8^+$ T-cells to attack the cancer cells [182]. This is the simplest form of this type of therapy. More advanced forms use dendritic cells from either the patient or elsewhere and fuse them with cancer cells also taken from the patient in the laboratory in the hope that this will stimulate the $CD8^+$ T-cells to attack the cancer. In advanced cancer cases, the immune system may be weakened to the point where it cannot respond effectively to a vaccine. Under these circumstances, the antibodies that seek out the antigens on the surface of the cancer cells are grown in the laboratory [183]. Such monoclonal antibodies may then be injected into the patient to attack the cancer cells. Alternatively, they may either have an anticancer drug or a radioactive isotope attached to them so that these may also be delivered to the cancer cell and help destroy it.

Increasingly cancer is being viewed as a stem cell based disease, a view that suggests new approaches to therapy [184]. However, despite the ingenuity shown by cancer researchers, a long struggle remains before cancer is brought under effective control. Fortunately, advances in cancer detection and location have aided earlier diagnosis that improves the chances of successful treatment [185].

6.25. Ageing: Causes and Consequences

In the later years of life a weakening of the immune system is thought to render humans more susceptible not only to cancer but also to invading pathogens. This is also a period when the slowing and increasingly imperfect regeneration of the body leads to malfunctions such as osteoporosis, a weakening of the bones, decreasing effectiveness of major organs such as the brain, heart, kidneys and liver, decreased musculature, eyesight and hearing, and loss of reproductive ability. The overall effect is that most of the elderly die of either arteriosclerosis, cancer or dementia [186]. However, humanity is resourceful in combating the effects of ageing as exemplified by joint replacement surgery where the insertion of replacement joints composed of metal, ceramic and plastic is almost routine. The replacement, or transplantation, of malfunctioning organs such as hearts, kidneys, livers and lungs with organs from donors is more complex as they must be connected to the blood supply and nervous system of the recipient. Because the donated organ invariably has a different genome from that of the recipient, the recipient's immune system attacks the cells of the donor organ and places it at the risk of being rejected as a consequence. This can usually be controlled by immune response suppressing drugs that the recipient has to take throughout life.

The success of the surgical replacement of a malfunctioning organ is less widespread than it could be because of shortages of donor organs. Except in the case of kidneys where a person may donate one kidney without great impairment, other organs only become available on the death of a donor with healthy organs. Consequently, there is great interest in the use of animal organs, particularly those of the pig, that are of similar size and function to those of the human. Such xenotransplantation also requires the use of immunosuppressor drugs to prevent organ rejection [187]. It also raises the danger of unwittingly transferring an animal virus to the human recipient with the possibility of producing a hitherto unencountered disease against which the human immune system is ineffective. Fortunately, the possibility of growing complete replacement organs from stem cells taken from the patient requiring an organ transplant is becoming increasingly likely as is discussed in section 5.17 [188]. Such an achievement would both obviate the need for immunosuppressing drugs and the possibility of viral transfer. A more distant possibility is cell by cell repair and rejuvenation of malfunctioning organs using stem cell therapy.

Inevitably the question arises as to why ageing occurs. The answers appear to be several, and each has to account for the cells of most of tissues, the somatic cells, that replace themselves for much of life eventually ceasing to do so, at which point senescence is reached. Examples of this are the skeleton that replaces itself completely every seven years for the greater part of life, but begins to lose weight when senescence sets in, and can become porous and brittle if osteoporosis develops. Similarly, skin is continuously replaced until senescence occurs when the underlying cells no longer reproduce as rapidly as those lost from the surface so that skin becomes thinner and wrinkles. In contrast, germline cells, sperm and ova, do not experience senescence, as they possess telomerase, an enzyme capable of renewing the telomeres, that confers a kind of "immortality" on them. Thus, the length of life appears to be programmed by the length of the telomeres at either end of DNA. If the telomers could be regenerated in senescent cells through treatment with telomerase it might be possible to extend the human lifespan [189]. However, some organs and tissues appear to age at different rates that suggests that the genome possesses other controls over ageing in addition to the telomere.

6.26. Future Patterns of Health

A major theme to emerge from this discussion of humanity's struggle with disease is a preoccupation with prolonging life through developing increasingly sophisticated defences against old and new diseases. Important as these developments are, viewing health in terms of lifespan alone does not give a full picture of the impact of disease on humanity. Quite apart from broad assessments of the quality of life, it is possible to assess quantitatively the effects of disease and other agencies on health through the *years* of *l*ife affected by a *d*isability, or

YLD. The diseases affecting this aspect of health are generally very different from the major causes of death as is seen from Fig. 6.18 [190,191]. Thus, four of the top ten causes of disability were mental illness together with iron deficiency, falls and alcohol consumption in 1990. Alcohol consumption was the leading cause of disability in men in developed nations and the fourth largest cause of disability in men in developing countries.

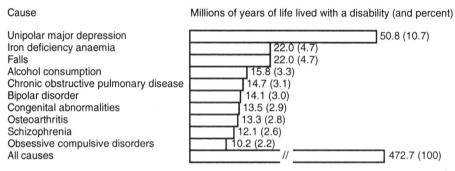

Cause   Millions of years of life lived with a disability (and percent)

Unipolar major depression — 50.8 (10.7)
Iron deficiency anaemia — 22.0 (4.7)
Falls — 22.0 (4.7)
Alcohol consumption — 15.8 (3.3)
Chronic obstructive pulmonary disease — 14.7 (3.1)
Bipolar disorder — 14.1 (3.0)
Congenital abnormalities — 13.5 (2.9)
Osteoarthritis — 13.3 (2.8)
Schizophrenia — 12.1 (2.6)
Obsessive compulsive disorders — 10.2 (2.2)
All causes — 472.7 (100)

Fig. 6.18. The major causes of disability globally in 1990. Data from [191].

To assess what the global threats to health are, how they are likely to change over time and how to intervene to ameliorate the burden of disease, the Harvard School of Public Health, the WHO and the World Bank set up the Global Burden of Disease Study in 1992 [190,192,193]. Using as a measure of the impact of disease and other causes on health, the number of healthy years of life lost either through premature death or through years lived with a disability weighted by the severity of that disability, or *d*isability *a*djusted *l*ife *y*ears (DALY), the burdens posed by the fifteen leading causes were ranked in descending order of importance for 1990 as shown in Fig. 6.19. By making estimates of the effects of changes in education and wealth, changes in societal behaviour and advances in medicine, a similar ranking was predicted for 2020. On this basis it is seen that the relative effects of disease and other agencies on health will change greatly over the period 1990 to 2020. At the beginning of the period one in three deaths globally resulted from either infectious, maternal, perinatal or nutritional causes, mostly in the developing nations. Within this group most of the deaths were those of infants and the basic cause was insufficient funds for healthcare and hygiene. However, even in 1990 fewer deaths arose from these causes than from non infectious diseases in developing nations indicating that the tide had turned against infectious diseases that exerted dominance in earlier times. That this is so is a tribute to public health programs that have eliminated smallpox globally, poliomyelitis from many nations and decreased measles, rubella, tetanus,

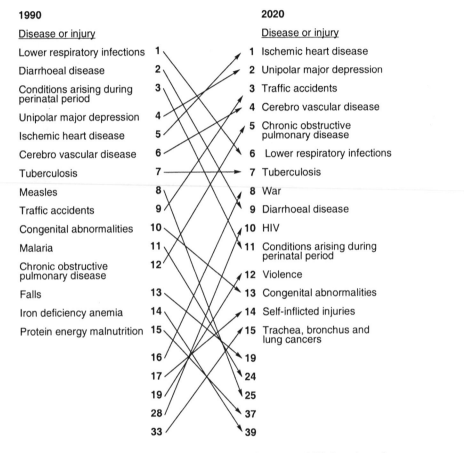

1990

Disease or injury

Lower respiratory infections 1
Diarrhoeal disease 2
Conditions arising during perinatal period 3
Unipolar major depression 4
Ischemic heart disease 5
Cerebro vascular disease 6
Tuberculosis 7
Measles 8
Traffic accidents 9
Congenital abnormalities 10
Malaria 11
Chronic obstructive pulmonary disease 12
Falls 13
Iron deficiency anemia 14
Protein energy malnutrition 15
16
17
19
28
33

2020

Disease or injury

1 Ischemic heart disease
2 Unipolar major depression
3 Traffic accidents
4 Cerebro vascular disease
5 Chronic obstructive pulmonary disease
6 Lower respiratory infections
7 Tuberculosis
8 War
9 Diarrhoeal disease
10 HIV
11 Conditions arising during perinatal period
12 Violence
13 Congenital abnormalities
14 Self-inflicted injuries
15 Trachea, bronchus and lung cancers
19
24
25
37
39

Fig. 6.19. The fifteen major causes of the number of healthy years of life lost through premature death or to years lived with a disability weighted by the severity of that disability, or disability adjusted life years (DALY), in 1990 and the predicted change in order of importance in 2020. Adapted from [191].

diphtheria and meningitis to a few cases each year in many nations and saved many millions of lives and billions of dollars [192]. However, it has been argued that the expenditure of more public health funds to dissuade humanity from smoking, excessive alcohol consumption and other disease and disability generating habits may save more lives and decrease disability to a greater extent than expenditure on medical research in other areas. Nevertheless, the challenges

to humanity's health are subject to considerable variation and those now causing
increasing concern are climate change, ozone depletion and biological warfare.

6.27. Climate Change: An Agent for Changing Disease Patterns

As humanity has increased tropospheric greenhouse gases and depleted the
stratospheric ozone layer, there has been a resulting warming of Earth and
increased exposure to ultraviolet-B radiation, respectively, as discussed in detail
in Chapters 8 and 9. The effect of these changes could have substantial effects on
human health [194]. Increased exposure to ultraviolet-B radiation has the
potential to increase the incidence of skin cancer, melanoma and eye damage in
addition to weakening the immune system and thereby increasing the incidence
of a wide range of diseases. Many such diseases are those found in the tropics
that may become endemic over a larger range as Earth warms and climate
changes to provide the pathogens and their vectors such as mosquitoes with a
more favourable climate and environment. Evidence for increasing disease
ranges is already present. Malaria is a prime example, and is expected to double
its range by 2020, as is another mosquito borne disease, dengue fever. Bilharzia,
or schistosomiasis, a parasitic worm that infects humans through contaminated
water is also likely to undergo a substantial extension in range with a warming
climate and an extension of irrigation. At present, as many as two hundred
million people are thought to be infected with bilharzia, an infection that leads to
death if untreated. A range of other water borne pathogens such as the *Giardia
lamblia* and *Cryptosporidium* protozoa that cause diarrhoea, the *Naegleria
fowleri* protozoan that causes meningitis, the *Salmonella enterica* bacterium that
causes severe gastric upsets and dysentery, and the rotavirus that causes
diarrhoea, are also expected to increase their range as Earth warms. While such
global expectations are speculative to some extent, the substantial effects of
climate change on health and disease distribution in the Pacific region are
already familiar through the alternation between the El Niño and La Niña
climatic effects stemming from changes in the magnitude and displacement of
water temperature across the Pacific Ocean.

With global warming, heat waves are expected to become more prevalent and
kill many more people through heat exhaustion. Extreme weather events are also

expected to become more frequent and to increase the loss of life. However, these rather gloomy predictions do not take into account humanity's response to these threats that will almost certainly result in improved disease control, healthcare and changes in lifestyle that will nullify much if not all of the effects of climate change and ozone depletion. While these challenges to health came about unintentionally, the rational approaches to them are, unfortunately, not so readily applied to the next challenge, biological warfare, a product of irrational minds.

6.28. Biological Warfare: The Great Betrayal

Given the devastating consequences of many diseases it is unfortunately inevitable that their use as weapons should be contemplated [195,196]. This represents an addition to humanity's capacity for violence that is different from other violent abilities. The difference lies in the recruitment and aiding of humanity's pathogenic evolutionary enemies and their toxins that maintain an unremitting attack, irrespective of a human state of war or otherwise, and thereby constitutes a great betrayal. Early instances of biological warfare date back to the Romans who infected their enemies' water supplies with dead animals. In 1347 a Tartar army catapulted bubonic plague victims into the besieged town of Kaffa on the Black Sea in the hope of infecting their enemy, and Sir Jeffrey Amherst used gifts of infected blankets to spread smallpox among hostile native Americans in 1758 during the Franco-British war in North America. While the League of Nations unsuccessfully sought to ban biological warfare in 1925, there are no certain twentieth and twenty-first century instances of biological weapons being used in combat, although the Japanese killed thousands of prisoners of war with anthrax, cholera, plague, dysentery and typhus during the war against China [197]. They also released plague carrying fleas over the Chinese province of Chekiang. These instances of biological warfare were not without cost to the perpetrators, for the Japanese suffered substantial casualties from the plague they released. In 1942, the Soviet Union may have deliberately infected German troops at Stalingrad with the *Francisella tularensis* bacterium to give them tularaemia, a particularly infectious disease that kills about twenty percent of its victims [196]. Whatever the truth, more than 100,000 cases of the disease appeared in the German and Soviet armies.

After the Second World War, the United States and the Soviet Union, in particular, engaged in extensive research and development of biological warfare agents but encountered major problems in weaponising these agents to the point where they would be a lesser threat to the attacker than the attacked. In 1969, the United States unilaterally renounced biological warfare and destroyed all of its biological weapons by May 1972. This was almost coincident with the formulation of the Biological and Toxin Weapons Convention, BTWC, of April 1972 that came into force in March 1975 and banned the development and production of biological weapons. By 2001, the convention had been signed by 160 nations and ratified by 143, including China, the United Kingdom, the United States, the Russian Federation, Iraq, Iran, Libya and North Korea. Despite this, some nations maintained their development of biological weapons. The greatest transgressor of the convention was the Soviet Union that, from 1972 to its collapse in 1990, maintained a massive biological warfare effort employing some 60,000 personnel at forty facilities in the development of anthrax, plague, smallpox, Marburg and Ebola weaponising programs. However, this incurred a penalty, for in 1979 about one hundred people died and extensive livestock were lost through an accidental and major release of anthrax spores at Ekaterinburg, formerly Sverdlovsk [198].

It is a great irony that the work of three of the giants of bacteriology, Robert Koch, Louis Pasteur and Joseph Lister, who respectively isolated *Bacillus anthracis* that causes anthrax, produced a vaccine against it and developed antiseptic precautions that, in combination, greatly reduced the livestock toll of this disease, should have generated the knowledge that facilitated the potential use of anthrax as a weapon. Anthrax is normally a disease of sheep and cattle and is rarely transferred to humans unless the bacterium, *Bacillus anthracis*, either enters a wound to cause cutaneous anthrax or is consumed in contaminated meat to cause intestinal anthrax. However, *Bacillus anthracis* can remain dormant as a spore for hundreds of years and then change into its infectious form in a warm and moist environment. Breathing in ten thousand or so spores (which sounds a lot but would scarcely cover the area of a period in this text) can result in inhalation anthrax that kills more than eighty percent of its victims unless they have either been immunized against anthrax or are treated with large doses of antibiotics within the first day of infection. This high death rate is only equalled by hemorrhagic fevers such as Ebola and is much greater than the thirty percent of epidemic typhus.

Sadly, the elimination of smallpox as an infectious disease, the consequent cessation of immunization programs and the loss of immunity in the global population have made it a potentially devastating agent for biological warfare. Fear of this is a significant contributor to the reluctance to destroy the only two known remaining samples of the variola virus, one each in the United States and the Russian Federation [199]. In 2004, a WHO recommendation to allow a genetic modification of the variola virus to aid research in smallpox treatment caused disquiet because it might give the impression that the United States and the Russian Federation were recommencing bioweapon research.

At present, the most likely to be encountered biological warfare agents are anthrax, smallpox, plague, tularaemia and Q fever, an incapacitating, but seldom fatal infection, caused by *Coxiella burnetii rickettsia*. The 1991 Gulf war drew attention to Iraq's maintenance of a substantial biological warfare program involving anthrax and botulinum toxin produced by the *Clostridium botulinum* bacterium, the most deadly toxin known [200]. Quite apart from nations who ignore the BTWC, terrorist groups are increasingly causing concern as potential users of biological weapons. The variola virus and *Bacillus anthracis* are readily cultured by competent microbiologists and are robust in handling as shown in the latter case by the deaths from inhalation anthrax resulting from mailed anthrax spores in the United States in late 2001 [201]. Earlier, in 1984, a terrorist group caused 750 people to become very ill by spreading *Salmonella enterica* bacteria in restaurants in the United States [196].

Gene technology raises the possibility of inserting antibiotic and vaccine resistant genes into pathogens to render infected victims defenceless against them and potentially for the attackers to develop a vaccine to render them immune to such a weapon. Should such a pathogen escape control, the consequent loss of life could be incalculable. This renders it even more important that the BTWC should be strengthened by making mandatory the regular inspection of facilities capable of producing biological warfare agents [202].

References

1. World Health Report 2004, WHO, Geneva, Switzerland, 2004.
2. A. A. Puca *et al.*, A genome-wide scan for linkage to human exceptional longevity identifies a locus on chromosome 4, *Proc. Natl. Acad. Sci. USA*, **2001**, *98*, 10505.
3. a) P. W. Ewald, The evolution of virulence, *Sci. Am.*, **1993**, *268*, April, 56. b) R. M. Nesse and G. C. Williams, Evolution and the origins of disease, *Sci. Am.*, **1998**, *279*, Nov., 86. c) D. Pimentel *et al.*, Ecology of increasing disease, *BioScience*, **1998**, *48*, Oct., 817. d) B. R. Levin, M. Lipsitch and S. Bonhoeffer, Population biology, evolution, infectious disease: convergence and synthesis, *Science*, **1999**, *283*, 806.
4. A. H. Mokdad *et al.*, Actual causes of death in the United States, 2000, *JAMA*, **2004**, *291*, 1238.
5. E. Marshall, Public enemy number one: tobacco or obesity? *Science*, **2004**, *304*, 804.
6. a) W. H. McNeill, *Plagues and Peoples*, Doubleday, New York, USA, 1976. b) R. M. Krause, The origin of plagues: old and new, *Science*, **1992**, *257*, 1073.
7. a) P. Hale *et al.*, HIV a "9" on Richter scale of viral diseases, *Nature*, **2001**, *412*, 271. b) P. R. Lamptey, Reducing heterosexual transmission of HIV in poor countries, *Brit. Med. J.*, **2002**, *324*, 207.
8. R. S. Gottfried, *The Black Death*, Macmillan, London, UK, 1986.
9. C. McEvedy, The Bubonic plague, *Sci. Am.*, **1988**, *258*, Feb., 74.
10. C. L. Mee, How a mysterious disease laid low Europe's masses, *Smithsonian*, **1990**, *20*, Feb., 67.
11. M. J. Keeling and C. A. Gilligan, Metapopulation dynamics of bubonic plague, *Nature*, **2000**, *407*, 903.
12. J. F. D. Shrewsbury, *A History of Bubonic Plague in the British Isles*, Cambridge University Press, Cambridge, UK, 1970.
13. a) A. G. Carmichael, Bubonic plague, in *The Cambridge World History of Human Disease*, K. F. Kiple, ed., Cambridge University Press, Cambridge, UK, 1993. b) G. L. Campbell and J. M. Hughes, Plague in India: a new warning from an old nemesis, *Annals Internat. Med.*, **1995**, *122*, 151. c) V. Ramalingaswami *et al.*, Plague in India, *Nature Med.*, **1995**, *1*, 1237. d) C. T. Fisher, Bubonic plague in modern China: an overview, *J. Orient. Soc. Aust.*, **1995-96**, *27* and *28*, 57.
14. a) S. T. Cole and C. Buchrieser, A plague o' both your hosts, *Nature*, **2001**, *413*, 467. b) J. Parkhill *et al.*, Genome sequence of *Yersinia pestis*, the causative agent of plague, *ibid.*, 523.
15. a) F. L. Black, Why did they die? *Science*, **1992**, *258*, 1739. b) W. M. Denevan, *The Population of The Americas in 1492*, 2nd edn., University of Wisconsin Press, Madison, USA, 1992. c) B. L. Turner and K. W. Burzer, The Columbian encounter and land-use change, *Environment*, **1992**, *34*, Oct., 16.
16. C. Zimmer, Can genes solve the syphilis mystery? *Science*, **2001**, *292*, 1091.
17. C. E. Bartecchi, T. D. MacKenzie and R. W. Schrier, The global tobacco epidemic, *Sci. Am.*, **1995**, *272*, May, 26.
18. a) J. Lederberger, Infectious history, *Science*, **2000**, *288*, 287. b) C. Zimmer, Genetic trees reveal disease origins, *Science*, **2001**, *292*, 1090. c) A. Osterhaus, Catastrophes after crossing species barriers, *Phil. Trans. Roy. Soc. B*, **2001**, *356*, 791. d) C. G. N. Mascie-Taylor and E. Karim, The burden of chronic disease, *Science*, **2003**, *302*, 1921.
19. L. H. Taylor, S. M. Latham and M. E. J. Woolhouse, Risk factors for human disease emergence, *Phil. Trans. Roy. Soc. B*, **2001**, *356*, 983.
20. S. Cleaveland, M. K. Laurenson and L. H. Taylor, Diseases of humans and their domestic mammals: pathogen characteristics, host range and the risk of emergence, *Phil. Trans. Roy. Soc. B*, **2001**, *356*, 991.
21. R. A Weiss, The Leeuwenhoek Lecture 2001. Animal origins of human infectious disease, *Phil. Trans. Roy. Soc. B*, **2001**, *356*, 957.

22. J. Parkhill *et al.*, Complete genome sequence of a multiple drug resistant *Salmonella enterica* serovar Typhi CT18, *Nature*, **2001**, *413*, 848.
23. a) B. Geenwood, The epidemiology of pneumococcal infection in children in the developing world, *Phil. Trans. Roy. Soc. B*, **1999**, *354*, 777. b) H. Tettelin *et al.*, Complete genome sequence of a virulent isolate of *Streptococcus pneumoniae*, *Science*, **2001**, *293*, 498.
24. C. Gerard, For whom the bell tolls, *Nature*, **1998**, 395, 217.
25. M. Brookes, Day of the mutators, *New Sci.*, **1998**, 14 Feb., 38.
26. a) R. Nathan, Japan's *E. coli* outbreak elicits fear, anger, *Nature Med.*, **1996**, *2*, 956. b) A. Coghlan, Killer strain raises urgent questions, *New Sci.*, **1997**, 25 Jan., 7. c) F. R. Blattner *et al.*, The complete genome sequence of *Escherichia coli* K-12, *Science*, **1997**, *277*, 1453. d) G. M. Pupo, R. Lan and P. R. Reeves, Multiple independent origins of Shigella clones of Escherichia coli and convergent evolution of many of their characteristics, *Proc. Natl. Acad. Sci. USA*, **2000**, *97*, 10567. e) J. A. Eisen, Gastrogenomics, *Nature*, **2001**, *409*, 463. f) J. Wain *et al.*, Acquisition of virulence-associated factors by the enteric pathogens *Escherichia coli* and *Salmonella enterica*, *Phil. Trans. Roy. Soc. B*, **2001**, *356*, 1027. g) N. T. Perna *et al.*, Genome sequence of enterohaemorrhagic *Escherichia coli* O157:H7, *Nature*, **2001**, *409*, 529. h) M. McClelland *et al.*, Complete genome sequence of *Salmonella enterica* serovar Typhimurium LT2, *Nature*, **2001**, *413*, 852. i) V. Souza, A. Castillo and L. E. Eguiarte, The evolutionary ecology of *Escherichia coli*, *Am. Sci.*, **2002**, *90*, 332.
27. D. A. Relman, The search for unrecognised pathogens, *Science*, **1999**, *284*, 1309.
28. a) S. B. Prusiner, The prion diseases, *Sci. Am.*, **1995**, *272*, Jan., 48. b) F. Edenhoffer *et al.*, Chemistry and molecular biology of transmissible spongiform encephalopathies, *Angew, Chem. Int. Edn.*, **1997**, *36*, 1674. c) S. B. Prusiner, Prion diseases and the BSE crisis, *Science*, **1997**, *278*, 245. d) A. Aguzzi and C. Weissmann, Prion research: the next frontiers, *Nature*, **1997**, *389*, 795. e) S. B. Prusiner, Nobel Lecture: Prions, *Proc. Natl. Acad. Sci. USA*, **1998**, *95*, 13363.
29. a) D. R. Brown, BSE: a post-industrial disease, *Chem. Indust.*, **2001**, 5 Feb., 73. b) D. MacKenzie, North America needs tougher BSE measures, *New Sci.*, **2003**, 5 July, 5. c) B. Hileman, Guarding against mad cow disease, *Chem. Eng. News.*, **2003**, *81*, 4 Aug., 22. d) S. B. Prusiner, Detecting mad cow disease, *Sci. Am.*, **2004**, *291*, July, 60. e) *Anon.*, It's a mad cow world, *New Sci.*, **2004**, 7 Aug., 35. f) P. Brown, Mad-cow disease in cattle and human beings, *Am. Sci.*, **2004**, *92*, 334.
30. a) M. E. Bruce, "New Variant" Creutzfeld-Jakob disease and bovine spongiform encephalopathy, *Nature Med.*, **2000**, *6*, 258. b) P. Brown, Mad-cow disease in cattle and human beings, *Am. Sci.*, **2004**, *92*, 334.
31. P. Jones, Smart proteins, *New Sci.*, *Inside Science 139*, **2001**, 17 March, 1.
32. a) A. C. Ghani *et al.*, Predicting vCJD mortality in Great Britain, *Nature*, **2000**, *406*, 583. b) P. Yam, Mad cow's human toll, *Sci. Am.*, **2001**, *284*, May, 10. c) N. M. Ferguson *et al.*, Estimating the human health risk from possible BSE infection of the British sheep flock, *Nature*, **2002**, *415*, 420.
33. a) S. Borman, Prion research accelerates, *Chem. Eng. News.*, **1998**, *76*, 9 Feb., 22. b) D. E. Koshland, Conformational changes: how small is big enough? *Nature Med.*, **1998**, *4*, 1112. c) S. W. Liebman, The shape of the species barrier, *Nature*, **2001**, *410*, 161. d) B. Hileman, The "mad" disease has many forms, *Chem. Eng. News*, **2001**, *79*, 9 April, 24. e) B. Caughey and D. A. Kocisko, Prion diseases: a nucleic acid accomplice? *Nature*, **2003**, *425*, 673. f) N. R. Deleault, R. W. Lucassen and S. Supattapone, RNA molecules stimulate prion protein conversion, *ibid.*, 717. g) J. Cozin, An end to the prion debate? Don't count on it, *Science*, **2004**, *305*, 589. h) G. Legname, *et al.*, Synthetic mammalian prions, *ibid.*, 673.
34. a) D. C. Gajdusek, C. J. Gibbs and M. Alpers, Transmission and passage of experimental "kuru" to chimpanzees, *Science*, **1967**, *155*, 212. b) D. C. Gajdusek, Nobel Prize acceptance speech. Unconventional viruses and the origin and disappearance of kuru, *Science*, **1977**, *197*, 943. c) A. Underwood, Cannibals to cows: the path of a deadly disease, *Newsweek*, **2001**, March 12, 41.

35.   a) A. Coghlan, Prion purge, *New Sci.*, **2001**, 28 July, 6. b) M. Enari, E. Fleshchig and C. Weissmann, Scrapie protein accumulation by scrapie infected neuroblastoma cells abrogated by exposure to a prion antibody, *Proc. Natl. Acad. Sci. USA*, **2001**, *98*, 9295. c) F. L. Heppner *et al.*, Prevention of scrapie pathogenesis by transgenic expression of anti-prion protein antibodies, *Science*, **2001**, *294*, 178.

36.   a) D. J. Selkoe, Amyloid protein and Alzheimer's disease, *Sci. Am.*, **1991**, *265*, Nov., 68. b) P. J. Thomas, B.-H. Qu and P. L. Pedersen, Defective folding as a basis of human disease, *Trends Biomed. Sci.*, **1995**, *20*, 456. c) G. Taubes, Misfolding the way to disease, *Science*, **1996**, *271*, 1493. d) J. Marx, New gene tied to common form of Altzheimer's, *Science*, **1998**, *281*, 507. e) D. Blacker *et al.*, Alpha-2 macroglobulin is genetically associated with Altzheimer disease, *Nature Genet.*, **1998**, *19*, 357. f) J. Hope, Breech-birth prions, *Nature*, **1999**, *402*, 7373. g) N. Iwata *et al.*, Metabolic regulation of brain Aβ by neprilysin, *Science*, **2001**, *292*, 1550. h) R. J. Ellis and T. J. T. Pinheiro, Danger – misfolding proteins, *Nature*, **2002**, *416*, 483. i) B. De Strooper and J. Woodgett, Mental plaque removal, *Nature*, **2003**, *423*, 392. j) A. Aguzzi and C. Haass, Games played by rogue proteins in prion disorders and Alzheimer's disease, *Science*, **2003**, *302*, 814. k) D. J. Selkoe, Folding proteins in fatal ways, *Nature*, **2003**, *426*, 900. l) M. P. Mattson, Pathways towards and away from Alzheimer's disease, *Nature*, **2004**, *430*, 631. m) M. Beckman, Untangling Alzheimer's by paring plaques bolsters amyloid theory, *Science*, **2004**, *305*, 762.

37.   a) V. Morell, Bacteria diversify through warfare, *Science*, **1997**, *278*, 575. b) A. F. Read and L. H Taylor, The ecology of genetically diverse infections, *Science*, **2001**, *292*, 1099. c) E. Gulbins and F. Lang, Pathogens, host-cell invasion and disease, *Am. Sci.*, **2001**, 89, 406.

38.   a) B. R. Bloom, On the particularity of pathogens, *Nature*, **2000**, *406*, 760. b) B. R. Levin and R. Antia, Why we don't get sick: the within-host population dynamics of bacterial infections, *Science*, **2001**, *292*, 1112. c) P.Cossart and P. J. Sansonetti, Bacterial invasion: the paradigms of enteroinvasive pathogens, *Science*, **2004**, *304*, 242.

39.   a) G. J. V. Nossal, Life, death and the immune system, *Sci. Am.*, **1993**, *269*, Sept., 20. b) I. L. Weissman and M. D. Cooper, How the immune system develops, *ibid.*, 32. c) C. A. Janeway, How the immune system recognises invaders, *ibid.*, 40. d) P. Marrack and J. W. Kappler, How the immune system recognizes the body, *ibid.*, 49. e) W. E. Paul, Infectious diseases and the immune system, *ibid.*, 56. f) Double agent. The cunning of our immune systems came at a price, *New Sci.*, **1998**, 5 Sept., 21. g) R. Medzhitov and C. Janeway, Innate immunity, *N. Engl. J. Med.*, **2000**, *343*, 338. h) J. K. Whitmire and R. Ahmed, The economy of T-cell memory: CD4$^+$ recession in times of CD8$^+$ stability?*Nature Med.*, **2001**, 7, 892. i) R. Medzhitov and C. A. Janeway, Decoding the patterns of self and nonself by the innate immune system, *Science*, **2002**, *296*, 298. j) P. Matzinger, The danger model: a renewed sense of self, *ibid.*, *301*. k) M. C. Carroll, The complement system in regulation of adaptive immunity, *Nature Immunol.*, **2004**, *5*, 981. l) D. H. Raulet, Interplay of natural killer cells and their receptors with the adaptive immune response, *ibid.*, 996. m) K. Hoebe, E. Janssen and B. Beutler, The interface between innate and adaptive immunity, *ibid.*, 971.

40.   a) B. Beutler, Inferences, questions and possibilities in toll-like receptor signalling, *Nature*, **2004**, *430*, 257. b) D. N. Cook, D. S. Pisetsky and D. A. Schwartz, Toll-like receptors in the pathogenesis of human disease, *Nature Immunol.*, **2004**, *5*, 975. c) A. Iwasaki and R. Medshitov, Toll-like receptor control of the adaptive immune responses, *ibid.*, 987.

41.   L. A. J. O'Neil, Immunity's early warning system, *Sci. Am.*, **2005**, *292*, Jan., 38.

42.   a) E. Samuel, Beating the superbug, *New Sci.*, **2001**, 28 April, 17. b) H. Muir, Killer's genome reveals its vicious arsenal, *New Sci.*, **2001**, 28 April, 17. c) D. A. A. Ala'Aldeen and H. Grundmann, Unveiling of genetic basis of resistance of *S. aureus* to antibiotics, *Lancet*, **2001**, *357*, 1219. d) M. Kuroda *et al.*, Whole genome sequence of meticillin resistant *Staphylococcus aureus*, *ibid.*, 1225. e) D. Ferber, Triple-threat microbe gained powers from another bug, *Science*, **2003**, *302*, 1488.

43.   a) H. M. Johnson, J. K. Russell and C. H. Pontzer, Superantigens in human disease, *Sci. Am.*, **1992**, *266*, April, 42. b) L. M. Steinman, Autoimmune disease, *Sci. Am.*, **1993**, *269*, Sept., 75. c) L. M. Lichtenstein, Allergy and the immune system, *Sci. Am.*, **1993**, *269*, Sept.,

85. d) P. Marrack, J. Kappler and B. L. Kotzin, Autoimmune disease: why and where it occurs, *Nature Med.*, **2001**, *7*, 899.

44. a) G. Weissmann, Aspirin, *Sci. Am.*, **1991**, *264*, Jan., 58. b) K. N. Allen, Aspirin - now we can see it, *Nature Med.*, **1995**, *1*, 882. c) S. J. Shiff and B. Rigas, Aspirin for cancer, *Nature Med.*, **1999**, *5*, 1348.

45. L. M. Lichtenberger *et al.*, Non-steroidal anti-inflammatory drugs (NSAIDs) associate with phospholipids: insight into the mechanism and reversal of NSAID-induced gastrointestinal injury, *Nature Med.*, **1995**, *1*, 154.

46. a) R. J. Wurtman and R. L. Bettiker, The slowing of treatment discovery, 1965-1995, *Nature Med.*, **1995**, *1*, 1122. b) J. Drews, Drug discovery: a historical perspective, *Science*, **2000**, *287*, 1960.

47. *The Pharmaceutical Century. Ten Decades of Drug Discovery*, J. Ryan, A. Newman and M. Jacobs, eds., ACS Publications, Washington DC, USA, 2000. http://pubs.acs.org/journals/pharmcent/html

48. T. Stephens and R. Brynner, *Dark Remedy: The Impact of Thalidomide and its Revival as a Vital Medicine*, Perseus, Boulder, USA, 2001.

49. W. McBride, Thalidomide and congenital anomalies, *Lancet*, **1961**, *2*, 1358.

50. a) M. Cimons, Thalidomide resurfaces for FDA consideration, *Nature Med.*, **1997**, *3*, 8. b) B. F. Hales, Thalidomide on the comeback trail, *Nature Med.*, **1999**, *5*, 489. c) T. Stephens, Reinventing thalidomide, *Chem. Br.*, **2001**, *37*, Nov., 38.

51. a) E. Jenner, An enquiry into the causes and effects of the variolae vaccinae, a disease discovered in some of the western counties of England, particularly Gloucestershire, and known by the name of cow pox, London, 1798, in *Classis of Medicine and Surgery*, C. N. B. Camac, ed., Dover, New York, USA, 1959, p. 213. b) A. Mantovani, Edward Jenner 200 years on, *Nature Med.*, **1996**, *2*, 256. c) H. Bazin, *The Eradication of Smallpox: Edward Jenner and the First and Only Eradication of a Human Infectious Disease*, Academic Press, London, UK, 2000.

52. F. Fenner *et al.*, *Smallpox and Its Eradication*, WHO, Geneva, Switzerland, 1988.

53. R. A. Seder and A. V. S. Hill, Vaccines against intracellular infections requiring cellular immunity, *Nature*, **2000**, *406*, 793.

54. a) R. I. Glass *et al.*, Rotavirus vaccines at the threshold, *Nature Med.*, **1997**, *3*, 1324. b) B. R. Bloom and R. Widdus, Vaccine visions and their global impact, *Nature Med.*, *Vaccine Supplement*, **1998**, *4*, 480. c) K. Birmingham, New vaccines aimed at the rich says UNICEF, *ibid.*, 477. d) P. Brown, Finally, a plan to reduce measles deaths, *Nature Med.*, **2000**, 6, 1305.

55. a) H. Wigzell, The immune system as a therapeutic agent, *Sci. Am.*, **1993**, *269*, Sept., 95. b) T. Beardsley, Better than a cure, *Sci. Am.*, **1995**, *272*, Jan., 88. c) M. R. Hilleman, Six decades of vaccine development - a personal history, *Nature Med.*, *Vaccine Supplement*, **1998**, *4*, 507. d) M. A. Liu, Vaccine developments, *ibid.*, 515.

56. a) L. Schlein, Hunting down the last of the poliovirus, *Science*, **1998**, *279*, 168. b) S. Kumar and M. Day, Off target, *New Sci.*, **1998**, 25 July, 18. c) T. Clarke, Polio's last stand, *Nature*, **2000**, *409*, 278. d) S. Blume and I. Geesink, A brief history of polio vaccines, *Science*, **2000**, *288*, 1593. e) L. Greensfelder, Polio outbreak raises questions about vaccine, *Science*, **2000**, *290*, 1867. f) D. MacKenzie, Polio strikes back, *New Sci.*, **2000**, 9 Dec., 14. g) Editorial, Polio eradication: the endgame, *Nature Med.*, **2001**, *7*, 131. h) M. Watanabe, Polio outbreak threatens eradication program, *Nature Med.*, **2001**, *7*, 135. i) N. Nathanson and P. Fine, Poliomyelitis eradication - a dangerous end game, *Science*, **2002**, *296*, 269. j) D. Butler, WHO prepares for final push to rid the world of polio, *Nature*, **2003**, *424*, 604. k) L. Roberts, Polio eradication effort adds new weapon to its armory, *Science*, **2005**, *307*, 190. l) C. Soares, Polio postponed, *Sci. Am.*, **2005**, *292*, Jan., 18.

57. a) G. Taubes, Salvation in a snippet of DNA, *Science*, **1997**, *278*, 1711. b) D. B. Weiner and R. C. Kennedy, Genetic vaccines, *Sci. Am.*, **1999**, *281*, July, 50. c) N. P. Restifo, Vaccines to die for, *Nature Biotechnol.*, **2001**, *19*, 527. d) A. Hunt and G. Evan, Till death us do part, *Science*, **2001**, *293*, 1785. e) G. J. Nossal, The double helix and immunology, *Nature*, **2003**,

421, 440. f) M. K. Collins and V. Cerundolo, Gene therapy meets vaccine development, *Trends Biotechnol.*, **2004**, *22*, 623. g) M. Scarselli *et al.*, The impact of genomics on vaccine design, *Trends Biotechnol.*, **2005**, *23*, 84.

58.   D. M. Pardoli, Cancer vaccines, *Nature Med.*, *Vaccine Supplement*, **1998**, *4*, 525.

59.   a) C. J. Arntzen, Pharmaceutical foodstuffs - oral immunization with transgenic plants, *Nature Med.*, *Vaccine Supplement*, **1998**, *4*, 502. b) T. Arakawa and W. H. R. Langridge, Plants are not just passive creatures! *Nature Med.*, **1998**, *4*, 550. c) J. K.-C. Ma *et al.*, Characterization of a recombinant plant monoclonal secretory antibody and preventive immunotherapy in humans, *Nature Med.*, **1998**, *4*, 601. d) C. O. Tacket *et al.*, Immunogenicity in humans of a recombinant bacterial antigen delivered in a transgenic potato, *ibid*, 607.

60.   D. MacKenzie, Vaccine failure, *New Sci.*, **1997**, 11 Oct., 5.

61.   S. Gandon *et al.*, Imperfect vaccines and the evolution of pathogen virulence, *Nature*, **2001**, *414*, 751.

62.   Y. Nakaseko and M. Yanagida, Cytoskeleton in the cell cycle, *Nature*, **2001**, *412*, 291.

63.   a) V. Hook, Superbugs step up the pace, *Chem. Br.*, **1997**, *33*, May, 34. b) C. Ainsworth, They're everywhere, *New Sci.*, **2001**, 19 May, 5. c) S. R. Palumbi, Humans as the world's greatest evolutionary force, *Science*, **2001**, *293*, 786.

64.   L. P. Kotra *et al.*, Bacteria fight back, *Chem. Indust.*, **2000**, 22 May, 341.

65.   H. C. Neu, The crisis in antibiotic resistance, *Science*, **1992**, *257*, 1064.

66.   H. Pearson, "Superbug" hurdles key drug barrier, *Nature*, **2002**, *418*, 469.

67.   a) S. B. Levy, Antibiotic resistance: an ecological imbalance, in *Antibiotic Resistance: Origins, Evolution, Selection and Spread*, Ciba Foundation Symposium 207, Wiley, Chichester, UK, **1997**, p. 1. b) J. E. Davies, Origins, acquisition and dissemination of antibiotic resistance determinants, *ibid.*, p. 15. c) T. Sheldon, Antibiotic resistance soars in Europe, *Nature Med.*, **2004**, *10*, 6.

68.   a) J. Davies, Inactivation of antibiotics and the dissemination of resistance genes, *Science*, **1994**, *264*, 375. b) H. Nikaido, Prevention of drug access to bacterial targets: permeability barriers and active efflux, *ibid.*, 382, c) B. G. Spratt, Resistance to antibiotics mediated by target alterations, *ibid.*, 388. d) R. E. Lenski, The cost of antibiotic resistance - from the perspective of a bacterium, in *Antibiotic Resistance: Origins, Evolution, Selection and Spread*, Ciba Foundation Symposium 207, Wiley, Chichester, UK, **1997**, p. 131. e) R. A. Skurray and N. Firth, Molecular evolution of multiply-antibiotic-resistant staphlococci, *ibid.*, p. 167. f) S. J. Schrag, V. Perrot and B. R. Levin, Adaptation to the fitness costs of antibiotic resistance in *Escherichia coli*, *Proc. Roy. Soc. B*, **1997**, *264*, 1287. g) J. Knight, Too much of a good thing, *New Sci.*, **1997**, 27 Sept., 18. h) V. Morell, Antibiotic resistance: road of no return, *Science*, **1997**, *278*, 575. i) S. B. Levy, The challenge of antibiotic resistance, *Sci. Am.*, **1998**, *278*, March, 32. j) D. Ferber, New hunt for roots of resistance, *Science*, **1998**, *280*, 27. k) E. C. Böttger *et al.*, Fitness of antibiotic microorganisms and compensatory mutations, *Nature Med.*, **1998**, *4*, 1343. l) C. Walsh, Molecular mechanisms that confer antibacterial drug resistance, *Nature*, **2000**, *406*, 775. m) G. L. Archer and J. M. Bosilevac, Signalling antibiotic resistance in Staphylococci, *Science*, **2001**, *291*, 1915. n) A. George, March of the superbugs, *New Sci.*, *Inside Science 162*, **2003**, 19 July, 1.

69.   a) R. Gaynes and D. Mommet, The contribution of antibiotic use on the frequency of antibiotic resistance in hospitals, in *Antibiotic Resistance: Origins, Evolution, Selection and Spread*, Ciba Foundation Symposium 207, Wiley, Chichester, UK, 1997, p. 47. b) H. Giamarellou and A. Antoniadou, The effect of monitoring of antibiotic use on decreasing antibiotic resistance in the hospital, *ibid.*, 76. c) M. L. Cohen, Epidemiological factors influencing the emergence of antimicrobial resistance, *ibid.*, 223. d) Claire, They're everywhere. Superbugs may be more widespread than anyone had realised, *New Sci.*, **2001**, 19 May, 5.

70.   a) W. Witte, Impact of antibiotic use in animal feeding on resistance of bacterial pathogens in humans, in *Antibiotic Resistance: Origins, Evolution, Selection and Spread*, Ciba Foundation Symposium 207, Wiley, Chichester, UK, 1997, p. 61. b) A. Coghlan, That's us

stuffed then, *New Sci.*, **1997**, 6 Dec., 5. c) V. Perreten *et al.*, Antibiotic resistance spread in food, *Nature*, **1997**, *389*, 801. d) D. Ferber, Superbugs on the hoof? *Science*, **2000**, *288*, 792.

71. B. Hileman, Resistance is on the rise, *Chem. Eng. News.*, **2001**, *79*, 19 Feb., 47.

72. a) C. T. Walsh, Vancomycin resistance: decoding the molecular logic, *Science*, **1993**, *261*, 308. b) A. Coghlan, Animal antibiotics "threaten hospital epidemics", *New Sci.*, **1996**, 27 July, 7. c) M. S. Gilmore and J. A. Hoch, A vancomycin surprise, *Nature*, **1999**, *399*, 524. d) R. Novak *et al.*, Emergence of vancomycin tolerance in Streptococcus pneumoniae, *ibid.*, 590.

73. W. Witte, Medical consequences of antibiotic use in agriculture, *Science*, **1998**, *279*, 996.

74. C. Mlot, Antidote for antibiotic use on the farm, *BioScience*, **2000**, *50*, Nov., 955.

75. *Anon.*, A smoking gun, *New Sci.*, **1998**, 21 March, 13.

76. M. Wadman, Group urges survey of antibioics in animals, *Nature*, **2001**, *409*, 273.

77. a) A. Gibbons, Exploring new strategies to fight drug-resistant microbes, *Science*, **1992**, *257*, 1036. b) M. L. Cohen, Epidemiology of drug resistance: implications for a post-antimicrobial era, *ibid.*, 1050. c) R. P. Bax, Antibiotic resistance - what can we do? *Nature Med.*, **1998**, *4*, 545. d) R. M. Anderson, The pandemic of antibiotic resistance, *Nature Med.*, **1999**, *5*, 147.

78. a) J. Travis, Reviving the antibiotic miracle, *Science*, **1994**, *264*, 360. b) J. E. Gabay, Ubiquitous natural antibiotics, *ibid.*, 373. c) L. Valigra, Engineering the future of antibiotics, *New Sci.*, **1994**, 30 April, 25. d) R. E. Service, Antibiotics that resist resistance, *Science*, **1995**, *270*, 724. e) T. Beardsley, Resisting resistance, *Sci. Am.* **1996**, *274*, Jan., 15. f) S. J. Brickner, Multidrug-resistant bacterial infections: driving the search for new antibiotics, *Chem. Indust.*, **1997**, 17 Feb., 131. g) S. Nemecek, Beating bacteria, New ways to fend off antibiotic-resistant pathogens, *Sci. Am.*, **1997**, *276*, Feb., 24. h) F. Arigoni *et al.*, A genome-based approach for the identification of essential bacterial genes, *Nature Biotechnol.*, **1998**, *16*, 851. i) H. Breithaupt, The new antibiotics, *Nature Biotechnol.*, **1999**, *17*, 1165. j) M. Brookes, Could designer antibiotics hit bugs where it hurts? *New Sci.*, **1998**, 5 Sept., 20. k) C. Walsh, Deconstructing vancomycin, *Science*, **1999**, *284*, 442. l) M. Ge *et al.*, Vancomycin derivatives that inhibit peptidoglycan biosynthesis without binding D-Ala-D-Ala, *ibid.*, 507. m) J. Rosamond and A. Allsop, Harnessing the power of the genome in the search for new antibiotics, *Science*, **2000**, *287*, 1973. n) K. C. Nicolaou and C. N. C. Boddy, Behind enemy lines, *Sci. Am.*, **2001**, *284*, May, 48. o) C. F. Amábile-Cuevas, New antibiotics and new resistance, *Am. Sci.*, **2003**, *91*, 13.

79. M. Leeb, A shot in the arm, *Nature*, **2004**, *431*, 892. b) C. Nathan, Antibiotics at the crossroads, *ibid.*, 899.

80. M. L. Cohen, Changing patterns of infectious disease, *Nature*, **2000**, *406*, 762.

81. A. J. McMichael, Environmental and social influences on emerging infectious diseases: past, present and future, *Phil. Trans. Roy. Soc. B*, **2004**, *359*, 1049.

82. B. R. Bloom and C. J. L. Murray, Tuberculosis: commentary on a reemergent killer, *Science*, **1992**, *257*, 1055.

83. G. K. Folkers and A. S. Fauci, The role of the US Government agencies in vaccine research and development, *Nature Med.*, *Vaccine Supplement*, **1998**, 4, 491.

84. a) World Bank, *World Development Indicators 1999*, World Bank, Washington, DC, USA, 1999. b) C. Hertzman, Health and human society, *Am. Sci.*, **2001**, *89*, 538.

85. a) D. E. Bloom and J. G. Williamson, Demographic transitions and economic miracles in emerging Asia, *World Bank Economic Review*, **1998**, *12*, 419. b) D. E. Bloom and D. Canning, The health and wealth of nations, *Science*, **2000**, *287*, 1207.

86. a) World Bank, *Confronting AIDS: Public Priorities in a Global Epidemic*, Oxford University Press, New York, USA, 1997. b) *The Use of Essential Drugs* (Tenth Model List of Essential Drugs), WHO Technical Reports Series No. 882, Eighth Report of the WHO Expert Committee, WHO, Geneva, 1998. c) D. E. Bloom and A. Rosenfield, River Path Associates, *A Moment in Time: AIDS and Business*, American Foundation for AIDS

Research, New York, USA, 1999. d) B. Pécoul *et al.*, Access to essential drugs in poor countries, *JAMA*, **1999**, *281*, 361.

87.   a) L. C. Chen, F. Wittgenstein and E. McKeon, The upsurge of mortality in Russia: Causes and policy implications, *Popn. Dev. Rev.*, **1996**, *22*, 517. b) T. Lincoln, Death and the demon drink in Russia, *Nature*, **1997**, *388*, 723. c) R. Stone, Stress: the invisible hand in Eastern Europe's death rates, *Science*, **2000**, *288*, 1732.

88.   T. Fujisaki and M. R. Reich, *Assessments of TDR's Contribution to Product Development for Tropical Diseases: The Case of Ivermectin for Onchoceriasis*, UND/World Bank/WHO Special Programme for Research and Training in Tropical Diseases, Geneva, Switzerland, 1997.

89.   M. R. Reich, The global drug gap, *Science*, **2000**, *287*, 1979.

90.   P. Wehrwein, Pharmacophilanthropy, *Harvard Public Health Review*, **1999**, Summer, 32.

91.   K. J. Watkins, AIDS: managing a global problem, *Chem. Eng. News.*, **2001**, *79*, 4 June, 25.

92.   a) J. Fincher, America's deadly rendezvous with the "Spanish Lady", *Smithsonian*, **1989**, *19*, Jan., 130. b) A. Crosby, *America's Forgotten Pandemic*, Cambridge University Press, Cambridge, UK, 1989. c) G. Kolata, *The Story of the Great Influenza Pandemic of 1918 and the Search for the Virus that Caused It*, Farrar, Strauss and Giroux, New York, USA, 1999. d) J. S. Oxford *et al.*, Who's that lady? *Nature Med.*, **1999**, *5*, 1351. e) A. Noymer and M. Garenne, The 1918 influenza epidemic's effects on sex differentials in mortality in the United States, *Popn. Dev. Rev.*, **2000**, *23*, 565. f) D. Martindale, No Mercy, *New Sci.*, **2000**, 14 Oct., 26.

93.   a) W. Smith, C. H. Andrewes and P. P. Laidlaw, A virus obtained from influenza patients, *Lancet*, **1933**, *225*, 66. b) T. Francis, Transmission of influenza by a filterable virus, *Science*, **1934**, *80*, 457.

94.   R. G. Webster and E. J. Walker, Influenza, *Am. Sci.*, **2003**, *91*, March-April, 122.

95.   a) H. Saul, Flu vaccines wanted: dead or alive, *New Sci.*, **1995**, 18 Feb., 26. b) E. D. Kilbourne, What are the prospects for a universal influenza vaccine? *Nature Med.*, **1999**, *5*, 1119. c) J. S. Oxford and R. S. Daniels, The Holy Grail of influenza virologists, *ibid.*, 484. d) M. Enserink, Looking the pandemic in the eye, *Science*, **2004**, *306*, 392. e) J. Kaiser, Facing down pandemic flu, the world's defenses are weak, *ibid.*, 394. f) K. Stöhr and M. Esveld, Will vaccines be available for the next influenza pandemic? *Ibid.*, 2195.

96.   a) V. Glaser, Neuraminidase inhibitors take bite out of influenza, *Nature Biotechnol*, **1998**, *16*, 1002. b) W. G. Laver, N. Bischofberger and R. G. Webster, Disarming flu viruses, *Sci. Am.*, **1999**, *280*, Jan., 56. c) G. Laver and E. Garman, The origin and control of pandemic influenza, *Science*, **2001**, *293*, 1776.

97.   a) J. K. Taubenberger *et al.*, Initial genetic characterisation of the 1918 "Spanish" influenza virus, *Science*, **1997**, *275*, 1793. b) J. K. Taubenberger, A. H. Reid and T. G. Fanning, The 1918 influenza virus: a killer comes into view, *Virology*, **2000**, *274*, 241. c) J. A. H. Reid *et al.*, Characterization of the 1918 "Spanish" influenza virus neuraminidase gene, *Proc. Natl. Acad. Sci. USA*, **2000**, *97*, 6785. d) J. Lederberg, H1N1-influenza as Lazarus: genomic reconstruction from the tomb of an unknown, *ibid.*, **2001**, *98*, 2115. e) C. F. Basler *et al.*, Sequence of the 1918 pandemic influenza virus nonstructural gene (NS) segment and characterisation of recombinant viruses bearing the 1918 NS genes, *ibid.*, 2746. f) M. J. Gibbs, J. S. Armstrong and A. J. Gibbs, Recombination in the hemagglutinin gene of the 1918 "Spanish Flu", *Science*, **2001**, *293*, 1842. g) E. C. Holmes, 1918 and all that, *Science*, **2004**, *303*, 1787. h) S. J. Gamblin *et al.*, The structure and receptor binding properties of the 1918 influenza hemagglutinin, *ibid.*, 1838. i) J. Stevens *et al.*, Structure of the uncleaved human H1 hemagglutinin from the extinct 1918 influenza virus, *ibid.*, 1866. j) J. K. Taubenberger *et al.*, Capturing a killer flu virus, *Sci Am.*, **2005**, *292*, Jan., 62.

98.   a) C. Scholtissek and E. Naylor, Fish farming and influenza pandemics, *Nature*, **1988**, *331*, 215. b) R. G. Webster *et al.*, Evolution and ecology of influenza A viruses, *Microbiol. Rev.*, **1992**, *56*, 152. c) E. C. J. Claas *et al.*, Infection of children with avian-human reassortment influenza virus from pigs in Europe, *Virology*, **1994**, *204*, 453. d) J. Cohen, The flu

pandemic that might have been, *Science*, **1997**, *277*, 1600. e) M. Day, In fear of flu, *New Sci.*, **1997**, 13 Dec., 5.

99.   a) J. C. de Jong *et al.*, A pandemic warning? *Nature*, **1997**, *389*, 554. b) M. Day, A killer waits, *New Sci.*, **1998**, 31 Jan., 18. c) K. Y. Yuen *et al.*, Clinical features and rapid viral diagnosis of human disease associates with avian influenza A H5N1 virus, *Lancet*, **1998**, *351*, 467. d) E. C. J. Claas *et al.*, Human influenza A H5N1 virus related to a highly pathogenic avian influenza virus, *ibid.*, 472. e) E. C. J. Claas and A. D. M. E. Osterhaus, New clues to the emergence of flu pandemics, *Nature Med.*, **1998**, *4*, 1122. f) M. Day, Sweating it out, *New Sci.*, **1999**, 30 Oct., 4. g) Y. P. Lin *et al.*, Avian-to-human transmission of H9N2 subtype influenza A viruses: relationship between H9N2 and H5N1 human isolates, *Proc. Natl. Acad. Sci. USA*, **2000**, *97*, 9654. h) T. Horimoto and Y. Kawaoka, Pandemic threat caused by avian influenza A viruses, *Clin. Microbiol. Rev.*, **2001**, *14*, 129. i) D. Cyranoski, Outbreak of chicken flu rattles Hong Kong, *Nature*, **2001**, *412*, 261. j) R. G. Webster, A molecular Whodunit, *Science*, **2001**, *293*, 1773. k) M. Hatta *et al.*, Molecular basis for high virulence of Hong Kong H5N1 influenza A viruses, *Science*, **2001**, *293*, 1840. l) A. Abbot, Human fatality adds fresh impetus to fight against bird flu, *Nature*, **2003**, *423*, 5. m) R. J. Webby and R. G. Webster, Are we ready for pandemic influenza? *Science*, **2003**, *302*, 1519. n) E. C. Holmes, J. K. Taubenberger and B.T. Grenfell, Heading off an influenza pandemic, *Science*, **2005**, *309*, 989. o) I. M. Longini *et al.*, Containing pandemic influenza at the source, *ibid.*, 1083.

100. W. L Heyward and J. W. Curran, The epidemiology of AIDS in the U.S., *Sci. Am.*, **1988**, *259*, Oct., 52.

101. a) G. Barre-Sinoussi *et al.*, Isolation of a T-lymphotropicc retrovirus from a patient at risk for acquired immune deficiency syndrome, *Science*, **1983**, *220*, 868. b) M. Popovic *et al.*, Detection, isolation and continuous production of cytopathic retroviruses (HTLV-III) from patients with AIDS and pre-AIDS, *Science*, **1984**, *224*, 497. c) R. C. Gallo and L. Montagnier, The chronology of AIDS research, *Nature*, **1987**, *326*, 435. d) R. C. Gallo, Human retroviruses in the second decade: a personal perspective, *Nature Med.*, **1995**, *1*, 753. e) S. B. Prusiner, Discovering the cause of AIDS, *Science*, **2002**, *298*, 1726. f) L. Montagnier, A history of HIV discovery, *ibid.*, 1727, g) R. C. Gallo, The early years of HIV/AIDS, *ibid.*, 1728.

102. a) R. R. Redfield and D. S. Burke, HIV infection: the clinical picture, *Sci. Am.*, **1988**, *259*, Oct., 70. b) D. Butler, Making global blood safety a priority, *Nature Med.*, **1995**, *1*, 7. c) D. Spurgeon, Canadian enquiry calls for "safety first" blood agency, *Nature*, **1997**, *390*, 432.

103. a) J. M. Mann *et al.*, The international epidemiology of AIDS, *Sci. Am.*, **1988**, 259, Oct., 60. b) R. A. Weiss, Gulliver's travels in HIVland, *Nature*, **2001**, *410*, 963. c) P. Piot *et al.*, The global impact of HIV/AIDS, *Nature*, **2001**, *410*, 968.

104. *AIDS Epidemic Update*, UNAIDS/WHO, Geneva, Switzerland, 2004.

105. a) J. Cohen, Ground zero: AIDS research in Africa, *Science*, **2000**, *288*, 2150. b) P. Piot, Global AIDS epidemic: time to turn the tide, *ibid.*, 2176. c) B. Schwartländer *et al.*, AIDS in a new millennium, *Science* **2000**, *289*, 64. d) K. Birmingham, UN acknowledges HIV/AIDS as a threat to world peace, *Nature Med.*, **2000**, *6*, 117. e) D. Normile, China awakes to fight projected AIDS crisis, *ibid.*, 2312. f) J. Cohen, HIV gains foothold in key Asian groups, *Science*, **2001**, *294*, 282. g) J. Kaufman and J. Jing, China and AIDS – the time to act is now, *Science*, **2002**, *296*, 2339. h) J. Cohen, Malawi: a suitable case for treatment, *Science*, **2002**, *297*, 927.

106. a) *Anon.*, The Durban Declaration, *Nature*, **2000**, *406*, 15. b) D. Butler, In and out of Africa, *Nature*, **2000**, *408*, 901. c) H. P. Binswanger, HIV/AIDS treatment for millions, *Science*, **2001**, *292*, 221. d) K. Birmingham, South Africa vs. big pharma, *Nature Med.*, **2001**, *7*, 390. e) B. G. Williams and E. Gouws, The epidemiology of human immunodeficiency virus in South Africa, *Phil. Trans. Roy. Soc. B*, **2001**, *356*, 1077.

107. a) J. G. Bartlett and R. D. Moore, Improving HIV therapy, *Sci. Am.*, **1998**, *279*, July, 65. b) T. Jardetsky, Conformational camoflage, *Nature*, **2002**, *420*, 623. c) A. McKnight and R. A.

Weiss, Blocking the docking of HIV-1, *Proc. Natl. Acad. Sci. USA*, **2003**, *100*, 10581. d) P. Cohen, Nowhere to hide, *New Sci.*, **2004**, 3 April, 34.

108.   a) W. C. Greene, AIDS and the immune system, *Sci. Am.*, **1993**, *269*, Sept., 67. b) M. J. Bevan and T. J. Braciale, Why can't cytotoxic T cells handle HIV? *Proc. Natl. Acad. Sci. USA*, **1995**, *92*, 5765. c) R. C. Desrosiers, Strategies used by human immunodeficiency virus that allow persistent viral replication, *Nature Med.*, **1999**, *5*, 273. d) J. M. McCune, The dynamics of CD4$^+$ T-cell depletion in HIV disease, *Nature*, **2001**, *410*, 974. e) A. J. McMichael and S. L. Rowland-Jones, Cellular immune responses to HIV, *ibid.*, 980. f) D. C. Douek *et al.*, HIV preferentially infects HIV-specific CD4$^+$ T cells, *Nature*, **2002**, *417*, 95. g) D. N. Levy *et al.*, Dynamics of HIV-1 recombination in its natural target cells, *Proc. Natl. Acad. Sci. USA*, **2004**, *101*, 4204.

109.   a) N. Short, A fight to the death, *Nature Med.*, **1995**, *1*, 122. b) D. D. Richman, HIV therapeutics, *Science*, **1996**, *272*, 1886. c) R. J. Wurtman, What went right: why is HIV a treatable infection, *Nature Med.*, **1997**, *3*, 715. d) C. Ezzell, Swallow this: three days of antiretroviral drugs, *Sci. Am.*, **2000**, *283*, Nov., 82. e) M. Altfield and B. D. Walker, Less is more? STI in acute and chronic HIV-1 infection, *Nature Med.*, **2001**, *7*, 881. f) P. Cohen, Nowhere to hide, *New Sci.*, **2004**, *3* April, 34.

110.   a) D. D. Richman, How drug resistance arises, *Sci. Am.*, **1998**, *279*, July, 68. b) A. Persidis, Progress against AIDS, *Nature Biotechnol.*, **2000**, *18*, 466. c) J. Cohen, Confronting the limits of success, *Science*, **2002**, *296*, 2320. d) P. Cohen, HIV wages war on "miracle" drugs, *New Sci.*, **2002**, 17 Aug., 22. e) R. J. Pomerantz, Cross-talk and viral reservoirs, *Nature*, **2003**, *424*, 136.

111.   a) C. Ezzell, AIDS drugs for Africa, *Sci. Am.*, **2000**, *283*, Nov., 80. b) J. P. Moatti *et al.*, Antiretroviral treatment for HIV infection in developing countries: an attainable new paradigm, *Nature Med.*, **2003**, *9*, 1449.

112.   a) B. R. Bloom, A perspective on AIDS vaccines, *Science*, **1996**, *272*, 1888. b) D. Baltimore and C. Heilman, HIV vaccines: prospects and challenges, *Sci. Am.*, **1998**, *279*, July, 78. c) C. Holden, First AIDS vaccine launched, *Science*, **1998**, *280*, 1697. d) D. R. Burton and J. P. Moore, Why do we not have an HIV vaccine and how can we make one? *Nature Med., Vaccine Supplement*, **1998**, *4*, 495. e) C. A. Heilman and D. Baltimore, HIV vaccines - where are we going, *ibid.*, 532. f) R. P. Johnson, Live attenuated AIDS vaccines: hazards and hopes, *Nature Med.*, **1999**, *5*, 154. g) A. J. McMichael and T. Hanke, Is an HIV vaccine possible? *Ibid.*, 612. h) D. C. Krakauer and N. Nowak, T-Cell induced pathogenesis in HIV: bystander effects and latent infection, *Proc. Roy. Soc. B*, **1999**, *266*, 1069. i) E. Niiler, Companies begin tests of HIV vaccines, *Nature Med.*, **2001**, *7*, 390. j) G. J. Nabel, Challenges and opportunities for development of an AIDS vaccine, *Nature*, **2001**, *410*, 1002. k) J. Cohen, *Shots in the Dark: The Wayward Search for an AIDS Vaccine*, Norton, New York, USA, 2001. l) D. Baltimore, Steering a course to an AIDS vaccine, *Science*, **2002**, *296*, 2297. m) B. Gashen *et al.*, Diversity considerations in HIV-1 vaccine selection, *ibid.*, 2354. n) J. D. Lifson and M. A. Martin, One step forward, one step back, *Nature*, **2002**, *415*, 272. o) J. W. Shiver *et al.*, Replication-incompetent adenoviral vaccine vector elicits effective antiimmunodeficiency-virus immunity, *ibid.*, 331. p) B. Gaschen *et al.*, Diversity considerations in HIV-1 vaccine selection, *Science*, **2002**, *296*, 2354. q) R. D. Mugerwa *et al.*, First trial of the HIV-1 vaccine in Africa: Ugandan experience, *Brit. Med. J.*, **2002**, *324*, 226. r) J. Cohen, AIDS vaccine trial produces disappointment and confusion, *Science*, **2003**, *299*, 1290. s) R. D. Klausner *et al.*, The need for a global HIV vaccine enterprise, *Science*, **2003**, *300*, 2036. t) R. C. Desrosiers, Prospects for an AIDS vaccine, *Nature Med.*, **2004**, *10*, 221.

113.   a) M. Mata *et al.*, Evaluation of a recombinant *Listeria monocytogenes* expressing in an HIV protein that protects mice against viral challenge, *Vaccine*, **2001**, *19*, 1435. b) A. Coghlan and M. Day, Bugs keep HIV at bay, *New Sci.*, **2001**, 3 Feb., 18.

114.   a) G. S. Shearer and M. Clerici, Protective immunity against HIV infection: has Nature done the experiment for us? *Immunol. Today*, **1996**, *17*, 21. b) B. Laumbacher and R. Wank, Recruiting HLA to fight HIV, *Nature Med., Vaccine Supplement*, **1998**, *4*, 505. c)

M. W. Makgoba, N. Solomon and T. J. P. Tucker, The search for an HIV vaccine, *Brit. Med. J.*, **2002**, *324*, 211. d) J. Clayton, Beating the odds, *New Sci.*, **2003**, 14 Feb., 34. e) C. Wilson, World without AIDS, *ibid.*, 38.

115.   a) T. J. Coates and C. Collins, Preventing HIV infection, *Sci. Am.*, **1998**, *279*, July, 76. b) N. Nathanson and J. D. Auerbach, Confronting the HIV pandemic, *Science*, **1999**, *284*, 16. c) P. Jha *et al.*, Reducing HIV transmission in developing countries, *Science*, 2001, *292*, 224. d) B. Schwartländer *et al.*, Resource needs for HIV/AIDS, *ibid.*, 2434. e) P. Hale and colleagues, Mission now possible for AIDS fund, *Nature*, **2001**, *412*, 271. f) A. Stone, Protect and survive, *New Sci.*, **2003**, 8 Feb., 42. g) R. L. Stoneburner and D. Low-Beer, Population-level HIV declines and behavioral risk avoidance in Uganda, *Science*, **2004**, *304*, 714.

116.   a) M. Essex and P. J. Kanki, The origins of the AIDs virus, *Sci. Am.*, **1988**, *259*, Oct., 44. b) S. Wain-Hobson, More ado about HIV's origins, *Nature Med.*, **1998**, *4*, 1001. c) R. A. Weiss and R. W. Wrangham, From *Pan* to pandemic, *Nature* **1999**, *397*, 385. d) F. Gao *et al.*, Origin of HIV-1 in the chimpanzee *Pan troglodytes troglodytes*, *ibid.*, 436. e) J. Cohen, AIDS virus traced to chimp subspecies, *Science*, **1999**, *283*, 772. f) Editorial, Half the story, *New Sci.*, **1999**, 6 Feb., 3. g) D. M. Hillis, Origins of HIV, *Science*, **2000**, *288*, 1757. h) J. Cohen, Searching for the epidemic's origins, *ibid.*, 2164. i) K. M. De Cock, Epidemiology and the emergence of human immunodeficiency virus and the acquired immune deficiency syndrome, *Phil. Trans. Roy. Soc. B*, **2001**, *356*, 795. j) P. M. Sharp *et al.*, The origins of acquired immune deficiency syndrome viruses: where and when? *ibid.*, 867.

117.   a) J. C. Caldwell and P. Caldwell, The African AIDS epidemic, *Sci. Am.*, **1996**, *274*, March, 40. b) S. Wain-Hobson, 1959 and all that, *Nature*, **1998**, *391*, 531. c) M. Day, A plague unleashed, *New Sci.*, **2000**, 30 Sept., 2000. d) P. A. Marx, P. G. Alcabes and E. Drucker, Serial human passage of simian immunodeficiency virus by unsterile injections and emergence of epidemic human immunodeficiency virus in Africa, *Phil. Trans. Roy. Soc. B*, **2001**, *356*, 911. e) D. Low-Beer, The distribution of early acquired immune deficiency syndrome cases and conditions for the establishments of new epidemics, *ibid.*, 927. f) J. Randerson, WHO accused of huge HIV blunder, *New Sci.*, **2003**, 6 Dec., 8.

118.   a) T. Zhu *et al.*, An African HIV-1 sequence from 1959 and implications for the origin of the epidemic, *Nature*, **1998**, *391*, 594. b) M. Balter, Virus from 1959 sample marks early years of HIV, *Science*, **1998**, *279*, 801.

119.   a) S. S. Frøland *et al.*, HIV-1 infection in Norwegian family before 1970, *Lancet*, **1988**, *1*, 1344. b) T. O. Jonassen *et al.*, Sequence analysis of HIV-1 group O from Norwegian patients infected in the 1960's, *Virology*, **1997**, *231*, 43.

120.   D. Vangroenweghe, The earliest cases of immunodeficiency virus type 1 group M in Congo-Kinshasa, Rwanda and Burundi and the origin of acquired immune deficiency syndrome, *Phil. Trans. Roy. Soc. B*, **2001**, *356*, 923.

121.   a) R. A. Weiss, Polio vaccines exonerated, *Nature*, **2001**, *410*, 1035. b) P. Blanco *et al.*, Polio vaccine samples not linked to AIDS, *ibid.*, 1045. c) N. Berry *et al.*, Analysis of oral polio vaccine CHAT stocks, *ibid.*, 1046. d) A. Rambaut *et al.*, Phylogeny and the origin of HIV-1, *ibid.*, 1047. e) J. Cohen, Disputed AIDS theory dies its final death, *Science*, **2001**, *292*, 615. f). H. Poinar, M. Kuch and S. Pääbo, Molecular analyses of oral polio vaccine samples, *ibid.*, 743. g) M. Worobey *et al.*, Contaminated polio vaccine theory refuted, *Nature*, **2004**, *428*, 820.

122.   a) J. Cribb, *The White Death*, Harper Collins, New York, USA, 1996. b) K. Ott, *Fevered Lives*, Harvard University Press, Cambridge, USA, 1997. c) D. B. Young, Blueprint for the white plague, *Nature*, **1998**, *393*, 515. d) C. Dye *et al.*, Global burden of tuberculosis: estimated incidence, prevalence and mortality by country, *JAMA*, **1999**, *282*, 677. e) D. G. Russell, TB comes to a sticky beginning, *Nature Med.*, **2001**, *7*, 894. f) J. Lee, The white plague, *New Sci.*, *Inside Science 155*, **2002**, 9 Nov., 1.

123.   J. Bunyan, *The Death of Mr Badman*, Dent, London, UK, 1928, p. 282.

124.   a) L. Spinney, Global emergency as TB toll mounts, *New Sci.*, **1996**, 30 March, 8. b) M. Day, "Poor" vaccine ruled out in TB puzzle, *New Sci.*, **1996**, 13 July, 14. c) R. M. May, The

348 Challenged Earth

rise and fall and rise of tuberculosis, *Nature Med.*, **1995**, *1*, 752. d) D. B. Young and D. B. Robertson, TB vaccines: global solutions for global problems, *Science* **1999**, *284*, 1479. e) A. Coghlan, A sure shot, *New Sci.*, **2002**, 16 Feb., 17. f) H. Hoag, New vaccines enter fray in fight against tuberculosis, *Nature Med.*, **2004**, *10*, 6.

125. S. Bhatia, T. Dranyi and D. Rowley, Tuberculosis among Tibetan refugees in India, *Soc. Sci. Med.*, **2002**, *54*, 423.
126. a) K. Kleiner, Incurable TB threatens millennium, *New Sci.*, **1997**, 1 Nov., 14. b) D. Butler, New fronts in an old war, *Nature*, **2000**, *406*, 672. c) D. Walton and P. Farmer, The new white plague, *JAMA*, **2000**, *284*, 2789. d) E. Stokstad, Drug-resistant TB on the rise, *Science*, **2000**, *287*, 2391. e) M. V. Shivlova and C. Dye, The resurgence of tuberculosis in Russia, *Phil. Trans. Roy. Soc. B*, **2001**, *356*, 1069. f) A. Coghlan, Lives before profit, *New Sci.*, **2001**, 7 July, 6. g) A. Coghlan and D. Concar, Coming home, *ibid.*, 28. h) C. Dye *et al.*, Erasing the world's slow stain: Strategies to beat multidrug-resistant tuberculosis, *Science*, **2002**, *295*, 2042.
127. a) T. Beardsley, Paradise lost? *Sci. Am.*, **1992**, *267*, Nov., 12. b) B. R. Bloom, Back to a frightening future, *Nature*, **1992**, *358*, 538. c) Y. Zhang *et al.*, The catalase-peroxidase gene and isoniazid resistance of *Mycobacterium tuberculosis*, *ibid.*, 591. d) P. Aldous, Genetic basis found for resistance to TB drug, *Science*, **1992**, *257*, 1038.
128. a) S. Sreevatsan *et al.*, Restricted structural gene polymorphism in the Mycobacterium tuberculosis complex indicates evolutionarily recent global dissemination, *Proc. Natl. Acad. Sci. USA*, **1997**, *94*, 9869. b) S. T. Cole *et al.*, Deciphering the biology of Mycobacterium tuberculosis from the complete genome sequence, *Nature*, **1998**, *393*, 537. c) D. B. Young, A post-genomic perspective, *Nature Med.*, **2001**, *7*, 11.
129. a) S. T. Cole and P. M. Alzari, TB – A new target, a new drug, *Science*, **2005**, *307*, 214. b) K. Andries *et al.*, A diarylquinoline drug active on the ATP synthase of *Mycobacterium tuberculosis*, *ibid.*, 223.
130. a) J. Rennie, Proteins 2, malaria, 0, *Sci. Am.*, **1991**, *265*, July, 10. b) P. H. Martin and M. G. Lefebvre, Malaria and climate: sensitivity of malaria potential transmission to climate, *Ambio*, **1995**, *24*, 200. c) A. M. Richman and F. C. Kafotos, Malaria: in the belly of the beast, *Nature Med.*, **1998**, *4*, 552. d) R. Carter and L. Ranford-Cartwright, Has the ignition key been found? *Nature*, **1998**, *392*, 227. e) F. Pearce, Malariasphere, *New Sci.*, **2000**, 15 July, 32. f) S. Budiansky, Creatures of our own making, *Science*, **2002**, *298*, 80. g) B. Greenwood and T. Mutabingwa, Malaria in 2002, *Nature*, **2002**, *415*, 670. h) L. H. Miller and B. Greenwood, Malaria – a shadow over Africa, *Science*, **2002**, *298*, 121.
131. a) L. H. Miller, M. F. Good and G. Milon, Malaria pathogenesis, *Science*, **1994**, *264*, 1878. b) L. H. Miller and J. D. Smith, Motherhood and malaria, *Nature Med.*, **1998**, *4*, 1244. c) P. Druilhe and J. L. Pérignon, Malaria from Africa blows hot and cold, *Nature Med.*, **1999**, *5*, 272. d) L. H. Miller *et al.*, The pathogenic basis of malaria, *Nature*, **2002**, *415*, 673. e) B. Greenwood, Between hope and a hard place, *Nature*, **2004**, *430*, 926
132. a) L. Pauling, *et al.*, Sickle cell anaemia: a molecular disease, *Science*, **1949**, *110*, 543. b) J. Flint, *et al.*, The population genetics of the haemoglobinopathies, *Clin. Haematol*, **1993**, *6*, 215. c) O. S. Platt, Sickle cell paths converge on hydroxyurea, *Nature Med*, **1995**, *1*, 307. d) D. J. Weatherall and J. B. Clegg, Thalassemia - a global public health problem, *Nature Med.*, **1996**, *2*, 847. e) C. Thompson, There's life in the old drug yet, *New Sci*, **1996**, 30 March, 15. f) S. A. Tishkoff *et al.*, Haplotype diversity and linkage disequilibrium at human *G6PD*: recent origin of alleles that confer malarial resistance, *Science*, **2001**, *293*, 455. g) E. Pennisi, Genetic change wards off malaria, *Science*, **2001**, *294*, 1439. h) D. Modiano *et al.*, Haemoglobin C protects against clinical *Plasmodium falciparum* malaria, *Nature* **2001**, *414*, 305.
133. a) E. Marshall, Gene *gemisch* cures sickle cell in mice, *Science*, **2001**, *294*, 2268. b) R. Pawliuk *et al.*, Correction of sickle cell disease in transgenic mouse models by gene therapy, *ibid.*, 2368.
134. L. Luzzatto and R. Notaro, Protecting against bad air, *Science*, **2001**, *293*, 442.

135. a) A. F. G. Slater and A. Cerami, Inhibition by chloroquine of a novel haem polymerase enzyme activity in malaria trophozoites, *Nature*, **1992**, *355*, 167. b) Editorial, The wrong resistance, *New Sci.*, **1996**, 13 July, 3. c) M. Day, Malaria falls to herbal remedy, *New Sci.*, **1996**, 13 July, 4. d) R. G. Ridley, Medical need, scientific opportunity and the drive for antimalarial drugs, *Nature*, **2002**, *415*, 686. e) K. Wengelnik *et al.*, A class of potent antimalarials and their specific accumulation in infected erythrocytes, *Science*, **2002**, *295*, 1311. f) I. M. Hastings, P. G. Bray and S. A. Ward, A requiem for chloquine, *Science*, **2002**, *298*, 74. g) T. E. Wellems, *Plasmodium* chloroquine resistance and the search for a replacement antimalarial drug, *ibid.*, 124. h) R. G. Ridley, To kill a parasite, *Nature*, **2003**, *424*, 887.

136. a) P. M. O'Neill, A worthy adversary for malaria, *Nature*, **2004**, *430*, 838. b) J. L. Vennerstrom *et al.*, Identification of an antimalarial synthetic trioxolane drug development candidate, *ibid.*, 900.

137. a) K. S. Jayaraman, India plans $200 million attack on malaria, *Nature*, **2000**, *386*, 536. b) A. Attaran, *et al.*, Balancing risks on the backs of the poor, *Nature Med.*, **2000**, *6*, 729.

138. a) M. Enserink, Bed nets prove their mettle against malaria, *Science*, **2001**, *294*, 2271. b) G. Vogel, An elegant but imperfect tool, *Science*, **2002**, *298*, 94. c) J. Hemingway, L. Field and J. Vontas, An overview of insecticide resistance, *ibid.*, 96. d) J. Randerson, Resistance to pesticides goes global in a flash, *New Sci.*, **2002**, 5 Oct., 15.

139. a) B. Holmes, In the bag, *New Sci.*, **1997**, 13 Dec., 11. b) X. Su, *et al.*, Complex polymorphisms in an ~330 kDa protein are linked to chloroquine-resistant *P. falciparum* in Southeast Asia and Africa, *Cell*, **1997**, *91*, 593.

140. a) D. Butler, Global agencies hold the financial key, *Nature*, **1997**, *386*, 10. b) J. Kaiser, Raising the stakes in the race for new malaria drugs, *Science*, **1998**, *281*, 1930. c) D. N. Nabarro and E. M. Tayler, The "Roll Back Malaria" campaign, *Science*, **1998**, *280*, 2067. d) M. Enserink, Malaria researchers wait for industry to join fight, *Science*, **2000**, *287*, 1956. e) J. Sachs and P. Malaney, The economic and social burden of malaria, *Nature*, **2002**, *415*, 680. f) J. D. Sachs, A new global effort to control malaria, *Science*, **2002**, *298*, 122. g) A. Attaran, Where did it all go wrong? *Nature*, **2004**, *430*, 932. h) G. Yamey, Roll back malaria: a failing global health campaign, *Br. Med. J.*, **2004**, *328*, 1086.

141. D. MacKenzie, The comeback killer, *New Sci.*, **1999**, 18 Sept., 13.

142. a) R. Ménard, Medicine: Knockout malaria vaccine? *Nature*, **2005**, *433*, 113. b) A.-K. Mueller *et al.*, Genetically modified *Plasmodium* parasites as a protective experimental malaria vaccine, *ibid.*, 164.

143. a) *Anon.*, Vaccines: a roller-coaster of hopes, *Nature*, **1997**, *386*, 535. b) A. Dove, Slow progress in malaria vaccine development, *Nature Med.*, *Vaccine Supplement*, **1998**, *4*, 479. c) L. H. Miller and S. L. Hoffman, Research toward vaccines against malaria, *ibid.*, 520. d) R. Carter, *et al.*, Malaria transmission blocking vaccines - how can their development be supported? *Nature Med.*, **2000**, *6*, 241. e) S. L. Hoffman and D. L. Doolan, Malaria vaccines - targeting infected hepatocytes, *ibid.*, 1218. f) T. L. Richie and A. Saul, Progress and challenges for malaria vaccines, *Nature*, **2002**, *415*, 694. g) D. Fox, Endgame, *New Sci.*, **2003**, 5. July, 34.

144. a) S. L. Hoffman, *et al.*, From genomics to vaccines: malaria as a model system, *Nature Med.*, **1998**, *4*, 1351. b) D. Wirth, Malaria: a 21st century solution for an ancient disease, *ibid.*, 1360. c) E. Pennisi, Malarial genome comes into view, *Science*, **1999**, *286*, 1263. d) X. Su *et al.*, A genetic map and recombination parameters of the human malaria parasite *Plasmodium falciparum*, *Science*, **1999**, *286*, 1351. e) Z. Lai *et al.*, A shotgun optical map of the entire *Plasmodium Falciparum* genome, *Nature Genet.*, **1999**, *23*, 309. f) E. Pennisi, Malaria's beginnings: on the heels of hoes? *Science*, **2001**, *293*, 416. g) S. Volkman *et al.*, Recent origin of *Plasmodium falciparum* from a single progenitor, *ibid.*, 482. h) S. L. Hoffman *et al.*, *Plasmodium*, human and *Anopheles* genomics and malaria, *Nature*, **2002**, *415*, 702. i) M. Enserink and E. Pennisi, Researchers crack malaria genome, *Science*, **2002**, *295*, 1207. j) E. Pennisi, Parasite genome sequenced, scrutinized, *Science*, **2002**, *298*, 33. k)

A. F. Cowman and B. S. Crabb, The *plasmodium falciparum* genome – a blueprint for erythrocyte invasion, *ibid.*, 126.

145. a) C. F. Curtis, The case for de-emphasizing genomics in malaria control, *Science*, **2000**, *290*, 1508. b) S. L. Hoffman, Research (genomics) is crucial to attacking malaria, *ibid.*, 1508. c) C. A. Long and S. L. Hoffman, Malaria – from infants to genomics to vaccines, *Science*, **2002**, *297*, 345.

146. a) D. Butler, Transgenic mosquitoes: a new solution, *Nature*, **1997**, *386*, 538. b) C. M. Morel *et al.*, The mosquito genome – a breakthrough for public health, *Science*, **2002**, *298*, 79. c) T. C. Kaufman, D. W. Severson and G. E. Robinson, The *Anopheles* genome and comparative insect genomics, *Science*, **2002**, *298*, 97. d) T. W. Scott *et al.*, The ecology of genetically modified mosquitoes, *ibid.*, 117. e) L. Alphey *et al.*, Malaria control with genetically manipulated insect vectors, *ibid.*, 119. f) R. A. Holt *et al.*, The genome of the malaria mosquito, *Anopheles gambiae*, *ibid.*, 129. g) O. Morton, Splat! *New Sci.*, **2003**, 22 March, 30.

147. A. Spielman and M. D'Antonio, *Mosquito: The Story of Mankind's Deadliest Foe*, Faber and Faber, London, UK, 2001.

148. a) B. Le Guenno, Emerging viruses, *Sci. Am.*, **1995**, *273*, Oct., 30. b) V. Morell, Chimpanzee outbreak heats up search for Ebola origin, *Science*, **1995**, *268*, 974. c) M. Day, Did ancient Athenians catch Ebola? *New Sci.*, **1996**, 29 June, 5. d) I. Wickelgren, A method in Ebola's madness, *Science*, **1998**, *279*, 983. e) G. J. Nabel, Surviving Ebola virus infection, *Nature Med.*, **1999**, 5, 373. f) Z. Yang *et al.*, Distinct cellular interactions of secreted and transmembrane Ebola virus glycoproteins, *Science*, **1998**, *279*, 1034. g) D. R. Burton and P. W. H. I. Parren, Fighting the Ebola virus, *Nature*, **2000**, *408*, 527. h) M. Balter, On the trail of Ebola and Marburg viruses, *Science*, **2000**, *290*, 923. i) H.-D. Klenk, Will we have and why do we need Ebola vaccine? *Nature Med.*, **2000**, 6, 1322. j) T. Clarke and J. Knight, Fast vaccine offers hope in battle with Ebola, *Nature*, **2003**, *424*, 602. k) N. J. Sullivan *et al.*, Accelerated vaccination for Ebola virus, haemorrhagic fever in non-human primates, *ibid.*, 681. l) A. Surendran, Ebola rears ugly head as vaccine enters trials, *Nature Med.*, **2004**, *10*, 9. m) M. Enserink, A puzzling outbreak of Marburg disease, *Science*, **2005**, *308*, 31.

149. a) D. MacKenzie, Powerless to stop the spread, *New Sci.*, **2003**, 12 April, 6. b) D. MacKenzie, Can we contain SARS? *ibid.*, 7. c) D. MacKenzie, Where did this deadly pneumonia come from? *Ibid.*, 8. d) M. Enserink, Clues to the animal origins of SARS, *Science*, **2003**, *300*, 1351. e) K. V. Holmes and L. Enjuanes, The SARS coronavirus: a postgenomic era, *ibid.*, 1377. f) J. Knight, Researchers get to grips with cause of pneumonia epidemic, *Nature*, **2003**, *422*, 547. g) D. Cyranoski and A. Abbot, Virus detectives seek source of SARS in China's wild animals, *Nature*, **2003**, *423*, 467. h) H. Pearson *et al.*, SARS. What have we learned? *Nature*, **2003**, *424*, 121. i) J. Bull, Epidemics in waiting, *Nature*, **2003**, *426*, 609. j) M. Enserink, Calling all coronavirologists, *Science*, **2003**, *300*, 413. k) K. V. Holmes and L. Enjuanes, The SARS coronavirus: a postgenomic era, *ibid.*, 1377.

150. a) P. Brown, Cholera's deadly hitchhiker, *New Sci.*, **1996**, 6 July, 14. b) S. H. Richardson and D. J. Wozniak, An ace up the sleeve of the cholera bacterium, *Nature Med.*, **1996**, *2*, 853.

151. a) J. C. Peters, Early history of Asiatic cholera, in India as known to Europeans A.D. 1503-1800, in *Asiatic Cholera*, E. C. Wendt, ed., William Wood, New York, USA, 1985. b) R. R. Colwell, Global climate and infectious disease: the cholera paradigm, *Science*, **1996**, *274*, 2025. c) M. Pascual *et al.*, Cholera dynamics and El Niño-Southern Oscillation, *Science*, **2000**, *289*, 1766.

152. P. J. Sansonetti, Slaying the Hydra all at once or head by head? *Nature Med.*, *Vaccine Supplement*, **1998**, *4*, 499.

153. a) M. Enserink, The enigma of West Nile, *Science*, **2000**, *290*, 1482. b) D. MacKenzie, Beware the birds, *New Sci.*, **2000**, 8 July, 4. c) M. Enserink, New York's deadly virus may stage a comeback, *Science*, **2000**, *287*, 2129. d) M. Enserink, West Nile drugs, vaccine still

years away, *Science*, **2000**, *290*, 1483. e) M. Enserink, West Nile's surprisingly swift continental sweep, *Science*, **2002**, *297*, 18. f) J. Ginsburg, The new polio? *New Sci.*, **2003**, 7 June, 41. g) S. Mukhopadhyay *et al.*, Structure of West Nile virus, *Science*, **2003**, *302*, 248.

154. a) A. G. Barbour and D. Fish, The biological and social phenomenon of Lyme disease, *Science*, **1993**, *260*, 1610. b) K. Schmidt, If you go down to the woods today, *New Sci.*, **1997**, 15 Nov., 44. c) A. G. Barbour and W. R. Zükert, New tricks of tick-borne pathogen, *Nature*, **1997**, *390*, 553. d) J. Kaiser, Of mice and moths - and Lyme disease, *Science*, **1998**, *279*, 984. e) C. G. Jones *et al.*, Chain reactions linking acorns to Gypsy Moth outbreaks and Lyme disease risk, *ibid.*, 1023.

155. a) W. O. C. M. Cookson and M. F. Moffat, Alchemy for asthma, *Nature Med.*, *Vaccine Supplement*, **1998**, *4*, 500. b) R. Doyle, Asthma world-wide, *Sci. Am.*, **2000**, *280*, June, 19.

156. J. Peto, Cancer epidemiology in the last century and the next decade, *ibid.*, 390.

157. a) R. A. Weinberg, How cancer arises, *Sci. Am.*, **1996**, *275*, Sept., 32. b) E. Ruoslahti, How cancer spreads, *ibid.*, 42. c) J. Rennie and R. Rusting, Twelve major cancers, *ibid.*, 92. d) B. A. J. Ponder, Cancer genetics, *Nature*, **2001**, *411*, 336. e) G. I. Evan and K. H. Vousden, Proliferation, cell cycle and apoptosis in cancer, *ibid.*, 342. f) J. H. J. Hoeijmakers, Genome maintenance mechanisms for preventing cancer, *ibid.*, 366. g) D. M. Parkin, F. I. Bray and S. S. Devas, Cancer burden in the year 2000. The global picture, *Eur. J. Cancer*, **2001**, *37*, Supplement 8, 4. h) K. Weissman, Life and cell death, *Chem. Br.*, **2003**, *39*, Aug., 19

158. a) D. Grimm, Disease backs cancer origin theory, *Science*, **2004**, *306*, 389. b) H. Rajagopalan and C. Lengauer, Aneuploidy and cancer, *Nature*, **2004**, *432*, 338.

159. R. A. DePinho, The age of cancer, *Nature*, **2000**, *408*, 248.

160. a) J. Massagué, G1 cell-cycle control and cancer, *Nature*, **2004**, *432*, 298. b) S. W. Lowe, E. Cepero and G. Evan, Intrinsic tumour suppression, *ibid.*, 307. c) M. B. Kastan and J. Bartek, Cell-cycle checkpoints and cancer, *ibid.*, 316. d) C. Dernicourt and S. F. Dowdy, Targeting apoptotic pathways in cancer cells, *Science*, **2004**, *305*, 1411. e) L. D. Walensky *et al.*, Activation of apoptosis in vivo by a hydrocarbon-stapled BH3 helix, *ibid.*, 1466. f) L. Li *et al.*, A small molecule Smac mimic potentiates TRAIL- and TNFa-mediated cell death, *ibid.*, 1471.

161. a) C. W. Greider and E. H. Blackburn, Telomeres, telomerase and cancer, *Sci. Am.*, **1996**, *274*, Feb., 80. b) N. Axelrod, Of telomeres and tumors, *Nature Med.*, **1996**, *2*, 158. c) V. A. Zakian, Life and cancer without telomerase, *Cell*, **1997**, *91*, 1. d) T. de Lange, Telomeres and senescence: Ending the debate, *Science*, **1998**, *279*, 334. e) T. de Lange and R. A. DePinho, Unlimited mileage from telomerase? *Science*, **1999**, *283*, 947. f) S. E. Artandi and R. A. DePinho, Mice without telomerase: what can they teach us about human cancer? *Nature Med.*, **2000**, *6*, 852. g) J. A. Baur *et al.*, Telomere position effect in human cells, *Science*, **2001**, *292*, 2075. h) R. Hodes, Molecular targeting of cancer: telomeres as targets, *Proc. Natl. Acad. Sci. USA*, **2001**, *98*, 7649.

162. a) D. Trichopoulos, F. P. Li and D. J. Hunter, What causes cancer? *Sci. Am.*, **1996**, *275*, Sept., 50. b) J. M. Bishop, Cancer: what should be done? *Science*, **1997**, *278*, 995. c) E. E. Calle, *et al.*, Body-mass index and mortality in a prospective cohort of U.S. adults, *N. Engl. J. Med.*, **1999**, *341*, 1097. d) D. S. Michaud *et al.*, Physical activity, obesity, height and the risk of pancreatic cancer, *JAMA*, **2001**, *286*, 921.

163. W. C. Willett, G. A. Colditz and N. E. Mueller, Strategies for minimising cancer risk, *Sci. Am.*, **1996**, *275*, Sept., 59.

164. a) R. Peto, *et al.*, *Mortality from Tobacco in Developed Countries, 1950-2000*, Oxford University Press, Oxford, UK, 1994. b) B.-Q. Liu *et al.*, Emerging tobacco hazards in China: 1. Retrospective proportionality mortality study of one million deaths, *Brit. Med. J.*, **1998**, *317*, 1411. c) R. Peto, Z.-M. Chen and J. Boreham, Tobacco - the growing epidemic, *Nature Med.*, **1999**, *5*, 15. d) M. J. Thun, L. F. Apicella and S. J. Henley, Smoking vs other risk factors as the cause of smoking-attributable deaths, *JAMA*, **2000**, *284*, 706. e) M. Raw, Fighting tobacco dependence in Europe, *Nature Med.*, **2001**, *7*, 13.

165.	a) J. Parsonnet, ed., *Microbes and Malignancy*, Oxford University Press, New York, USA, 1999. b) M. A. Epstein, Historical background, *Phil. Trans. Roy. Soc. B*, **2001**, *356*, 413. c) O. G. Pybus *et al.*, The epidemic behaviour of hepatitis C virus, *Science*, **2001**, *292*, 2323.

166.	a) D. Ganem, The X files - one step closer to closure, *Science*, **2001**, *294*, 2299. b) M. J. Bouchard, L.-H. Wang and R. J. Schneider, Calcium signalling by HBx protein in hepatitis B virus DNA replication, *ibid.*, 2376. c) G. Gong *et al.*, Human hepatitis C virus NS5A protein alters intracellular calcium levels, induces oxidative stress, and activates STAT-3 and NF-κB, *Proc. Natl. Acad. Sci. USA*, **2001**, *98*, 9599.

167.	a) A. Covacci *et al.*, *Helicobacter pylori* virulence and genetic geography, *Science*, **1999**, *284*, 1328. b) M. J. Blaser, Linking *Helicobacter pylori* to gastric cancer, *Nature Med.*, **2000**, 6, 376. c) D. Ferber, Cracking gut bugs' cell-skewing strategy, *Science*, **2001**, *294*, 2269.

168.	a) S. Mirsky, A host with infectious ideas, *Sci. Am.*, **2001**, *284*, May, 26. b) C. Zimmer, Do chronic diseases have an infectious root, *Science*, **2001**, *293*, 1974.

169.	a) D. J. Leffell and D. E Brash, Sunlight and skin cancer, *Sci. Am.*, **1996**, *275*, 38. b) H. N. Ananthaswamy *et al.*, Sunlight and skin cancer: inhibition of p53 mutations in UV-irradiated mouse skin by sunscreens, *Nature Med.*, **1997**, *3*, 510.

170.	a) R. Doll, Mortality from lung cancer in asbestos workers, *Brit. J. Indust. Med.*, **1955**, *12*, 81. b) J. C. Wagner, C. A. Sleggs and P. Marchand, Diffuse pleural mesothelioma and asbestos exposure in the North Western Cape Province, *Brit. J. Indust. Med.*, **1960**, *17*, 260. c) J. Peto, *et al.*, The European mesothelioma epidemic, *Brit. J. Cancer*, **1999**, *79*, 666.

171.	C. Cislaghi and P. L. Nimis, Lichens, air pollution and lung cancer, *Nature*, **1997**, *387*, 463.

172.	a) J. U. Ahmed, Radon in the human environment: assessing the picture, *IAEA Bull.*, **1994**, 2, 32. b) P. Phillips, T. Denman and S. Barker, Silent but deadly, *Chem. Br.*, **1997**, *33*, Jan., 35. c) F. Pearce, Undermining our lives? *New Sci.*, **1998**, 14 March, 20.

173.	a) J. M. Bishop, Enemies within: the genesis of retrovirus oncogenes, *Cell*, **1981**, *23*, 5. b) R. A. Weinberg, Tumor suppressor genes, *Science*, **1991**, *254*, 1138. c) D. Haber and E. Harlow, Tumor suppressor genes: evolving definitions in the genomic age, *Nature Genet.*, **1997**, *16*, 320. d) A. Oliff, J. B. Gibbs and F. McCormick, New molecular targets for cancer therapy, *Sci. Am.*, **1996**, *275*, Sept., 110. e) K. W. Kinzler and B. Vogelstein, Gatekeepers and caretakers, *Nature*, **1997**, *386*, 761. f) G. Evan and T. Littlewood, A matter of life and cell death, *Science*, **1998**, *281*, 1317. g) M. Park in *The Genetic Basis of Human Cancer*, eds., B. Vogelstein and K. W. Kinder, McGraw-Hill, New York, USA, 1998, p. 205. h) R. Takimoto and W. S. El-Deiry, DNA replication blockade impairs p53-transactivation, *Proc. Natl. Acad. Sci. USA*, **2001**, *98*, 781.

174.	a) G. Ferbeyre and S. W. Lowe, The price of tumour supression? *Nature*, **2002**, *415*, 26. b) S. D. Tyner *et al.*, p53 mutant mice that display early ageing-associated phenotypes, *ibid.*, 45. c) A. Surendran, China approves world's first gene therapy drug, *Nature Med.*, **2004**, *10*, 9. d) D. P. Lane and P. M. Fischer, Turning the key on p53, *Nature*, **2004**, *427*, 789. e) L. T. Vassilev *et al.*, In vivo activation of the p53 pathway by small-molecule antagonists of MDM2, *ibid.*, 844.

175.	a) E. Marshall, Cancer warriors claim a victory, *Science*, **1998**, *279*, 1842. b) Cancer incidence and mortality, 1973 - 1995, *Cancer*, **1998**, *82*, 1197. c) E. Levy-Lahad and S. E. Plon, A risky business – assessing breast cancer risk, *Science*, **2003**, *302*, 574.

176.	a) S. Hellman and E. E. Vokes, Advancing current treatments for cancer, *Sci. Am.*, **1996**, *275*, Sept., 84. b) K. Antman, When are bone marrow transplants considered? *ibid.*, 90.

177.	a) J. Folkman, Fighting cancer by attacking its blood supply, *Sci. Am.*, **1996**, *275*, Sept., 116. b) F. McCormick, New-age drug meets resistance, *Nature*, **2001**, 412, 281. c) J. Reedjik, New clues for platinum antitumor chemistry: kinetically controlled metal binding to DNA, *Proc. Natl. Acad. Sci. USA.*, **2003**, *100*, 3611.

178.	a) L. J. Old, Immunotherapy for cancer, *Sci. Am.*, **1996**, *275*, Sept., 102. b) D. M. Pardolli, Cancer vaccines, *Nature Med.*, *Vaccine Supplement*, **1998**, *4*, 525. c) W. F. Anderson, Gene therapy scores against cancer, *Nature Med.*, **2000**, *6*, 862. d) S. A. Rosenberg, Progress in human tumour immunology and immunotherapy, *Nature*, **2001**, *411*, 380. e) P. Moingeon, Cancer vaccines, *Vaccine*, **2001**, *19*, 1305.

179. J. Bonner, Cancer vaccines, *Chem. Indust.*, **2001**, 7 May, 277.
180. a) B. Vogelstein and K. W. Kinzler, Achilles' heel of cancer? *Nature*, **2001**, *412*, 865. b) K. Raj, P. Ogston and P. Beard, Virus-mediated killing of cells that lack p53 activity, *Nature*, **2001**, *412*, 914.
181. D. Kirn, R. L. Martuza and James Zwiebel, Replication-selective virotherapy for cancer: biological principles, risk management and future directions, *Nature Med.*, **2001**, *7*, 781.
182. a) D. W. Kufe, Smallpox, polio and now a cancer vaccine? *Nature Med.*, **2000**, *6*, 252. b) F. O. Nestle, J. Banchereau and D. Hart, Dendritic cells: on the move from bench to bedside, *Nature Med.*, **2001**, *7*, 761.
183. a) N. Houghton and D. A. Scheinberg, Monoclonal antibody therapies - a "constant" threat to cancer, *ibid.*, 373. b) J. M. Reichert, Monoclonal antibodies in the clinic, *Nature Biotechnol.*, **2001**, 19, 819. c) G. J. Nabel, Genetic, cellular and immune approaches to disease therapy: past and future, *Nature Med.*, **2004**, *10*, 135
184. a) M. F. Clarke, At the root of brain cancer, *Nature*, **2004**, *432*, 281. b) C. Sawyers, Targeted cancer therapy, *ibid.*, 294. c) P. A. Beachy, S. S. Karhadkar and D. M. Berman, Tissue repair and stem cell renewal in carcinogenesis, *ibid.*, 324. d) S. K. Singh *et al.*, Identification of human brain tumour initiating cells, *ibid.*, 396.
185. a) D. Sidransky, Advances in cancer detection, *Sci. Am.*, **1996**, *275*, Sept., 71. b) M. L. Giger and C. A. Pelizzari, Advances in tumor imaging, *ibid.*, 76.
186. a) L. Guarente, What makes us tick? *Science*, **1997**, *275*, 943. b) C. E. Finch and R. E. Tanzi, Genetics of aging, *Science*, **1997**, *278*, 407. c) J. H. Morrison and P. R. Hof, Life and death of neurons in the aging brain, *ibid.*, 412. d) S. W. J. Lamberts, A. W. van den Beld and A.-J. van der Lely, The endrocrinology of aging, *Science*, **1997**, *278*, 419. e) D. Kipling and R. G. A. Faragher, Ageing hard or hardly ageing? *Nature*, **1999**, *398*, 191. f) J. Kingsland, Age-old story, *New Sci.*, *Inside Science 117*, **1999**, 23 Jan., 1. g) T. B. Kirkwood and S. N. Austad, Why do we age? *Nature* **2000**, *408*, 233. h) T. Finkel and N. J. Holbrook, Oxidants, oxidative stress and the biology of ageing, *ibid*, 239.
187. a) D. H. Sachs, Xenographs, cloning and the immune system, *Nature Med.*, **1997**, *3*, 951. b) L. E. Chapman and E. T. Bloom, Clinical xenotransplantation, *JAMA*, **2001**, *285*, *2304*.
188. a) R. P. Lanza, J. B. Cibelli and M. D. West, Human therapeutic cloning, *Nature Med.*, **1999**, *5*, 975. b) G. J. Annas, A. Caplan and S. Elias, Stem cell politics, ethics and medical progress, *ibid.*, 1339.
189. a) A. G. Bodnar *et al.*, Extension of life-span by introduction of telomerase into normal human cells, *Science*, **1998**, *279*, 349. b) M. R. R. Rose, Can human aging be postponed? *Sci. Am.*, **1999**, *281*, Dec., 68. c) M. R. R. Rose, No easy fixes, *ibid.*, 72.
190. C. C. J. L. Murray and A. D. Lopez, Evidence based health policy; lessons from the global burden of disease study, *Science*, **1996**, *274*, 740.
191. A. D. Lopez and C. C. J. L. Murray, The global burden of disease, 1990 - 2020, *Nature Med.*, **1998**, *4*, 1241.
192. G. Lewis and R. Araya, Classification, disability and the public health agenda, *Brit. Med. Bull.*, **2001**, *57*, 3.
193. B. R. Bloom, The future of public health, *Nature*, **1999**, *402*, C63.
194. a) J. Lee, Our health in flux, *New Sci.*, *Inside Science 116*, **1998**, 12 Dec., 1. b) P. R. Epstein, Climate and health, *Science*, **1999**, *285*, 347. c) P. Martens, How will climate change affect human health? *Am. Sci.*, **1999**, *87*, 534. d) P. Martens *et al.*, Climate change and future populations at risk of malaria, *Glob. Clim. Change*, *Supplementary Issue*, **1999**, *9*, S89. e) P. R. Epstein, Is global warming harmful to health? *Sci. Am.*, **2000**, *283*, Aug., 36.
195. a) L. A. Cole, The specter of biological weapons, *Sci. Am.*, **1996**, *275*, Dec., 30. b) M. B. A. Oldstone, *Viruses, Plagues and History*, Oxford University Press, Oxford, UK, 1998. c) D. Mackenzie, Bioarmageddon, *New Sci.*, **1998**, 19 Sept., 42. d) M. Murphy and R. Butler, Disease as a weapon of war, *Chem. Indust.*, **2001**, 1 Oct., 604.

354 Challenged Earth

196. S. M. Block, The growing threat of biological weapons, *Am. Sci.*, **2001**, *89*, 29.
197. a) P. Hadfield, Wartime skeletons return to haunt Japan, *New Sci.*, **1995**, 25 Feb., 12. b) P. Hadfield, Lethal legacy, *New Sci.*, **2001**, 3 Feb., 5.
198. a) P. J. Jackson *et al.*, PCR analysis of tissue samples from the 1979 Sverdlovsk anthrax victims: the presence of multiple *Bacillus anthracis* strains in different victims, *Proc. Natl. Acad. Sci. USA*, **1998**, *95*, 1224. b) J. Guillemin, *Anthrax: The Investigation of a Deadly Outbreak*, University of California Press, Berkely, USA, 1999.
199. a) J. Maurice, Virus wins stay of execution, *Science*, **1995**, *267*, 450. b) E. Marshall, President revokes plan to destroy smallpox, *Science*, **1999**, *284*, 718. c) R. Stone, WHO puts off destruction of U.S., Russian caches, *Science*, **2002**, *295*, 598. d) M. Enseink and R. Stone, Dead virus walking, *Science*, **2002**, *295*, 2001. e) E. Check, Unanimous vote approves tweak to smallpox genome, *Nature*, **2004**, *432*, 263.
200. C. Seelos, Lessons from Iraq on bioweapons, *Nature*, **1999**, *398*, 187
201. a) D. A. Henderson, The looming threat of bioterrorism, *Science*, **1999**, *283*, 1279. b) C. Dennis, The bugs of war, *Nature*, **2001**, *411*, 232. c) D. MacKenzie, Trail of terror, *New Sci.*, **2001**, 27 Oct., 5. d) M. Enserink, This time it was real: knowledge of anthrax put to the test, *Science*, **2001**, *294*, 490. e) J. Cohen and E. Marshall, Vaccines for biodefense: a system in distress, *ibid.*, 498. f) R. Dalton, Genetic sleuths rush to identify anthrax strains in mail attacks, *Nature*, **2001**, *413*, 657. g) D. MacKenzie, A shot in the dark, *New Sci.*, **2001**, 3 Nov., 4. h) K. Brown, A "sure killer" yields to medicine, *Science*, **2001**, *294*, 1813.
202. a) D. Dickson, Iraq crisis spurs new bioweapons moves, *Nature*, **1998**, *391*, 831. b) D. MacKenzie, Deadly secrets, *New Sci.*, **1998**, 16. c) D. MacKenzie, Unleash the bugs of war, *New Sci.*, **2001**, 12 May, 4. d) M. Wheelis, Deterring bioweapons development, *Science*, **2001**, *291*, 2089. e) J. Knight, US rejects bioweapon inspections, *Nature*, **2001**, *412*, 365. f) R. Stone, Down to the wire on bioweapons talks, *Science*, **2001**, *293*, 414.

Chapter 7

Energy: The Basis of Modern Civilization

"The discovery and production of mineral and energy resources have always been shrouded in the physically unknown and the economically uncertain."

C. D. Masters, D. H. Root and E. D. Attanasi, *Science*, **1991**, *253*, 146.

7.1. Civilization and Energy

During 4-10 January 1998, a moisture laden jet stream heading north from the Gulf of Mexico collided with air from the Arctic to produce a great ice storm [1]. Some 1,300 kilometres in length, it enveloped eastern Ontario, southern Quebec and parts of the maritime provinces of Canada together with parts of New York and Maine in the United States as seen in Fig. 7.1. It was caused by snow from the jet stream melting as it fell through the warmer upper air of a temperature inversion into its freezing lower air to form super-cooled rain. This rain built up layers of ice to a depth of eighty millimetres on electricity transmission lines

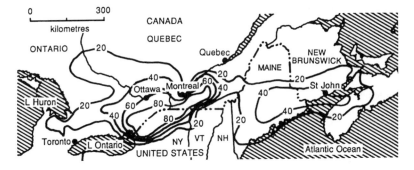

Fig. 7.1. The great ice storm of 4-10 January 1998 that enveloped the northeastern American seaboard. The thickness of ice build up within the contours is shown in millimetres. Adapted from [1].

that collapsed under the weight. As a result, three million people in Quebec, 600,000 in Ontario, 4,000 in New Brunswick and many more in New York and Maine lost all electrical power for periods ranging from several hours to as long as a month. The storm was a potent reminder of the dependency of sophisticated societies on readily available energy. Its severity raised fears that it and the seemingly increasing frequency of extreme weather events might reflect changes in climate caused by increasing atmospheric carbon dioxide levels resulting largely from fossil fuel combustion. Thus, the storm and its effect neatly encapsulated concerns about the security and source of future energy supplies and the effect of increasing energy use on climate and the environment [2-6].

The great dependency of the developed nations on huge supplies of dominantly fossil fuels grew swiftly during the twentieth century and with it came great increases in prosperity, health and life expectancy. Inevitably, the developing nations aspire to the same quality of life that the developed nations enjoy, and many are increasing their energy consumption in a trend that is expected to extend well into the twenty-first century. According to the International Energy Agency, humanity used 416.9 exajoules (Ej) of energy in 2000 as shown in Fig. 7.2 [7,8]. Of this energy, 79.4 percent came from the fossil

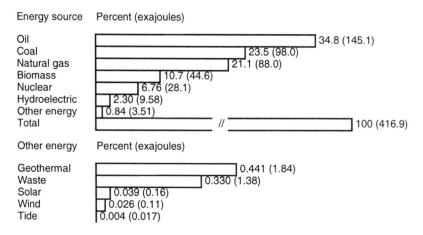

Fig. 7.2. The pattern of global energy use in 2000 according to the International Energy Agency [7,8]. The first value is the percentage of total energy use and the second value is the amount of energy used in exajoules. The scale used in the lower section of the figure is larger than that used in the upper section.

fuels oil, coal and natural gas. When the 10.7 and 0.33 percent of energy gained from combustible biomass and waste, respectively, were taken into account, 90.4 percent of all energy used resulted from combustion with only 9.6 percent coming from alternative energy resources. Thus, it is hardly surprising that atmospheric carbon dioxide levels are rising as is discussed in detail in Chapter

8. In an effort to address these concerns through an international agreement, the Kyoto Protocol that came into force on 16 February 2005, has been formulated to at least slow the rise in carbon dioxide levels [9]. Simultaneously, increasing demand for more energy has intensified scrutiny of Earth's remaining fossil fuel reserves and alternative energy resources. It is this that underlies most of the discussion in this chapter.

7.2. Total Primary Energy: The Sum of all Energies

Total primary energy is the sum of all of the energy consumed by humanity apart from that contained in food. There are a confusing variety of ways in which it is assessed that varies from something as simple as the amount of natural gas burned in an oven to roast a turkey to the number of tonnes of enriched uranium used in nuclear power plants to provide France with almost forty percent of its energy supply in 2000. There is also a confusing variety of measures of energy peculiar to the particular energy source. These are exemplified by millions of barrels of oil equivalent (Mboe), millions of tonnes of oil equivalent (Mtoe), millions of British Thermal Units (MBtu), kilowatt hours (kWh), kilocalories (kcals) and megajoules (MJ). Amidst this variety of measures, the joule (J) is becoming the most frequently used international measure of energy. One joule is the energy required to raise the temperature of one gram of water slightly less than 0.25°C, or alternatively, 4,184 joules or 4.184 kilojoules (kJ) is the energy required to raise the temperature of one kilogram of water by 1°C. The kilojoule is the unit used to inform consumers of the amount of energy in processed food that is usually printed on the back of food packets. It is also the unit used to calculate the energy contained in the average food consumption needed to keep an adult healthy and to provide the energy for everyday activities. This is about 11,500 kilojoules daily. However, these amounts of energy are small by comparison with the energy produced by either a fossil fuel powered or a nuclear powered electricity plant and accordingly prefixes such as kilo (k), one thousand times, mega (M), one million times, giga (G), one billion times, tera (T), one thousand billion (or one trillion) times, peta (P) one million billion times (or one quadrillion) and exa (E), one billion billion times are often used. This is exemplified by the 416.9 exajoules (EJ) of total primary energy used in 2000.

Because of the wide range of sources from which information about energy is available it is convenient to be able to readily convert one measure of energy to another as shown in Table 7.1.

Conversion of Energy Units					
To:	MJ	Mboe	Mtoe	MBtu	kWh
From:	Multiply by:				
MJ	1	1.75×10^{-10}	2.39×10^{-11}	9.48×10^{-4}	0.278
Mboe	5.71×10^{9}	1	0.136	5.40×10^{6}	1.586×10^{9}
Mtoe	4.184×10^{10}	7.33	1	3.97×10^{7}	1.163×10^{10}
MBtu	1.055×10^{3}	1.846×10^{-7}	2.52×10^{-8}	1	2.88×10^{2}
kWh	3.597	6.30×10^{-10}	8.60×10^{-11}	3.412×10^{-3}	1

Table 7.1. The interconversion of megajoules (MJ), million barrels of oil equivalent (Mboe), million tonnes of oil equivalent (Mtoe), millions of British Thermal units (Mbtu) and kilowatt-hours (kWh).

Energy usage is expected to continue to increase rapidly in the twenty-first century, as shown in Fig. 7.3, mainly because of the expansion of the economies of developing nations [10,11]. According to this projection, fossil fuel use will

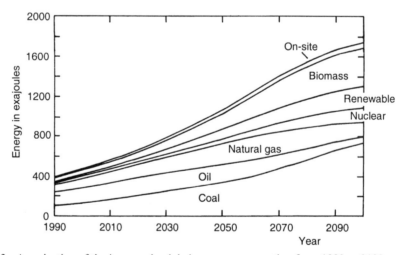

Fig. 7.3. A projection of the increase in global energy consumption from 1990 to 2100 according to the International Institute for Applied Systems Analysis. (One exajoule (EJ) is 10^{18} joules (J)). Adapted from [10].

grow to about two and a half times its present use and will be dominated by coal as oil and gas reserves diminish and these fuels become increasingly expensive.

However, by 2100 almost half of the energy used will be derived from nuclear power and renewable resources such as hydroelectricity and geothermal, wind, solar thermal, photovoltaic and tidal power in addition to biomass burning and on-site energy generation from roof mounted thermal solar and photovoltaic panels. Over all, it is projected that by 2100 there will be a four and a half fold increase in energy consumption by comparison with the 1990 consumption.

Such projections inevitably involve a range of assumptions that strongly influence the outcome. For the projections shown in Fig. 7.3 it has been assumed that the six billion 2000 population will reach about eight and a half and ten to eleven billion by 2050 and 2100, respectively. It is also assumed that an annual gross domestic product (GDP) global growth rate of 2.9 percent and 2.4 percent will apply in the periods 1990-2050 and 2050-2100, respectively, and that a 1.4 percent annual increase in global energy consumption will be accompanied by continuing improvements in the efficiency of energy use. As a consequence of their relatively low cost, fossil fuels are assumed to dominate energy supply for much of the twenty-first century. As China and India develop, and because of their large coal reserves, coal is assumed to become an increasingly important energy resource through direct combustion in the early twenty-first century with liquid fuels produced from coal assuming increasing importance thereafter.

The energy use projections shown in Fig. 7.3 are in the middle of the range of thirty-four energy projections produced by the International Institute for Applied Systems Analysis (IIASA) and the World Energy Council (WEC) that range from a little below 880 exajoules to almost 3,000 exajoules for global energy usage by 2100 [12,13]. Independently, the United States Department of Energy has made projections based on annual global GDP growth rates of 2.3, 3.0 and 3.6 percent that result in the 1990 total global energy usage increasing to 571.8, 657.2 and 749.1 exajoules, respectively, by 2025 [14]. Other projections show similar increases in energy and fossil fuel consumption for the first thirty years of the twenty-first century [6,8,15]. While any projection of energy availability and use for the twenty-first century is fraught with uncertainty, it is highly probable that most of this energy will come from the sun either directly as solar energy, as hydroelectric and wind energy derived from sun driven weather patterns, or as stored energy in biomass and fossil fuels.

7.3. The Solar Origin of Energy

Most of the energy used by humanity is gained from the sun through photosynthesis of carbohydrates from carbon dioxide and water that occurs in the great diversity of green plants, the dominate lifeforms of the landmasses and oceans, as shown in Fig. 7.4a. Apart from plants being the origin of all food and

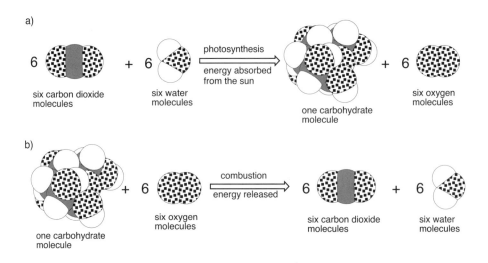

Fig. 7.4. a) Photosynthesis utilizes sunlight in a complex sequence of reactions to produce a carbohydrate molecule from six molecules each of carbon dioxide and water. b) Biomass burning reverses this process to produce heat carbon dioxide and water although it is dominantly the cellulose in plant fibres that has a similar composition to carbohydrates that is burnt. The shaded, clear and chequered spheres represent carbon, hydrogen and oxygen atoms, respectively.

oxygen for air-breathing animals, their combustion in the form of wood, grass and agricultural waste, or biomass, represents a major energy source, particularly for many developing countries. Both the digestion of food and biomass combustion release energy and represent reversals of the photosynthetic process to produce carbon dioxide and water from carbohydrates and the plethora of complex molecules derived from carbohydrates that make up living organisms as shown in Fig. 7.4b. Thus, the growth of green plants and their combustion to release energy and to produce carbon dioxide and water that is assimilated to produce more green plants represents a renewable energy cycle.

In contrast to decaying plant and animal matter exposed to air that is broken down to water and carbon dioxide, buried plant and animal matter out of contact with air is converted to natural gas, which is mainly methane, by bacterial action within as short a period as a year, and into oil and coal over hundreds of thousands and millions of years through combinations of bacterial action, pressure and heat. This represents a storing of solar energy that is released from fossil fuels by combustion to produce carbon dioxide and water. The burning of a fossil fuel produces heat, one form of energy, in direct proportion to the number of oxygen molecules that combine with each carbon atom in the fossil fuel as is shown in Fig. 7.5a. Methane produces most energy as it requires two oxygen molecules to convert one methane molecule to carbon dioxide and water. Oil, that on average has two hydrogen atoms bound to each carbon atom, requires one and a half oxygen molecules to convert it to carbon dioxide and water. The composition of coal varies substantially, but the dominant combustible material in it is carbon that requires only one oxygen molecule to convert it to carbon dioxide and water. Thus, to convert a carbon atom in methane, oil and coal to carbon dioxide and water requires two, one and a half, and one molecules of oxygen, respectively. Clearly, methane produces more energy for the same amount of the greenhouse gas carbon dioxide produced, an important consideration in view of current concern about increasing atmospheric carbon dioxide levels. (It is seen from Fig. 7.4b that biomass burning requires one molecule of oxygen for each carbon atom converted to carbon dioxide if it retains the same carbon:hydrogen:oxygen atomic ratio as carbohydrates.)

In a simple sense, it is not surprising that Earth is warming as the great increase in fossil fuel use that has occurred since the Industrial Revolution represents a release of solar energy and the corresponding carbon dioxide over a very much shorter time than that over which it was stored. So great has been the magnitude of this release that Earth's combined geological and photosynthetic assimilation of the carbon dioxide have been unable to adjust sufficiently rapidly to prevent the steady increase in atmospheric carbon dioxide levels and the accompanying enhancement of the greenhouse effect. (The greenhouse effect is discussed in detail in Chapter 8.) It is therefore not surprising that a fuel that is attracting increasing attention is hydrogen that only produces water on burning, as seen from Fig. 7.5d, and is discussed in section 7.11.

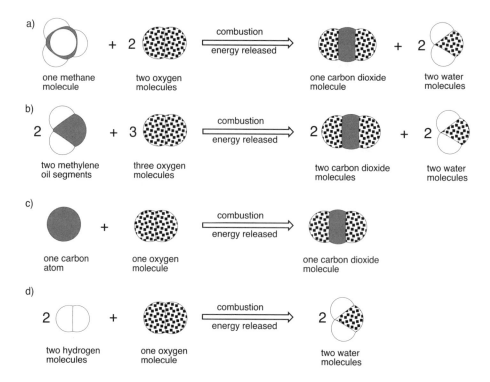

Fig. 7.5. The combustion of three fossil fuels and hydrogen. a) One methane molecule burns with two molecules of oxygen to produce one molecule of carbon dioxide and two molecules of water. b) Two methylene segments of oil burn with three molecules of oxygen to produce two molecules of carbon dioxide and two molecules of water. c) One atom of carbon, which is what coal is largely composed of, burns with one molecule of oxygen to produce one molecule of carbon dioxide. Because the heat liberated is proportional to the number of oxygen molecules consumed per carbon atom of each fuel, the amount of heat liberated on burning methane, oil and coal is in the ratio 2:1.5:1, respectively. d) The burning of two molecules of hydrogen produces heat and two water molecules. The shaded, clear and chequered spheres represent carbon, hydrogen and oxygen atoms, respectively.

While the molecular basis for the relative amounts of energy and carbon dioxide produced on burning a fuel is informative, in the everyday world far greater quantities are considered as is seen in Fig. 7.6 [16]. The superiority of methane over oil and coal in terms of the amount of energy and carbon dioxide produced is clearly seen. The relatively low amounts of energy released from dry wood and straw reflect the substantial amounts of water that they contain. The promise of hydrogen as a potential fuel of the future is emphasized by its high energy content and its zero carbon dioxide production. Uranium-235 is an important fuel that provides an enormous amount of energy and will probably remain an important energy source in the future. The heat liberated from these fuels may be directly used for heating, to boil water, to cook food, smelt iron or weld steel and so on. Alternately, it may be converted into mechanical motion either as in the expansion of volume of the high temperature air, carbon dioxide and water produced in piston engines and gas turbines, or it may be used to produce high pressure steam that turns turbines to produce electricity.

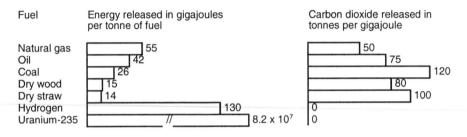

Fig. 7.6. The energy released on combustion of a range of fuels and the amount of carbon dioxide released. The composition of coal varies greatly and the figures quoted refer to a high quality coal. Adapted from [16].

7.4. The Wealth of Nations and Energy

The many and varied ways in which nations generate wealth are reflected in their gross domestic product, or GDP. Among nations there is a very large range of GDPs, as shown in Fig. 7.7, where those of the economic colossi represented by the United States and Japan are dominant [8,11]. The next largest GDPs are produced by the major western European nations that are followed by China, the most populous of nations (1.275 billion population in 2000 [17]). However, the correspondence between GDP and energy consumption [18] is brought into sharper focus when the GDP and energy consumption per person is examined as shown in Fig. 7.8. In broad terms the more energy consumed per person the greater is the GDP per person with the two most populous nations, India and China being at the low end of the scale while Japan, the United States and the western European nations are at the high end of the scale.

The 1997 global gross domestic product of US$33 trillion was disproportionately shared between the developed, or Organization for Economic

Nation   2000 Gross domestic product in billions of US$

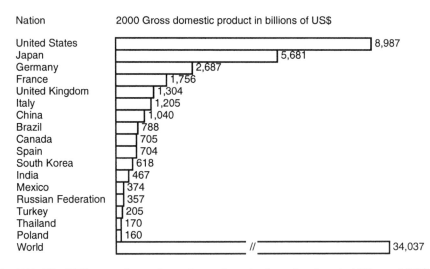

Fig. 7.7. The 2000 gross domestic products of a selection of nations in billions of US$ (1995 $US). Data from [8].

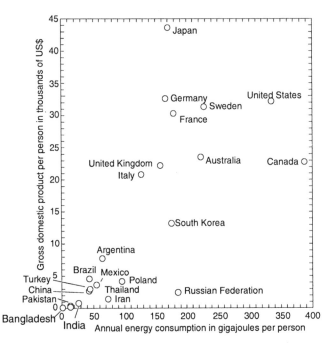

Fig. 7.8. The variation of the gross domestic product (GDP in 1995 US$) per person with the energy consumed per person in 2000 for a selection of nations calculated from data in [17,18]. Biomass energy gained from wood, grass, dung and other waste combustion is not taken into account and constitutes up to fifty percent of the energy consumed in some of the developing nations.

Cooperation and Development (OECD), nations with a fifty-four percent share leaving the developing and transitional economy nations (those of the former Soviet bloc) that represent the bulk of Earth's population, with forty-six percent as shown in Fig. 7.9 [19]. According to these estimates, the global GDP is expected to increase to US$ 67.3 trillion by 2020, with a decline in the developed nations' share to forty-two percent and a growth of the other nations' share to fifty-eight percent, largely as a result of China's gross domestic product share increasing from 13.2 percent to 20.8 percent. This coincided with the OECD nations using fifty-four percent of the energy consumed in 1997, a share that is expected to decline to forty-four percent by 2020. Globally, energy consumption is expected to rise by fifty-seven percent from a total primary energy use equivalent to that produced from 8,610 million tonnes of oil in 1997 to that produced from 13,529 million tonnes of oil in 2020, as shown in Fig. 7.10 [19].

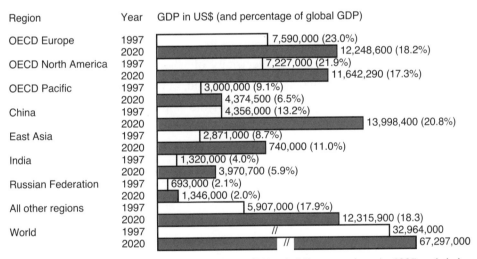

Region	Year	GDP in US$ (and percentage of global GDP)
OECD Europe	1997	7,590,000 (23.0%)
	2020	12,248,600 (18.2%)
OECD North America	1997	7,227,000 (21.9%)
	2020	11,642,290 (17.3%)
OECD Pacific	1997	3,000,000 (9.1%)
	2020	4,374,500 (6.5%)
China	1997	4,356,000 (13.2%)
	2020	13,998,400 (20.8%)
East Asia	1997	2,871,000 (8.7%)
	2020	740,000 (11.0%)
India	1997	1,320,000 (4.0%)
	2020	3,970,700 (5.9%)
Russian Federation	1997	693,000 (2.1%)
	2020	1,346,000 (2.0%)
All other regions	1997	5,907,000 (17.9%)
	2020	12,315,900 (18.3)
World	1997	32,964,000
	2020	67,297,000

Fig. 7.9. The gross domestic products (GDP) in $US of different regions in 1997 and their predicted values in 2020. Data from [19].

In addition to the higher energy consuming nations tending to have the higher gross domestic products per person, there is a striking coincidence of the wealthy nations being found in the temperate regions of Earth while the poorer nations largely occupy a band on either side of the equator [20]. There is also a strong coincidence between the wealthier nations either possessing substantial coastlines and good access to the oceans or having good water access to the oceans within one hundred kilometres of the coast. Although there is little that nations can do about their geographical location, improving access to energy and the wealth that flows from it is a preoccupation of most governments and within it access to oil supplies is usually the one that induces most concern.

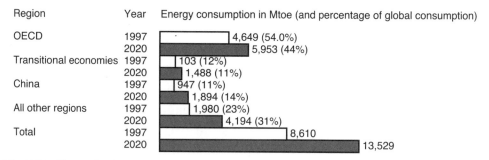

Fig. 7.10. The total primary energy consumption of different regions in millions of tonnes of oil equivalent in 1997 and their projected values in 2020. Biomass energy is not included. Data from [19].

While it is informative to discuss national gross domestic products and access to energy, this can obscure the effect this has on individuals. Thus, in 2002, 1.6 billion people had no access to electricity of whom more than eighty percent lived in South Asia and sub-Saharan Africa and, at the then rate of development, it was estimated that eighteen percent of the population would still have no access to electricity by 2030 assuming a global population of 8.3 billion [11]. Although this is a lesser proportion than in 2002, the lives of these 1.5 billion unfortunate people will be diminished by this deprivation.

7.5. Oil: The Great Addiction

The liquid crude oil found in deposits scattered about Earth's surface has become such a convenient and heavily used source of energy that current predictions suggest that it is likely to be close to exhaustion by 2050, some two hundred years from the beginning of its commercial exploitation. After the drilling of the first American commercial oilwell in Titusville, Pennsylvania in 1859, the global consumption of oil grew at an astonishing rate, as shown in Fig. 7.11, and with it the search for oilfields. The price of oil is very sensitive to supply from the oilfields, the demand from customers and events that affect oil supply as is also seen from Fig. 7.11 that shows the fluctuation of the price of oil from 1861 to

2001 in 2001 US$ [16,18]. After the drilling of the Titusville oilwell the demand for oil grew quickly as did its price. As more oilwells came into production prices fell until demand outstripped production to cause major price fluctuations that evened out as other oil producers appeared and Texas overtook Pennsylvania as the major oil producer in the United States.

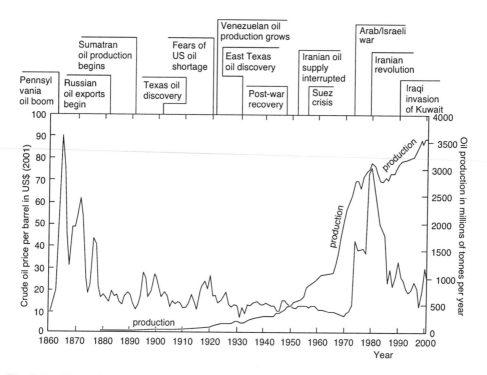

Fig. 7.11. The production of oil from 1861 to 2001 and the fluctuations in its price (in 2001 US$) from 1861 to 2001. Based on [16,18].

Oil prices remained fairly stable from 1930 to 1970. However, in 1973, the outbreak of the Yom Kippur Arab-Israeli war saw the price of oil soar to almost US$40 a barrel as the Organization of Petroleum Exporting Countries (OPEC), dominated by Iran, Iraq, Kuwait, Saudi Arabia and the United Arab Emirates, and producing fifty percent of global oil, cut off oil supplies to those nations considered friendly to Israel. This was countered to some extent by increasing oil production from elsewhere, a switch to other fuels and an increased efficiency of oil usage. However, the major alternative oil supplier, the United States, was unable to increase production to lower prices to their pre-war level and oil prices remained high until the Iranian Islamic revolution in 1979. This stopped the supply of Iranian oil and prices soared to US$80 a barrel. Thereafter oil prices gradually declined to US$20 a barrel in 1986 as production from newly developed oilfields in Alaska, Mexico and the North Sea increased production. By this time the OPEC share of oil production had dropped to thirty-two percent of global oil production. From 1986 to 2000, oil prices fluctuated around US$25 a barrel and it appeared inevitable that because of the dominance of the huge Middle Eastern oil reserves, control of oil prices would eventually return to OPEC unless either other sources of oil become more economical to exploit or alternative energy sources were used. In 2003-2005, political instability and war in the Middle East combined with rapidly increasing demand from China and other developing nations caused oil prices to rise to US$60 a barrel with the possibility of US$100 being reached. This accentuated the search for new oil reserves and alternative energy sources.

Inevitably, speculation about the extent of Earth's oil reserves and the duration of the availability of oil increased as oil consumption grew. The best known prediction of oil supply was made in 1956 by a geologist with Shell Oil, M. K. Hubbert, who correctly and controversially predicted that oil production in the contiguous forty-nine American states would reach a maximum in about 1969 and thereafter decline [21,22]. He predicted this on the basis that for newly discovered oil bearing regions production starts slowly and then accelerates as the large oilfields holding most of the oil are discovered. In the next stage the production levels out as the production of oil from the large fields slows as the amount left in them diminishes. (The more numerous smaller oilfields discovered in the same region are usually more difficult to extract.) Thereafter, a maximum in oil production is reached as the rate of discovery of new fields slows and the extraction of oil from the proven fields declines. This maximum production is reached when about half of all of the oil has been extracted. Beyond this point production falls off quite rapidly in almost a mirror image of its earlier rise to give a bell-shaped curve for the growth and decline of unrestricted production from a particular oil producing region over time. While Hubbert's prediction of a bell-shaped curve for the production of oil was made for the United States, it appears to apply generally. On this basis, recent predictions for global oil production reaching a maximum have ranged from 2000 to beyond 2020, after which a decline will set in [21,22]. At the end of 2000, proven oil reserves were 142.7 billion tonnes, or 1046.2 billion barrels, that at 2000 production rates would be exhausted in 38.4 years according to a 2000 British Petroleum analysis and other analyses [18,23]. These reserves represented the oil that could be economically extracted at the level of 2000 demand and price and were dominated by the sixty-five percent found in the Middle East.

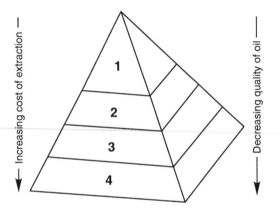

Fig. 7.12. The oil extraction pyramid where a relatively small proportion of the oil, 1, in an oilfield is of high quality and cheap to extract. Thereafter, successive proportions of the oil represented by slices 2-4 of the pyramid decline in quality and are increasingly costly to extract.

The exploitation of an oilfield may be viewed as a pyramid as shown in Fig. 7.12. The top of the pyramid represents the oil that is both of high quality and most economical to extract. Once this is pumped out, the quality of the oil tends to decrease and its extraction to become more costly as the pyramid is descended and the oilfield becomes increasingly depleted. The extent and rapidity with which the pyramid is descended and the oil is extracted depends substantially on the price of oil and advances in location and extraction technology. These technological advances include fully computerized seismic surveys for oilfield detection, the pumping of either carbon dioxide, natural gas or steam into largely depleted oilwells to force out the last of the oil, drilling that allows the drill head at the end of a flexible pipe to be steered through rock strata to the pockets of trapped oil and the development of more sophisticated techniques to extract oil from reserves found at increasing depths in the oceans [24].

Liquid crude oil is not the only source of oil. It is estimated that the tar-like oil in oil sands and oil shales in Australia, Brazil, Canada, China, Sweden, the United States and elsewhere amounts to several trillion barrels of oil; far more than the known reserves of liquid light crude oil [25]. Such oil is expensive to extract and presently it is only in the Canadian province of Alberta, that probably holds three hundred billion barrels of oil, that large scale commercial extraction is occurring. The bituminous oil coated grains of the Albertan oil sands look like coffee grounds. The gaining of the oil from them starts with mining of the oil sands followed by a complex extraction of the oil after which it is refining into petroleum and other oil products. Oil shales are hard rock materials that also have to be mined and require complex extraction methods for oil extraction. These sources of oil are usually not counted in oil reserve estimates because they are not presently considered economically viable.

Given the huge importance of a reliable energy supply to the global economy and living standards, the level of proven reserves of oil and other fossil fuels is a preoccupation of most governments and is explored next.

7.6. Fossil Fuel Reserves: The Anxious Estimates

Of the reserves of the most widely used fossil fuel, oil, 65.3 percent were in the Middle East in 2000 and were being depleted far less rapidly there than elsewhere as shown in Fig. 7.13 [18]. At the end of 2000, it appeared that Earth's proven oil reserves would last for about forty years at the then current rates of extraction. However, it might be expected that dwindling oil supplies will cause economic strain well before that. The twelve nations with the largest proven oil reserves at the end of 2000 are shown in Fig. 7.14. Of these, those of Saudi Arabia were by far the largest and made up 25.7 percent of proven reserves.

Until about 1965, coal was the dominant fossil fuel but a combination of the large amount of ash, soot and sulfur dioxide produced from its combustion and the increasing availability of more conveniently used oil and natural gas caused it to gradually lose this dominance. By 2000, coal accounted for 28.1 percent of the primary energy derived from fossil fuels globally compared with the 44.6 and 27.3 percent derived from oil and natural gas, respectively. However, the global reserves of coal greatly exceed those of oil and their geographical distribution

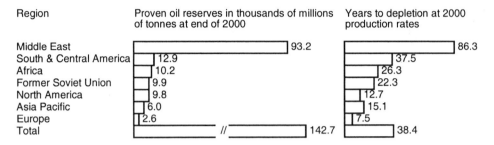

Fig. 7.13. Regional proven oil reserves at the end of 2000 and their expected times to depletion at 2000 production rates [18].

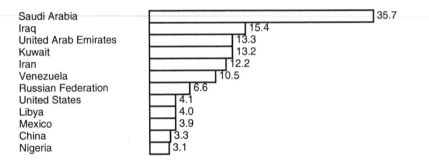

Fig. 7.14. The twelve nations possessing the largest proven reserves of oil at the end of 2000 represented 87.8 percent of all proven reserves [18].

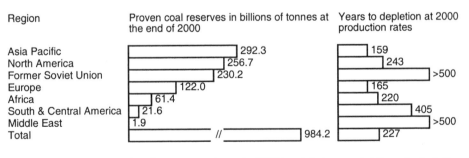

Fig. 7.15. Regional proved coal reserves at the end of 2000 and their expected times to depletion at 2000 production rates [18].

is quite different from that of oil as seen in Fig. 7.15. Thus, the regions of the Asia Pacific, North America, the former Soviet Union and Europe dominate the proven coal reserves whereas the oil rich regions of Africa, South and Central

America and the Middle East have only small coal reserves. At the 2000 rate of coal usage the global reserves would last for 227 years, substantially longer than those of oil. Of the individual nations, the United States possesses the largest coal reserves followed by the Russian Federation, China, Australia and India as seen from Fig. 7.16.

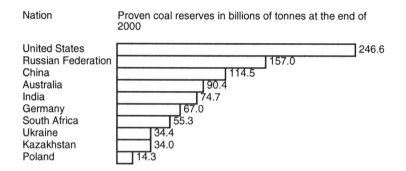

Nation	Proven coal reserves in billions of tonnes at the end of 2000
United States	246.6
Russian Federation	157.0
China	114.5
Australia	90.4
India	74.7
Germany	67.0
South Africa	55.3
Ukraine	34.4
Kazakhstan	34.0
Poland	14.3

Fig. 7.16. The ten nations possessing the largest proven reserves of coal at the end of 2000 represented ninety percent of all proven reserves [18].

A relative newcomer as a major fossil fuel is natural gas whose proven reserves in 2000 were dominated by the former Soviet Union and the Middle East that together possessed 72.7 percent of the global total as shown in Fig. 7.17. At the 2000 rate of usage, the proven natural gas reserves would last for 62

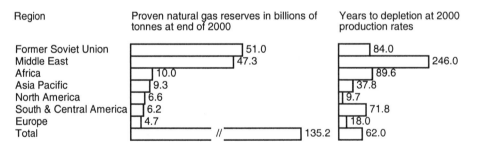

Region	Proven natural gas reserves in billions of tonnes at end of 2000	Years to depletion at 2000 production rates
Former Soviet Union	51.0	84.0
Middle East	47.3	246.0
Africa	10.0	89.6
Asia Pacific	9.3	37.8
North America	6.6	9.7
South & Central America	6.2	71.8
Europe	4.7	18.0
Total	135.2	62.0

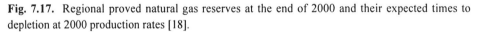

Fig. 7.17. Regional proved natural gas reserves at the end of 2000 and their expected times to depletion at 2000 production rates [18].

years. Among nations, the Russian Federation held 32.0 percent of proven natural gas reserves followed by Iran and Qatar with 15.3 and 7.4 percent, respectively, as shown in Fig. 7.18.

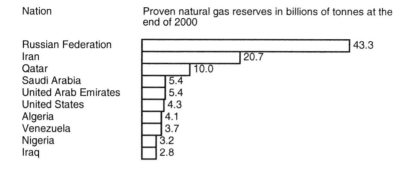

Fig. 7.18. The ten nations possessing the largest proven reserves of natural gas at the end of 2000 represented sixty-one percent of all proven reserves [18].

7.7. Patterns of Energy Consumption

There is a considerable variation in the relative fossil fuel wealth of geographical regions and nations that not only illustrates geological differences but also has economic and political implications. In 2000, oil represented 34.8 percent of all energy consumed (Fig. 7.2) and was the most traded energy source. Inevitably this placed the oil rich Middle East in a powerful trading position. However, the Middle East does not dominate the coal and natural gas reserves and this has a restraining effect. Should the price of oil rise too steeply, alternative fossil fuels and energy technologies would become economically competitive. The gasification of coal and the derivation of liquid fuel from coal is not presently economically competitive with the production of oil and natural gas, but a substantial rise in the price of the latter fuels and an improvement in technology could make coal derived gas and liquid fuels attractive. Quite apart from this, nations adjust their energy sources and consumption according to their own energy resources and wealth as shown for the major geographical regions in 2000 in Fig. 7.19 [18]. Six of the seven regions showed a strong dependence on oil with the exception of the former Soviet Union that derived 52.5 percent of its energy from natural gas of which it possessed 37.7 percent of Earth's proven reserves. The Middle East also showed a heavy dependence on natural gas at 44.4 percent that together with oil totalled 97.5 percent of that region's energy consumption as a consequence of plentiful reserves of these fuels. Europe showed the greatest dependence on nuclear power at 11.6 percent consistent with that region's relatively poor supply of fossil fuels. Of all nations, France was the most dependent on nuclear power at 36.7 percent of energy consumed in 2000 that was a reflection of the availability of significant national uranium deposits.

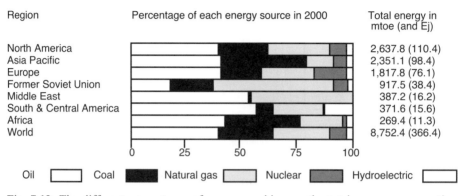

Fig. 7.19. The different percentages of sources making up the total energy consumption of geographic regions (8,752.4 millions of tonnes of oil equivalent) excluding biomass in 2000 according to British Petroleum [18].

Energy production and consumption within developed nations, in particular, presents a complex pattern as exemplified by those of the United States, a wealthy and technologically advanced nation, that generated about twenty percent of global GDP at the beginning of the twenty-first century. However, to achieve this in 2000 the United States consumed a colossal amount of energy that exceeded the energy produced within its national borders by twenty-eight percent as shown in Fig. 7.20 [26]. Of this energy, about sixty-three percent of the oil consumed was imported. This dependency on foreign oil inevitably affects the foreign and internal policies of the United States. This is exemplified by the controversial proposal to open the Alaskan Arctic National Wildlife Refuge to oil exploration that set conservationists and oil companies at loggerheads [27]. Most developed nations show variations on the pattern of energy consumption of the United States with most of them also being very dependent on imported oil, as are many developing nations although their individual energy consumption is much less.

In the United States and globally, almost sixty percent of the energy used is not converted into the action for which it was intended such as heating and illuminating buildings, and powering engines, elevators and computers, but is lost in the generation of unused heat, friction and noise and through transmission to the point of use. Transportation is particularly poor in this respect with only some twenty percent of the energy consumed being transformed into motion. The remainder is lost as heat and through the poor efficiency of engines and

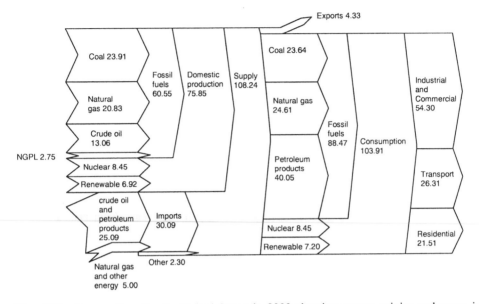

Fig. 7.20. Energy flow for the United States in 2000 showing energy origins and usage in exajoules. NGPL is natural gas plant liquids. "Other 2.30" represents small losses and gains of energy and minor energy sources including liquid hydrogen and methanol and net electricity imports. There are minor discrepancies in some of additions due to rounding errors and slight differences in accounting for energy flows. Based on [26].

transmissions. Even the more efficient electricity generating power plants convert only about forty percent of the energy gained from fossil fuel combustion to electricity. This is exemplified by the fluidised bed coal-fired power plant generating 600 megawatts of electricity per hour shown in Fig. 7.21. The hourly amounts of air, coal, water and steam passed through the power plant are enormous, as is the approximately 700 tonnes of carbon dioxide produced. Depending on the temperature at which the coal is burnt, a tonne or so of the oxides of nitrogen is produced and, depending on the sulfur content of the coal, up to twenty tonnes of sulfur dioxide are also produced. In modern coal-fired power plants oxides of nitrogen, sulfur dioxide, soot and other particulate matter are removed from the flue gases before they reach the atmosphere [28].

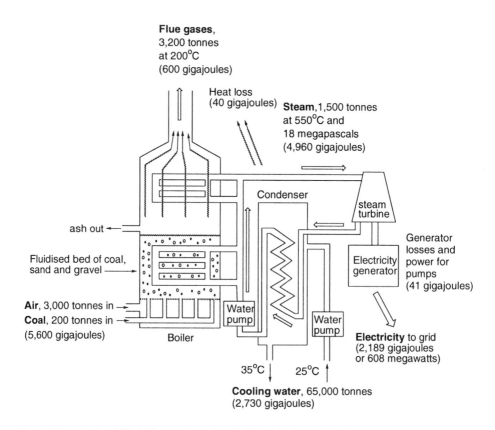

Flue gases,
3,200 tonnes
at 200°C
(600 gigajoules)

Fig. 7.21. A simplified illustration of a fluidised bed coal-fired 600 megawatt electricity generating power plant. The hourly quantities of air, coal and water used and the amounts of energy involved are based on data from [16]. The flue gases are 2,500 tonnes of nitrogen and 700 tonnes of carbon dioxide. The steam pressure of eighteen megapascals is equivalent to 180 times the pressure of the atmosphere.

7.8. Inflammable Ice: Methane Hydrate

The intense interest in the remaining reserves of fossil fuels inevitably results in a similar interest in the potential of any new source of fossil fuel. Recently attention has focused on methane hydrate. A vast amount of methane is trapped in the permafrost and deep ocean sediments as methane hydrate [29]. When extracted, it has the appearance of effervescent soft ice that is readily ignited to burn with a reddish flame leaving water behind. The effervescence is caused by combustible methane escaping as the temperature of the methane hydrate increases and the pressure that it experienced in the permafrost and ocean sediments is released. The possibility of using methane hydrate as another source of fossil fuel is attractive.

Methane hydrate was first discovered in the early nineteenth century but only became of practical importance in the early 1930s when it was encountered as an ice-like material that blocked natural gas pipelines under conditions resembling those in permafrost and deep ocean sediments. The first definite findings of naturally occurring methane hydrate were in permafrost in North America and Siberia in the 1960's, and seismological studies first detected oceanic methane hydrate off the North Carolina coast of the United States in the 1970's. The currently estimated size of methane hydrate deposits is ten to twenty trillion tonnes. This is one to two hundred times greater than the proven conventional methane reserves and two to four times the methane energy equivalent of all known recoverable and inaccessible deposits of coal, methane and oil. As methane hydrate deposits are widely spread around the coasts of the continents it is possible that their exploitation would lessen the dependence of several nations such as China, India and Japan on imported fossil fuels.

In its most common form, methane hydrate forms an elegant crystalline structure in which twenty water molecule form the apices of a dodecahedron that encloses a methane molecule as shown in Fig. 7.22 [30]. A water molecule is at the corner of each of the pentagonal faces of each dodecahedron through which a series of dodecahedra are linked to form an extended array. If each dodecahedron holds a methane molecule there are 5.75 water molecules in the methane hydrate for every methane molecule. In this form the methane and the water are so neatly packed together that one cubic metre of solid methane hydrate contains 0.8 cubic metres of liquid water and one hundred and sixty-four cubic metres of methane gas. Most of this methane is produced by bacteria metabolising buried organic matter in the absence of air. A smaller amount is produced directly from deeply buried organic matter at temperatures of $80\text{-}150\,^\circ\text{C}$. Some is also produced by the thermal decomposition of underlying oil and coal deposits. Surprisingly little of this methane escapes into the atmosphere and it has been suggested that archaea, closely related to the methane producing methanogens, oxidize methane to carbon dioxide and hydrogen where the latter is used by another group of microbes that reduce sulfate to hydrogen sulfide. This intermicrobial collaboration is known as "syntrophy" [31].

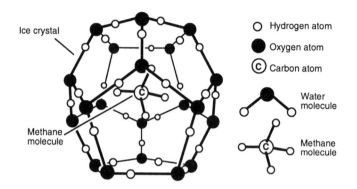

Fig. 7.22. A single crystal of methane hydrate where a methane molecule is enclosed in a dodecahedral ice crystal of twenty water molecules. Such crystals link together through their pentagonal faces to form an extended array. Adapted from [30].

Methane hydrate is normally only stable at $0^{\circ}C$ or lower at the pressures experienced at 150 metres depth or deeper in permafrost or below about 300 metres or more in the ocean. The heat from Earth's core sets a maximum depth of about 2,000 metres below the surface at which methane hydrate is stable. Should either the temperature rise, the pressure diminish, or both, methane hydrate will release methane gas with a huge increase in volume that is thought to have been responsible for major disruptions of ocean floors in the past. It has been speculated that some of the unexplained disappearances of ships may have been due to a sudden methane hydrate gas release that caused the ocean to become very frothy such that the water density fell below that needed to provide buoyancy [32]. It is also appears that massive rapid releases of methane hydrate may have been an important factor in ending ice ages as methane is a very potent greenhouse gas as is discussed in Chapter 8 [33].

The possibility of exploiting the onshore methane hydrate in permafrost and the globally spread deposits of methane hydrate in oceanic sediments is obviously attractive as methane is a readily transportable and clean fossil fuel. It is also attractive because the methane is two to five times more concentrated than it is in natural methane gas deposits and about ten times more concentrated than it is in the other potential sources of methane in sands, shales, coal seams and deep aquifers. There are a variety of possible methods for recovering methane from methane hydrate of which the simplest is the pumping of hot water into the deposit to break down the methane hydrate into methane gas and liquid water. As with other fossil fuel deposits, the amount of recoverable methane in the methane hydrate deposit must be present in economically viable amounts before any exploitation is attempted. So far no commercial exploitation has occurred with the possible exception of the Messoyakha gasfield in Siberia where methane hydrate in the permafrost may have broken down to add methane to the natural gas deposit [34].

7.9. Capturing Carbon Dioxide: First Attempts to Slow Global Warming

As discussed in section 7.2, humanity's energy use will increase greatly in the future and the bulk of this is likely to be supplied by fossil fuels until at least the middle of twenty-first century. Most of this increase will come from the developing nations as they aspire to higher living standards. However, the consequence could be that increasing atmospheric carbon dioxide levels arising from burning of fossil fuels produces global warming and climate change to an extent that partially frustrates the realization of these aspirations as a consequence of rising sea levels, crop failure and the accompanying spread of some diseases presently confined to the equatorial regions.

The Kyoto Protocol seeks to reduce carbon dioxide emissions through imposing targets and penalties, but its mandate is short and covers too few of the major carbon dioxide producing nations to have a large impact. Irrespective of these difficulties and driven by a combination of the desire to lessen the rate of atmospheric carbon dioxide increase and by the imposition of carbon taxes in some developed nations, public and commercially funded development of carbon dioxide capture, or sequestration, technologies is growing rapidly. An obvious method is to simply grow trees and allow them to convert carbon dioxide to cellulose [35]. This approach has an important role and is good for wildlife, but to rely on this alone would see humanity's already scarcely sufficient agricultural land increasingly taken over by trees. This is amply illustrated by a 2,000 square kilometre forest being required to absorb all of the carbon dioxide produced by a 500 megawatt coal-fired power plant during its lifetime. To absorb humanity's annual carbon dioxide production would require the planting of a forest the size of India each year. In any event, trees have a natural lifetime after which they decay and release their carbon as carbon dioxide to the atmosphere.

It has also been proposed that the great phytoplankton blooms induced in oceanic surface water through fertilization with ferrous sulfate increases the uptake of atmospheric carbon dioxide and represents another method of biological capture of carbon dioxide [36]. As the increased phytoplankton growth supports greater populations of zooplankton and other marine species that die to leave their skeletons to sink to the ocean floor, largely as calcium carbonate, this represents an enhancement of a natural process that permanently removes carbon dioxide from the atmosphere. However, there are concerns that substantial use of such oceanic fertilization could trigger unknown effects on deep ocean lifeforms and the closely related biogeochemical processes.

Geological processes have a much greater ability to retain carbon dioxide as both gas and as carbonates and already act as a colossal carbon store as seen in Fig. 7.23 [37]. These processes, both on the continents and in the oceans, are

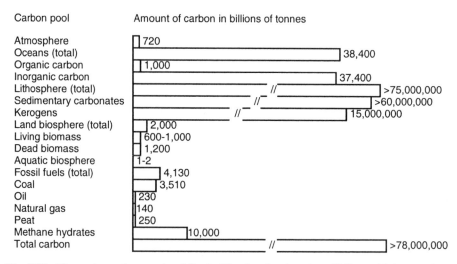

Fig. 7.23. The major carbon pools of Earth. The data is taken from [37] to which an estimate of 10,000 billion tonnes of carbon in methane hydrate is added. Kerogens are the precursors of oil and coal.

increasingly being examined and the first tentative steps to use them as carbon dioxide sinks have been taken.

The first practical large scale sequestration of carbon dioxide started in 1996 in the Sleipner oil and natural gas field in the North Sea 240 kilometres off the Norwegian coast [38]. One of the Sleipner reservoirs contains natural gas with a nine percent carbon dioxide content that has to be reduced to 2.5 percent or less to make it marketable. The excess carbon dioxide is isolated on the gas rig and is pumped under pressure at a rate of a million tonnes annually into the 32,000 square kilometre porous sandstone Utsira Formation that is two hundred metres thick some thousand metres below the bed of the North Sea. There the carbon dioxide is trapped as it dissolves in saline water where some of it is expected to slowly form carbonates and become part of the rock substrata. This increases the cost of the natural gas production by an amount that is small by comparison with the Norwegian carbon taxes that would be levied if the carbon dioxide was released into the atmosphere. A similar approach is being taken with the Natuna project. Here natural gas from the Natuna field in the South China Sea off Singapore is seventy-one percent in carbon dioxide. Other carbon dioxide sequestration processes are either planned or in operation in Algeria, Australia, Canada, Poland and the United States [39].

Somewhat ironically, and several decades before the prospects of global warming became a matter of concern, the combustion of fossil fuel was considered a good source of carbon dioxide for a variety of industrial uses as largely remains the case today. Carbon dioxide is captured by passing the flue gases through water containing monoethanolamine that loosely binds it in a chemical reaction that produces monoethanolaminecarbamic acid [40]. The captured carbon dioxide is then released by heating and is either stored in its gaseous form under pressure or is cooled to form solid carbon dioxide, or "dry ice", and the monoethanolamine is recycled back into the carbon dioxide capture process. Much of this gaseous carbon dioxide is sold to the carbonated drink industry and the "dry ice" to the food industry where it is used for cooling. An example of this is the 300 megawatt coal-fired power plant at Shady Point, Oklahoma, where all of the carbon dioxide is extracted from the flue gases and sold to a soft drink manufacturer [41]. Such usage defrays the cost of the carbon dioxide capture but ultimately releases the carbon dioxide to the atmosphere. The carbon dioxide separation from flue gases with monoethanolamine and subsequent permanent storage is likely to increase the cost of electricity by ten to thirty percent depending on the type of power plant if there is no defraying of cost through sale of the carbon dioxide. Other methods of carbon dioxide capture involving membrane technologies and the conversion of carbon dioxide into methanol, ethanol and polymers are under study and may lower this cost to some extent [42]. However, the amount of carbon dioxide produced through electricity generation by fossil fuelled power plants vastly exceeds any use envisaged for it and accordingly its permanent sequestration is of pressing importance.

When carbon dioxide is pumped into depleted oilwells and coalmines, saline aquifers and the deep ocean it is sequestered there for many hundreds of years at least and probably for thousands in some cases [43]. The estimates of the capacity of these reservoirs for carbon dioxide are very broad, as seen from Fig. 7.24, and the minimum capacity of 1,500 billion tonnes is much greater than humanity's 2000 carbon dioxide emissions from all sources of some thirty billion tonnes [41]. A more active use of carbon dioxide gas is to pump it into oilwells to force out ten to fifteen percent of the oil that would otherwise not be recoverable and thereby defray the cost of the carbon dioxide sequestration. Large amounts of methane trapped in coal seams can also be released by pumping carbon dioxide into them where for every methane molecule released two carbon dioxide molecules are captured, a good exchange ratio as each methane molecule subsequently burnt produces only one carbon dioxide molecule.

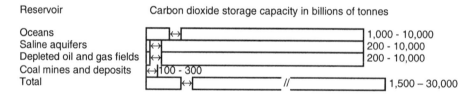

Fig. 7.24. Estimates of the capacity of the major reservoirs for storing carbon dioxide. Data from [41].

While the sequestration of carbon dioxide in underground reservoirs and the deep ocean is a promising option to offset increasing atmospheric carbon dioxide levels, there are a range of aspects that have to be considered. Not the least is the fact that carbon dioxide levels of greater than ten percent, or one hundred thousand parts per million, by volume in air are fatal to humans. Thus, the possibility that a sudden release of sequestered carbon dioxide could reach this level and asphyxiate people in the immediate vicinity is a serious matter. This has occurred naturally at least twice in the past through sudden releases of carbon dioxide from Lake Monoun and Lake Nyos in Cameroon [44]. Because carbon dioxide is heavier than air it forms a blanket on the ground that is only dispersed by wind. In the worst case, during the night of 21 August 1986, a sudden release of eighty million cubic metres of carbon dioxide from Lake Nyos asphyxiated 1,700 people and many cattle as they slept. The cause of this unusual event was that the lake water between fifty and two hundred metres deep became super-saturated with carbon dioxide from volcanic springs and an unknown disturbance caused its fast release. Subsequently, control systems have been installed to safely regulate the release of carbon dioxide from Lakes Monoun and Nyos. Nevertheless, there is an obvious parallel to be drawn between these events and the sequestration of carbon dioxide by pumping it into the deep ocean. Clearly, the deep ocean provides an immensely greater reservoir for carbon dioxide than do Lakes Monoun and Nyos. However, the potential exists for a localized build up of dissolved carbon dioxide that could result in a sudden massive carbon dioxide release. Detrimental leaks of volcanic carbon dioxide through rock strata also occur as exemplified by Mammoth Mountain in California where conifers have been dying and early signs of human asphyxia have been reported [45]. These occurrences indicate a need for careful monitoring and regulation of oceanic and underground carbon dioxide sequestration on a national and international basis.

Quite apart from the stability of carbon dioxide sequestration in the deep ocean, the effect of dissolved carbon dioxide on the pH of 8.1 of oceanic water and its impact on aquatic life must be considered [46]. In coastal waters the pH of water in tide pools varies daily with the rise and fall of tides and their biota have evolved to adapt to this, although even here pH changes to values below the natural range are likely to be deleterious. However, in the deep oceans pH appears to have been stable for millennia and their living organisms are probably adapted to this stability. It is estimated that to maintain atmospheric carbon dioxide levels at 550 parts per million, twice the preindustrial level, until 2100 by absorption of carbon dioxide in the oceans would lower their pH globally by 0.1 pH unit. The full impact of such acidification is not readily predicted but it should be recalled that at the height of the acid rain crisis of the 1970s-1990s the acidification of rivers and lakes had very detrimental consequences for aquatic life as discussed in section 3.16 [47].

Despite reservations about the effects of some forms of carbon dioxide sequestration on the environment, the potential to almost completely eliminate carbon dioxide emissions from the electricity, natural gas and oil industries is within sight and is likely to become cheaper as new technologies come into use. However, these sources of carbon dioxide are both stationary and large and only produce about thirty-five percent of carbon dioxide emissions. The major part of humanity's carbon dioxide emissions arise from individually small scale, but cumulatively very large, production as exemplified by the internal combustion engine in transportation where its mobility poses a much greater problem in capturing the carbon dioxide produced. One solution proposed is to increase the use of biomass in the form of "biofuels" for vehicular use. Another is to take the carbon out of fossil fuels before they are burnt. This approach involves the production of hydrogen from fossil fuels and the sequestering of the residual carbon dioxide to produce hydrogen for use in redesigned internal combustion engines and fuel cells. This is the basis for what is sometimes termed a "hydrogen economy" as discussed in section 7.11.

7.10. Biomass Energy: A Tried but Limited Solution

Since humans discovered fire, wood, straw and dried dung have been important sources of domestic heat and this is particularly the case in developing nations. In 2000, it was estimated that three billion people, dominantly in the developing nations, were mainly dependent on wood and charcoal for their energy supply [48]. Now, more sophisticated approaches to the sustainable use of biomass are increasingly being considered as a possible way of slowing the rise of atmospheric carbon dioxide levels. They are based on the growing of fuel crops whose direct combustion, or the combustion of fuels produced from them, produces energy and carbon dioxide that is photosynthetically sequestered to produce the next fuel crop. While this renewable energy approach is an attractive prospect, analysis shows that it can only offer a partial solution to the reduction of carbon dioxide emissions.

Of the developing nations highly dependent on fuelwood for simple needs such as cooking and warmth, many have reached the point where their wood supplies are close to exhaustion as they have largely relied on natural regrowth and have not supplemented this with replacement tree planting. It will take several decades to ameliorate this situation, particularly in sub-Saharan Africa where population growth creates increasing demands for food production increases so that land available for producing fuel crops is scarce. At a more advanced level where annual plantation of fuel crops is carried out a significant amount of non renewable energy is still required. This is illustrated by the Brazilian sugarcane ethanol industry where crushed cane, or bagasse, is used to fire boilers and steam engines used in the cane crushing and sugar extraction processes [49]. While some of the ethanol produced also supplies energy for its production, some energy supplementation by fossil fuels occurs for this purpose and also in the production and transportation of the extensively used fertilizers and pesticides. Thus, when ethanol is produced as an automotive fuel from sugar, as occurs on a large scale in Brazil (10.9 million tonnes equivalent to about 6.6 million tonnes of oil equivalent in 1996), the carbon dioxide produced in the bagasse burning and during fermentation and the fuel use of the ethanol is largely taken out of the atmosphere as the next sugarcane crop grows. However, the carbon dioxide derived from the supplementary fossil fuel used in producing the ethanol simply adds to the atmospheric level. For Brazil, that is the largest user of sugarcane ethanol as an automotive fuel, the fossil fuel used in ethanol production resulted in 3.0 million tonnes of carbon dioxide emissions in 1996. The carbon dioxide emissions from the fuel use of ethanol was 23.7 million tonnes so that 20.7 million tonnes of carbon dioxide emissions from ethanol replaced the same emissions from fossil fuels. Of the liquid biofuels, ethanol produced from sugar is the most energy efficient and, at the efficiency of ethanol production achieved by Brazil of 114 gigajoules of energy per hectare annually, more than twice the area of Earth's croplands would be required to replace all of the 1996 fossil fuel use.

The amount of sunlight falling on croplands affects both the types of crops grown and their efficiency in converting solar energy to a potential biofuel. Thus, while 220 watts per square metre (69,400 gigajoules per hectare) of sunlight falls on Brazil annually, only 170 watts per square metre (53,600 gigajoules per hectare) falls on the midwestern United States where ethanol is produced from maize (4.0 million tonnes equivalent to about 2.4 million tonnes of oil equivalent annually) [49]. The conversion of the maize biomass to ethanol at 2,200 kilograms per hectare is less productive than is the case with Brazilian sugarcane at 3,200 kilograms per hectare. On this basis, the replacement of ten percent of the United States' 1996 petroleum use by ethanol from maize would require twelve percent of the United State's cropland. However, it is estimated that the energy from the fossil fuel used in growing the maize and producing the ethanol is equal to seventy-five percent of the energy in the ethanol produced so that the overall replacement of petroleum by ethanol is only twenty-five percent. Thus, about forty-eight percent of United States cropland would be needed for ethanol produced from maize to replace the carbon dioxide produced by ten percent of the national petroleum use. It is apparent that the present methods of ethanol production do not offer a major replacement for fossil fuels simply because insufficient land is available to produce the fuel crops. At present ethanol is mainly produced from the carbohydrates in sugarcane and maize. It is possible that the overall yield of ethanol production could be increased by producing ethanol from the residual cellulose of both crops but this method of ethanol production is less efficient than that from carbohydrates [49,50].

Plant oils produced from soybean, sunflower and canola crops are triglycerides that can be used in diesel engines and are often collectively called "biodiesel". However, energy derived from them is presently more costly in the amount of petroleum used in their production than is the case for ethanol. Nevertheless, plant oils recovered from commercial cooking provides a secondary source of biodiesel fuel. Despite the inability of present biofuel technology to offer a major replacement for petroleum, its broader use will lessen the use of fossil fuels to a worthwhile extent if sustainably managed [51].

Despite the limitations of biofuels, they can make a significant contribution to energy supply. A prime example of a lesser dependence on fossil fuels is provide by Brazil where 35.4% of total primary energy was derived from renewable resources of which 20.4% came from fuel produced from sugar and plantation wood in 2000 as shown in Fig. 7.25 [52]. Brazil is able to achieve this high proportion of renewable energy production as a consequence of a moderate population density and abundant sunshine and water, characteristics that many nations do not enjoy.

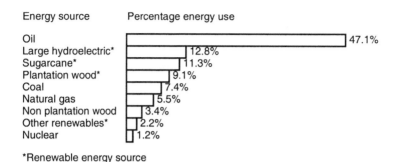

Fig. 7.25. The energy usage profile for Brazil in 2000 showing a 35.4% usage of renewable energy. Data from [52].

7.11. The Hydrogen Economy: A Distant Dream

It has long been a dream that water could be an abundant and pollution free fuel. To some extent hydroelectricity achieves this through the use of falling water to drive turbines and generate electricity. However, this is only a partial fulfilment of the dream as the availability of suitable water courses and their seasonal variation limits expansion of hydroelectricity generation a long way short of the capacity to replace fossil fuels in electricity generation. The alternative is to split water into its constituent elements, hydrogen and oxygen and then to recombine them either through combustion or in fuel cells to give energy and water as a by-product [53,54]. This concept is sometimes referred to as a "hydrogen economy" and its attraction is an inexhaustible fuel supply and the elimination of carbon dioxide emissions as shown in Fig. 7.5d. However, the splitting of water through the presently only viable method, electrolysis, as shown in Fig. 7.26a, requires more energy than the energy contained in the hydrogen produced and so represents an overall loss of energy. Usually, the electricity used in the electrolysis is generated by fossil fuelled power plants so that the hydrogen produced indirectly generates carbon dioxide emissions. There are no usable natural sources of hydrogen, although microorganisms such as the alga *Chlamydomas reinhartii* that uses sunlight to split water into hydrogen and oxygen show some promise of being able to provide some hydrogen as a fuel.

At the beginning of the twenty-first century, ninety-seven percent of hydrogen used in the United States was produced either from natural gas, oil or coal with an attendant production of carbon dioxide identical to that produced by burning the equivalent amount of these fossil fuels if the hydrogen production was one hundred percent efficient as a comparison of Fig. 7.5a-c and 7.26b-d shows. This does not take into account the additional carbon dioxide arising from the fossil fuels used to generate the heat and pressure required in the hydrogen production. In practice, the energy of the natural gas, oil and coal is transferred

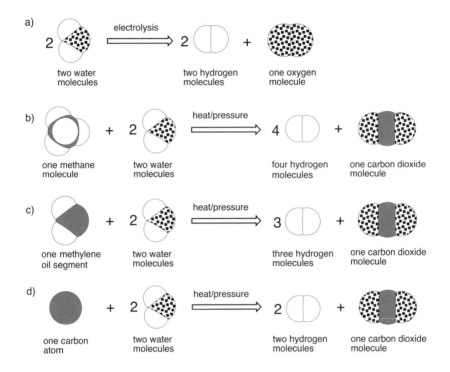

Fig. 7.26. a) The formation of hydrogen and oxygen by the electrolysis of water. The production of hydrogen from methane, that is the dominant constituent of natural gas, oil and coal is shown in b, c and d, respectively. It is seen that methane produces more hydrogen than does oil than does coal in a 2:1.5:1 ratio. The shaded, clear and chequered spheres represent carbon, hydrogen and oxygen atoms, respectively.

to the hydrogen produced with about seventy-two, seventy-six and fifty-seven percent efficiency [53]. Thus, hydrogen is not a primary energy source but is an agent of transfer, or a carrier, of energy from other sources. Most of the hydrogen produced is used in oil refining, in the food industry where it is used to hydrogenate edible oils to produce margarine, and in the chemical industry where it is used to produce many chemicals in every day use such as ammonia.

Similarly large proportions of the hydrogen used in other developed nations are also produced from fossil fuels.

Clearly, the production of hydrogen from fossil fuels will not assist in lowering carbon dioxide production, but it does offer a route to capturing transportation carbon dioxide emissions if hydrogen is used as a fuel. An alternative to the production of hydrogen from fossil fuels is to produce it by electrolysis of water using either hydroelectricity or nuclear or wind generated electricity. However, the magnitude of the change in the present energy supply structure required to achieve this is colossal. It is estimated that 230,000 tonnes of hydrogen would be required daily to replace the oil used in surface transportation in the United States in 2003 [55]. This is an optimistically low estimate as it assumes a presently unachievable sixty-four percent energy conversion into vehicular motion. If all of the hydrogen was generated through electrolysis of water it would require either the building of 500 natural gas fuelled power plants generating 800 megawatts each, 500 coal fuelled power plants generating 800 megawatts each, 200 Hoover Dams generating 2,000 megawatts each, 400 nuclear power plants generating 1,000 megawatts each, or a mix thereof, costing US$400 billion, or one twentieth of the 2003 gross domestic product of the United States. If the fossil fuelled power plant options were adopted this would increase the amount of carbon dioxide that would have to be sequestered if the advantages of using hydrogen as a fuel with respect to global warming were to be realized. The magnitude of the energy infrastructure changes required globally to switch to hydrogen for transportation appears even more daunting using current hydrogen production technologies.

If more modest targets are set for hydrogen production to complement rather than completely replace other fuels, an increase in hydroelectricity generation offers a sustainable source of hydrogen production as does electricity generated through wind power and photovoltaic cells. In the future it may be possible to produce hydrogen directly from water using solar powered devices. Nevertheless, for at least the next two to three decades it is probable that a major obstacle to any significant switch to hydrogen as a fuel without incurring major carbon dioxide emissions will be its low availability from non fossil fuel sources.

Despite these difficulties, experimentation with hydrogen powered devices is proceeding but assessments of their potential for bringing a hydrogen economy closer are mixed [55-57]. While the spectacular use of liquid hydrogen and liquid oxygen as rocket fuel is well-known, the combustion of gaseous hydrogen in engines quite similar to the conventional internal combustion engine is of more practical value for general use. However, it is the very different use of hydrogen as a fuel in fuel cells that has gradually found a small but increasing usage as power packs in spacecraft, emergency power generators and in prototype hydrogen fuel cell powered cars. To understand how a fuel cell works it is helpful to take a closer look at the splitting of water into hydrogen and oxygen during electrolysis (Fig. 7.26a). Water represents the least energetic states in which hydrogen and oxygen coexist. As a consequence, the splitting of water into hydrogen and oxygen requires energy. For almost two centuries it has been known that this can be achieved by supplying this energy as electricity through two electrodes immersed in water. The water first splits into hydrogen atoms which have each lost an electron to become hydrogen ions (H^+) and simultaneously forms oxygen molecules that have gained two electrons to become a peroxide ion, O_2^{2-}. The hydrogen ions regain their electrons at the positively charged cathode and join together to form hydrogen that bubbles off, and the peroxide ion loses two electrons at the negatively charged anode to form oxygen which also bubbles off. In 1839, the Welsh physicist Sir William Grove showed that this process could be reversed to produce electricity and water from hydrogen and oxygen. This is the basis of the modern hydrogen fuel cell shown in Fig. 7.27 [56,57].

The storage of hydrogen presents a significant obstacle to its use as a major fuel. By weight, hydrogen contains about three times the energy contained in petroleum. However, by volume, gaseous hydrogen contains only about ten percent of the amount of energy stored by petroleum even when the hydrogen gas is compressed at 20 megapascals (200 times the pressure exerted by the atmosphere). Thus, for a car powered by an internal combustion engine to travel the same distance on pressurized gaseous hydrogen as on petroleum a hydrogen tank ten times larger than a petroleum tank would be required. If liquid hydrogen was used instead the fuel tank could be reduced to about three times the size of the petroleum tank, but to retain the hydrogen in its liquid form requires refrigeration at –253°C that itself entails additional energy expenditure. Neither

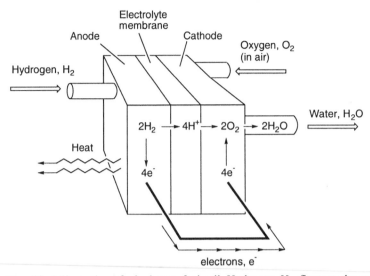

Fig. 7.27. Simplified illustration of a hydrogen fuel cell. Hydrogen, H_2, flows to the anode where each molecule splits into two hydrogen ions, H^+, and two electrons, e^-. The hydrogen ions cross the electrolyte membrane to the cathode, but the electrons cannot traverse the membrane and so pass along an external circuit to the cathode. Oxygen, O_2, flows to the cathode where it gains two electrons to form a peroxide ion, O_2^{2-}, that combines with the hydrogen ions to form water. Heat is released during these processes. The electrons passing along the external circuit represent an electrical current that can be used to power electromotors and other devices.

option is practical for general use. Alternative storage possibilities are the loose combination of hydrogen with metals to form metal hydrides and the absorption of hydrogen in metal-organic compounds and microporous forms of carbon called nanotubes [58,59]. These raise the possibility of storing hydrogen safely and in a dense form to allow the use of fuel tanks more closely approaching the size of present petroleum tanks. However, these technologies are still in their infancy and appear unlikely to provide safe storage of hydrogen for general use for several decades.

Concerns have been expressed that leakage of growing amounts of hydrogen in to the atmosphere with increased use of hydrogen as a fuel could be a new source of global warming and depletion of the ozone layer [60,61]. Hydrogen is second to methane as a trace gas in the atmosphere. It is produced biologically in soil and natural waters, through the oxidation of methane and other hydrocarbons to carbon dioxide and water, through biomass and fossil fuel burning, and escapes during oil refining, the hydrogenation of edible oils to produce

margarine and from other chemical technologies as shown in Fig. 7.28. Of the
seventy-seven million tonnes of molecular hydrogen produced annually, about a
quarter is converted to water by hydroxyl radicals in the atmosphere and the
remainder is absorbed in the soil by biological processes. Should an increase in
hydrogen emissions to the atmosphere occur, the reaction of hydrogen with
hydroxyl radicals to produce water and thereby decrease their availability to
destroy methane and other hydrocarbons could lead to a significant increase in
the levels of these greenhouse gases in the troposphere. This could result in
tropospheric warming and a concomitant cooling of the lower stratosphere.
Increased levels of hydrogen in the stratosphere would result in a moistening of
the stratosphere through its combination with oxygen to produce water.
Together, these effects could result in an increased formation of polar
stratospheric ice clouds that are responsible for the catalysis of ozone destruction
over the Arctic and Antarctic in early spring as is discussed in Chapter 9. While
the magnitude of such effects is uncertain they do emphasise the desirability of
greater efficiency in energy use.

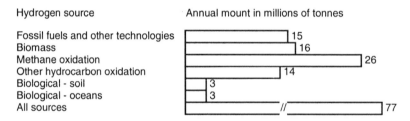

Fig. 7.28. Sources of atmospheric molecular hydrogen according to [60].

7.12. Nuclear Fission: The Divisive Energy Source

Amidst the concerns about humanity's dependence on fossil fuels, their climate
changing effects and the search for alternative energy sources, nuclear power
based on the fission of uranium-235 makes a substantial contribution to the
global energy supply. The development of this new energy source was
overshadowed by the race to produce a nuclear weapon and the awesomely
destructive explosive release of energy by the first such weapon at Alamogordo,
New Mexico, in 1944. However, this was preceded by the controlled and
sustained release of nuclear energy at the University of Chicago under the

direction of Enrico Fermi in 1942 [62]. This gaining of a new energy source was achieved in great secrecy under the West Stands at Stagg Field and is there commemorated by a bronze plaque which states:

"On December 2 1942 man achieved here the first self-sustaining chain reaction and thereby initiated the controlled release of nuclear energy."

Whilst this achievement was kept secret until the end of the Second World War in 1945, its subsequent revelation as a new source of energy to some extent offset the deep foreboding felt by much of humanity after witnessing the destructive power of the nuclear weapons dropped on Hiroshima and Nagasaki.

The first nuclear power plant to supply electricity for household and commercial use was Calder Hall in the United Kingdom in 1956. Many other nuclear power plants quickly followed in the United Kingdom, the United States and other nations, particularly during the period 1960-1980. to the point were nuclear generated electricity is an important component of the energy supplies of many nations [63]. It is seen from Fig. 7.29 that 442 civilian nuclear power plants were operating in thirty nations in November 2002 with a combined electricity generating capacity of 356,746 megawatts and providing seventeen percent of electricity globally according to the International Atomic Energy Agency, IAEA [64]. At that time, another thirty-five nuclear reactors were under construction with an estimated generating capacity of 27,743 megawatts.

From the inception of the construction of nuclear reactors for electricity generation the link between nuclear power and nuclear weapons has cast a shadow over public acceptance of this energy source [65-67]. In addition, the knowledge that nuclear power plants produce extremely dangerous radioactive fission products that require storage for many thousands of years is a concern for the public and governments alike. This concern was heightened by the catastrophic accident at Chernobyl in Ukraine in 1986 and the lesser accident at Three Mile Island in the United States in 1979 as discussed in section 1.13 [68]. In addition to the fission products from uranium-235 in a nuclear reactor core, the fact that substantial amounts of uranium-238 are converted to plutonium-239, the nuclear explosive used together with uranium-235 in nuclear fission weapons and as a trigger component of nuclear fusion weapons, has further increased anxiety at the terrorist afflicted beginning of the twenty-first century. Nevertheless, as the detrimental consequences of global warming have become apparent, proponents of nuclear power have advocated its increased use to slow the rate of increase in atmospheric carbon dioxide levels [69]. Simultaneously, safer and more advanced reactor designs are appearing [70].

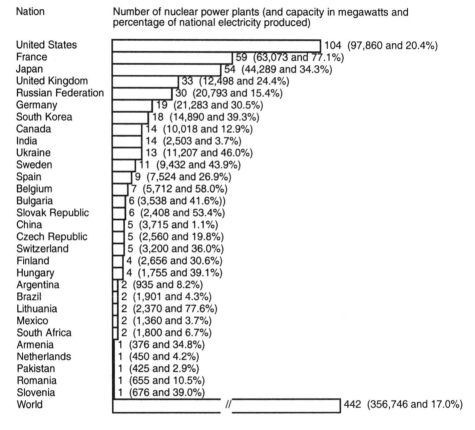

Nation	Number of nuclear power plants (and capacity in megawatts and percentage of national electricity produced)
United States	104 (97,860 and 20.4%)
France	59 (63,073 and 77.1%)
Japan	54 (44,289 and 34.3%)
United Kingdom	33 (12,498 and 24.4%)
Russian Federation	30 (20,793 and 15.4%)
Germany	19 (21,283 and 30.5%)
South Korea	18 (14,890 and 39.3%)
Canada	14 (10,018 and 12.9%)
India	14 (2,503 and 3.7%)
Ukraine	13 (11,207 and 46.0%)
Sweden	11 (9,432 and 43.9%)
Spain	9 (7,524 and 26.9%)
Belgium	7 (5,712 and 58.0%)
Bulgaria	6 (3,538 and 41.6%))
Slovak Republic	6 (2,408 and 53.4%)
China	5 (3,715 and 1.1%)
Czech Republic	5 (2,560 and 19.8%)
Switzerland	5 (3,200 and 36.0%)
Finland	4 (2,656 and 30.6%)
Hungary	4 (1,755 and 39.1%)
Argentina	2 (935 and 8.2%)
Brazil	2 (1,901 and 4.3%)
Lithuania	2 (2,370 and 77.6%)
Mexico	2 (1,360 and 3.7%)
South Africa	2 (1,800 and 6.7%)
Armenia	1 (376 and 34.8%)
Netherlands	1 (450 and 4.2%)
Pakistan	1 (425 and 2.9%)
Romania	1 (655 and 10.5%)
Slovenia	1 (676 and 39.0%)
World	442 (356,746 and 17.0%)

Fig. 7.29. The number of nuclear power plants, their electricity generating capacity and the percentage of national electricity generated for the thirty nuclear power generating nations in November 2002. Data from [64].

To aid a more informed assessment of nuclear power and its future as an energy source, a more detailed analysis of its origins and potential development is presented below.

7.13. Nuclear Fission

Uranium atoms are the heaviest found on Earth in more than trace amounts and were formed in a supernova explosion billions of years ago. Today, some of the energy stored from the supernova is released in nuclear power plants to produce electricity [70]. It is an extraordinarily powerful source of energy as shown by the ability of the uranium-235 in a six gram pellet of nuclear reactor fuel to produce the same amount of energy as a tonne of coal. However it requires a sophisticated technology to achieve this. Uranium comes in two forms, or isotopes, of which the scarce nuclear fuel fissile uranium-235 comprises 0.7 percent and the abundant uranium-238 99.3 percent. (The number 235 indicates that 92 positively charged protons and 143 uncharged neutrons make up the uranium-235 nucleus while 92 protons and 146 neutrons make up the uranium-238 nucleus.)

Uranium-235 and uranium-238 are naturally radioactive and slowly decay to lead-206 with respective half-lives of 0.71 and 4.5 billion years; the times it takes for half of all of the uranium-235 and uranium-238 now on Earth to decay to lead-206. As a consequence, about ninety-nine percent of all of the uranium-235, and half of the uranium-238 present when Earth formed 4.5 billion years ago now exists as lead-206, a familiar non radioactive soft metal. This reflects the fact that many of the nuclei produced in the original supernova explosion are unstable and are slowly converting to stable nuclei through radioactive decay. The heaviest stable nucleus is that of bismuth-209. However, none of this is directly related to the use of uranium-235 to generate electricity in nuclear power plants. This requires the splitting, or fission, of the uranium-235 nucleus into two smaller nuclei to release energy by hitting it with a neutron at a carefully controlled speed [16,70]. This neutron is described as a "thermal neutron" because it travels at the speed a neutron or small atom would normally travel at the temperature of the nuclear reactor core (about 300°C) around which a power plant is built.

There are a variety of ways in which uranium-235 fission occurs one of which is shown in Fig. 7.30 where the lighter krypton-92 and barium-141 nuclei and three neutrons are produced as fission products. The sum of the masses of krypton-92, barium-141 and the three neutrons is slightly less than the mass of uranium-235 and the colliding neutron. This difference is the mass that is converted into energy upon uranium-235 fission and is initially expressed as the kinetic energy of the krypton-92 and barium-141 nuclei and the three neutrons and the gamma radiation produced that generates heat which is the useful form of nuclear energy used in power plants as is discussed below. The source of this

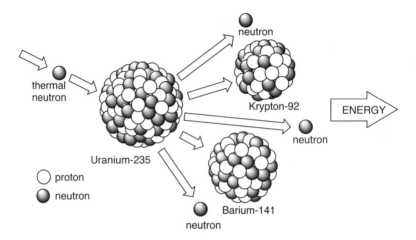

Fig. 7.30. A thermal neutron collides with a uranium-235 nucleus and splits it into krypton-92 and barium-141 nuclei and three neutrons to release energy as heat and gamma radiation. This is one of the several ways that uranium-235 splits into two lighter nuclei to release energy and neutrons. In this example the energy released is 3.2×10^{-11} joules for each uranium-235 nucleus, or 82 petajoules per tonne of uranium-235.

released energy is a small part of the force that bound the ninety-two protons and 143 neutrons, each weighing approximately one atomic mass unit, in the uranium-235 nucleus and is usually referred as the "binding energy". When all of the possible ways that uranium-235 can split are considered, one uranium-235 nucleus and the neutron that splits it produce several lighter nuclei fission products and 2.43 neutrons and 202 million electron volts of energy (MeV). This converts to eighty-two petajoules of energy released per tonne of uranium-235 that is the same amount of energy released by about three million tonnes of coal and two million tonnes of oil.

The amount of energy released by uranium-235 fission is calculated through the equation: $E = mc^2$, derived by the Nobel Laureate Albert Einstein. This shows that the energy released is equivalent to the difference between the sum of the mass of the uranium-235 nucleus and the neutron that hits it and the sum of the masses of the fission product nuclei and neutrons multiplied by the speed of light twice. This is the basis of the calculation of the energy produced by the nuclear fission shown in Fig. 7.30 and the several other pathways through which uranium-235 splits into pairs of other nuclei as fission products and either two or three neutrons in addition to that used to cause the fission. (The conversion of mass into energy is not unique to nuclear fission for when something is burnt mass is also lost but the proportion is very much smaller. Thus, when coal is burnt less than half a billionth of its mass is lost.)

The fission of a single uranium-235 nucleus produces a very small amount of energy. Accordingly, it is necessary to use the neutrons released by the fission process to cause other uranium-235 nuclei to undergo fission in a chain reaction to build up the amount of energy released. These neutrons must be slowed to a speed at which they possess the appropriate energy to split other uranium-235 nuclei on collision in a self-sustaining chain reaction. In a pressurized water reactor (PWR) this is achieved by circulating pressurized water through the reactor core, as shown in Fig. 7.31, so that the neutrons lose much of their energy through collision with the nuclei of the hydrogen atoms of water (and are said to be moderated) such that they participate in the uranium-235 chain reaction [70,71]. Natural uranium contains insufficient uranium-235 to give the "critical mass" required to sustain this chain reaction in this widely used reactor type and therefore the amount of uranium-235 is increased through processing to three percent or greater in the enriched uranium used. (In the less used deuterium oxide, or heavy water, and graphite moderated reactors natural uranium fuel is used [16].)

In a pressurized water reactor, the fuel is in the form of uranium oxide pellets enriched to three or more percent in uranium-235 and contained in metal tubes, or fuel rods, immersed in circulating water that is pressurized to one hundred times the pressure exerted by the atmosphere. If the uranium-235 chain reaction was not carefully controlled it could run so fast that the resulting increase in temperature would evaporate the moderating and cooling water with a consequent melting of the reactor core and rupture of the steel pressure vessel containing it. To avoid this, the level of neutrons, or the neutron flux, in the reactor core is automatically controlled by the lowering and raising of control rods made of neutron absorbing material. As the fast moving fission products from uranium-235 slow down through collisions with either the moderating water or uranium-235 in the fuel rods, they lose their energy as heat that raises the temperature of the circulating pressurized water to about 300°C. This hot water circulates through a heat exchanger in a steam generator that produces high pressure steam that turns a turbine to produce electricity. The steam from the turbine is condensed by circulating cooling water as it returns to the steam generator. To prevent the accidental escape of radioactive materials into the atmosphere the reactor and steam generator are housed inside a reinforced concrete containment shell.

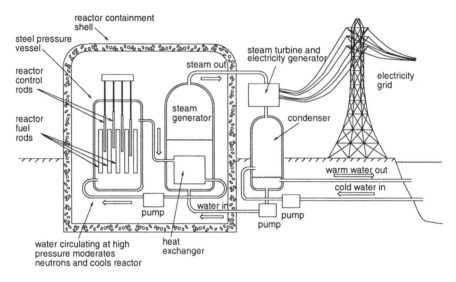

Fig. 7.31. A simplified diagram of a nuclear power plant built around a high pressure water moderated and cooled enriched uranium-235 nuclear reactor often referred to as a light water (as opposed to heavy water in which the two hydrogen atoms are replaced by the heavier deuterium isotope of hydrogen) pressurized reactor, or PWR.

While uranium-238 does not undergo fission in the nuclear reactor, it can absorb a neutron to produce plutonium-239. Some of the early nuclear reactors were designed to maximize the production of plutonium-239 for use in nuclear weapons. Today most nuclear reactors are designed to minimize the production of plutonium-239 but it is still produced in substantial amounts along with the uranium-235 fission products in the nuclear reactor fuel rods. Some of it undergoes fission and adds to the energy released in the reactor core. Ultimately, the amount of uranium-235 in the fuel rods drops below the level at which the uranium-235 chain reaction can be sustained and the "spent" fuel rods are then replaced with fresh fuel rods. The spent fuel rods are extremely radioactive and generate a large amount of heat such that they must be stored under water in cooling ponds on the reactor site for several years. As the shortest lived fission products decay away the fuel rods cool to the point where they can either be placed into permanent storage or chemically reprocessed to separate plutonium-239 and also unused uranium-235 for use in fresh fuel rods from the fission products.

At present there are only four non military reprocessing plants at Sellafield in the United Kingdom, La Hague in France, Tarapur in India and Tokai in Japan. Nevertheless, whether spent fuel rods are reprocessed or not, the problem of long term storage of the highly radioactive by-products of the nuclear power cycle, often referred to as high-level waste, remains. Apart from military facilities, there are no operating stores that are considered permanent and most nuclear power plants store their high-level waste on site. Foremost among the dangerous radioactive waste in the spent fuel rods after ten years of storage are strontium-90 and cesium-137 which have half-lives of around thirty years. Initially, these two radioactive isotopes generate most of the heat in nuclear waste but after several centuries they decay to levels where they are too small to be of concern. However, carbon-14 (with a 432.2 years half-life), technetium-99 (211,100 years), plutonium-239 (24,360 years), neptunium-237 (2.20 million years) and small amounts of other fission products require much longer term storage. It is obvious from this that quite apart from the fear of plutonium-239 falling into terrorists' hands, its radioactivity will have decreased to only about one eighth of its original value after 73,080 years, a length of time hugely greater than recorded history.

7.14. Yucca Mountain: The Forever Nuclear Store

The most common proposal for safe storage of nuclear waste is deep underground in stable rock formations away from water tables. The first requirement arises from the need to store the waste for many thousands of years to allow the longer lived fission products to decay to a safe level, and the second is to avoid radioactive contamination of water in aquifers that either are or may be drawn upon for human use. In 2002, there were 43,000 tonnes of high-level nuclear waste stored at a large number of locations in the United States and it was growing at 2,000 tonnes a year, and with it increasing pressure to find secure storage. By far the most discussed proposal for underground storage is Yucca Mountain in Nevada that is planned as the national high-level waste store in the United States [72]. Here, large tunnels have been drilled into Yucca Mountain three hundred metres below the surface and it is planned to extend them to fifty-six kilometres in length. Up to 70,000 tonnes of mainly spent fuel rods will be stored in the tunnels in as durable as possible containers essentially forever as there are no plans to retrieve them. A similar method of storage is envisaged at the nuclear waste repository at Äspö in Sweden where a tunnel spirals 450 metres down in bedrock granite [73]. The magnitude of the storage problem is illustrated by 150,000 tonnes of high-level waste that had accumulated globally by 2000 and was growing by 10,000 tones a year. An additional source of high-level nuclear waste is the nuclear reactors themselves that become very radioactive during operation and ultimately wear out over thirty years or so and must be dismantled and safely stored.

The past record of storage and processing of nuclear waste is not reassuring. Much of this waste accumulated during the nuclear weapons programs of the Cold War with regions of the Soviet Union, in particular, becoming badly contaminated [74]. A combination of deliberate and accidental dumping of Soviet high-level nuclear waste and naval nuclear reactors at sea in the Arctic and elsewhere has lead to substantial contamination. Similar contamination also exists on land as illustrated by probably the worst known accident arising with nuclear waste storage. This occurred at a nuclear weapons production site at Mayak in the southern Ural region of the now Russian Federation on 29 September 1957. A radioactive waste pit overheated and exploded to spread radioactive waste including long lived strontium-90, cesium-137 and plutonium-239 over twenty thousand square kilometres forcing the evacuation of sixteen thousand people [75]. In the United States, contamination of groundwater and soil has occurred in a number of places, particularly at Hanford in Washington State, a major site used in the nuclear weapons program, where high-level liquid waste leaked from storage tanks [76]. Fortunately, major decontamination and storage programs are now underway. The main British high-level waste reprocessing and storage facility at Sellafield has been plagued by controversy that was increased by the release of radioactive waste including technetium-99, iodine-129 and cesium-137 into the Irish Sea, some of which found its way to the Arctic and northern Canada [77]. Fortunately, the once widespread dumping of radioactive waste in the oceans is now prohibited although proposals for storage of nuclear waste under the deep ocean floor have been made [78].

7.15. The Limits to the Use of Nuclear Fission Power

Irrespective of the concerns discussed above, the use of nuclear fission power is limited by the paucity of the known and anticipated recoverable deposits of uranium which amount to 3.4 and seventeen million tonnes, respectively [5]. This is sufficient to fuel uranium-235 fission reactors for sixty-eight and 340 years, respectively, at the 2000 annual rate of consumption of 50,000 tonnes. If uranium-235 fission was expected to provide completely for the expected doubling of electricity demand by 2050 the uranium supply would be rapidly exhausted, unless the approximately 4.4 billion tonnes of uranium dissolved in the oceans could be efficiently extracted.

It appeared for several decades that it might be possible to use uranium-238 that makes up 99.3 percent of natural uranium to generate nuclear power and thereby increase the energy produced by uranium-235 reactors by almost one hundred percent. As discussed in section 7.13, absorption of a neutron by uranium-238 produces plutonium-239 that can also undergo nuclear fission in a chain reaction to produce fission products, neutrons and energy in a similar way to uranium-235. Thus, a nuclear fuel is "bred" from uranium-238. This led to the fast breeder reactor, so called because fast neutrons are absorbed by uranium-238 to "breed" plutonium-239 that in turn is split by a fast neutron to release energy. However, fast neutrons are less effective in causing nuclear fission and as a consequence the fast breeder reactor fuel rods must contain up to twenty percent of either uranium-235 or plutonium-239. In addition, water cannot be used as a coolant as it slows down neutrons and impedes the plutonium-239 breeding process. Consequently, sodium metal that becomes liquid at 98°C and boils at 883°C is used as the coolant and heat transfer agent in fast breeder reactors much as water is used in the reactor shown in Fig. 7.31. However, sodium burns in air and reacts explosively with water to produce hydrogen gas that forms an explosive mixture with air so that a sodium leak could cause a catastrophic explosion.

At first, the fast breeder reactor seemed almost too good to be true as it produces its own plutonium-239 fuel which undergoes fission to give abundant energy and fission products that, although requiring long term storage as high-level waste, cannot be used in nuclear fission weapons. However, disappointment was to come. The first fast breeder reactor to produce electricity was built in Idaho in the United States in 1951 and was followed by fast breeder reactors at Dounreay in Scotland, at Morestel in France and the Monju reactor in Japan, amongst others. Almost without exception problems arose with the liquid sodium cooling systems so that most fast breeder reactors have been closed down and appear not to be a viable power source for the near future [79]. Nevertheless, it has been proposed that fast breeder reactors should be redesigned to operate solely on the large global weapons grade plutonium-239 stockpile in a mode that also destroys the long lived fission products, thereby decreasing the possibility of nuclear weapon proliferation and also requirements for high-level waste storage [80].

Another possibility for nuclear fission power is based on the absorption of a neutron by non fissile thorium-232 to give fissile uranium-233 that can be split by a neutron to give fission products, neutrons and energy in a similar way to uranium-235. This has the advantage that all natural thorium is thorium-232 that is three times more abundant than uranium. At present this alternative source of nuclear power is attracting little commercial attention, apart from in India, despite a range of new fission reactor designs becoming available [81].

While the extensive debate about nuclear power has continued, some developed nations future energy plans have been prepared that decrease dependence on nuclear power. This is exemplified by the 2003 energy policy for the United Kingdom that envisaged no new nuclear power plants being built [82].

7.16. Nuclear Fusion: The Elusive Power Source

The provision of energy from nuclear fusion has the unfortunate reputation of being available within thirty or so years ever since it was first considered in the middle of the twentieth century and as a consequence is now regarded with some scepticism. Nevertheless, its distant feasibility is there for all to see as it is the process that occurs in the sun and other stars. The major obstacle to be overcome to make this energy source viable is the attainment of the conditions where controlled nuclear fusion can occur [83]. There are a range of nuclear fusion reactions which can occur and the one most considered for the generation of nuclear fusion energy is the fusion of nuclei of hydrogen-2 (deuterium) and hydrogen-3 (tritium) to produce helium-4, a neutron and energy as shown in Fig. 7.32 [84]. However, this does not produce a chain reaction as in the uranium-235 fission process discussed in section 7.13. Instead, the energy generated has to continually cause more fusion reactions of deuterium and tritium nuclei. While deuterium is in plentiful supply, tritium is not. Thus, it is envisaged that in a

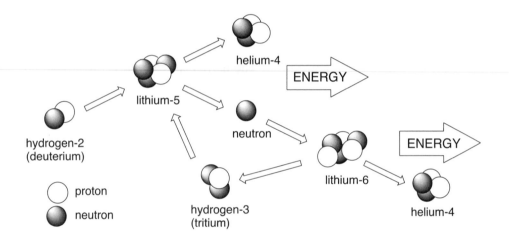

Fig. 7.32. The fusion of the deuterium and tritium nuclei into unstable lithium-5 that disintegrates into helium-4 and a neutron and energy. The neutron subsequently collides with a lithium-6 nucleus that release further energy as it fissions into helium-4 and tritium nuclei where the latter is available to fuse with a deuterium nucleus to continue the fusion cycle.

fusion reactor the very fast moving neutrons produced by the fusion process would be absorbed by lithium-6 to produce energy and helium-4 and tritium nuclei where the latter can fuse with another deuterium nucleus to continue the fusion cycle. Lithium-6 constitutes 7.42 percent of all lithium and is in plentiful supply. It is anticipated that ultimately some of the energy can be tapped from the fusion cycle to produce steam to drive electricity generating turbines.

The fusion of deuterium and tritium nuclei can only occur at a temperature of 100,000,000°C or greater, at which temperature everything becomes a plasma [85]. Conventional materials cannot be used to contain the fusion reaction but a container generated by a magnetic field can and nuclear fusion has been achieved for tiny fractions of a second in this way. However, the energy needed to produce this magnetic field and the extremely high temperature required has greatly exceeded the amount of energy produced by nuclear fusion so far. In 1985, President Ronald Regan of the United States and Chairman Mikhail Gorbachev of the Soviet Union called for a joint development of nuclear fusion energy. From this came a 1987 agreement to commence the design and construction of an International Thermonuclear Experimental Reactor (ITER) with the aim of maintaining a nuclear fusion reaction for ten minutes or more at a cost of US$10 billion. The design is based on a tokamak that produces a toroidal, or doughnut, shaped magnetic container for the plasma in which the nuclear fusion process occurs. After a long and tortuous round of discussions about the technical, funding and political aspects of the project it was decided in June 2005 to build the ITER at Cadarache in France [86] The participants in the ITER project are the European Community, Japan, the Russian Federation and the United States. Should the ITER achieve its goal it will represent the first stage towards building electricity generating nuclear fusion power plants. While such plants will not produce highly radioactive fission products from their fuel, in contrast to uranium-235 fission reactors, they will still suffer the consequences of intense neutron fluxes that produce highly radioactive by-products from the reactor structural materials and embrittle the reactor structure itself.

7.17. Non Carbon Renewable Energy

Non carbon renewable energy such as hydroelectricity, solar, wind, tidal, wave and geothermal energy is attractive because it does not directly generate carbon dioxide as a by-product and is derived from constantly renewed sources. Presently, non carbon renewable energy is dominated by hydroelectricity that supplied nineteen and 2.3 percent of global electricity and energy, respectively, in 2000 [6]. Most of this was supplied by more than 45,000 large dams as discussed in section 3.8. The largest dam under construction so far is the Three Gorges Dam on the Yangtze River at Sandouping in China [87]. Due to be completed in 2009, it will have an electricity generating capacity of 19,200 megawatts. To this will be added other large dams either under construction or planned, dominantly in Asia and Africa [88]. While global water run-off greatly exceeds the amount used to generate hydroelectricity, it shows seasonal variability depending on location and it is impractical to access most of it for hydroelectricity generation. This and political considerations lead to estimates of possible increases in global hydroelectricity generation ranging from two to threefold that could supply the anticipated large increase in electricity demand, particularly in developing countries. However, the benefits of hydroelectricity are offset by some disadvantages. Millions of people have been displaced by large dams that greatly change water flow with detrimental effects on river channels, aquatic biota and estuarine fisheries. And, depending upon the rock strata over which the catchment water flows, the accumulation of silt behind the dam can diminish the water storage capacity to the point where the hydroelectricity generated by the dam is much decreased.

The other non carbon renewable energy sources are geothermal, solar, wind and tidal that supplied 0.441 0.039, 0.026 and 0.004 percent of global energy, respectively, in 2000 [7]. After nuclear power, geothermal energy is only the second energy source not derived from the sun and arises from the heat of Earth's molten core. The use of hot springs and geothermal high pressure steam from deep below the surface to provide direct heating and steam to drive electricity generating turbines is a well-established and increasingly utilized source of geothermal energy [89]. More advanced systems under construction and planned pump water several kilometres below the surface to percolate through hot substrata to produce high pressure steam that is piped back to the surface. While there is considerable scope for expansion of geothermal energy usage, it appears unlikely to be able to supply more than a small percentage of future energy demand.

Sunlight reaching the surface of the continents amounts to a long term average of 850,000 exajoules per year when diurnal and seasonal variations are taken into account [2]. This represents about two thousand times humanity's 2000 energy consumption. However, the problem is to capture this energy in a useful form close to where it is needed. Photosynthesis captures by far the major proportion of sunlight transformed into energy but is only about one percent efficient in using this energy to produce carbohydrates. While not as sophisticated as photosynthesis, a range of technologies is either in practical use or being developed to derive energy from sunlight [90]. Silicon based photovoltaic cells that convert sunlight directly into electricity have efficiencies of up to twenty percent and have proved effective in roof mounted panels to provide household electricity. Together with roof mounted direct water heating units, such usage is likely to grow although it is dependent on diurnal and seasonal cycles and its effectiveness diminishes as distance from the equator increases. Large scale electricity supply from solar energy is feasible in regions were sunshine is plentiful and is usually based either on large banks of photovoltaic cells or on parabolic mirrors that focus sunlight to produce steam to drive electricity generating turbines.

An unusual solar energy project is the German designed solar thermal power plant planned for Australia at Buronga in New South Wales [91]. It is based on the generation of a powerful airflow in an 8,500 hectare circular greenhouse-like structure during the day as the sun heated air rises to a central one kilometre high tower through which it passes as shown in Fig. 7.33. The hot airflow generated will be sufficient to turn air turbines and generate 200 megawatts of electricity. At the same time, solar heat absorbers will absorb heat during the day and release it at night to continue the airflow and the generation of electricity at night. The electricity generated should be sufficient to power 200,000 homes and will replace fossil fuel burning by an amount equivalent to that producing 750,000 tonnes of carbon dioxide annually. The project is based on a prototype fifty megawatt solar thermal power plant operated at Manzanares in Spain from 1982 to 1989.

The use of wind to propel sailing vessels and to drive water pumping and grain milling has been in use for a long time and the growth of interest in wind generating electricity represents a refinement of this renewable energy resource [92]. The global production of electricity from wind turbines doubled every three years during 1990 to 2000. At the end of 2000 their electricity producing capacity was 17,600 megawatts of which over 6,000 and 2,000 megawatts was produced in Germany and Denmark, respectively. By 2003, about fifteen percent of Denmark's electricity consumption was met by wind power. Increasingly, wind turbines are appearing both onshore and offshore in small groups and in wind farms containing one hundred turbines or more. Most such turbines generate one megawatt of electricity but increasingly those generating two and more megawatts are being installed. However, due to wind variability, wind turbines can only operate effectively for about one third of the time and in some regions a paucity of wind makes it impractical to establish wind farms. It is likely that the use of wind power will continue to increase into the foreseeable future with the major objections to such expansion being based on visual and noise pollution, bird kills and interference with radar.

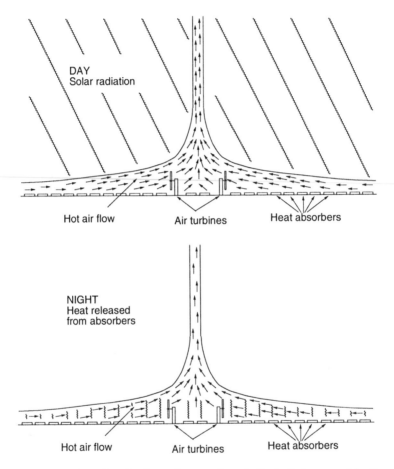

Fig. 7.33. A simplified view of a solar thermal power plant similar to that proposed for Buronga in Australia. During the day the air in a giant greenhouse-like structure is heated by solar radiation creating a massive hot airflow following the rise of the roof to a one kilometre tall central tower. This airflow turns air turbines that generate 200 megawatts of electricity. Simultaneously, large amounts of heat are absorbed by extensive heat absorbers. At night the heat absorbers release their heat to continue the airflow to turn the wind turbines.

The gravitational effects of the moon produces the rise and fall of tides that potentially provide a water flow to drive electricity generating turbines if the height of the tide is sufficient. Tidal power offers a moderate source of energy but this and the potentially large amount of energy available from ocean currents and waves are largely untapped [93]. Thus, in addition to hydroelectricity generation that is already an important energy source, other non carbon alternative energy sources have the potential to make increasing contributions to the overall energy supply but at their present stage of development and adoption are unlikely to provide a major alternative to fossil fuels for several decades.

7.18. Future Energy Use and Supply

Predictions of humanity's future patterns of global energy use are subject to a wide range of factors and are uncertain as a consequence. Nevertheless, there are two factors that provide some guidance. The first is that, even if viable technologies were now available to completely replace the use of fossil fuels, the enormity of the infrastructure change required to sustain present energy demand and future growth would take decades to achieve. The second factor is that, despite their drawbacks, uranium-235 fission and hydroelectric power are presently the most feasible options to sustain present and future levels of energy use and significantly slow the rate of increase in fossil fuel consumption over the next several decades. It is sobering to note that despite the promise of renewable energy technologies, their contribution to global electricity generation decreased from twenty-four percent in 1970 to fifteen percent in 2001 according to the International Energy Agency [94].

In view of these considerations, it appears that if the rates of global climate change and sea level rise already set in train for the next several centuries [95] are not to be increased, there is little alternative but to embark on a substantial global carbon dioxide sequestration program. While carbon dioxide could become a valuable feed stock for chemical industries, most of it will have to be permanently sequestered. Inevitably, this will increase the cost of energy and encourage greater efficiency in energy use. Under these circumstances, the several types of available renewable energy should become more price competitive. The signatories to the Kyoto Protocol have little option but to embark on such endeavours and there are signs that some non signatory nations are also engaging in similar ventures.

The adoption of new patterns of energy use is likely to vary from nation to nation depending on their national energy resources and the extent of the dependency on imported energy supplies that they find acceptable. Thus, in 2003 the United States proposed to greatly expand its nuclear power program to simultaneously increase electricity supply, produce hydrogen as a fuel and lessen its dependence on imported fossil fuels [69]. In contrast, Europe declared its intention to gain much of its energy from renewable resources by 2050 [96] and the United Kingdom indicated its aspiration to reduce its carbon dioxide emissions by sixty percent by 2050 through the measures indicated in Fig. 7.34 [82].

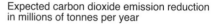

Source of carbon dioxide emission reduction | Expected carbon dioxide emission reduction in millions of tonnes per year

Increased household energy efficiency — 4-6
Increased industrial energy efficiency — 4-6
Increased use of renewable energy sources — 3-5
Reduced vehicle emissions — 2-4
Carbon trading — 2-4

Fig. 7.34. A governmental proposal for reducing the United Kingdom's carbon dioxide emissions by 15-25 million tonnes annually by 2020 to gain a 60% cut by 2050. The open and shaded boxes indicate the uncertainty in the estimated savings. Data from [82].

An economic insight into the adoption of various energy sources was offered in a 2004 report on the costs of generating electricity in the United Kingdom by the Royal Academy of Engineering. This showed that biomass, wind, wave and tidal generated electricity was not price competitive with fossil fuel and uranium-235 generated electricity as indicated in Fig. 7.35 [94,97]. Wind farms suffered

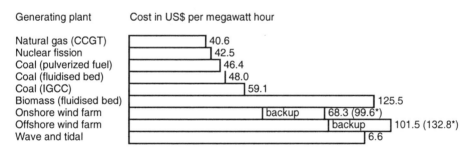

Generating plant | Cost in US$ per megawatt hour

Natural gas (CCGT) — 40.6
Nuclear fission — 42.5
Coal (pulverized fuel) — 46.4
Coal (fluidised bed) — 48.0
Coal (IGCC) — 59.1
Biomass (fluidised bed) — 125.5
Onshore wind farm — backup 68.3 (99.6*)
Offshore wind farm — backup 101.5 (132.8*)
Wave and tidal — 6.6

CCGT = Combined cycle gas turbine; IGCC = Integrated combined cycle
* Including auxiliary backup power plant.

Fig. 7.35. Variation of electricity generating costs with type of generating plant in the United Kingdom in 2004. Data from [97].

from the disadvantage of requiring an auxiliary backup power plant to produce electricity during the sixty-five percent of time when the wind was too light or too strong. However, on the optimistic side increased efficiency in current energy production and usage technologies combined with decreases in deforestation, increases in reforestation and improved farming practice may well slow the rate of increase in atmospheric carbon dioxide levels [98]. However, it appears inevitable that fossil fuels will continue to be the major and increasing source of energy at least until 2050 and probably well beyond that unless major non carbon energy sources become economically competitive.

References

1. M. Kerry *et al.*, Glazed over: Canada copes with the ice storm of 1998, *Environment*, **1999**, *41*, 6.
2. M. Hoffert *et al.*, Energy Supply in Engineering Response to Global Climate Change, ed. G. Watts, CRC, Boca Raton, USA, **1997**, p. 205.
3. D. Anderson, Renewable energy technology and policy for development, *Annu. Rev. Energy. Environ.*, **1997**, *22*, 187.
4. V. Smil, Energy in the twentieth century: resources, conversions, costs, uses, and consequences, *Annu. Rev. Energy Environ.*, **2000**, *25*, 21.
5. M. I. Hoffert *et al.*, Advanced technology paths to global climate stability: energy for a greenhouse planet, *Science*, **2002**, *298*, 981.
6. J. Chow, R. J. Kopp and P. R. Portney, Energy resources and global development, *Science*, **2003**, *302*, 1528.
7. International Energy Agency, *Renewables in Global Energy Supply*, IEA, Paris, France, 2002.
8. Key World Energy Statistics, International Energy Agency, Paris, 2002.
9. a) J. Depledge and R. Lamb, *Caring for Climate: A Guide to the Climate Change Convention and the Kyoto Protocol*. UNFCCC, Bonn, Germany, 2003. A frequent update on the Kyoto protocol appears at: www.unfcc.int.resource.convkp.html. b) *Anon.*, Climate respite? *New Sci.*, **2005**, 19 Feb., 4.
10. A. McDonald, Combating acid deposition and climate change, *Environment*, **1999**, *41*, April, 4.
11. *Anon.*, Energy and poverty, *IAEI Bull.*, **2002**, 44/2 25.
12. N. Nakicenovic and K. Riahi, *An assessment of technological change across selected energy scenarios*, World Energy Council, 2001. (http://www.iiasa.ac.at/Publications/ Documents/ PR-02-005.pdf)
13. N. Nakicenovic, Energy scenarios, in *Global Perspectives*, International Systems for Applied Systems Analysis/World Energy Council, (http://www.iiasa.ac.at/Research/ ECS/docs/book_st/wec) A 2004 update may be found at http://wwwundp.org/energy/ docs/WEAOU_full.pdf/weaover2004.htm
14. *International Energy Outlook*, US Department of Energy, Washington, DC, USA, 2004. (www.eia.doe.gov/oiaf/ieo/index.html)
15. J. T. Houghton *et al.*, eds., *Climate Change 2001: The Scientific Basis*, IPCC, Cambridge University Press, Cambridge, UK, 2001. Also at www.ipcc.ch/pub/reports.htm
16. J. Ramage, *Energy. A Guidebook*, Oxford University Press, Oxford, UK, 1997.

17. *World Population Prospects. The 2002 Revision.* Population Division of the United Nations, New York, USA, **2002**.
18. *Statistical Review of World Energy*, British Petroleum, London, UK, 2001.
19. *World Energy Outlook*, International Energy Authority, Paris, France, 2000.
20. J. D. Sachs, A. D. Mellinger and J. L. Gallup, The geography of poverty and wealth, *Sci. Am.*, **2001**, *284*, March, 62.
21. a) M. K. Hubbert, Nuclear energy and the fossil fuels, Shell Development Company Publication 95, 1956, p. 40. b) M. K. Hubbert, Degree of advancement of petroleum exploration in the United States, *Am. Assoc. of Petrol. Geol. Bull.*, **1967**, *51*, 2207. c) M. K. Hubbert, The world's evolving energy system, *Am. J. Phys.*, **1981**, *49*, 1007.
22. a) M. A. Adelman and M. C. Lynch, Fixed view of resource limits creates undue pessimism, *Oil Gas J.*, **1997**, 7 April, 56. b) P. J. McCabe, Energy resources – cornucopia or empty barrel? *Am. Assoc. of Petrol. Geol. J.*, **1998**, *82*, 2110. c) R. A. Kerr, The next oil crisis looms large – and perhaps close, *Science*, **1998**, *281*, 1128.
23. a) C. L. Campbell and J. H. Laherrère, The end of cheap oil, *Sci. Am.*, **1998**, *278*, March, 60. b) N. Selley, The future for oil – predicting the global production peak, *Energy World*, **2000**, *280*, June, 8. c) B. Holmes and N. Jones, Brace yourself for the end of cheap oil, *New Sci.*, **2003**, 2 Aug., 9.
24. R. N. Anderson, Oil production in the 21st century, *Sci. Am.*, **1998**, *278*, March, 68.
25. a) R. L. George, Mining for oil, *Sci. Am.*, **1998**, *278*, March, 66. b) N. Jones and B. Holmes, Can heavy oil avert an energy crisis? *New Sci.*, **2003**, 2 Aug., 11. c) I. M. Head, D. M. Jones and S. R. Larter, Biological activity in the deep subsurface and the origin of heavy oil, *Nature*, **2003**, *426*, 344.
26. G.V. Kaiper, U.S. Energy Flow – 2002, U.S. Department of Energy, 2002, UCRL-ID-129990-00.
27. a) W. W. Gibbs, The Arctic oil and wildlife refuge, *Sci. Am.*, **2001**, *284*, May, 54. b) J. Pelley, Will drilling for oil disrupt the Arctic National Wildlife Refuge? *Environ. Sci. Technol.*, **2001**, *35*, 240 A. c) T. Burnhill, Wishing well, *New Sci.*, **2002**, 5 Jan., 16.
28. a) E. Corcoran, Cleaning up coal, *Sci. Am.*, **1991**, *264*, May, 71. b) S. B. Alpert, Clean coal technology and advanced coal-based power plants, *Annu. Rev. Energy and Environ.*, **1991** 16, 1. c) A. Rose, T. Torries and W. Labys, Clean coal technologies and future prospects for coal, *ibid.* 90. d) C. O Bauer, Developing clean coal technology, *Environ. Sci. Technol.*, **2003**, 37, 27 A. e) A. J. Minchener, Fluidized bed combustion systems for power generation and other industrial applications, *Proc. Inst. Mech. Eng. A*, **2003**, 217, 9.
29. a) K. A. Kvenvolden, A primer on the geological occurrence of gas hydrate, in *Gas Hydrate: Relevance to World Margin Stability and Climate Change*, eds., J.-P. Henriet and J. Mienert, Geological Society, Special Publications, London, UK, 1998, p. 9. b) D. Pendick, The power below, *New Sci.*, **1998**, 30 May, 36. c) B. U. Haq, Gas hydrates: greenhouse nightmare? Energy panacea or pipe dream? *GSA Today*, **1998**, Nov., 1. d) E. Suess *et al.*, Flammable ice, *Sci. Am.*, **1999**, *77*, Nov., 52. e) K. A. Kvenvolden, Potential effects of gas hydrates on human welfare, *Proc. Natl. Acad. Sci. USA*, **1999**, *96*, 3420. f) K. A. Kvenvolden, Gas hydrate and humans, in *Challenges for the Future*, eds. G. D. Holder and P. R. Bishnoi, New York Academy of Sciences, New York, USA, 2000. g) R. L. Kleinberg and P. G. Brewer, Probing gas hydrate deposits, *Am. Sci.*, **2001**, *89*, 244. h) Q. Schierimeier, Expedition trawls seabed for energy-rich gas crystals, *Nature*, **2002**, *415*, 355. i) E. D. Sloan, Fundamentals and applications of natural gas hydrates, *Nature*, **2003**, *426*, 353. j) R. A. Kerr, Gas hydrate resource: Smaller but sooner, *Science*, **2004**, *303*, 946.
30. C. A. Koh, Towards a fundamental understanding of natural gas hydrates, *Chem. Soc. Rev.*, **2002**, *31*, 157.
31. a) E. F. Delong, Resolving a methane mystery, *Nature*, **2000**, *407*, 577. b) A. Boetius *et al.*, A marine microbial consortium apparently mediating anaerobic oxidation of methane, *Nature*, **2000**, *407*, 623.

32.   a) R. D. McIver, Role of naturally occurring gas hydrates in sediment transport, *Am. Assoc. Petrol. Geol. Bull.*, **1982**, *789*, 66. b) R. Corfield, Close encounters with crystalline gas, *Chem. Br.*, **2002**, *38*, May, 23.

33.   a) R. A. Kerr, A smoking gun for an ancient methane discharge, *Science*, **1999**, *286*, 1465. b) D. Normile, Ocean project drills for methane hydrates, *Science*, **1999**, *286*, 1456. c) S. Simpson, Methane fever, *Sci. Am.*, **2000**, *282*, Feb., 12. d) T. Blunier, "Frozen" methane escapes from the sea floor, *Science*, **2000**, *288*, 68. e) J. P. Kennett *et al.*, Carbon dioxide isotope evidence for methane hydrate instability during quaternary interstadials, *Science*, **2000**, *288*, 128. f) S. B. Jacobsen, Gas hydrates and deglaciations, *Nature* **2001**, *412*, 691. g) I. A. Pecher, Gas hydrates on the brink, *Nature*, **2002**, *420*, 622. h) W. T. Wood *et al.*, Decreased stability of methane hydrates in marine sediments owing to phase-boundary roughness, *ibid*, 656.

34.   T. S. Collett and G. D. Ginsberg, Gas hydrates in the Messoyakha gas field of the West Siberian Basin - a re-examination of the geological evidence, *Internatl. J. Off-Shore Polar Engn.*, **1998**, *8*, 22.

35.   a) D. Martindale, Plant a tree, *Sci. Am.*, **2000**, *282*, Feb., 59. b) R. J. Scholes and I. R. Noble, Storing carbon on land, *Science*, **2001**, *294*, 1012. c) B. A. McCarl and E. A. Schneider, Greenhouse mitigation in U.S. agriculture and forestry, *Science*, **2001**, *294*, 2481. d) J. Pelley, Taking credit for forest carbon sinks, *Environ. Sci. Technol.*, **2003**, *37*, 58 A.

36.   a) S. W. Chisholm, P. J. Falkowski and J. J. Cullen, Dis-crediting ocean fertilization, *Science*, **2001**, *294*, 309. b) G. H. Rau and K. Caldeira, Minimizing the effects of CO_2 storage in oceans, *Science*, **2002**, *295*, 275.

37.   P. Falkowski *et al.*, The global carbon cycle: A test of our knowledge of Earth as a system, *Science*, **2000**, *290*, 291.

38.   a) C. Hanisch, Exploring options for CO_2 management, *Environ. Sci. Technol.*, **1999**, *33*, 66 A. b) H. Herzog, B. Eliasson and O. Kaarstad, Capturing greenhouse gases, *Sci. Am.*, **2000**, *282*, Feb., 54. c) H. J. Herzog, What future for carbon capture and sequestration? *Environ. Sci. Technol.*, **2001**, *35*, 149 A.

39.   R. F. Service, The carbon conundrum, *Science*, **2004**, *305*, 962.

40.   a) H. J. Herzog and E. M. Drake, Carbon dioxide recovery and disposal from large energy systems, *Annu. Rev. Energy. Environ.*, **1996**, *21*, 145. b) P. Freund, Making deep reductions in CO_2 emissions from coal-fired power plant using capture and storage of CO_2. *Proc. Inst. Mech. Eng. A*, **2003**, *217*, 1.

41.   a) E. A. Parson and D. W. Keith, Fossil fuels without CO_2 emissions, *Science*, **1998**, *282*, 1053. b) R. Corfield, Lest sleeping giants rise..., *Chem. Br.*, **2003**, *39*, March, 30.

42.   a) D. Penman, Mineral sieve filters out carbon from flue gases, *New Sci.*, **2003**, 4 Oct., 26. b) Y. Cui, H. Kita and K. Okamoto, Preparation and gas separation properties of zeolite T membrane, *Chem. Commun.*, **2003**, 2154.

43.   a) C. Hanisch, The pros and cons of carbon dioxide dumping, *Environ. Sci. Technol.*, **1998**, *32*, 20 A. b) P. Reimer, Out of the air, into the ground, *Chem. Indust.*, **1998**, July, 563. c) B. Omerod, A salty solution for carbon dioxide, *ibid.*, 569. d) P. G. Brewer *et al.*, Direct experiments on the ocean disposal of fossil fuel CO_2, *Science*, **1999**, *284*, 943. e) D. W. Keith and E. A Parson, A breakthrough in climate change policy? *Sci. Am.*, **2000**, *282*, Feb., 60. f) P. G. Brewer and F. M. Orr, CO_2: the burning issue, *Chem. Indust.*, **2000**, Sept., 567. g) R. L. Rawls, Sequestering CO_2, *Chem. Eng.*, **2000**, *78*, 18 Sept., 66. h) S. Holloway, Storage of fossil fuel-derived carbon dioxide beneath the surface of the Earth, *Annu. Rev. Energy Environ.*, **2001**, *26*, 145. i) B. Hileman, How to reduce greenhouse gases, *Chem. Eng. News*, **2002**, *80*, 27 May, 37. j) R. G. Bruant *et al.*, Safe storage of CO_2 in deep saline aquifers, *Environ. Sci. Technol.*, **2002**, *36*, 240 A. k) E. J. Wilson, T. L. Johnson and D. W. Keith, Regulating the ultimate sink: managing the risks of geological CO_2 storage, *Environ. Sci. Technol.*, **2003**, *37*, 3476.

44.   a) G. W. Kling *et al.*, The 1986 Lake Nyos gas disaster in Cameroon, West Africa, *Science*, **1987**, *236*, 169. b) Y. Nojiru *et al.*, Gas discharge at Lake Nyos, *Nature*, **1990**, *346*, 322. c)

S. J. Freeth *et al.*, Conclusions from Lake Nyos disaster, *Nature*, **1990**, 348, 201. d) G. W. Kling *et al.*, Degassing Lake Nyos, *Nature*, **1994**, *368*, 405. e) T. Clarke, Taming Africa's killer lake, *Nature*, **2001**, *409*, 554. f) K. Krajick, Efforts to tame second African "killer lake" begin, *Science*, **2003**, *299*, 805.

45.   C. D. Farrar *et al.*, Forest-killing diffuse CO_2 emission at Mammoth Mountain as a sign of magmatic unrest, *Nature*, **1995**, *376*, 675.

46.   a) B. A. Siebel and P. J. Walsh, Potential impacts of CO_2 on deep-sea biota, *Science*, **2001**, *294*, 319. b) B. A. Seibel and P. J. Walsh, We are faced with serious decisions, *Science*, **2002**, *295*, 276. c) L. D. D. Harvey, Impact of deep-ocean carbon sequestration on atmospheric CO_2 and on surface-water chemistry, *Geophys. Res. Lett.*, **2003**, *30*, 41.

47.   a) D. W. Schindler, Effects of acid rain on freshwater ecosystems, *Science*, **1988**, *239*, 149. b) A. Henriksen, *et al.*, Northern European lake survey, 1995, *Ambio*, **1998**, *27*, 80. c) B. L. Skjelkvåle and R. F. Wright, Mountain lakes; sensitivity to acid deposition and global climate change, *Ambio*, **1998**, *27*, 280. d) T. Hesthagen, I. H. Sevaldrud and H. M. Berger, Assessment of damage to fish populations in Norwegian lakes due to acidification, *Ambio*, **1999**, *28*, 112.

48.   a) J. S. Levine *et al.*, Biomass burning. A driver for global change, *Environ. Sci. Technol.*, **1995**, *29*. 120 A. b) H. Schulte-Bisping, M. Bredemeier and F. Beese, Global availability of wood and energy supply from fuelwood and charcoal, *Ambio*, **1999**, *28*, 592. c) K. Betts, How charcoal fires heat the world, *Environ. Sci. Technol.*, **2003**, *37*, 160 A.

49.   H. S. Kheshigi, R. C. Prince and G. Marland, The potential of biomass fuels in the context of global climate change: Focus on transportation needs, *Annu. Rev. Energy. Environ.*, **2002**, *25*, 199.

50.   L. R. Lynd, Overview and evaluation of fuel ethanol from cellolosic biomass: Technology, economics, the environment and policy, *Annu. Rev. Energy. Environ.*, **1996**, *21*, 403.

51.   a) A. Kendall, A. McDonald and A. Williams, The power of biomass, *Chem. Indust.*, **1997**, 5 May, 342. b) Y.-M. Choo and A.-N Ma, Plant power, *Chem. Indust.*, **2000**, 21 Aug., 530. c) I. M. Arbon, Worldwide use of biomass in power generation and combined heat and power schemes, *Proc. Inst. Mech. Eng. A: Power and Energy*, **2002**, 216, 41. d) R. Newton, Biofuels are the future, *Chem. Indust.*, **2003**, 2 June, 14. e) H. R. Bungay, Confessions of a bioenergy advocate, *Trends Biotechnol.*, **2004**, *22*, 67. f) L. T. Angenent *et al.*, Production of bioenergy and biochemicals from industrial and agricultural wastewater, *ibid.*, 477. g) J. R. Rostrup-Nielsen, Making fuels from biomass, *Science*, **2005**, *308*, 1421. h) G. W. Huber *et al.*, Production of liquid alkanes by aqueous-phase processing of biomass derived carbohydrates, *ibid.*, 1446.

52.   J. Goldemberg *et al.*, How adequate policies can push renewables, *Energy Policy*, **2004**, *32*, 1141.

53.   C. J. Winter and J. Nitsch, eds., *Hydrogen as an Energy Carrier*, Springer-Verlag, New York, USA, 1988.

54.   a) J. M. Ogden, Prospects for building a hydrogen energy infrastructure, *Annu. Rev. Energy Environ.*, **1999**, *24*, 227. b) S. Dunn, Hydrogen futures: toward a sustainable energy system, *Int. J. Hydrogen Energy.*, **2002**, *27*, 235. c) C. C. Elam *et al.*, Realizing the hydrogen future: the International Energy Agency's efforts to advance hydrogen energy technologies, *Int. J. Hydrogen Energy*, **2003**, *28*, 601. d) R. Winder, Hydrogen fuels, *Chem. Indust.*, **2003**, 17 Feb., 19. e) M. Z. Jacobson, W. G. Colella and D. M. Golden, Cleaning the air and improving health with hydrogen fuel-cell vehicles, *Science*, **2005**, *308*, 1901.

55.   P. M. Grant, Hydrogen lifts off – with a heavy load, *Nature*, **2003**, *424*, 129.

56.   a) M. Schrope, Which way to energy utopia? *Nature*, **2001**, *414*, 682. b) D. W. Keith and A. E. Farrell, Rethinking hydrogen cars, *Science*, **2003**, *301*, 315. c) S. Ashley, On the road to fuel-cell cars, *Sci. Am.*, **2005**, *292*, March, 50.

57.   L. D. Burns, J. B. McCormick and C. E. Borroni-Bird, Vehicle of change, *Sci. Am.*, **2002**, *286*, Oct., 40.

58.   a) L. Schlapbach and A. Züttel, Hydrogen-storage materials for mobile applications, *Nature*, **2001**, *414*, 353. b) E. Willcocks, Fuel cells go mobile, *Chem. Br.*, **2003**, Jan., 27. c)

M. D. Ward, Molecular fuel tanks, *Science*, **2003**, *300*, 1104. d) N. L. Rossi *et al.*, Hydrogen storage in microporous metal-organic frameworks, *ibid,*, 1127. e) M. Jacoby, Filling up with hydrogen, *Chem. Eng. News*, **2005**, *83*, 22 Aug., 42.

59.   B. McEnaney, Go further with H_2, *Chem. Brit.*, **2003**, Jan., 24.

60.   P. C. Novelli *et al.*, Molecular hydrogen in the troposphere: global distribution and budget. *J. Geophys. Res.*, **1999**, *104*, 30427.

61.   a) T. K. Tromp *et al.*, Potential environmental impact of a hydrogen economy on the stratosphere, *Science*, **2003**, *300*, 1740. b) M. J. Prather, An environmental experiment with H_2? *Ibid.*, 581. c) M. G. Schultz, T. Diehl. G. P. Brasseur and W. Zittel, Air pollution and climate-forcing impacts of a global hydrogen economy, *ibid.*, 624. d) A. Ananthaswamy, Reality bites for the dream of a hydrogen economy, *New Sci.*, **2003**, 15 Nov., 6.

62.   H. Blix, The dual challenge of a nuclear age, *IAEA Bull.*, **1993**, *1*, 33.

63.   J. Johnson, New life for nuclear power, *Chem. Eng. News*, **2000**, *78*, 2 Oct., 39.

64.   IAEA, International datafile, *IAEA Bull.*, **2002**, *44* (2), 59.

65.   W. Häfel, Energy from nuclear power, *Sci. Am.*, **1990**, *275*, Sept., 91.

66.   The Royal Society. *Nuclear Energy-The Future Climate*, The Royal Society. Document 11/99, London, UK, 1999.

67.   P. W. Beck, Nuclear energy in the twenty-first century: Examination of a contentious subject, *Annu. Rev. Energy Environ.*, **1999**, *24*, 113.

68.   A. Borovoi and S. Bogatov, Consequences of Chernobyl. A view ten years on, *Adv. Nucl. Sci. Technol.*, **1997**, *25*, 171.

69.   a) W. C. Sailor *et al.*, A nuclear solution to climate change, *Science*, **2000**, *288*, 1177. b) D. Chandler, America steels itself to take the nuclear plunge, *New Sci.*, **2003**, 9 Aug., 10. c) M. El Baradei, Nuclear power. An evolving scenario, *IAEA Bull.*, **2004**, *46* (1) 4.

70.   a) C. W. Forsberg and A. M. Weinberg, Advanced reactors, passive safety, and acceptance of nuclear energy, *Annu. Rev. Energy Environ.*, **1990**, *15*, 133. b) M. W. Golay and N. E Todreas, Advanced light-water reactors, *Sci. Am.*, **1990**, 262, April, 58. c) J. A. Lake *et al.*, Next-generation nuclear power, *Sci. Am.*, **2002**, *286*, Jan., 71. d) D. Butler, Energy: Nuclear power's new dream, *Nature*, **2004**, *429*, 238. e) E. P. Loewen, Heavy-metal nuclear power, *Am. Sci.*, **2004**, *92*, 522.

71.   D. Sang, Atoms unleashed, *New Sci.*, *Inside Science 157*, **2003**, 18 Jan., 1.

72.   a) J. Flynn and P. Slovic, Yucca Mountain: A crisis for policy: Prospects for America's high-level nuclear waste program, *Annu. Rev. Energy Environ.*, **1995**, *20*, 83. b) C. G. Whipple, Can nuclear waste be stored safely at Yucca Mountain? *Sci. Am.*, **1996**, *274*, June, 56. c) J. Flynn *et al.*, Overcoming tunnel vision: Redirecting the U.S. high-level nuclear waste program, *Environment*, **1997**, *39*, April, 6. d) P. P. Craig, High-level nuclear waste: The status of Yucca Mountain, *Annu. Rev. Energy. Environ.*, **1999**, 24, 461. e) R. C. Ewing and A. Macfarlane, Yucca Mountain, *Science*, **2002**, *296*, 659. f) C. Macilwain, Out of sight out of mind? *Nature*, **2001**, *412*, 850. g) J. Johnson, Yucca Mountain, *Chem. Eng. News*, **2002**, *80*, 8 June, 20.

73.   a) S. Banwart, P. Wikberg and O. Olsson, A testbed for underground nuclear repository design, *Environ. Sci. Technol.*, **1997**, *31*, 510 A. b) R. Stone, Deep repositories: out of sight, out of terrorists' reach, *Science*, **2004**, *303*, 161. c) C. Thegerström, Down to earth and below. Sweden's plans for nuclear waste, *IAEA Bull.*, **2004**, *46* (1) 36.

74.   a) N. Precoda, Radionuclides in Russia's resource-rich Arctic regions, *Ambio*, **1995**, *24*, 316. b) R. Edwards, Hot ice could contaminate fish, *New Sci.*, **1996**, 2 March, 7. c) D. Schneider, Not in my backyard, *Sci. Am.*, **1997**, *276*, March, 14.

75.   a) R. Edwards, Russia's toxic shocker, *New Sci.*, 6 Dec., **1997**. b) R. Stone, Retracing Mayak's radioactive cloud, *Science*, **1999**, *283*, 164.

76.   a) G. Zorpette, Hanford's nuclear wasteland, *Sci. Am.*, **1996**, *274*, May, 72. b) R. Renner, U.S. Nuclear cleanup shows signs of progress, *Environ. Sci. Technol.*, **1997**, *31*, 134 A. c) J. Johnson, Nuclear waste cleanup stirs, *Chem. Eng. News*, **1998**, *76*, June, 26. d) J. A. MacDonald, Cleaning up the nuclear weapons complex, *Environ. Sci. Technol.*, **1999**, *33*, 314 A.

77. a) R. Edwards, Irish concern over safety of Sellafield waste tanks, *New Sci.*, **1996**, 15 June, 6. b) F. Pearce, Sellafield leaves its mark on the frozen north, *New Sci.*, **1997**, 10 May, 14. c) R. Edwards, High-level risk, *New Sci.*, **1998**, 20 June, 13. d) R. Edwards, Dirty business, *New Sci.*, **2000**, 9 Dec., 2000.

78. a) K.-L. Sjöblom and G. Lindsley, Sea disposal of radioactive wastes: the London Convention 1972, *IAEA Bull*, **1994**, *36* (2), 13. b) C. D. Hollister and S. Nadis, Burial of radioactive waste under the seabed, *Sci. Am.*, **1998**, *278*, Jan., 60.

79. F. von Hippel and S. Jones, The slow death of the fast breeder, *Bull. Atomic Scientists*, **1997**, Sept./Oct., 46.

80. G. Zorpette, Return of the breeder, *Sci. Am.*, **1996**, *274*, Jan., 34.

81. a) J. A. Lake, R. G. Bennett and J. F. Kotek, Next-generation nuclear power, *Sci. Am.*, **2002**, *286*, Jan., 70. b) M. S. Kazami, Thorium fuel for nuclear energy, *Am. Sci.*, **2003**, *91*, 408.

82. a) D. Bachtold, Britain to cut CO_2 without relying on nuclear power, *Science*, **2003**, *299*, 1291. b) R. Edwards, Britain's global energy vision, *New Sci.*, **2003**, 1 March, 12.

83. H. Muir, Celestial Fire, *New Sci.*, *Inside Science 161*, **2003**, 21 June, 1.

84. a) U. Colombo and U. Farinelli, Progress in Fusion Energy, *Annu. Rev. Energy Environ.*, **1992**, 17, 123. b) D. Ward, Fusion power – energy for the future? *Energy World*, **2000**, March, 15.

85. a) R. W. Conn *et al.*, The international thermonuclear experimental reactor, *Sci. Am.*, **1992**, *266*, April, 75. b) H. P. Furthe, Fusion, *Sci. Am.* **1995**, *273*, 140. c) G. Brunfield, Just around the corner, *Nature*, **2005**, *436*, 318.

86. a) D. Clery and D. Normile, ITER rivals agree to terms; site said to be Cadarache, *Science*, **2005**, *308*, 934. b) D. Butler, Japan consoled with contracts as France snares fusion project, *Nature*, **2005**, *435*, 1142.

87. a) F. Pearce, The biggest dam in the world, *New Sci.*, **1995**, 28 Jan., 25. b) C. Kwai-cheong, The Three Gorges Project of China: Resettlement prospects and problems, *Ambio*, **1995**, *24*, 98.

88. World Commission on Dams, Dams and Development. *A New Framework for Decision Making*, EarthScan, London, UK, 2000.

89. a) J. E. Mock, J. W. Tester and P. M. Wright, Geothermal energy from the Earth: Its potential impact as an environmentally sustainable resource, *Annu. Rev. Energy Environ.*, **1997**, *22*, 305. b) P. M. Wright, Geothermal energy – harnessing heat from the center of the Earth, *Chem. Indust.*, **1998**, 16, March, 208. c) N. Fell, Deep heat, *New Sci.*, **2003**, 22 Feb., 40.

90. a) C. J. Weinberg and R. H. Williams, Energy from the sun, *Sci. Am.*, **1990**, 263, Sept., 99. b) I. Dostrovsky, Chemical fuels from the sun, *Sci. Am.*, **1991**, *265*, Dec., 102. c) W. Hoagland, Solar energy, *Sci. Am.*, **1995**, *273*, Sept., 136. d) D. Anderson, Renewable energy technology and policy for development, *Annu. Rev. Energy Environ.*, **1997**, *22*, 187. e) M. Hammonds, Getting power from the sun, *Chem. Indust.*, **1998**, *16*, March, 219. f) J. Wolfe, Capitalising on the sun, *ibid.*, 224. g) J. Johnson, U.S. Photovoltaic exports jump, *Chem. Eng. News*, **2000**, *78*, 9 Oct., 45. h) A. Steinfeld and M. Epstein, Light years ahead, *Chem. Br.*, **2001**, 37, May, 30. i) R. F. Service, Is it time to shoot for the sun? *Science*, **2005**, *309*, 548.

91. a) *Time*, Tower of power, **2002**, 18 November, 78. b) M. Burke, Australia eyes a piece of the sun, *Env. Sci. Technol.*, **2002**, *36*, 403 A. c) M. Murphy, Out of thin air, *Chem. Indust.*, **2003**, 17 March, 14. d) R. Nowak, Power tower, *New Sci.*, **2004**, 31 July, 42.

92. a) D. Milborrow, Catching the breeze, *Chem. Indust.*, **1998**, 16 March, 214. b) J. G. McGowan and S. R. Connors, Windpower: A turn of the century review, *Ann, Rev. Energy Environ*, **2000**, 25, 147. c) K. S. Betts, U.S. begins to explore offshore wind energy, *Env. Sci. Technol.*, **2001**, 35, 358 A. d) D. Milborrow, Wind energy technology – the state of the art, *Proc. Inst. Mech. Eng., A*, **2002**, *216*, *A: Power and Energy*, 23. e) J. Johnson, Blowing green, *Chem. Eng. News*, **2003**, *81*, 24 Feb., 27. f) J. Randerson, Turbines get stealthy, *New*

Sci., **2003**, 9 Aug., 6. g) N. Fleming, Crunch time looms for offshore wind power, *New Sci.*, **2003**, 6 Dec., 30. h) J. Knight, Breezing into town, *Nature*, **2004**, *430*, 12.

93.   a) F. Pearce, Catching the tide, *New Sci.*, **1998**, 20 June, 38. b) P. L. Fraenkel, Power from marine currents, *Proc. Inst. Mech. Eng., A: Power and Energy*, **2003**, *216*, 1. c) T. Setoguchi *et al.*, A performance study of a radial turbine for wave energy conversion, *ibid.*, 15. d) J. Johnson, Power from moving water, *Chem. Eng. News*, **2004**, 4 Oct, 23.

94.   J. Hogan and P. Cohen, Is the green dream doomed to fail? *New Sci.*, **2004**, 17 July, 6.

95.   a) K. Hasselmann *et al.*, The challenge of long-term climate change, *Science*, **2003**, *302*, 1923. b) T. M. L. Wigley, The climate change commitment, *Science*, **2005**, *307*, 1766. c) G. A. Meehl *et al.*, How much more global warming and sea level rise? *Ibid.*, 1769.

96.   J. Randerson, The clean green energy dream, *New Sci.*, **2003**, 16 Aug., 8.

97.   *The Cost of Generating Electricity*, The Royal Academy of Engineering, London, UK, 2004, (http:/www.raeng.org.uk/news/temp/cosy_generation.report.pdf)

98.   S. Pacala and R. Socolow, Stabilization wedges: solving the climate problem for the next 50 years with current technologies, *Science*, **2004**, *305*, 968.

Chapter 8

Greenhouse Earth and Climate Change

"Human beings have in recent years discovered that they may have succeeded in achieving a momentous but rather unwanted accomplishment. Because of our numbers and our technology, it now seems likely that we have begun altering the climate of our planet."

T. R. Karl, N. Nicholls and J. Gregory in Scientific American, May, 1997.

8.1. The Greenhouse Effect: Earth's Solar Energy Balance

Of all of the factors that affect life on Earth, climate is one of the most influential. Yet, unless a particularly disturbing event occurs such as a flood, a drought, a hurricane or an ice storm, climate tends to be very much taken for granted. However, when the cycle of the seasons that provides humanity with sustenance and replenishes the landmasses with fresh water as rain and snow is considered it becomes clear that the origin of this largesse warrants closer study. A good starting point is a comparison of the equable climate and the 15°C average surface temperature of Earth with conditions on the airless moon that has an average surface temperature of -18°C despite being the same distance from the sun. This difference arises because Earth has a thin layer of air, the atmosphere, that retains warmth from the sun and provides insulation from the cold of space [1,2]. This has been likened to the effect of a greenhouse in retaining warmth and is often referred to as the "natural greenhouse effect". Despite its name, this effect is different from that of a garden greenhouse where the glass lets in sunlight and prevents the warmed air from escaping. Earth's natural greenhouse effect is substantially more complex and has operated for several billions years. It arises predominantly from absorption of energy as infrared radiation by the greenhouse gases water, carbon dioxide and methane in the lower part of the atmosphere, the troposphere. These gases are minor components of dry air at ground level that by volume is 78.09 percent nitrogen, N_2, 20.94 percent oxygen, O_2, 0.93 percent argon, Ar, 0.038 percent carbon dioxide, CO_2, and 0.00017 percent methane, CH_4. However, in the atmosphere these amounts are slightly decreased by the presence of water vapour, H_2O, that varies between 0.5 and 2 percent depending upon local conditions and seasonal variations. (Nitrogen and oxygen are often referred to as dinitrogen and dioxygen because they are diatomic.)

To understand Earth's natural greenhouse effect the role of the sun must first be considered. The sun has a surface temperature of 5,700°C and emits energy as light that ranges from short wavelength high energy ultraviolet radiation at about 0.1 micrometres wavelength to the lower energy infrared radiation at about 3.5 micrometres wavelength as seen in Fig. 8.1. Much of the sunlight falls in the visible wavelength range between 0.4 and 0.7 micrometres and the sun appears as a white star. In accord with the wavelength of radiation increasing as temperature decreases, Earth emits its energy as infrared radiation at longer wavelengths dominantly in the 4.0 to 60 micrometres range that is often referred

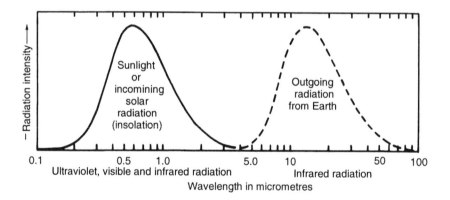

Fig. 8.1. The wavelengths of radiation emitted by the sun are consistent with a surface temperature of 5,700°C, and those emitted by Earth are consistent with a temperature of -18°C. The smooth curve shown for Earth's outgoing radiation does not take into account the natural greenhouse effect. Adapted from [1].

to as thermal infrared radiation as it is felt as heat. This radiation is invisible to the human eye and is consistent with Earth behaving as if its temperature is -18°C, the air temperature at about six kilometres altitude in the mid latitudes. The beautiful colours of Earth seen in pictures from space arise from reflected sunlight, not from radiation emitted by Earth. This becomes clear from pictures of the night side of Earth where the only light seen arises from lightening and city illuminations [3].

The average global surface temperature of Earth is maintained at 15°C by the natural greenhouse effect. Clearly, if the amount of energy arriving as sunlight was simply matched by the combination of sunlight reflected back into space and the energy emitted by Earth as longer wavelength infrared radiation the surface temperature would be similar to that of the moon. To understand why this is not so the role of the atmosphere and the natural greenhouse effect in maintaining Earth's solar energy balance has to be taken into account as shown in Fig. 8.2 [4,5]. The sunlight, or incoming solar radiation, that enters the atmosphere

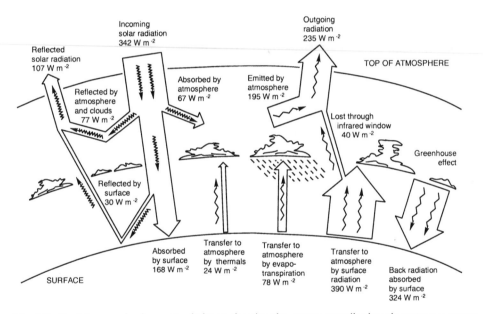

Fig. 8.2. Earth's annual solar energy balance showing the energy contributions in watts per square metre, W m^{-2}. Adapted from [4].

and the energy released back into space from Earth is measured in watts per square metre (W m^{-2}). (Incoming solar radiation is often referred to as insolation.) This measure of energy is readily understood when it is recalled that domestic electrical heaters are often labelled as being either 1,000 watt or 2,000 watt heaters and so on. If all of the heat produced by a 1,000 watt heater falls on an area of one square metre the heat produced is equivalent to 1,000 W m^{-2} over the length of time the heater is on. The averaged annual amount of incoming solar radiation reaching Earth is 342 W m^{-2}. Of this, 77 W m^{-2} is reflected back into space by the atmosphere and the clouds, aerosols, dust and smoke floating in it. Earth's surface reflects a further 30 W m^{-2} of the sunlight so that a total of 31.3 percent of sunlight, or 107 W m^{-2}, is reflected back into space; a reflecting action referred to as Earth's albedo. This leaves 19.6 percent, or 67 W m^{-2}, of the incoming solar radiation to be absorbed by the atmosphere and the residual 49.1 percent, 168 W m^{-2}, to be absorbed at the surface where part of it powers photosynthesis in green plants. Of this solar energy absorbed by the atmosphere and the surface, 235 W m^{-2} escapes back into space as infrared radiation so that, together with the reflected sunlight, it balances the energy from the sun and Earth maintains a steady temperature. However, a major part of the infrared radiation released from the surface is recycled in the atmosphere and is returned to the surface through the natural greenhouse effect that maintains the average global surface temperature of Earth at about 33°C warmer than that of the moon.

The escaping 235 W m^{-2} of infrared radiation is much less than the total amount of energy entering the atmosphere. A substantial 390 W m^{-2} of infrared radiation is emitted directly from the surface, 78 W m^{-2} is released into the atmosphere when the water vapour transpired by plants and from evaporation condenses to form rain and snow and 24 W m^{-2} escapes the surface in thermals that, when added to the 67 W m^{-2} entering the atmosphere directly as sunlight, amounts to a total of 559 W m^{-2} entering the atmosphere. Of this energy, 324 W m^{-2} is absorbed by the greenhouse gases that either retain it in the atmosphere or re-radiate it to the surface while 235 W m^{-2} is radiated back into space, 40 W m^{-2} of which passes through the "infrared window" in the wavelength range in which the greenhouses gases do not absorb infrared radiation. Other slightly different estimates of Earth's solar energy balance have also appeared but the broad picture remains the same [6].

8.2. The Greenhouse Gases

The natural greenhouse effect is largely a consequence of the major greenhouse gases in the troposphere, water, carbon dioxide and methane, absorbing much of Earth's emitted infrared radiation. This occurs because the atoms in these molecules are joined by chemical bonds that may be likened to springs linking the atoms together. When stretched and released, strong springs vibrate quickly while weak springs vibrate slowly. The chemical bonds in molecules behave in a similar way so that if there is a strong bond between atoms in a molecule it vibrates quickly while a weak bond vibrates slowly. Because these greenhouse gases consist of three or more linked atoms they act as a collection of vibrating chemical bonds and the angle subtended by three of their linked atoms rapidly increases and decreases so that their shapes change continuously. Consequently, a molecular property of the greenhouse gas molecules, called a dipole moment, that changes at the same frequencies at which the bonds vibrate and their shapes change enables them to absorb infrared radiation. In contrast, diatomic nitrogen and oxygen cannot absorb infrared radiation because they have no dipole moments to be either generated or changed by their bond vibrations and therefore are not greenhouse gases. It is for this reason that the low levels of water, carbon dioxide, methane and other greenhouse gases in the atmosphere assume such importance in maintaining Earth's temperature. (The wavelength of infrared radiation is equal to the speed of light divided by the infrared frequency. Thus, high frequencies correspond to short wavelengths and *vice versa* so that rapidly vibrating strong chemical bonds absorb shorter wavelength infrared radiation and slowly vibrating weaker bonds absorb longer wavelength infrared radiation.)

In the troposphere, the greenhouse gas molecules absorb a large portion of the infrared radiation emitted from Earth's surface as seen in Figure 8.2. Through collisions, they transfer some of the energy from this radiation to the plentiful diatomic nitrogen and oxygen molecules that rebound vibrating and tumbling to a greater extent than before. Further collisions cause transfer of this energy to other molecules and so on. The greenhouse gas molecules also re-radiate infrared energy to Earth's surface. These interactions effectively store a major part of Earth's emitted infrared energy that warms the troposphere and Earth's surface. This is the molecular basis of the natural greenhouse effect and is the reason for calling water, carbon dioxide and methane, and other gases that have a similar warming effect, greenhouse gases. Because of the abilities of the these gases to absorb infrared radiation, the amount of Earth's emitted infrared radiation does not follow the smooth curve that might have been anticipated from Fig. 8.1. Instead, as much of the infrared radiation is absorbed by the greenhouse gases, the residual amount leaving Earth is equal to the area under the dog-eared curve shown in Fig. 8.3 [2]. This dog-eared curve may be looked upon as Earth's infrared signature. The area between the smooth curve shown by the broken line and the dog-eared curve represents the amount of infrared radiation retained by the atmosphere through the natural greenhouse effect.

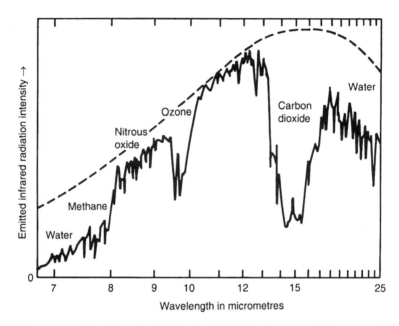

Fig. 8.3. The amount of Earth's emitted infrared radiation lost into space is the area under the dog-eared curve. The amount that would be lost if the greenhouse gases were absent is the amount under the dashed curve. The major wavelength ranges in which the various greenhouse gases absorb infrared radiation are labelled. Adapted from [2].

Carbon dioxide absorbs infrared radiation particularly strongly between the wavelengths of 12 to 18 micrometres as is seen from Fig. 8.3. Water is the most abundant greenhouse gas, although molecule for molecule it is less effective than carbon dioxide. It absorbs infrared radiation strongly between the wavelengths 4.5 to 8.5 micrometres and 18 micrometers and longer wavelengths. Three other naturally occurring greenhouse gases, methane, nitrous oxide (N_2O) and ozone (O_3), also absorb some of Earth's emitted infrared radiation. However, there remains a "window" where little of Earth's emitted infrared radiation is absorbed in the wavelength ranges 8.0 to 9.5 micrometres and 10 to 13 micrometres. If either the levels of water, carbon dioxide, methane and nitrous oxide increase, or new greenhouse gases that absorb some of the infrared radiation passing through the infrared window enter the atmosphere, such as either the chlorofluorocarbons or the hydrochlorofluorocarbons and hydrofluorocarbons that are replacing them (see Chapter 9), the troposphere would be expected to warm because of an "enhanced greenhouse effect". This is what drives the current debate on global warming and climate change [7-10].

8.3. The Enhanced Greenhouse Effect

If greenhouse gases levels increase the capacity of the troposphere to retain heat will also increase. However, precise climatic consequences are difficult to predict because of the complexity of the interactions that control Earth's climate and tropospheric and surface temperatures. Nevertheless, it is well established that the levels of carbon dioxide and other greenhouse gases in the atmosphere have increased substantially in the last fifty years or so. This has resulted in increased global temperatures as a manifestation of an "enhanced greenhouse effect" and discernable changes in climate and rises in sea level. Given the large and increasing population, much of which lives in low lying coastal areas, sea level rises of tens of centimetres or more will cause distress through large scale flooding, loss of arable land and all of the attendant consequences.

This realization of the potential effects of an enhanced greenhouse effect has caused extensive debate over several decades and has lead to major internationally sponsored studies. In particular, the Intergovernmental Panel on Climate Change (IPCC) was jointly established by the World Meteorological Organization and the United Nations Environment Program in 1988 to assess the extent of human induced, or anthropogenic, climate change occurring, to assess the environmental and socioeconomic impacts of climate change and to formulate response strategies. It is clear from the massive IPCC reports that the extent of anthropogenic climate change is significant and that the agricultural, economic, health and sociological effects of this change will be variable and in many cases, damaging [4,5].

The biosphere, in which humanity not only resides but is also part of, is extremely complex with all of its facets interacting in diverse ways through feedback mechanisms. Despite many sophisticated attempts at modelling the consequences of an enhanced greenhouse effect, a degree of guessing is still involved because understanding of all of the factors affecting climate and their interactions is incomplete [7,11-13]. This is exemplified by two attempts in 2000 to predict soil moisture changes in the United States by 2100 as a consequence of global warming [14]. One model predicted drier soil, the other wetter soil. Because of such contradictions in climate observations and predictions some polarization of views about the likely future of Earth's climate has arisen [15]. Some of the uncertainty in understanding global temperature changes was caused by satellite temperature measurements that appeared to show that the troposphere at about 3.5 kilometres altitude was cooling despite the expectation that it should be warming as increasing greenhouse gas levels trapped more infrared radiation. This apparent contradiction arose because the temperature measuring satellites orbiting at 800-870 kilometres altitude slowly lost twenty kilometres of altitude over seventeen years and this was not taken into account in the temperature measurements so that the lower middle troposphere appeared to be cooling contrary to expectation if global warming was occurring [16]. When this and other refinements of satellite data were considered the lower middle troposphere was found to have been warming at about $0.07°C$ per decade.

The concerns about increasing amounts of carbon dioxide and other greenhouse gases in the atmosphere and their potential to cause climate change are not new [17]. In 1827, the Frenchman Jean-Baptiste Fourier wrote about the warming effect of the atmosphere on ground temperatures. Similar discussions were published by the Irishman J. Tyndall in 1861 and by the American Samuel Langley in 1884. However, it was the Swedish Nobel Laureate Svante Arrhenius who first made quantitative estimates of the effect of changes in atmospheric carbon dioxide levels on temperature in 1896. His estimation of a temperature rise of $5-6°C$ for a doubling of carbon dioxide levels is similar to those reported to be likely by the IPCC in its 2001 assessment. The IPCC considers it probable that, unless a significant curbing of anthropogenic greenhouse gas emissions occurs, average global temperature will rise by $1.4-5.8°C$ by 2100 coupled to an increase in sea level of between nine to eighty-eight centimetres through a combination of the effects of the expansion of ocean water as it warms and increasing amounts of melt water from thawing glaciers and ice caps [4,5,8]. The wide range estimated for the global temperature rise reflects different assumptions about the rate of increase in fossil fuel usage and uncertainties in its contributions to global warming. (It is salutary to recall that since the depths of the last ice age nineteen to thirty thousand years ago some fifty million cubic metres of water from melting ice have flowed into the oceans to increase their level by about 130 metres [18].)

The known temperature rises are not uniform. Northwestern North America, the Beaufort Sea, the Siberian plateau, the Antarctic Peninsula and the Bellinghausen Sea have warmed to the greatest extent so far [19]. According to one analysis, by 2100 Canada and the Russian Federation can expect temperature rises of more than 6°C; China, several Central Asian, Middle Eastern, northwestern and sub-Saharan African nations 5.0-5.9°C; the United States, northern South America, most of Europe and Scandinavia, much of the remaining African states, India and Australia 4.0-4.9°C; Greenland, Southeast Asia, Japan, Argentina and Peru 3.0-3.9°C; and Ireland, New Zealand and Uruguay 2.9°C or less [20]. A number of Middle Eastern, central Asian, northwestern and sub-Saharan African nations that already regularly experience temperatures above 40°C, and frequently suffer drought, will be particularly hard hit by some of the largest temperature rises despite their production of greenhouse gases being among the smallest. In the colder northern continental regions of Earth rising temperatures will result in increased permafrost melting and consequent decay of previously frozen buried plant matter and release of carbon dioxide and methane, settling of buildings and coastlines, and flooding [21].

Already, some extreme weather events are being interpreted in terms of global warming. This was exemplified by the heatwave that occurred during 1-15 August in the European summer of 2003 that was the hottest since the commencement of temperature records in 1851, particularly in central and southwestern Europe [22,23]. One of the effects of this heatwave was a substantial increase in death rate in the German state of Baden-Würtemberg amounting to 900–1,300 deaths beyond those expected under normal climatic conditions in a population of 10.7 million as is shown in Fig. 8.4. In France, the death rate increased by fifty-four percent nation-wide, or 14,000 deaths beyond those expected during the period of the heatwave. In Portugal, crop losses generally amounted to US$12.3 billion and forest fires attributed to the heatwave caused losses of US1.6 billion.

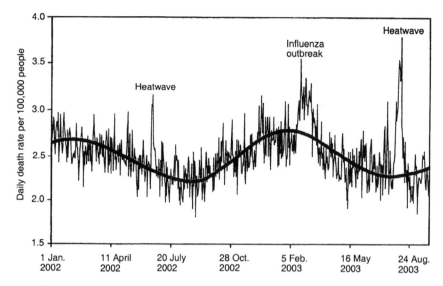

Fig. 8.4. Seasonal variation of the death rate in Baden-Würtemberg showing the normal higher mortality in winter. Superimposed on this are the increased death rates arising from a small heatwave in June 2002 and a much larger one in August 2003. An influenza outbreak in February-March 2003 caused a substantial increase in death rate. Adapted from [23].

Heatwaves and their effects on health and crops have been quite common phenomena throughout humanity's existence but now the fear is that their frequency and duration will increase with global warming. The cause of the 2003 European heatwave was a persistent anticyclone over northwest Europe. Anticyclones are common meteorological occurrences and it is impossible to prove that this particular anticyclone and the resulting heatwave were a consequence of global warming. However, one 2004 analysis of European climate showed that anthropogenic global warming had increased the likelihood of an anticyclone causing a heatwave similar to that of 2003 fourfold with a ninety percent probability that more than half the risk of such a heatwave occurring arose from this cause [24]. Another analysis of Europe and North America showed that heatwaves were likely to become more intense, more frequent and of greater duration in the second half of the twenty-first century as a consequence of anthropogenic climate change [25].

While attempts to separate the human impact on climate from natural variability are important in understanding the changing climate, they are taking on another significance as exemplified by eight American states and New York

City initiating lawsuits in 2004 to force five power companies to reduce their carbon dioxide emissions because of the deleterious climate change associated with them [22]. It appears inevitable that more such lawsuits and corresponding legislation to reduce greenhouse gas emissions will follow as the actual and potential damaging effects of climate change are increasingly recognised.

8.4. Increasing Carbon Dioxide and Other Greenhouse Gas Levels

Continuous monitoring of carbon dioxide, methane and nitrous oxide over several decades has shown that the levels of these greenhouse gases in the atmosphere are steadily rising [26]. The gathering of the two impressive sets of measurements of atmospheric carbon dioxide levels shown in Fig. 8.5

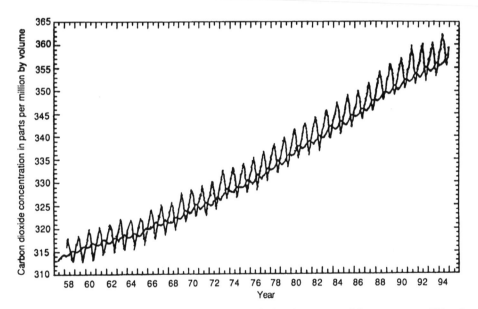

Fig. 8.5. Variations of atmospheric carbon dioxide levels expressed in parts per million by volume. The curve showing the large annual fluctuations represents measurements made at Mauna Loa, Hawaii, and the other line represents those made at the South Pole. Adapted from [26].

started in the International Geophysical Year 1957-1958. The first set o f measurements from the observatory 3,400 metres high on Mauna Loa, Hawaii, show a steady increase in carbon dioxide from 1957 onwards, and is sometimes referred to as the Keeling record after the atmospheric scientist C. D. Keeling who initiated the measurements. Superimposed on the steady increase in atmospheric carbon dioxide levels are annual fluctuations reflecting the northern seasonal cycle. Thus, when photosynthesis reaches a maximum in summer the carbon dioxide level falls to a low point and when photosynthesis diminishes in winter, and microbial breakdown of fallen leaves releases carbon dioxide, the level rises to a high point. The second set of measurements from the South Pole also shows a steady rise in carbon dioxide level but the annual fluctuation is six months out of sequence with that seen for the Mauna Loa measurements because the southern summer coincides with the northern winter and *vice versa*. The annual fluctuations in the South Pole carbon dioxide levels are smaller because the bulk of Earth's landmasses is in the northern hemisphere as is its terrestrial vegetation.

The seasonally adjusted average of the two sets of carbon dioxide increases shown in Fig. 8.4 closely coincide with the increases in the rate at which carbon dioxide enters the atmosphere largely from fossil fuel combustion and to a lesser extent from cement production over the same period. While the production of carbon dioxide from fossil fuel combustion is readily understood its generation through cement production is perhaps not so obvious. Central to cement production is the heating of limestone that is dominantly calcium carbonate, $CaCO_3$, to produce quick lime, or calcium oxide, CaO, that liberates carbon dioxide to add to that produced by the combustion of the fossil fuel that produced the heating [27]. The same process occurs in volcanoes whose lavas are alkaline. Such a volcano is Mount Etna in Italy that belches twenty-five million tonnes of carbon dioxide into the atmosphere annually [28].

On a longer time scale than that shown in Fig. 8.5, the level of carbon dioxide in the atmosphere rose from a preindustrial level of about 280 parts per million by volume in 1865 to 380 parts per million by volume in 2004, an increase of thirty-five percent, and continues to grow. The carbon dioxide emissions for some of the countries with larger emissions in 2001 are shown in Fig. 8.6 [29]. The United States contributed about twenty-four percent of the annual carbon dioxide emissions, Europe twenty-six percent, China thirteen percent, the Russian Federation six percent, Japan five percent and India four percent. High per capita emissions characterize the developed nations, particularly the Australia, Canada and the United States, while for the developing nations, exemplified by China and India, the per capita emissions are relatively low. This pattern is changing as the developed nations become more energy efficient and the developing nations increase their fossil fuel use [30].

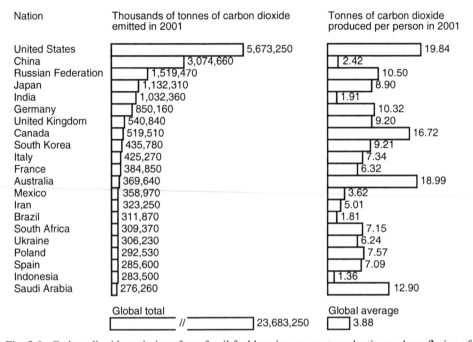

Nation	Thousands of tonnes of carbon dioxide emitted in 2001	Tonnes of carbon dioxide produced per person in 2001
United States	5,673,250	19.84
China	3,074,660	2.42
Russian Federation	1,519,470	10.50
Japan	1,132,310	8.90
India	1,032,360	1.91
Germany	850,160	10.32
United Kingdom	540,840	9.20
Canada	519,510	16.72
South Korea	435,780	9.21
Italy	425,270	7.34
France	384,850	6.32
Australia	369,640	18.99
Mexico	358,970	3.62
Iran	323,250	5.01
Brazil	311,870	1.81
South Africa	309,370	7.15
Ukraine	306,230	6.24
Poland	292,530	7.57
Spain	285,600	7.09
Indonesia	283,500	1.36
Saudi Arabia	276,260	12.90
Global total	23,683,250	Global average 3.88

Fig. 8.6. Carbon dioxide emissions from fossil fuel burning, cement production and gas flaring of some of the larger emitting nations in 2001 [29]. Gas flaring is the burning of gaseous by-products from oilfields and refineries.

The annual anthropogenic production of carbon dioxide is very small compared with the total amount circulating in the biosphere as shown in Fig. 8.7 [5,9,13,26,29-34]. Because of the global nature of the quantities assessed in Fig. 8.7, significant uncertainty exists in some of their magnitudes. This is particularly so in the assessments of some of the flows of carbon dioxide from one reservoir to another, and in some cases seeming discrepancies between the amount of carbon dioxide absorbed and the capacities of the reservoirs, or sinks, to absorb it have led to the postulation of "missing sinks". These uncertainties should be borne in mind in the discussion that follows.

Land plants use about 440 billion tonnes of carbon dioxide in photosynthesis of which half is returned to the atmosphere through plant respiration and the other half through decay. The oceans absorb about 370 billion tonnes of carbon

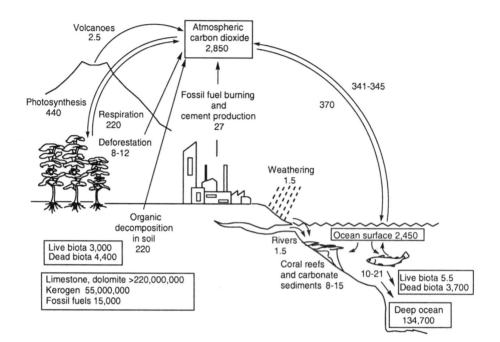

Fig. 8.7. Estimated annual circulation of carbon dioxide in billions of tonnes annually for 2004 shown unboxed. The boxed quantities refer to the carbon dioxide in the major carbon reservoirs after conversion to their corresponding amounts as carbon dioxide. The fossil fuels include oil shales and sands and are at the high end of estimates of total fossil fuels. The figure is mainly based on data in [5,9,13,26,29-34] with adjustments for increased carbon dioxide emissions up to 2004.

dioxide of which 341-345 billion tonnes are returned to the atmosphere leaving twenty-five to twenty-nine billion tonnes in the ocean. Of this, eight to fifteen billion tonnes are incorporated into carbonate sediments, coral reefs and other aquatic biota. The residual ten to twenty-one billion tonnes are retained in the deep ocean as carbon dioxide resulting mainly from the metabolising of dead plankton. Although humanity's contribution of carbon dioxide to the total circulation is relatively small, the biological and associated geological processes adjust too slowly to accommodate all of it. Thus, since 1865, the beginning of the industrial age, to 2004 the total amount of carbon dioxide in the atmosphere increased by thirty-five percent from 2,100 billion tonnes to 2,850 billion tonnes. This compares with 140,850 billion tonnes existing as dissolved carbon dioxide and carbonate salts and dead biota in the oceans. At present, atmospheric carbon dioxide is increasing at about 1.6 parts per million by volume, or twelve billion tonnes, annually. It is sobering to reflect that if all of the known reserves of fossil fuel were burnt, a further 15,000 billion tonnes or more of carbon dioxide would be released.

8.5. Biological Responses to Increasing Carbon Dioxide Levels

To some extent Earth redresses the carbon dioxide imbalance caused by humanity pouring twenty-seven billion tonnes of carbon dioxide into the atmosphere annually through fossil fuel combustion and cement production. Eight to twelve billion tonnes are added to this through topsoil erosion and deforestation and the resulting burning and decay of trees and other plants. However, only about twelve billion tonnes of this thirty-five to thirty-nine billion tonnes remain in the atmosphere [33]. Of the residual twenty-five to twenty-nine billion tonnes, some is incorporated into coral growth and the major portion is transported to the deep ocean largely as dead plankton that is metabolised to carbon dioxide [13] some of which forms carbonate sediments. (The role of the oceans is considered in more detail in section 8.13.)

A comprehensive global ground and satellite analysis of plant growth between 1982 and 1999 showed that the amount of carbon incorporated in plant growth increased by six percent, or 3.4 billion tonnes, equivalent to 12.5 billion tonnes of carbon dioxide (or 0.7 billion tonnes annually), with forty-two percent of this increase occurring in the Amazonian rainforests [34]. Direct regional measurements show that pristine Amazonian forests absorb more carbon dioxide through photosynthesis than they lose through respiration [35]. It is estimated that the Amazon basin probably absorbs two billion tonnes of carbon dioxide annually in this way and that South American forests as a whole removed much more during 1986-1987. Increased growth in the northern forests, most of which are in the Russian Federation and adjacent regions, was estimated to remove more than two billion tonnes of atmospheric carbon dioxide annually in the late 1980s and early 1990s [36]. Thus, it appears that despite massive deforestation in some regions, increased plant growth and reforestation is providing a significant compensation.

A combination of northern hemisphere carbon dioxide level seasonal variations and satellite monitoring of photosynthesis shows that the northern growing season increased in length by about twelve days over the period 1981-1991 [37]. This was particularly the case between 45°N and 70°N where in the higher latitudes there was a ten percent decrease in snow cover from 1973 to 1992 and an increasingly earlier disappearance of snow in spring. This coincided with a general warming in large areas of Alaska, northwestern Canada and northern Eurasia during the same period. Measurements of growth in European forests and tree ring analysis in Mongolia also indicate that photosynthetic activity has increased as atmospheric carbon dioxide levels have risen [38]. It is possible that this increased photosynthesis and growing season length arises from natural variations rather than an enhanced greenhouse effect, but it does show that small temperature variations on a global scale may be very significant in their effect on forest growth on a regional scale. However, the effect varies with the type of plant and geographic location as shown by the Alaskan white spruce for which annual growth rings collected for a ninety year period showed a decrease in width towards the end of the twentieth century as temperature rose [39]. This occurred because the rise in temperature was not matched by increased rainfall with the result that the spruce became stressed and growth decreased.

The overall increase in terrestrial photosynthesis may not arise solely from an increase in the length of the northern growing season. It seems probable that plants grow faster in carbon dioxide-rich environments provided that water and nutrients are adequate, an effect sometimes referred to as carbon dioxide fertilization. Green plants largely fall into two classes, C_3 and C_4, according to the way in which they convert carbon dioxide and water into carbohydrates. In C_3 plants the first molecule into which carbon dioxide is incorporated during photosynthesis contains three carbon atoms, while in C_4 plants it contains four carbon atoms. Almost all forest trees are C_3 plants as are many important crops such as wheat, rice, potatoes and beans. The C_4 plants include many grasses from the hot arid tropical and sub-tropical regions and also maize, sorghum and sugarcane. Under controlled carbon dioxide rich conditions both C_3 and C_4 plants show increased growth [40]. This, combined with an increase in the length of the growing season, could moderate the predicted increase in global temperatures as carbon dioxide and other greenhouse gas levels rise. It may also alter the balance between the populations of C_3 and C_4 plants whose metabolisms of carbon dioxide in photosynthesis differ.

Estimates of the amounts of carbon dioxide absorbed regionally and globally through photosynthesis vary significantly [41]. The reliability of these estimates has become controversial both scientifically and also in the international negotiations over the Kyoto Protocol that seeks to decrease the production of carbon dioxide and other greenhouse gases as is discussed in section 8.17 [42]. The amounts of carbon dioxide absorbed annually as carbon from the air over the United States are as shown in Fig. 8.8 and add up to 0.30-0.58 billion tonnes, or 1.1-2.1 billion tonnes of carbon dioxide, equivalent to four to eight percent of that produced from burning fossil fuels globally, according to a 2001 estimate [43]. It is seen that there is considerable uncertainty in these estimates of the amount of carbon dioxide sequestered by individual sinks. By comparison, seven to twelve percent of Europe's carbon dioxide emissions were absorbed through photosynthesis [44].

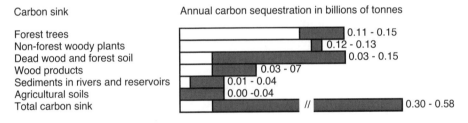

Carbon sink   Annual carbon sequestration in billions of tonnes

Forest trees   0.11 - 0.15
Non-forest woody plants   0.12 - 0.13
Dead wood and forest soil   0.03 - 0.15
Wood products   0.03 - 07
Sediments in rivers and reservoirs   0.01 - 0.04
Agricultural soils   0.00 -0.04
Total carbon sink   0.30 - 0.58

Fig. 8.8. Annual absorption of carbon by sinks in the United States excluding Alaska and Hawaii in the period 1980-1990 [43]. The shaded areas show the extent of uncertainty in the estimate of the size of each sink with the second value indicating the maximum size of each sink.

While negotiating the Kyoto Protocol, the United States and some other nations sought to offset the carbon dioxide emission reductions set for them against the amount of carbon dioxide absorbed by their forests. This was unacceptable to the majority of signatories to the protocol and, because of this and economic considerations, Australia and United States did not ratify it. In the short term it may be possible to absorb the increasing amounts of carbon dioxide that humanity pumps into the atmosphere by reforestation, but in the medium to longer term there seems little alternative to decreasing carbon dioxide emissions if the climate changes already evident are not to be exacerbated.

Apart from the direct biological responses to rising carbon dioxide levels, indirect responses are also likely. Thus, some modelling suggests that rainfall will increase substantially as the amount of atmospheric water vapour increases with global warming with increased risks of flooding and economic loss [45]. Coincidentally, increasing temperature and rainfall are expected to change breeding, growth and population patterns of land and oceanic animals and plants and the diseases that they are subject to [46]. However, the picture is complex as there is also evidence that the aerosols and other particulate matter produced by fossil fuel burning suppress rainfall over urban and industrial areas [47].

8.6. Rising Methane and Nitrous Oxide Levels

Much of the discussion of the enhanced green house effect centres on carbon dioxide because it represents a particularly large proportion of the greenhouse gases in the atmosphere and because its level is steadily rising. However the level of another greenhouse gas, methane, rose from 700 parts per billion by volume in preindustrial times to 1,721 parts per billion in 1994 and continues to rise [4,48]. Since 1970 there has been an annual increase of one percent in methane levels. Most of this methane is produced by microbes called methanogens that live in oxygen starved, or anaerobic, environments such as those found at the bottom of swamps and wet rice paddies, in animal wastes, sewage and landfills, and in termite and ruminant digestive tracts. This, together with deforestation, natural gas leakage, and gas, oil and coal production releases about 650 million tonnes of methane into the atmosphere annually. Increases in intensive farming and waste production also contribute to the rising methane levels to which some researchers have attributed a major part of global warming to [49].

Nitrous oxide, N_2O, levels have been increasing by about 0.3 percent per year, and its preindustrial level of 275 had increased to 311 parts per billion by volume in 1994 and have also continued to rise [4,5]. This greenhouse gas is generated by microbial action in the soil. The growing application of nitrogenous fertilizers increases the total amount of nitrogen in the soil and as a consequence the amount of nitrous oxide produced is also increased and is probably largely responsible for the rising nitrous oxide levels in the atmosphere.

The lifetime of carbon dioxide in the atmosphere is between fifty and two hundred years depending on where it is absorbed in the carbon cycle, that of methane is about twelve years and that of nitrous oxide is about 120 years [4]. If the global warming effect of each carbon dioxide molecule is defined as unity over a twenty-year period, those of methane and nitrous oxide are thirty-five and 260, respectively. However, because the levels of methane and nitrous oxide are much lower than that of carbon dioxide the latter greenhouse gas has the largest overall effect.

As the atmospheric levels of greenhouse gases increase the amount of infrared radiation absorbed by them will not increase proportionately. This is because Earth emits a fixed amount of infrared radiation at each wavelength. The physical law (the Beer-Lambert Law) governing the absorption of infrared radiation shows that at first the amount of infrared radiation absorbed increases linearly as the level of the greenhouse gas increases from a level of zero. However, as the proportion of infrared radiation absorbed grows the increase in the level of greenhouse gas required to double the amount absorbed becomes increasingly larger than the level required to absorb the first amount of infrared radiation. Thus, a doubling of the present level of carbon dioxide would only increase the amount of infrared radiation absorbed by forty percent and it would take four times the present carbon dioxide level to double the amount of infrared radiation absorbed. Similar considerations apply to methane and nitrous oxide.

The synthetic halocarbon greenhouse gases that include the chlorofluoro-carbons, their hydrochlorofluorocarbons and hydrofluorocarbon replacements, and a range of other carbon compounds containing one or more atoms of fluorine, chlorine, bromine and iodine are presently at very low atmospheric levels where a doubling of their levels would double the amount of infrared radiation they absorb. They also absorb infrared radiation in the infrared window where the major greenhouse gases do not absorb.

8.7. Clouds, Aerosols, Smoke and Soot

It is seen from the solar energy balance in Fig. 8.2 that clouds, aerosols, dust and smoke reflect sunlight back into space and decrease the amount of solar energy absorbed by the troposphere and Earth's surface. As a result Earth is cooler than would otherwise be the case and this cooling effect is often referred to as a negative climate forcing [4,50]. The warming effect of the greenhouse gases is referred to as a positive climate forcing. If the anthropogenic changes in climate forcings could all be either accurately determined or calculated it would be possible to more confidently assess their likely effects on global temperatures and climate. Such accuracy has yet to be achieved and is at the centre of much of the debate, research and climate modelling that has occurred over recent decades.

While the extra warming effect, or positive forcing, arising from the greenhouse gas absorption of Earth's emitted infrared radiation can be calculated with a high degree of accuracy, estimating the climate forcings for the other effects that accompany the production of carbon dioxide through fossil fuel burning is far less certain. Fossil fuels contain varying amounts of sulfur that is converted into sulfur dioxide (SO_2) when the fuel is burnt. In the atmosphere sulfur dioxide is converted to sulfur trioxide (SO_3) and then to sulfuric acid (H_2SO_4) that adds to the sulfuric acid aerosol already present as a consequence of volcanic activity and biological processes. Forest and grass fires and the burning of wood and grass as fuels, or biomass burning, also produce aerosols. Collectively, these aerosols increase the reflection of sunlight back into space and so have a cooling, or direct negative forcing, effect. In addition to this, aerosol particles affect the condensation of atmospheric water and the formation of clouds. Overall, the increased aerosol levels appear to produce greater cloud cover that reflects more sunlight in an indirect negative climate forcing and, coincidentally, may decrease rainfall [47]. Considerable uncertainty exists over the size of this indirect aerosol effect. Thus, clouds formed of water droplets appear to both reflect sunlight and emit infrared radiation into space from their upper surfaces whilst also absorbing infrared radiation [10,11,51]. However, ice clouds formed at higher altitude may absorb more infrared radiation than they emit and the sunlight they reflect and thereby have a warming effect. As the extent to which anthropogenic aerosols affect cloud formation is highly uncertain the inclusion of the impact of anthropogenic aerosols in the modelling of future climate results in similar uncertainty.

Soot is produced by incomplete combustion of fossil fuels and biomass and its level in the atmosphere has increased along with that of the anthropogenic aerosols. Soot contributes a negative forcing by decreasing the sunlight reaching the surface but also produces a positive climate forcing as a consequence of its infrared radiation absorbing properties. Again substantial uncertainty exists about the size of these effects that appear to be localized over the more highly industrialized areas of Earth. However, the effect of soot, smoke and aerosols can be visually very obvious as shown by the huge brown haze that cut out fifteen percent of sunlight over southern Asia in 2002. This was one of the more obvious pollution episodes that have contributed to the progressive decrease in sunlight by two to three percent and maximum summer temperature by 0.6°C each decade experienced by China since the mid 1950s as coal burning has increased [52]. Such decreases in the sunlight reaching Earth's surface through the combined effects of aerosols and soot are sometimes referred to as global dimming and fears have been expressed that their cooling effects will be lost as their levels decrease with the introduction of cleaner energy use [53,54].

It is possible to estimate the sum of the various anthropogenic global climate forcings, albeit with considerable uncertainity, as is shown in Fig. 8.9 [4]. The positive forcings of carbon dioxide, methane and nitrous oxide and the halocarbons are estimated with a high level of confidence. The negative forcing attributed to decreased levels of stratospheric ozone occurs because it is a greenhouse gas and its warming effect is proportionately decreased.

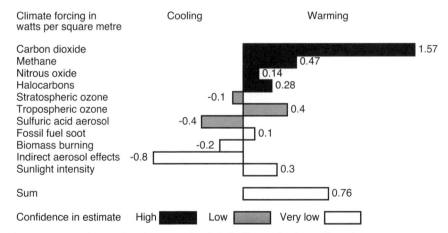

Fig. 8.9. Estimates of annual anthropogenic global climate forcing effects in watts per square metre from the beginning of the industrial revolution to 1992. Data from [4,9].

(Stratospheric ozone is discussed in more detail in Chapter 9.) Conversely, tropospheric ozone levels have increased due to human activities and have generated a positive forcing. The magnitudes of the latter two forcings, together with that for the sulfuric acid aerosol, are estimated with a low level of confidence. A very low level of confidence is attached to the estimates of the forcings from the soot and aerosols formed from fossil fuel and biomass burning. The indirect climate forcing effects of aerosols on the formation of clouds and the frequency of rain showers and a variety of other processes affecting temperature are so difficult to estimate that only the mid range value of the forcing is shown. The final contributor to changes in climate forcing is an increase in sunlight intensity that is independent of human activities.

The sum of the forcings shown in Fig. 8.9 is a positive annual forcing of 0.76 watts per square metre that would have a global warming effect only about thirty percent of that expected from the warming effects of the greenhouse gases alone. If the upper estimate of -1.5 watts per square metre is used for the indirect aerosol climate forcing, the additive effect of all of the climate forcings becomes zero. Although a very low level of confidence is attached to these summations, Fig. 8.9 clearly shows why anthropogenic effects on global temperature are less than those anticipated from the infrared absorbing properties of the greenhouse gases alone with the consequence that the global temperature rise since the industrial revolution is only 0.6°C.

The most abundant greenhouse gas, water, does not feature in Fig. 8.9 because approximately seventy percent of Earth's surface is covered by water and the variation of the natural atmospheric water vapour level is substantial and regional so that any apportioning of humanity's influences is almost impossible. However, as the atmosphere warms it contains more water vapour that further enhances greenhouse warming but may also increase cloud cover that reflects sunlight back into space.

8.8. Global Warming and Climate Change

In 1995 the IPCC stated that "the balance of evidence suggests that there is a discernible human influence on climate" and has expressed similar views since [4,5]. The variations in global temperature over the last 150 years shown in Fig. 8.10 indicate a rise in temperature of about 0.6°C during that time. This coincides with the increase in atmospheric carbon dioxide and greenhouse gas levels over the same period [55]. However, the small general temperature rise has been

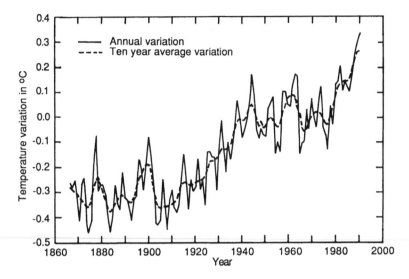

Fig. 8.10. Variations in global temperature shown as an annual (solid line) and a ten-year (broken line) average. Temperatures have been corrected for the El Niño effect, a change in Pacific weather patterns that lowers the average global temperature. Adapted from [55].

accompanied by temperature fluctuations that largely reflect natural climate variability arising from shifts in wind patterns and cloud cover. A report by the National Climatic Data Centre in the United States also showed that global surface temperatures gradually increased during the twentieth century with the 1980s and 1990s being particularly warm [56,57]. Nevertheless, a small global cooling was experienced in 1982-1984 and 1991-1992 due to the eruption of the volcanoes El Chichón and Pinatubo [58]. These particularly powerful eruptions injected large quantities of sulfur dioxide into the stratosphere that subsequently formed sulfuric acid aerosols that rapidly spread globally to increase the density of those already in the stratosphere from other sources. The resulting increased reflection of sunlight back into space produced small global coolings that illustrate the complex set of factors that affect global temperature.

While it is clear that temperature rose globally in the twentieth century, its significance for humanity becomes more apparent when it is placed in perspective with respect to global temperature over a longer period. For some time it was thought that from 950 to 1250 the average surface temperatures were about 1 °C warmer than in 1990 in Western Europe and this time became known as the Medieval Warm Period [59]. However, recalibration of the data on which this initial estimation was based and collection of much more data indicates that although the period 800 to 1400 was warmer than those immediately preceding and following it, its warmth was exceeded by that of the late twentieth century as shown in Fig. 8.11 [60]. The period 1400 to 1900 was a comparatively cold period and is often referred to as the Little Ice Age. In Europe, the summer decade 1994-2003 was the hottest in the last 500 years [61]. Less temperature data is available for the Southern Hemisphere and assessments of temperature over the past 2,000 years are consequently less reliable. Nevertheless, the steep rise in temperature during the twentieth century is clearly seen in Fig. 8.11.

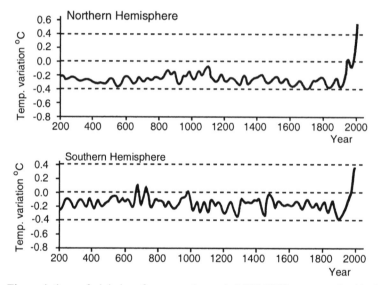

Fig. 8.11. The variations of global surface over the period 200-2000, compared with the 1961-1990 mean temperature. Adapted from [60].

(Reliable direct temperature measurements before 1880 are scarce and accordingly temperatures prior to that are estimated from proxies such as the hydrogen/deuterium, $^1H/^2H$, and oxygen $^{16}O/^{18}O$ isotopic ratios of ice cores from glaciers, variations in the thickness of tree annual growth rings and the calcium to magnesium ratio in shells. Because it has been possible to calibrate these proxies against accurately measured temperatures during the twentieth century, these calibrations can be applied to reliably estimate temperatures for earlier times.)

8.9. The Temperature Record in the Ground

Another record of past temperatures lies in the ground and may be recaptured by measurement of the change in temperature in boreholes as depth increases [62]. The ground surface temperature changes quickly to follow daily, seasonal and annual temperature changes. As the warmth at the surface diffuses downwards in the ground the daily temperature fluctuations are filtered out below one metre depth. Annual variations are filtered out below twenty to thirty metres so that a smoothed record of temperature changes over the last several hundred years is obtained. A 1993 assessment of surface temperature records reconstructed from the temperature variations found in boreholes in Europe, Greenland and North America showed a cooling trend of about 1.5°C beginning in the 1400s and ending in the mid 1800s. This coincides with records of very cold European and North American winters during that period. A 1995 study showed that during the previous 100-150 years North American regional surface temperatures had risen by an average of 1°C with Alaska showing the greatest increase of 2°C to 4°C. This was followed in 2000 by a more extensive collection of underground temperatures from 616 boreholes on every continent except Antarctica that showed that Earth has warmed by 1°C during the last five hundred years with the twentieth century being the warmest by 0.5°C with warming being faster than for any of the previous four centuries [63].

8.10. Melting Ice and Rising Oceans

As Earth becomes warmer an obvious concern is that glaciers, polar ice caps and icebergs will melt so that rising sea levels will inundate large areas of low lying coastal regions and islands. In the depths of the last ice age, 21,000 years ago, ice sheets where up to four kilometres high and covered thirteen times as much land as they do now. As Earth warmed and the ice sheets melted, sea levels rose by more than one hundred metres to inundate three million square kilometres of land in a process that is slowly continuing [8]. For the last one hundred years glaciers have been shrinking in the Andes, Alaska, the Canadian Rockies, the European Alps, Greenland, the Himalayas, the New Zealand Alps, parts of central Asia and on Mount Kilimanjaro in Tanzania together with melting snowpacks in the western United States. This has produced most of the rise of about 2 millimetres a year in ocean levels with a smaller contribution coming from the increase in volume of water as the oceans warm [2,64,65]. Ninety percent of Earth's ice is in Antarctica were the ice sheets cover an area of fourteen million square kilometres, or roughly the area of the continental United States and Mexico combined. The total volume of the Antarctic ice is about thirty million cubic kilometres and if it all melted a seventy-four metre global rise in the level of the oceans would result. This has focused great scrutiny on Antarctica.

The Antarctic ice cap has an average depth of 2,500 metres and is formed as snow is compressed into ice under the weight of successive snowfalls. The ice cap is not stationary, however, and slowly flows outwards from the South Pole towards the coast to form great ice shelves that extend out into the ocean where pieces break off to form icebergs. While this is a normal process, some of the ice shelves have shrunk over several decades, as exemplified by the Wordie Ice Shelf that decreased from 2,000 to 700 square kilometres in area between 1966 and 1989, and this has raised concerns that this may be a result of anthropogenic global warming [66]. This concern heightened in January 1995 when 4,200 square kilometres of the Larsen Ice Shelf broke away to form icebergs and was followed by the disintegration of another 3,250 square kilometre section over thirty-five days from 31 January 2002 [67]. However, while these dramatic events occurred on the Antarctic Peninsula at the western end of the continent that has been slowly warming by 1.4°C over the last three decades together with the southern tip of Patagonia, a cooling of up to 1.1°C has occurred over the same period in central and eastern Antarctica where glaciers have thickened and sea ice has increased [68-70].

Overall, satellite measurements show that the area of Antarctic sea ice increased by 212,000 square kilometres between 1979 and 1998 [69]. This may be due to a variety of factors including changes in ocean and atmospheric circulation around and above Antarctica, respectively. The latter appears to be linked to the 10°C cooling of the lower stratosphere due to ozone depletion that has occurred since 1985 and the resulting increase in the Antarctic vortex, a powerful system of westerly rotating winds that encompass Antarctica and extend from high in the stratosphere down into the troposphere [70]. (Ozone depletion and the Antarctic vortex are discussed in Chapter 9.) However, satellite observations of the area of Antarctic sea ice have only been collected relatively recently so that a longer term knowledge has to rely on less direct observations. A collection of 42,258 whaling records for the periods 1931 to 1957 and 1972 to 1987 after which commercial whaling was banned, suggest that between the mid 1950s and the early 1970s the sea ice area contracted by twenty-five percent, or 5.65 million square kilometres [69]. These whaling records give the position of the catch and, because catches occurred at the ice edge, the boundary of the sea ice.

Considerable interest has arisen in the probable behaviour of the West Antarctic ice sheet, the largest ice sheet, during a time of global warming. This is because most of this ice sheet rests on land below the level of the ocean that makes it more vulnerable to melting if ocean temperatures rise than the bulk of the ice cap that rests on higher ground. However, the West Antarctic ice sheet has been slowly melting for the past ten thousand years and is likely to continue to do so irrespective of human activities [72]. Its complete melting would produce a six metre rise in ocean level. There is evidence that this has occurred in the very distant past and that sometimes the entire Antarctic ice cap has melted with massive increases in ocean levels resulting [73]. Antarctica was largely ice free eighty million years ago when Earth was more than 6°C warmer than the present 15°C global surface average. Thereafter, a fluctuating but continuous cooling occurred and the first transient Antarctic ice sheet formed about thirty-five million years ago [74]. This was followed by the formation of major ice sheets that showed substantial seasonal variation in area. Permanent ice sheets appeared fifteen million years ago and have been present to this day. Modelling suggests that these changes coincide with a fall in atmospheric carbon dioxide levels from as high as 1,000 parts per million to 280 part per million, the preindustrial value. This has substantial implications for Earth as present predictions indicate that the current rate of anthropogenic global warming and rising carbon dioxide levels could see temperatures similar to those at the extinction of the dinosaurs sixty-five million years ago and an almost ice free planet by 2300, and earlier if the warming rate increases.

8.11. The Temperature Record in the Ice

An extensive record of the changing amounts of carbon dioxide and methane in the atmosphere and the accompanying changes in air temperature over the last 420,000 years has been obtained at the Russian Antarctic research station, Vostok [75,76]. By drilling 3,632 metres into the ice, scientists effectively became time travellers as the deeper they drilled the older became the ice core samples they took so that they held in their hands the compressed snow of blizzards that swept Antarctica tens and hundreds of thousands of years ago. The age of the ice was determined by counting the annual layers of ice.

The air bubbles trapped in the ice represent a scientific treasure in the form of a continuous record of the composition of the air over Antarctica, and the ice itself yields the accompanying changes in temperature. Generally, the ice is older than the air it traps because air can diffuse upwards and downwards through the compacting snow layers for 2,500 to 4,300 years before it is totally sealed in to become an unchanging atmospheric record. The temperature record is obtained by measuring the ratios of the amounts of hydrogen, ^{1}H, and its heavier isotope, deuterium, ^{2}H, in water obtained from the ice core samples. Most water molecules contain two hydrogen atoms bound to an oxygen atom, while a very small number contain one atom of heavier deuterium instead of hydrogen. These differing water molecules effectively form an "ice thermometer" because in water vapour the ratio of those containing hydrogen only to those containing deuterium increases as the climate cools. This ratio is preserved when the water vapour falls as snow and so a continuous temperature record can be determined from the layers of snow compacted into ice. A second ice thermometer is based on the plentiful lighter oxygen isotope, ^{16}O, and the scarce heavier isotope, ^{18}O, that show a similar variation in ratio in water vapour as temperature changes.

All of this information comes to life in Fig. 8.12 that shows a complete record of the changes in atmospheric carbon dioxide and methane levels, determined from analysis of the air in the bubbles trapped in the ice, and the accompanying rises and falls in temperature over the last 420,000 years. A striking pattern emerges. When carbon dioxide and methane levels were high so was the temperature (relatively speaking, as the mean temperature at Vostok is -55°C) and *vice versa*. As temperature, carbon dioxide and methane levels fell in a fluctuating way from 410,000 years ago an ice age, or glacial period, set in until 330,000 years ago when a rise in the levels of both gases was accompanied by a

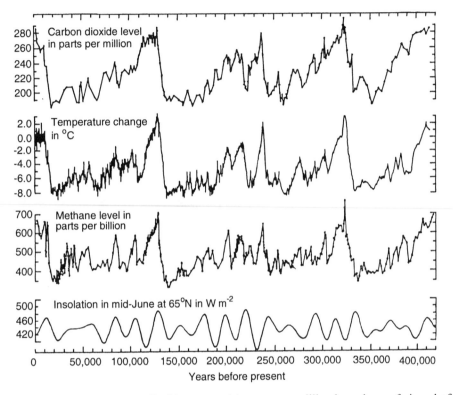

Fig. 8.12. The levels of carbon dioxide measured in parts per million by volume of air and of methane in parts per billion by volume of air. The corresponding changes in Antarctic air temperature by comparison with the average air temperature at Vostok, -55°C, are also shown together with the variation of incoming solar radiation, or insolation, at 65°N in mid-June resulting from the Milankovitch cycle. Adapted from [75].

rise in temperature when Earth enjoyed a warm, or interglacial, period. However, from 325,000 years ago the cycle began again and another ice age arrived only to end 240,000 years ago with another warm period that lasted 5,000 years that was followed by another glacial period that lasted until 130,000 years ago. Then another warm period arrived and lasted until 120,000 years ago before plunging into the last ice age that ended about 11,600 years ago at the end of the most recent cold period, the Younger Dryas, named after a tundra plant, *Dryas octopetala*, that was widely spread at that time. The climatically benign

Holocene epoch then began and persists to this day. A similar time sequence of high carbon dioxide and methane levels coinciding with interglacial periods, and low levels of these gases coinciding with glacial periods is also found in ice cores from Greenland and shows that this is a global phenomenon [77].

A coincidental, but fascinating, illustration of the adage that one discovery leads to another was the finding of the 280 kilometre long and 500 metre deep Lake Vostok almost four kilometres below the ice on which the Vostok research station stands [78]. The lake has been isolated for at least half a million years so that the bacteria in it has been isolated for a similar time and may provide interesting insights into the evolution of isolated life. Lake Vostok is the largest of the seventy or so lakes buried below the Antarctic ice.

An even longer record of glacial cycles has been derived from drilling 3,190 metres into the ice at Dome C in East Antarctica through the European Project for Ice Coring in Antarctica (EPICA) [79]. Analyses of the ice cores have provided a 740,000 year long record of climate encompassing eight glacial cycles by 2004 that is expected to extend to 800,000 years and to encompass a ninth glacial cycle when the ice core analyses are complete. The first 420,000 year data from Dome C are in good agreement with the Vostok data.

The striking correlation between increases in temperature and increases in carbon dioxide and methane levels seen in Fig. 8.12 is of great interest in view of the increasing levels of these greenhouse gases at the present time. It appears that a major proportion of the carbon dioxide level changes arise because warm oceans absorb less carbon dioxide than cold oceans with corresponding increases and decreases in atmospheric carbon dioxide as temperature rises and falls. A combination of the melting of glaciers and ice sheets as the climate warms causes an increase in the extent of swamps and the size of ruminant and termite populations as plant life increases. Correspondingly, this results in increased activity in methane producing bacteria, or methanogens, and an increase in methane production. Some of the increased methane levels arise from thawing of permafrost and the release of methane from ice-like methane hydrates [80]. (Methane hydrates are discussed in section 7.8.)

The quite regular repetition of the major temperature variations in Fig. 8.12 at about 100,000 years intervals, on which are superimposed other large variations at intervals of about 41,000 years and 23,000 years, suggests that a driving force characterised by similar time intervals triggers these variations. Such variations characterise Earth's orbit around the sun and accompanying variations in the tilt of Earth's axis from the vertical with respect to the plane of Earth's orbit and the rotation of that axis about the vertical. These variations are known as the Milankovitch cycles.

8.12. The Milankovitch Cycles and Climate Change

Eight times within the last 800,000 years snow built up over tens of thousands of years in the high northern latitudes and compacted into extensive glaciers and ice sheets that reached thicknesses of several kilometres and extended southward into midwestern America and central Europe. Simultaneously, extensive ice sheets built up in the southern hemisphere. These glacial periods ended quite abruptly and the ice sheets contracted over several thousands of years during interglacial periods to positions similar to those that they occupy today. An impressive coincidence exists between the timing of these periodic advances and retreats of the ice sheets and the accompanying climate changes and changes in Earth's orbit around the sun. These changes in the shape of Earth's orbit and in the tilt and orientation of Earth's axis, shown in Fig. 8.13, are caused by the gravitational pull of the moon and the planets, and are known as Milankovitch cycles after the Serbian Milutin Milankovich who proposed that the resulting

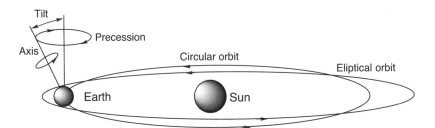

Fig. 8.13. The Earth moves in a slightly eccentric ellipse around the sun. At present the ellipse is 297 million kilometres across at its greatest extent with Earth being 146 and 151 million kilometres from the sun at either extreme. The axis of Earth is tilted at about 23.5° from the vertical about which it precesses. The orientation for a southern summer at the present time is shown.

changes in the sunlight reaching Earth's northern hemisphere control the size of the northern ice sheets [81,82]. Earlier, the Scotsman James Croll presented a similar theory in which the variation of incoming solar radiation at the poles with changes in Earth's orbit determined the coming and going of ice ages [83].

The shape of Earth's orbit around the sun changes from near circular to elliptical and back again over 100,000 years. As the orbit becomes more elliptical, Earth's distance from the sun at the orbit's nearest and farthest points increases so that the seasons in one hemisphere show greater extremes than in the other. This eccentric orbit is presently close to circular and as a result Earth is in an interglacial period that has been regularly preceded by others at about 100,000-year intervals. At present Earth reaches its greatest distance from the sun during southern winters so that they are a little colder than northern winters as a consequence. Conversely, Earth approaches closest to the sun during southern summers so that they are a little warmer than northern summers. Superimposed on this cycle is a second climate modifying effect, that of the regular variation of the tilt of Earth's axis about which it rotates.

The tilt of Earth's axis is now 23.5° from the vertical and over periods of about 41,000 thousand years it changes from 21.5° to 24.5° degrees and back again. Summers become hotter and winters become colder in both hemispheres as the tilt increases because sunlight enters the atmosphere at a smaller angle in summer and a greater angle in winter. A third climate modifying influence is the rotation of Earth's spin axis through a complete circle about the vertical every 23,000 years. This rotation decides whether summer in a particular hemisphere coincides with Earth being at a greater of lesser distance from the sun as it traverses its orbit. When the effects of Earth's orbit shape and axis rotation reinforce each other in one hemisphere they oppose each other in the other hemisphere. The combined effects of these two factors, together with changes in the tilt of Earth's axis, cause the amount of sunlight reaching the high northern and southern latitudes to vary over a range of twenty percent in a complex manner.

When the southern hemisphere experiences its maximum tilt towards the sun at the point where it most closely approaches the sun hot summers result and the corresponding winters are at their coldest because the south tilts away from the sun. However, in the northern hemisphere cool summers and mild winters occur. These are the conditions coinciding with the volume of ice reaching a maximum at periods about 100,000 years apart that approximately match the time over which Earth's orbit changes from near circular to elliptical and back again. Superimposed on this pattern are smaller expansions and shrinkages of ice volume in both hemispheres occurring at intervals of about 41,000 and 23,000 years that coincide with the duration of the periods over which a complete cycle of the change in the tilt of Earth's axis and its rotation about the vertical occurs, respectively. High global temperatures and high carbon dioxide and methane levels match periods of high northern summer sunlight and *vice versa*.

The summer sunlight reaching the southern hemisphere also varies with the Milankovitch cycles so that low summer sunlight in the southern hemisphere corresponds to high summer sunlight in the northern hemisphere and *vice versa*. However, the closely correlated variations in time in temperature, carbon dioxide and methane levels derived from the Vostok and Greenland ice cores when the northern hemisphere summer sunlight is high and southern hemisphere summer sunlight is low is consistent with the existence of an efficient thermal communication between the two hemispheres. This closely matched timing of the coming and going of ice ages in the two hemispheres has been the subject of much debate with the invoking of the coupling of the exchange of heat between the northern and southern hemispheres through the atmosphere that takes a year and the oceans that takes a thousand years, the variations of greenhouse gas levels, and coupling between the orbital variations themselves to explain the cause of the beginning or ending of ice ages in the two hemispheres differing by a relatively small, but significant, thousand years or so from the Milankovitch expectation of a 100,000 year interval [84]. Thus, while Earth's orbital motion about the sun is the initiator of the coming and going of ice ages, coincident changes in Earth's atmosphere and oceans amplify the changes in temperature and climate and both form thermal communication channels between the northern and southern hemispheres.

Typically during a 100,000 year cycle the amount of incoming solar radiation changes by about sixty watts per square metre during summer at 65°N and similar variations occur at 78°S, the latitude of the Vostok ice station, but with a different time sequence as expected from the Milankovitch cycles. However, the magnitudes of these variations of incoming solar radiation are insufficient to cause all of the 8°C temperature difference between the cold depths of a glacial period and the warmth of an interglacial period detected in either Antarctica or

Greenland. It appears that as the ice sheets start to shrink at the end of an ice age, reflection of sunlight from them decreases and contributes to the warming along with an increasing greenhouse effect as the levels of atmospheric carbon dioxide, methane and water rise due to increasing biological activity and release from melting ice, warming oceans and permafrost. This process reverses at the beginning of an ice age. As is discussed in section 8.14, the oceans probably play an additional and major part in establishing this synchrony between the hemispheres. (Small long and short term changes in solar irradiance superimpose on the variations of incoming solar radiation that Earth experiences through the Milankovitch cyles [85,86].)

At this point, it is of more than a passing interest to speculate about the length of the warm Holocene epoch that humanity currently enjoys and which has lasted for some eleven thousand years so far. A cursory examination of the temperature variation deduced from the Vostok ice cores and of the Milankovich cycle might lead to a worrying conclusion that the beginning of a new ice age is imminent as the two previous warm periods have lasted for about five and ten thousand years, respectively. This is a conclusion that contrasts greatly with the present concerns about global warming. Fortunately, the Milankovich cycle predicts only a small variation in incoming solar radiation over the next fifty thousand years that together with the increased greenhouse effect arising from the increasing carbon dioxide level, is likely to extend the Holocene [87]. Modelling suggests that the Holocene, which some have suggested might be better called the Anthropocene because of humanity's climate changing activities, may well continue for another fifty thousand years. One such long warm period has occurred in the past at a time when the incoming solar radiation and the coupling between the three components of the Milankovitch cycle were similar to that experienced today.

8.13. The Role of the Oceans

The rise in atmospheric carbon dioxide levels with increasing temperature seen in Fig. 8.12 appears to be mainly a consequence of the interactions between the atmosphere and the oceans that are central to the changeability of Earth's climate [12,88]. The amount of carbon dioxide in the atmosphere is only about two percent of that stored in the oceans where much of it exists as either bicarbonate

(HCO$_3^-$) or carbonate (CO$_3^{2-}$) that can be looked upon as molecular sized carbon dioxide reservoirs, or sinks, as shown in Fig. 8.14. In a continuous process,

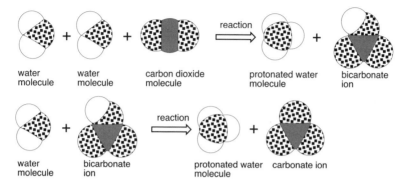

Fig. 8.14. The reactions of carbon dioxide (CO$_2$) dissolved in water (H$_2$O). In the upper section two water molecules react with carbon dioxide to form protonated water (H$_3$O$^+$), where a charged hydrogen atom or ion has added to a water molecule, and bicarbonate (HCO$_3^-$). In the lower section a water molecule reacts with bicarbonate to form a protonated water molecule and carbonate (CO$_3^{2-}$). The open, chequered and shaded atoms are hydrogen, oxygen and carbon, respectively.

carbon dioxide alternately dissolves in the ocean surface water and escapes back into the atmosphere as the balance between the amount of carbon dioxide in the atmosphere and the ocean surface is maintained. However, while the reversible reactions shown in Fig. 8.14 happen in a few milliseconds, the oceans consist of layers of water at different temperatures that can take a considerable number of years to mix depending on location. It would take a thousand years or so for all of the water in the oceans to completely mix between these layers. Because of this, the oceans adjust slowly to changes in the atmospheric carbon dioxide levels. Thus, although the capacity of the oceans to store carbon dioxide is large, their slow response has resulted in only forty to fifty percent of all anthropogenic carbon dioxide produced so far dissolving in them. This has substantial implications for predictions of future climate and for understanding the variation of carbon dioxide levels with the coming and going of ice ages.

While the dissolution of carbon dioxide in the oceans has slowed the rate of increase in atmospheric carbon dioxide levels, and thereby the rate of climate change, a penalty has been incurred. The reaction of carbon dioxide with water to form bicarbonate and carbonate releases protons to form protonated water, H$_3$O$^+$, as is shown in Fig. 8.14. As a consequence the acidity of the oceans has risen as the amount of carbon dioxide dissolved from the atmosphere has increased [89]. Depending upon the level of future anthropogenic carbon dioxide emissions, this acidification is predicted to continue. Although the acidification so far is quite small it is sufficient to cause some dissolution of calcium carbonate from the skeletons of coral and plankton and the shells of crustaceans with deleterious effects on their health and possibly consequent population depletion.

Compared with the major atmospheric gases, oxygen and nitrogen, carbon dioxide is quite soluble in water and it becomes more so as the temperature drops. As a result, the warmer surface water layers contain ten percent less carbon dioxide than do the deep cold water layers. Consequently, chilled water sinking into the ocean abyss carries with it a large amount of carbon dioxide, an action that is sometimes described as a solubility pump. Phytoplankton in oceanic surface water removes about 178 billion tonnes of carbon dioxide annually from the atmosphere as it photosynthesises organic matter as shown by a global satellite survey [13,90]. This 1998 value is about twice the estimate derived from earlier studies and compares with the 207 billion tonnes of carbon dioxide removed from the atmosphere by land plants to give a combined primary production of 105 billion tonnes of carbon incorporated into plant material annually. However, as the average lifetime of plankton is about a week it only constitutes 0.2 percent of global biomass at any given time whereas the terrestrial plants have lifetimes ranging from months to many years and constitute the bulk of global biomass. Nevertheless, phytoplankton has an important role in mediating atmospheric carbon dioxide levels. About twenty-five percent of phytoplankton is not consumed by grazing species and upon dying is metabolised to carbon dioxide by bacteria as its sinks towards the ocean abyss [12,13,88,90,91]. Some of this carbon dioxide forms carbonate salts but most of it dissolves in the cold deep ocean water. Eventually, much of the deep water carbon dioxide finds its way back to the surface through ocean circulation but the overall effect is to lower atmospheric carbon dioxide levels. This process of transferring atmospheric carbon dioxide to deep water is often referred to as a biological pump.

Most of Earth's ocean area is in the southern hemisphere and the biological pump appears to have its major impact south of 30°S in the Southern Ocean. There, the very cold and dense surface waters that represent about sixteen percent of the area of Earth's oceans, are outcrops of the much larger total volume of cold deep water. If the biological pump either slowed or stopped in the Southern Ocean surrounding Antarctica, carbon dioxide would escape into the atmosphere from the carbon dioxide rich deep waters and atmospheric carbon dioxide levels would rise. As it is, the biological pump captures much of the carbon dioxide escaping from the deep water and returns it to the depths. This is a very important part of the interaction of the oceans with the atmosphere that is examined on a larger scale in section 8.14.

8.14. The Thermohaline Cycle

The oceans have a very strong influence on global temperature and climate through surface currents carrying warm and less salty water and deep currents carrying cold and very salty water that act as a powerful global pumping system linking the Atlantic, Indian and Pacific Oceans. This is the thermohaline cycle shown in Fig. 8.15 [92,93]. Weather patterns cause winds to carry more moisture from the North Atlantic than is replaced by rain, snow and runoff from the

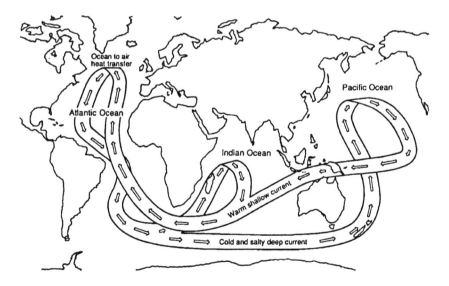

Fig. 8.15. A simplified illustration of thermohaline cycle that links the major oceans and is responsible for the temperate climate of Europe.

surrounding continents so that there is an overall loss of water that concentrates salt in the ocean's surface water. The warm Atlantic surface water flowing northward as a powerful surface current carries heat equivalent to a million billion watts, or a petawatt, at 24°N. In the vicinity of Greenland this surface water is chilled by about 8°C as it warms the eastward blowing winds. This results in a 5°C warming of the air over Europe as the ocean releases approximately one hundred times the present global anthropogenic energy consumption in this gigantic warming effect. The chilling and evaporation of the surface water makes it denser and higher in salt and carbon dioxide concentration as it sinks into the depths. Now, two to three kilometres down, it flows southwards through the Atlantic abyss and veers eastward around the southern tip of Africa and south to Antarctica. There, warmer and less dense than the near freezing surface water, it rises to the surface only to be chilled to near freezing point and sinking back into the depths. This deep Antarctic water floods northward into the Atlantic, Indian and Pacific Oceans. While the deep water flowing into the Atlantic is soon entrained back into the much stronger deep southern current, that flowing into the Indian and Pacific Oceans heads towards the tropics where it warms and rises to form warm surface currents in the tropics. These currents flow to the north of Australia and westward through the Indian Ocean and thence around the tip of Africa to complete the thermohaline cycle. It is estimated that this water circulation that produces the enormous transfer of heat northwards in the Atlantic Ocean is equivalent to a hundred times the flow of the Amazon. If the thermohaline cycle stopped, Dublin would experience a climate similar to that of Spitzbergen nearly 1,000 kilometres north of the Arctic Circle and London's climate would become similar to that of Irkutsk in Siberia.

The possibility of such dramatic climate changes occurring may seem a little far-fetched. However, dust and methane variations found in cores taken from the Greenland ice cap and from North Atlantic sediments are consistent with rapid temperature rises of about 10°C occurring repeatedly over as little as ten years at the beginning of cold periods that lasted for a thousand or so years in North America and Europe [94,95]. At least twenty-one of these Dansgaard-Oeschger events, named after their discoverers, occurred between 75,000 and 15,000 years ago at regular intervals [96]. Similar events have occurred nine times during the Holocene epoch at about 1,500 year intervals, albeit with lesser climate modifying effects than those occurring during the last ice age [97]. The Little Ice Age, that was at its coldest in the seventeenth century, is thought to have arisen from the most recent Dansgaard-Oeschger event. It appears that small increases in sunlight intensity magnified by accompanying changes in atmospheric circulation produce a rise in temperature that may be the trigger for these occurrences. The rising temperature probably releases substantially increased numbers of icebergs from Canada, Greenland and Iceland that dilute the salty North Atlantic with fresh water as they melt. This weakens the thermohaline cycle and a cold period sets in. Evidence for this comes from North Atlantic sediments that contain coarse rock particles picked up on land by flowing ice sheets and subsequently dropped into the ocean by melting icebergs.

Analyses of rock fragments in North Atlantic sediments also suggest that at least six times in the past 65,000 years at intervals of five to fifteen thousand years armadas of icebergs larger than those of the Dansgaard-Oeschger events broke away from the margins of the Canadian ice sheets to drift into the North Atlantic [98]. These occurrences coincide with particularly cold periods during which ice sheets may have built to three to four kilometres in height and became unstable and collapsed at their edges to produce sudden releases of iceberg armadas from Hudson Bay. These Heinrich events, also named after their discoverer, probably produced enough fresh water as the icebergs melted to either greatly slow or stop the thermohaline cycle and initiate a cold period in Europe. Each Heinrich event is followed by a particularly warm Dansgaard-Oeschger event with subsequent Dansgaard-Oeschger events becoming progressively cooler. This is consistent with the North American ice sheet growing in height until a partial collapse causes the next Heinrich event to start the cycle over again. These events illustrate the volatility of Earth's past climate on a millennial scale that is superimposed on the coming and going of ice ages associated with the much longer scale Milankovitch cycles.

Despite lingering uncertainties about some details of the Dansgaard-Oeschger and Heinrich events, it appears that the last very cold period of the last ice age, the Younger Dryas that started 12,900 years ago and lasted until 11,600 years ago, was triggered in a different way from these iceberg initiated events [99]. The North American the ice sheets started retreating 14,000 years ago and released massive volumes of melt water. Initially, much of this water flowed down the Mississippi to the Gulf of Mexico. At the same time the huge Lakes Agassiz and Ojibway (to the northwest of the present Great Lakes) formed as water filled the great depressions left by the ice sheets and was trapped by them as they slowly retreated. Eventually, the rising fresh water of the lakes broke through the ice barrier blocking the St. Lawrence River and flowed into the North Atlantic at rates of up to twenty-five cubic kilometres a day for a year or more before slowing to half this flow rate for many more years. This great dilution of the North Altantic salt water either greatly slowed or stopped the thermohaline cycle to usher in a great chilling of North America around the mouth of the St Lawrence, Iceland, the British Isles, Scandinavia, and North and Central Europe. The last major and sudden release of fresh water from Lakes Agassiz and Ojibway into the North Atlantic occurred 8,200 years ago [100]. As the North American ice sheets continued their retreat the ice barrier between the expanding Lakes Agassiz and Ojibway to the south and Hudson Bay to the north thinned and finally burst. This released 200,000 cubic kilometres of fresh water into Hudson Bay and from thence into the Labrador Sea and the North Atlantic, whereupon the temperature dropped by 4°C to 8°C in central Greenland 1.5°C to 3°C around the northeastern North Atlantic consistent with another slowing or cessation of the thermohaline cycle. There is also evidence that lesser releases of fresh water from the melting North American ice sheets caused smaller cooling episodes during the later period of the Ice Age. Coincidentally, such slowings and cessations of the thermohaline cycle in the past appear to have resulted in accumulation of warm surface water in the southern hemisphere that warmed the southern air while cooling occurred in the north and *vice versa* when the thermohaline cycle strengthened or restarted with a time lag of about a thousand years consistent with this oceanic connection being dominant [101].

The above discussion shows that the thermohaline cycle is vulnerable to sudden additions of large amounts of fresh water to the North Atlantic. This has attracted considerable attention as it shows what could happen should human induced global warming greatly dilute the North Atlantic with fresh water [101]. At present, the amount of rain and snow that falls in the Atlantic basin, together with the inflowing river water, appears to be exceeded by the amount of water that evaporates. This, combined with the capacity of the thermohaline cycle to entrain and salinize the incoming fresh water, maintains the saltiness of the North

Atlantic. Should this capacity be exceeded, the density of the surface water might decrease to the point where its sinking as it is chilled lessens to such an extent that the thermohaline cycle stops. (This is illustrated by a comparison with the North Pacific whose surface water is not as salty and dense as that of the North Atlantic and does not sink to form a deep cold current.) Interestingly, the current global warming induced melting and contraction of the Arctic ice sheets that might eventually lead to a considerable cooling in Europe promise the short term benefit of a month long opening of the Northwest Passage between the Atlantic and the Pacific Oceans along the Canadian and Alaskan coasts [102]. This is the seaway sought by Sebastian Cabot in 1508 and Henry Hudson in 1611 and finally navigated with difficulty by Roald Amundsen in 1906. A longer opening of a northern sea route between the Atlantic and Pacific Oceans along the Russian coast may also occur.

Clearly, the thermohaline cycle acts as a powerful heat pump between the northern and southern hemispheres with a thousand year cycle time. This may be one reason why the temperature variations of the south closely follow the variation of summer sunlight in the north with a small lag on the geological timescale. The second, and much faster, heat transfer channel is the atmosphere. From this emerges a plausible picture of the coming and going of ice ages. An ice age is initiated by cool summers and mild winters in the northern hemisphere triggered by the Milankovitch cycles. The spread of the northern ice sheets is greater than in the south because the northern ice sheets mainly form on better insulating land while those in the south extend out over the oceans to a greater extent. As the ice sheets grow, the ocean levels drop by a hundred metres or more, the atmosphere dries and deserts form from which winds carry dust into the atmosphere. An immediate effect of the drying of the atmosphere is a decrease in the level of the most important greenhouse gas, water, and a consequent cooling of the oceans that dissolve more carbon dioxide as a consequence. This decreased atmospheric carbon dioxide level produces a further decrease in the greenhouse effect and further cooling to which the cooling effect of decreasing methane levels adds as wetlands either freeze or dry out and ruminant and termite populations diminish. Coincident with this onset of an ice age, the thermohaline cycle stops possibly because changing weather patterns cause the North Atlantic to be less salty.

The end of an ice age commences with an increase in northern summer sunlight whose warming effect may be initially amplified by accumulated dust settled on ice and snow absorbing the increasing sunlight and decreasing the amount of sunlight reflected back into space [103]. This initial warming is further reinforced by increasing amounts of atmospheric water, carbon dioxide and methane as oceans warm and biological activity increases to produce further global warming. The rate of increase in atmospheric methane levels appears to be too rapid to arise solely from the increased activity of methanogens in newly thawed swamps and it is probable that two other sources of methane come into play. The first is the melting of permafrost to release methane and the second is the warming of oceans to release huge amounts of methane from the methane hydrates on the ocean floor [104]. If the North Atlantic weather patterns now return to those where the strengthening thermohaline cycle begins to further warm the north the rate of retreat of the ice sheets will increase and a return to interglacial conditions occurs.

While the above pictures of climate change do not include human influences, the impact of changes in the greenhouse gas levels on global temperatures and climate incorporated in them cannot be ignored. As anthropogenic carbon dioxide and other greenhouse gas levels increase the probability that weather patterns will change as a consequence becomes greater. One particularly frequently expressed concern is that North Atlantic weather patterns might change to a point where either an increase in surface water temperature or an increased inflow of fresh water or both might trigger a cessation of the thermohaline cycle and return glacial conditions to the North Atlantic. Such a great human induced climate change is unprecedented and is hopefully avoidable. Even so, lesser climate changes are not particularly attractive as is evident from the well-known El Niño phenomenon that occurs naturally with some regularity and causes sudden climate changes that are quite small by comparison with those that rising greenhouse gas levels might induce.

8.15. El Niño and Sudden Climate Change

The strong influence of changes in winds and ocean currents on climate is dramatically illustrated in the southern hemisphere by the sudden ENSO climatic change events that have near global consequences and are second only to the changing seasons in affecting weather patterns [105]. The acronym, ENSO, is derived from the El Niño phenomenon, first observed as an oceanic change, and the Southern Oscillation phenomenon, first observed as an atmospheric change, that together represent a major climatic variation over a very large area. Some of Earth's most prolific fisheries lie along the semi-desert coasts of southern Ecuador and Peru. Normally, strong offshore winds, the trade winds, blow steadily and push the warm surface water of the Pacific westward so that cold nutrient rich water wells up to replace it along the western coast of South America. A small fish, the anchoveta, flourishes under these conditions and provides huge catches for fishermen, seabirds and mammals alike. The trade winds, supported by a high pressure region over the eastern Pacific, produce a pooling of warm water on the western side of the Pacific where a low pressure region, the Indonesian Low, develops and is the centre of tropical rainfall. Here, warm water evaporates in huge quantities to produce high atmospheric moisture levels that result in the monsoons of Southeast Asia and the Indian sub-continent and plentiful rainfalls over northern and eastern Australia. These rains lead to good harvests for one of Earth's most populous areas.

At intervals of two to ten years the trade winds weaken and the low pressure region moves eastwards so that westerly anti-trade winds develop and the warm surface water of the Pacific moves eastwards. As the warm water moves inshore, and the upwelling of the nutrient rich cold waters along the South American coast ceases, the anchoveta vanish to leave the seabirds and mammals to starve in great numbers and the fishermen to face hardship. Simultaneously, the moisture laden air results in torrential rainstorms in Ecuador and Peru that produce crop destroying and life threatening floods and landslides. Because these hardships begin near Christmas, the Peruvian fishermen long ago named the climatic change accompanying them, El Niño, the Christ Child. On the western side of the Pacific, the occurrence of an El Niño event usually causes a weakening of the monsoons in the Asian equatorial regions and droughts in Australia, and most of the missing rain falls in mid Pacific storms. The much looked for end of an El Niño and the return of the anchoveta with the upwelling of cold water on the South American coast, and of more equable weather patterns generally, has become known as La Niña, the Little Girl. These alternating events have occurred throughout the Holocene epoch. However, the trigger for the onset of the El Niño and La Niña events is not known with certainty but it appears to be associated with changes in water flow in the Pacific abyss.

Particularly strong El Niño events occurred in 1891, 1911, 1925, 1941, 1953, 1957-58, 1982-83, a lengthy one in 1991-94 and again in 1997-98 and 2002-2003 [106,107]. The onset of an El Niño event can be predicted from temperature changes in the surface water of the Pacific, a slackening of the trade winds and falling atmospheric pressure across the Pacific, but the strength, duration and consequences of the event varies and are less predictable, although the accurate prediction of the onset of the 2002-2003 El Niño suggests that forecasting is improving. The particularly strong El Niño in 1982-83 resulted in six hundred deaths and US$900 million worth of damage arising from floods in Peru alone. This, and other El Niño events have also produced extensive floods in Ecuador, Argentina, Uruguay and Brazil and corresponding droughts and failed monsoons on the western side of the Pacific. The last and worst El Niño of the twentieth century occurred in 1997-98, and 23,000 deaths and US$33 billion worth of damage were attributed to the weather extremes it created amid fears that the frequency and strengths of El Niño events may be increasing as a consequence of global warming [108].

The climatic reach of an El Niño event is not confined to the Pacific region. Often, climatic conditions in the Pacific and Indian Oceans are connected [109]. Thus, when pressure is high over the Pacific it is low over the Indian Ocean and strong monsoons occur. Conversely, when pressure is low over the Pacific and high over the Indian Ocean, as is usually the case during an El Niño event, the monsoons often either weaken or fail and drought occurs in countries bordering the Indian Ocean as happened in India for half of the El Niño events in the twentieth century. This oscillating atmospheric phenomenon is the Southern Oscillation. While the climatic effects of ENSO are strongly felt in the southern landmasses, more northerly regions do not escape as shown by floods and storms in California and the American midwest and hard winters in Europe that have been attributed to the effects of El Niño [106]. The extraordinarily heavy July rains over central Europe and the extensive flooding of the Oder and Niese rivers in July 1997, coincided with the onset of drought in Australia and the 1997-98 El

Niño event. Some typical weather patterns arising under ENSO conditions are shown in Fig. 8.16 [110].

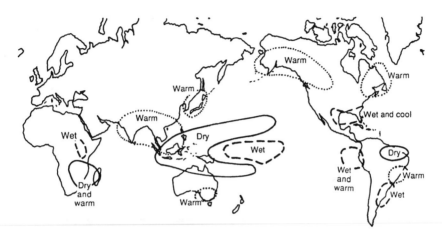

Fig. 8.16. Rainfall and temperature variations associated with ENSO conditions in the southern hemisphere summer. Adapted from [110].

While ENSO atmospheric pressure variations over the Indian and Pacific Oceans have by far the largest effect on climate, less extensive atmospheric pressure oscillations, known as the North Atlantic Oscillation, occur over the North Atlantic Ocean where they affect North American, European and northern Asian rainfall, storm tracks and temperatures at about ten year intervals [111].

8.16. Warming Oceans and Bleaching Coral Reefs

The impact of global warming on coral reefs has attracted great attention as these beautiful and colourful formations that cover several hundreds of thousands of square kilometres in shallow oceanic waters of the tropics have become bleached to varying extents as the oceans warm [112]. This was particularly noticeable in the Indian Ocean during the 1997-1998 El Niño event when ocean temperatures rose substantially. Coral reefs are made from calcium carbonate from successive generations of dead coral polyps. The varied and vivid colours of the reefs arise from the microscopic photosynthesizing zooxanthellae that live symbiotically inside the living coral polyps that provide a stable environment for them on the reef surfaces. Above about 30°C, the zooxanthellae disappear from the coral polyps that then become transparent so that the white calcium carbonate coral skeleton shows through and the reef appears bleached. The disappearance of the zooxanthellae is either a consequence of their death or their leaving the coral for cooler environments. Coral can slowly recover from such bleaching but there is a fear that should oceanic temperatures continue to rise many reefs will die. This is not only because of the loss of zooxanthellae on which the coral depends, but also because increasing atmospheric carbon dioxide levels increase dissolved carbon dioxide concentrations. This causes the ocean surface water to become more acidic and decreases the availability of calcium carbonate from which successive generations of coral polyps build reefs. There are other threats to coral reefs, notably pollution, eroded soil runoff and bacterial infection, but oceanic warming and increasing carbon dioxide levels appear to be the biggest threats.

Coral reefs have disappeared many times in the last billion years and have then recovered over thousands of years. However, such recovery rates are too slow to provide reassurance to those dependent on coral reefs for their livelihoods through fishing and tourism and for coastal protection. In the Indian Ocean alone it is estimated that over a twenty year period of moderate bleaching and rapid recovery the cost would be US$700 million whereas if severe bleaching occurs and recovery is either slow or non-existent the cost could be US$8 billion. Should the coral reefs be lost their niche in nature would probably be slowly taken over by other reef forming organisms such as the rudists, types of molluscs, that formed dominant and extensive reefs in the ocean shallows millions of years ago when the corals where largely absent [113]. The ecological and economic consequences of such a change are unpredictable.

8.17. The Future

What is to be concluded about the likely effects of human activity on future climate? A summary on the basis of the geological time scale would say that given that past and future climate is substantially controlled by the Milankovitch cycles, substantial temperature and climate changes will occur and that, although the Holocene epoch may continue for another fifty thousand years, it is inevitable that new a ice age will eventually arrive. However, this is not a very helpful observation as recorded history is very brief by comparison with geological timescales and most human concerns are based on the much shorter period of several generations. Nevertheless, it is well established that human activities have increased the level of greenhouse gases in the last one hundred years. There is unequivocal evidence that this is causing global warming that is transforming the Holocene epoch into an Anthropocene epoch during which the anthropogenic climate changes already in train will continue. While it is clear that feedback mechanisms that dampen the effects of increasing atmospheric greenhouse gas levels on climate exist, they ought not be overstressed lest climate changes that imperil substantial sections of humanity occur. Accordingly, it seems common sense to reduce energy demands, fossil fuel burning and generation of carbon dioxide and other greenhouse gases by developing more energy efficient technologies. Although population growth is slowing, population will continue to grow until the end of the century (as discussed in Chapter 2) and generate increasing energy demand.

In an attempt to address some of these concerns the United Nations Framework Convention on Climate Change, signed by 162 nations at the Earth Summit in Rio de Janeiro in June 1992, called for the reduction of anthropogenic greenhouse gas emissions to 1990 levels by 2008 to 2012 by developed nations and sought less stringent requirements for developing nations [114]. Article 2 of the treaty contains the statement:

"The ultimate objective of this Convention and any related legal instruments that the Conference of the Parties may adopt is to achieve, in accordance with the relevant provisions of the Convention, stabilization of greenhouse gas concentrations in the atmosphere at a level that would prevent dangerous anthropogenic interference with the climate system. Such a level should be achieved within a time frame sufficient to allow ecosystems to adapt naturally to climate change, to ensure that food production is not threatened, and to enable economic development to proceed in a sustainable manner."

While few could quibble with the general sentiments of this statement, the two sentences of which it is composed carry different messages. The first addresses the desire to avoid dangerous climate change. The second is equivocal in that it accepts that climate change is underway and seeks to incorporate the continuance of food supply and development while sustaining a liveable climate. It also seeks to satisfy all interest groups simultaneously, in particular those who are persuaded that climate change threatens their future, and also those who may be less fearful of climate change and see proposals for its amelioration as a threat to development in developing nations and to industry in developed nations. Given the differing interests both within individual nations and among the members of the United Nations, the blandness of the statement is not surprising.

Subsequently, it became clear that achievement of the required emission reductions was unlikely. Efforts to strengthen commitments to reducing greenhouse gas emissions through a legally binding treaty generated substantial dissension prior to and during the second and third conventions held in Geneva in January 1996 and Kyoto in December 1997, respectively [115]. Even so, the Kyoto conference did produce the Kyoto Protocol the intent of which was to reduce global emissions of greenhouse gases by five percent below 1990 emission levels by thirty-eight developed nations by 2008 to 2012 and the signatories were expected to ratify the treaty by 15 March 1999. [116]. These reductions were to apply to carbon dioxide, methane and nitrous oxide and also to hydrochlorofluorocarbons, hydrofluorocarbons, perfluorocarbons and sulfur hexafluoride that have no natural source. Despite the much lesser threat that hydrochlorofluorocarbons and hydrofluorocarbons pose to the ozone layer compared with that of the chloroflurocarbons that they are replacing, they are effective greenhouse gases. The perfluorocarbons, that are composed of carbon and fluorine only, are in a similar category. Sulfur hexafluoride, that has an atmospheric concentration of about three parts per trillion by volume, is a very effective greenhouse gas with an atmospheric lifetime of 3,200 years and is 24,900 times as effective as carbon dioxide in global warming potential [117]. Sulfur hexafluoride has been used in heavy electrical insulation and in the gas filled cushions of training shoes and, perversely, has been released into the biosphere in environmental studies of water and effluent flow.

However, substantial disputes occurred over the fairness of the requirements of the protocol and their economic consequences [118]. The United States, in particular, wished absorption of carbon dioxide by forests and other vegetation to greatly lessen the amount by which it would have to reduce its carbon dioxide emissions to comply with the protocol. Other nations disputed the validity of such a strategy and also the magnitude of the carbon sinks that the United States claimed to possess [42]. The consequence was that the United States, by far the largest carbon dioxide producer, declined to ratify the protocol in company with Australia. However, one hundred and forty-one nations ratified the protocol on 16 February 2005, but it only became legally binding on the ratifying developed nations [119]. Without the participation of the United States and rapidly industrialising China and India, who as developing nations were not required to abide by the protocol, reduction of carbon dioxide and other greenhouse emissions will be much less effective. Nevertheless, broad consideration of methods of cutting carbon dioxide emissions and countering their climate changing effects is underway in the United States and elsewhere [120].

Adherence to the provisions of the Kyoto Protocol should reduce the rate of increase of carbon dioxide and other greenhouse gas emissions and the rate of climate change should also be slowed [121]. However, unless the growth of emissions ceases by 2025, the hoped for stabilizing of atmospheric carbon dioxide at 550 parts per million by mid century will not occur and the extent of climate change will increase as carbon dioxide levels increase. It was an important achievement to bring the Kyoto Protocol into force but there is as yet no provision to replace it when its mandate expires in 2012. It is very probable that the developing nations will require that a replacement treaty should also be binding on developing nations. Tortuous and extended negotiations are probable.

The 2001 IPCC report is unequivocal about humanity's impact on climate through assertions such as "there is new and stronger evidence that most of the warming observed over the last 50 years is attributable to human activities" and "Human influences will continue to change atmospheric composition throughout the 21st century" [5]. To what extent temperatures will rise and climate change is girt with some uncertainty [122]. However, this is no reason to delay attempts to minimize these changes by seeking to be less profligate in the use of fossil fuels, to diminish the emission of greenhouse gases and maximize the use of alternative energy sources. Failure to do so will probably generate some unpleasant climatic surprises, one of which could be the cessation of the thermohaline cycle in the North Atlantic and chilling of Europe, the northeast of the United States' and eastern Canada. Alternatively, a general warming of all parts of Earth might occur to present quite different climate changes. It is salutary to reflect that present atmospheric carbon dioxide levels are thirty percent higher than they were 130,000 years ago when it was 2-3°C warmer globally, sea levels were six metres higher, and hippopotamuses and crocodiles inhabited tropical wetlands in what is now southern England [31].

These alternative visions of possible future climates carry with them great challenges to humanity. Already the climate changes that have occurred will take several centuries for Earth to fully adjust to even if the amounts of greenhouse gases now in the atmosphere were frozen at their current level, and their effects are likely to be disruptive [123].

References

1. J. Gribbin and M. Gribbin, The greenhouse effect, *New Sci.*, *Inside Science 92*, **1996**, 13 July, 1.
2. E. S. Nisbet, *Leaving Eden*, Cambridge University Press, Cambridge, UK, 1991.
3. K. Southwell, Night lights, *Nature*, **1997**, *390*, 21.
4. J. T. Houghton *et al.*, eds, *Climate Change 1995: The Science of Climate Change*, IPCC, Cambridge University Press, Cambridge, UK, 1996.
5. J. T. Houghton *et al.*, eds., *Climate Change 2001: The Scientific Basis*, IPCC, Cambridge University Press, Cambridge, UK, 2001.
6. a) S. H. Schneider, The changing climate, *Sci. Am*, **1989**, *261*, Sept., 38. b) L. Bengtsson, Modelling and prediction of the climate system, *Alexander von Humbolt Magazine*, **1997**, *69*, July, 3.
7. a) R. M. White, The great climate debate, *Sci. Am.*, **1990**, *263*, July, 18. b) G. A. McBean, Global change models - a physical perspective, *Ambio*, **1994**, *23*, 13. c) R. G. Prinn, The interactive atmosphere: global atmospheric-biospheric chemistry, *ibid.*, 50. d) P. P. Tans and P. S. Baldwin, Climate change and carbon dioxide, *Ambio*, **1995**, *24*, 376. e) B. Hileman, Climate observations substantiate global warming models, *Chem. Eng. News*, **1995**, *73*, 27 Nov., 18. f) T. R. Karl, N. Nicholls and J. Gregory, The coming climate, *Sci. Am.*, **1997**, *276*, May, 54. g) B. Hileman, Global climate change, *Chem. Eng. News*, **1997**, *75*, 17 Nov., 8.
8. a) D. Schneider, The rising seas, *Sci. Am.*, **1997**, *276*, March, 96. b) P. U. Clark and A. C. Mix, Ice sheets by volume, *Nature*, **2000**, *406*, 689. c) Y. Yokoyama *et al.*, Timing of the last glacial maximum from observed sea-level minima, *ibid.*, 713. d) K. Lambeck and J. Chapell, Sea level change through the last glacial cycle, *Science*, **2001**, *292*, 679. e) J. A. Church, How fast are sea levels rising? *Science*, **2001**, *294*, 802. f) C. Cabanes, A. Cazenave and C. Le Provost, Sea level rise during past 40 years determined from satellite and in situ observations, *ibid.*, 840. g) W. Munk, Ocean freshening, sea level rising, *Science*, **2003**, *300*, 2041. h) D. A. King, Climate change science: adapt, mitigate, or ignore? *Science*, **2004**, *303*, 176. i) O. H. Pilkey and J. A. G. Cooper, Society and sea level rise, *Science*, *ibid.*, 1781.
9. J. Houghton, *Global Warming: The Complete Briefing*, 2nd edn., Cambridge University Press, Cambridge, UK, 1997.
10. F. S. Rowland, Climate change and its consequences, *Environment*, **2001**, *43*, March, 28.
11. a) N. E. Graham, Simulation of recent global temperature trends, *Science*, **1995**, *267*, 666. b) R. A. Kerr, Greenhouse forecasting still cloudy, *Science*, **1997**, *276*, 1040. c) K. E.

Trenberth, The use and abuse of climate models, *Nature*, **1997**, *386*, 131. d) K. Hasselmann, Are we seeing global warming? *Science*, **1997**, *276*, 914. e) P. J. Sellers *et al.*, Modelling the exchanges of energy, water, and carbon between continents and the atmosphere, *Science*, **1997**, *275*, 502. f) M. A. Cane *et al.*, Twentieth-century sea surface temperature trends, *Science*, **1997**, *275*, 957. g) T. M. L. Wigley and S. C. B. Raper, Interpretation of high temperature projections for global-mean warming, *Science*, **2001**, *293*, 451. h) T. M. Smith, T. R. Karl and R. W. Reynolds, How accurate are climate simulations? *Science*, **2002**, *296*, 483. i) R. T. Pierrehumbert, The hydrological cycle in deep-time climate problems, *Nature*, **2002**, *419*, 191. j) M. R. Allen and W. J. Ingram, Constraints on future changes in climate and the hydrological cycle, *ibid.*, 224. k) R. A. Kerr, Climate modellers see scorching future as a real possibility, *Science*, **2005**, *307*, 497.

12. a) J. L. Sarmiento and C. Le Quéré, Oceanic carbon dioxide uptake in a model of century-scale global warming, *Science*, **1996**, *274*, 1346. b) C. W. Mullineaux, The plankton and the planet, *Science*, **1999**, *283*, 801. c) K. Caldeira and P. B. Duffy, The role of the Southern Ocean in uptake and storage of anthropogenic carbon dioxide, *Science*, **2000**, *287*, 620. d) D. Bakker and A. Watson, A piece in the CO_2 jigsaw, *Nature*, **2001**, *410*, 765.

13. P. G. Falkowski *et al.*, The global carbon cycle: a test of our knowledge of Earth as a system, *Science*, **2000**, *290*, 291.

14. R. A. Kerr, Dueling models: future U.S. climate uncertain, *Science*, **2000**, *288*, 2133.

15. a) P. E. Kauppi, The United Nations Climate Convention: Unattainable or irrelevant, *Science*, **1995**, *270*, 1454. b) R. A. Kerr, Sky-high findings drop new hints of greenhouse warming, *Science*, **1996**, *273*, 34. c) F. Pearce, Sit tight for 30 years, argues climate guru, *New Sci.*, **1996**, Jan., 7. d) B. D. Santer *et al.*, A search for human influences on the thermal structure of the atmosphere, *Nature*, **1996**, *382*, 39. e) P. J. Michaels and P. C. Knappenberger, Human effect on global climate? *Nature*, **1996**, *384*, 522. f) G. R. Weber, Human effect on global climate? *ibid.*, 523. g) B. D. Santer, J. S. Boyle and D. E. Parker, Human effect on global climate? *Ibid.*, 524. h) F. Pearce, Greenhouse wars, *New Sci.*, **1997**, 19, July, 38.

16. a) D. J. Gaffen, Falling satellites, rising temperatures, *Nature*, **1998**, *394*, 615. b) F. J. Wentz and M. Schabel, Effects of orbital decay on satellite-derived lower-tropospheric temperature trends, *Nature*, **1998**, *394*, 661. c) J. E Hansen *et al.*, Global climate data and models: A reconciliation, *Science*, **1998**, *281*, 930. d) J. Hecht, The heat is on, *New Sci.*, **1998**, 15 Aug., 4. e) K. Y. Vinnikov and N. C. Grody, Global warming trend of mean tropospheric temperature observed by satellites, *Science*, **2003**, *302*, 269. f) Z. Merali, Sceptics forced into climate climbdown, *NewSci.*, **2005**, 20 Aug., 10.

17. a) J. Uppenbrink, Arrhenius and global warming, *Science*, **1996**, *272*, 1122. b) S. Arrhenius, On the influence of carbonic acid in the air upon the temperature of the ground, *Phil. Mag.*, **1896**, *41*, 273. c) M. Heiman, A review of the contemporary global carbon cycle and as seen a century ago by Arrhenius and Högbom, *Ambio*, **1997**, *26*, 17. d) H. Rodhe, R. Charlson and E. Crawford, Svante Arrhenius and the greenhouse effect, *ibid.*, 2.

18. K. Lambeck, T. M. Esat and E.-K. Potter, Links between climate and sea levels for the past three million years, *Nature*, **2002**, *419*, 199.

19. a) J. Hansen *et al.*, GISS analysis of surface temperature change, *J. Geophys. Res.*, **1999**, *104*, 30997. b) J. Hansen *et al.*, Global warming continues, *Science*, **2002**, *295*, 275.

20. F. Pearce, A searing future, *New Sci.*, **2000**, 11 Nov., 4.

21. a) F. E. Nelson, O. A. Anisimov and N. I. Shiklomanov, Subsidence risk from thawing permafrost, *Nature*, **2001**, *410*, 889. b) E. Goldman, Even in the High Arctic nothing is permanent, *Science*, **2002**, *297*, 1493. c) E. Stokstad, Defrosting the carbon freezer of the North, *Science*, **2004**, *304*, 1618. d) F. Pearce, Climate warming as Siberia melts, *New Sci.*, **2005**, 13 Aug., 12.

22. M. R. Allen and R. Lord, The blame game, *Nature*, **2004**, *432*, 559.

23. C. Schär and G. Jendritzky, Hot news from summer 2003, *Nature*, **2004**, *432*, 559.

24. P. A. Stott, D. A. Stone and M. R. Allen, Human contribution to the European heat wave of 2003, *Nature*, **2004**, *432*, 610.

25.   G. A. Meehl and C. Tebaldi, More intense, more frequent, and longer lasting heat waves in the 21st century, *Science*, **2004**, *305*, 994.

26.   C. D. Keeling *et al.*, Interannual extremes in the rate of rise of atmospheric carbon dioxide since 1980, *Nature*, **1995**, *375*, 666.

27.   E. Worrell *et al.*, Carbon dioxide emissions from the global cement industry, *Annu. Rev. Environ. Energy*, **2001**, *26*, 303.

28.   C. S Powell, Greenhouse gusher, *Sci. Am.*, **1991**, *265*, Oct., 14.

29.   *Key World Energy Statistics*, International Energy Agency, Paris, 2002.

30.   a) T. A. Siddiqi, Carbon dioxide emissions from the use of fossil fuels in Asia: An overview, *Ambio*, **1996**, *25*, 229. b) E. Masood, Asian economies lead increase in carbon dioxide emissions, *Nature*, **1997**, *388*, 213. c) J. Romm *et al.*, A road map for U.S. carbon reductions, *Science*, **1998**, *279*, 669.

31.   T. J. Goreau, Balancing atmospheric carbon dioxide, *Ambio*, **1990**, *19*, 230.

32.   B. Moore and B. H. Braswell, Planetary metabolism: understanding the carbon cycle, *Ambio*, **1994**, *23*, 4.

33.   a) P. P. Tans, I. Y. Fung, and T. Takahashi, Observational constraints of the global atmospheric CO_2 budget, *Science*, **1990**, *247*, 1431. b) T. Beardsley, Tracking the missing carbon, *Sci. Am.*, **1991**, *264*, April, 9. c) H. A. Mooney and G. W. Koch, The impact of rising CO_2 concentrations on the terrestrial biosphere, *Ambio*, **1994**, *23*, 74. d) T. Takahashi, The fate of industrial carbon dioxide, *Science*, **2004**, *305*, 252. e) C. L. Sabine *et al.*, The oceanic sink for CO_2, *ibid.*, 267.

34.   R. R. Nemani *et al.*, Climate-driven increases in global terrestrial net primary production from 1982 to 1999, *Science*, **2003**, *300*, 1560.

35.   J. Grace *et al.*, Carbon dioxide uptake by an undisturbed tropical rainforest in southwest Amazonia, 1992 to 1993, *Science*, **1995**, *270*, 778.

36.   T. P. Kolchugina and T. S. Vinson, Role of Russian forests in the global carbon balance, *Ambio*, **1995**, *24*, 258.

37.   a) C. D. Keeling, J. F. S. Chin, and T. P. Whorf, Increased activity of northern vegetation inferred from atmospheric CO_2 measurements, *Nature*, **1996**, *382*, 146. b) I. Fung, A greener north, *Nature*, **1997**, *386*, 659. c) R. B. Myneni *et al.*, Increased plant growth in the northern latitudes from 1981 to 1991, *Nature*, **1997**, *386*, 689.

38.   a) P. E. Kauppi, K. Mielikainen, and K. Kuusela, Biomass and carbon budget of European forests from 1971-1990, *Science*, **1992**, *256*, 70. b) G. C. Jacoby, R. D. D'Arrigo, and T. Davaajamts, Mongolian tree rings and 20th-century warming, *Science* **1996**, *273*, 771. c) A. S. Moffat, Temperate forests gain ground, *Science*, **1998**, *282*, 1253.

39.   V. A. Barber, G. P. Juday and B. F. Finney, Reduced growth of Alaskan white spruce in the twentieth century from temperature-induced drought stress, *Nature*, **2000**, *403*, 668.

40.   a) F. A. Bazzaz and E. D. Fajer, Plant life in a CO_2-rich world, *Sci. Am.*, **1992**, *266*, Jan., 68. b) G. D. Farquhar, Carbon dioxide and vegetation, *Science*, **1997**, *278*, 1411.

41.   a) O. Phillips *et al.*, *Science*, Changes in the carbon balance of tropical forests: Evidence from long-term plots, **1998**, *282*, 439. b) S. Fan *et al.*, A large terrestrial carbon sink in North America implied by atmospheric and oceanic carbon dioxide data and models, *ibid.*, 442. c) D. W. Schindler, The mysterious missing sink, *Nature*, **1999**, *398*, 105. d) K. J. Nadelhofer *et al.*, Nitrogen deposition makes a minor contribution to carbon sequestration in temperate forests, *ibid.*, 145. e) M. Battle *et al.*, Global carbon sinks and their variability inferred from atmospheric O_2 and $\delta^{13}C$, *Science*, **2000**, *287*, 2467. f) J. Fang *et al.*, Changes in forest biomass carbon storage in China between 1949 and 1998, *Science*, **2001**, *292*, 2320. g) J. Randerson, No easy answer, *New Sci.*, **2002**, 13 April, 16. h) J. Grace and Y. Malhi, Carbon dioxide goes with the flow, *Nature*, **2002**, *416*, 594. i) J. E. Richey *et al.*, Outgassing from Amazonian rivers and wetlands as a large tropical source of atmospheric CO_2, *ibid.*, 617. j) R. A. Lovett, Rain might be a leading carbon sink factor, *Science*, **2002**, *296*, 1787. k) J. Kaiser, Satellites spy more forest than expected, *Science*, **2002**, *297*, 919. l) F. Achard, Determination of deforestation rates of the world's humid tropical forests, *ibid.*,

999. m) C. L. Goodale and E. A. Davidson, Uncertain sinks in the shrubs, *Nature*, **2002**, *418*, 593.

42. a) J. Kaiser, Possibly vast greenhouse gas sponge ignites controversy, *Science*, **1998**, *282*, 386. b) C. B. Field and I. Y. Fung, The not-so-big U.S. carbon sink, *Science*, **1999**, *285*, 544. c) E.-D. Schulze, C. Wirth and M. Heimann, Managing forests after Kyoto, *Science*, **2000**, *289*, 2058. d) R. J. Scholes and I. R. Noble, Storing carbon on land, *Science*, **2001**, *294*, 1012. e) F. Pearce, Tree farms won't save us after all, *New Sci.*, **2002**, 26 Oct., 10. f) C. Körner, Slow in, rapid out – carbon flux studies and Kyoto targets, *Science*, **2003**, *300*, 1242.

43. S. W. Pacala *et al.*, Consistent land- and atmosphere-based U.S. carbon sink estimates, *Science*, **2001**, *292*, 2316.

44. I. A. Janssens *et al.*, Europe's terrestrial biosphere absorbs 7 to 12% of European anthropogenic CO_2 emissions, *Science*, **2003**, *300*, 1538.

45. R. Schur, The investment forecast, *Nature*, **2002**, *415*, 483. b) F. Pearce, Europe's wake-up call, *New Sci.*, **2002**, 24 Aug., 4.

46. a) C. D. Harvell *et al.*, Climate warming and disease risks for terrestrial and marine biota, *Science*, **2002**, *296*, 2158. b) Ecological effects of climate fluctuations, *Science*, **2002**, *297*, 1292. c) G.-R. Walther *et al.*, Ecological responses to recent climate change, *Nature*, **2002**, *416*, 389. d) E. Sohn, Now mammals are feeling the heat, *New Sci.*, **2002**, 5 Oct., 9. e) T. L. Root *et al.*, Fingerprints of global warming on wild animals and plants, *Nature*, **2003**, *421*, 57. f) K. Krajick, All downhill from here? *Science*, **2004**, *303*, 1600. g) C. Both *et al.*, Large-scale geographical variations confirms that climate change causes birds to lay early, *Proc. Roy. Soc. B*, **2004**, *271*, 1657.

47. a) O. B. Toon, How pollution suppresses rain, *Science*, **2000**, *287*, 1763, b) D. Rosenfeld, Suppression of rain and snow by urban and industrial air pollution, *ibid.*, 1793.

48. a) D. R. Blake and F. S. Rowland, Continuing worldwide increase in tropospheric methane, *Science*, **1988**, *238*, 45. b) T. E. Graedel and P. J. Crutzen, The changing atmosphere, *Sci. Am.*, **1989**, *261*, Sept., 28. c) P. J. Crutzen, On the role of CH_4 in atmospheric chemistry: sources, sinks and possible reductions in anthropogenic sources, *Ambio*, **1995**, *24*, 52. d) L. Milich, The role of methane in global warming: Where might mitigation strategies be focussed? *Glob. Environ. Change*, **1999**, *9*, 179.

49. J. Hansen *et al.*, Global warming in the twenty-first century: An alternative scenario, *Proc. Natl Acad. Sci. USA*, **2000**, *97*, 9875.

50. a) S. E. Schwartz and M. O. Andreae, Uncertainty in climate change caused by aerosols, *Science*, **1996**, *272*, 1121. b) T. Skodvin and J. S. Fuglestvedt, A comprehensive approach to climate change: Political and scientific considerations, *Ambio*, **1997**, *26*, 351. c) Y. J. J Kaufman and R. S. Fraser, The effect of smoke particles on clouds and climate forcing, *Science*, **1997**, *277*, 1636. d) J. Hansen *et al.*, The missing climate forcing, *Phil. Trans.Roy. Soc. B*, **1997**, *352*, 231. e) J. H. Seinfeld, Clouds, contrails and climate, *Nature*, **1998**, *39*, 837. f) J. T. Kiehl, Solving the aerosol puzzle, *Science*, **1999**, *283*, 1273. g) J. E. Penner, D. Hegg and R. Leaitch, Unraveling the role of aerosols in climate change, *Env. Sci. Technol.*, **2001**, *35*, 332 A. h) R. A. Kerr, Rising global temperature, rising uncertainty, *Science*, **2001**, *292*, 192. i) Y. J. Kaufman, D. Tanré and O. Boucher, A satellite view of aerosols in the climate system, *Nature*, **2002**, *419*, 215. j) U. Lohmann and G. Lesins, Stronger constraints on the anthropogenic indirect aerosol effect, *Science*, **2002**, *298*, 1012. k) E. Pallé *et al.*, Changes in Earth's reflectance over the past two decades, *Science*, **2004**, *304*, 1299.

51. a) J. E. Penner, The cloud conundrum, *Nature*, **2004**, *432*, 926. b) A. S. Ackerman *et al.*, The impact of humidity above stratiform clouds on indirect aerosol climate forcing, *ibid.*, 1014.

52. a) S. Menon *et al.*, Climate effects of black carbon aerosols in China and India, *Science*, **2002**, *297*, 2250. b) F. Pearce and R. Edwards, Forest fires fuel pollution crisis, *New Sci.*, **2002**, 17 Aug., 8. c) F. Pearce, Pollution is plunging us into darkness, *New Sci.*, **2002**, 14 Dec., 6.

53.   a) J. K. Angell, Variation in United States cloudiness and sunshine duration between 1950 and the drought year of 1988, *J. Clim.*, **1990**, *3*, 296. b) B. G. Liepert, Recent changes in solar radiation under cloudy conditions in Germany, *Int. J. Climatol.*, **1997**, *17*, 1581. c) G. Stanhill and S. Cohen, Global dimming: a review of the evidence for a widespread and significant reduction in global radiation with discussion of its probable causes and possible agricultural consequences, *Agricult. Forest Meteorol.*, **2001**, *112*, 2075. d) S. Cohen, B. Liepert and G. Stanhill, Global dimming comes of age, *EOS*, **2004**, *85*, 362.

54.   a) F. Pearce, Heat will soar as haze fades, *New Sci.*, **2003**, 7 June, 7. b) Q. Schiermeier, Clear skies raise global-warming estimates, *Nature*, **2005**, *435*, 1142.

55.   P. D. Jones and T. M. L. Wigley, Global warming trends, *Sci. Am.*, **1990**, *263*, Aug., 66.

56.   R. A. Kerr, Hottest year, by a hair, *Science*, **1998**, *279*, 315.

57.   a) G. Vogel and A. Lawler, Hot year, but cool response in Congress, *Science*, **1998**, *280*, 1684. b) B. Hileman, Case grows for climate change, *Chem. Eng. News*, **1999**, *77*, 9 Aug., 16.

58.   a) J. Horgan, Volcanic disruption, *Sci. Am.*, **1992**, *266*, March, 16. b) M. R. Rampino and S. Self, The atmospheric effects of El Chichón, *Sci. Am.*, **1994**, *250*, Jan., 34. c) D. Schneider, A river (of mud) still runs through it, *Sci. Am.*, **1995**, *273*, July, 17.

59.   R. S. Bradley, M. K. Hughes and H. F. Diaz, Climate in medieval time, *Science*, **2003**, *302*, 404.

60.   a) F. Pearce, Climatologists hit back at greenhouse sceptics, *New Sci.*, **2003**, 12 July, 5. b) M. E. Mann and P. D. Jones, Global surface temperatures over the past two millennia, *Geophys. Res. Lett.*, **2003**, *30*, 1820, doi:1029/**2003**GL017814.

61.   J. Luterbacher *et al.*, European seasonal and annual temperature variability, trends and extremes since 1500, *Science*, **2004**, *303*, 1499.

62.   a) H. N. Pollack and D. S. Chapman, Underground records of changing climate, *Sci. Am.*, **1993**, *268*, June, 16. b) D. Deming, Climatic warming in North America: analysis of borehole temperature, *ibid.*, 1576.

63.   a) J. T. Overpeck, The hole record, *Nature*, **2000**, *403*, 714. b) S. Huang, H. N. Pollack and P.-Y. Shen, Temperature trends over the past five centuries reconstructed from borehole temperatures, *Nature*, **2000**, *403*, 756.

64.   a) W. Haeberli and M. Beniston, Climate change and its impact on glaciers and permafrost in the Alps, *Ambio*, **1998**, *27*, 258. b) W. Krabill *et al.*, Rapid thinning of parts of the southern Greenland ice sheet, *Science*, **1999**, *283*, 1522. c) D. Dahl-Jensen, The Greenland ice sheet reacts, *Science*, **2000**, *289*, 404. d) R. Thomas *et al.*, Mass balance of the Greenland ice sheet at high elevations, *ibid.*, 426. e) W. Krabill *et al.*, Greenland ice sheet; High-elevation balance and peripheral thinning, *ibid.* 428. f) C. S. Hvidberg, When Greenland ice melts, *Nature*, **2000**, *404*, 551. g) R. Irion, The melting snows of Kilimanjaro, *Science*, **2001**, *291*, 1690. h) B. Hileman, Climate change, *Chem. Eng. News*, **2001**, *79*, 10 April, 9. i) K. Alverson *et al.*, A global paleoclimate observing system, *Science*, **2001**, *293*, 47. j) N. McDowell, Melting ice triggers Himalayan flood warning, *Nature*, **2002**, *416*, 776. k) H. J. Zwally *et al.*, Surface melt-induced acceleration of Greenland ice-sheet flow, **2002**, *297*, 218. l) M. F. Meier and M. B. Dyurgerov, How Alaska affects the world, *ibid.*, 350. m) A. A. Arendt *et al.*, Rapid wastage of Alaska glaciers and their contribution to rising sea level, *ibid.*, 382. n) F. Gasse, Kilimanjaro's secrets revealed, *Science*, **2002**, *298*, 548. o) L. G. Thompson *et al.*, Kilimanjaro ice core records: Evidence of Holocene climate change in tropical Africa, *ibid.*, 589. p) F. Pearce, Meltdown, *New Sci.*, **2002**, 2 Nov., 44. q) R. F. Service, As the West goes dry, *Science*, **2004**, *303*, 1124.

65.   a) S. Levitus *et al.*, Warming of the world ocean, *Science*, **2000**, *287*, 2225. b) M. Fukasawa *et al.*, Bottom water warming in the North Pacific Ocean, *Nature*, **2004**, *427*, 825. c) G. C. Hegerl and N. Bindoff, Warming the world's oceans, *Science*, **2005**, *309*, 254. d) T. P. Barnett *et al.*, Penetration of human-induced warming into the world's oceans, *ibid.*, 284.

66.   C. S. M. Doake and D. G. Vaughan, Rapid disintegration of Wordie Ice Shelf in response to atmospheric warming, *Nature*, **1991**, *350*, 328.

67.   a) M. Fahnestock, An ice shelf breakup, *Science*, **1996**, *271*, 775. b) H. Rott, P. Skvarca and T. Nagler, Rapid collapse of northern Larsen ice shelf, Antarctica, *ibid.*, 788. c) C. S. M. Doake *et al.*, Breakup and conditions for stability of the northern Larsen Ice Shelf, Antarctica, *Nature*, **1998**, *391*, 778. d) G. Walker, Southern exposure, *New Sci.*, **1999**, 14 Aug., 42. e) D. G. Vaughan *et al.*, Devil in the detail, *Science*, **2001**, *293*, 1777. f) D. Malakoff, Antarctic Ice Shelf collapses, *Science*, **2002**, *295*, 2359. g) B. Hileman, Giant chunk of Antarctic ice shelf collapses, *Chem. Eng. News*, **2002**, *80*, 25 March, 12. h) R. A. Kerr, A single climate mover for Antarctica, *Science*, **2002**, *296*, 825. i) J. Kaiser, Breaking up is far too easy, *Science*, **2002**, *297*, 1494, 1996. j) J. Kaiser, Warmer ocean could threaten Antarctic ice shelves, *Science*, **2003**, *302*, 759. k) A. Shepherd *et al.*, Larsen Ice Shelf has progressively thinned, *Science*, **2003**, *302*, 856. l) B. Holmes, Melting ice, global warning, *New Sci.*, **2004**, 2 Oct., 8. m) E. Domack *et al.*, Stability of the Larsen B ice shelf on the Antarctic Peninsula during the Holocene epoch, *Nature*, **2005**, *436*, 681.

68.   a) D. G. Vaughan, How does the Antarctic ice sheet affect sea level rise? *Science*, **2005**, *308*, 1877. b) C. H. Davis *et al.*, Snowfall-driven growth in East Antarctic ice sheet mitigates recent sea-level rise, *ibid.*, 1898.

69.   a) F. Pearce, The icehouse effect, *New Sci.*, **2002**, 1 June, 6. b) H. J. Zwally *et al.*, Variability of Antarctic sea ice 1979-1998, *J. Geophys. Res.*, **2005**, *107* (C5), 3041, doi:10.1029/2000JC000733.

70.   a) D. W. J. Thompson and S. Solomon, Interpretation of recent southern hemisphere climate change, *Science*, **2002**, *296*, 895. b) D. J. Karoly, Ozone and climate change, *Science*, **2003**, *302*, 236. c) N. P. Gillett and D. W. J. Thompson, Simulation of recent southern hemisphere climate change, *Science*, **2003**, *302*, 273.

71.   a) E. Murphy and J. King, Icy message from the Antarctic, *Nature*, **1997**, *389*, 20. b) W. K. de la Mare, Abrupt mid-twentieth-century decline in Antarctic sea-ice extent from whaling records, *ibid.*, 57. c) D. MacKenzie, Sea ice meltdown, *New Sci.*, **1997**, 6 Sept., 4. d) E. W. Wolff, Whither Antarctic sea ice? *Science*, **2003**, *302*, 1164. e) M. A. J. Curran *et al.*, Ice core evidence for Antarctic sea ice decline since the 1950's, *ibid.* 1203.

72.   a) H. Conway *et al.*, Past and future grounding-line retreat of the West Antarctic ice sheet, *Science*, **1999**, *286,* 280. b) S. Simpson, Melting away, *Sci. Am.*, **2000**, *282*, Jan., 14. c) R. A. Bindschadler and C. R. Bentley, On thin ice, *Sci. Am.*, **2002**, *287*, Dec., 66. d) R. A. Ackert, An ice sheet remembers, *Science*, **2003**, *299*, 57. e) J. O. Stone, Holocene deglaciation of Marie Byrd Land, West Antarctica, *ibid.*, 99.

73.   a) J. Horgan, Antarctic meltdown, *Sci. Am.*, **1993**, *268*, March, 7. b) C. R. Bentley, Rapid sea-level rise soon from West Antarctic ice sheet collapse, *Science*, **1997**, *275*, 1077. c) W. W. Nicholls, Predicted reduction in basal melt rates of an Antarctic ice shelf in a warmer climate, *Nature*, **1997**, *388*, 460. d) R. A. Kerr, Signs of past collapse beneath Antarctic ice, *Science*, **1998**, *281*, 17. e) R. P. Schere *et al.*, Pleistocene collapse of the West Antarctic ice sheet, *ibid.*, 82. f) R. Bindschadler, Future of the West Antarctic ice sheet, *Science*, **1998**, *282*, 428. g) C. R. Bentley, Ice on the fast track, *Nature*, **1998**, *394*, 21.

74.   a) P. Barrett, Cooling a continent, *Nature*, **2003**, *421*, 221. b) R. M. DeConto and D. Pollard, Rapid Cenozic glaciation of Antarctica induced by declining atmospheric CO_2, *ibid.*, 245.

75.   J. R. Petit *et al.*, Climate and atmospheric history of the past 420,000 years from the Vostok ice core, Antarctica, *Nature*, **1999**, *399*, 429.

76.   D. M. Sigman and E. A. Boyle, Glacial/interglacial variations in atmospheric carbon dioxide, *Nature*, **2000**, *407*, 859.

77.   a) C. Lorius *et al.*, The ice-core record: climate sensitivity and future greenhouse warming, *Nature*, **1990**, *347*, 139. b) W. Dansgaard *et al.*, Evidence for general instability of past climate from a 250-kyr ice-core record, *Nature*, **1993**, *364*, 218. c) J. Chappellaz *et al.*, Synchronous changes in atmospheric CH_4 and Greenland climate between 40 and 8 kyr BP, *Nature*, **1993**, *366*, 443. d) C. Lorius and H. Oeschger, Palaeo-perspectives: reducing uncertainties in global change? *Ambio*, **1994**, *23*, 30. e) E. J. Brook, T. Sowers and J. Orchardo, Rapid variations in atmospheric methane concentration during the last 110,000

years, *Science*, **1996**, *273*, 1087. f) M. Legrand and P. Mayewski, Glaciochemistry of polar ice cores: A review, *Rev. Geophys.*, **1997**, *35*, 219. g) J. R. Petit *et al.*, Four climate cycles in Vostok ice-core, *Nature*, **1997**, *387*, 359. h) W. R. Howard, A warm future in the past, *Nature*, **1997**, *388*, 418. i) R. B. Alley and M. L. Bender, Greenland ice cores: Frozen in time, *Sci. Am.*, **1998**, *278*, Feb., 66.

78.   a) F. D. Carsey and J. C. Horvath, The lake that time forgot, *Sci. Am.*, **1999**, *281*, Oct., 42. b) M. J. Siegert, Antarctica's Lake Vostok, *Am. Sci.*, **1999**, *87*, Nov.-Dec., 510. c) W. F. Vincent, Icy life on a hidden lake, *Science*, **1999**, *286*, 2094. d) C. R. Bentley, Stirring the icy waters, *Nature*, **2000**, *403*, 610. e) H. Gavaghan, Life in the deep freeze, *Nature*, **2002**, *415*, 828.

79.   a) G. Walker, Frozen time, *Nature*, **2004**, *429*, 596. b) J. F. McManus, A great grand-daddy of ice cores, *ibid.*, 611. c) EPICA community members, Eight glacial cycles from an Antarctic ice core, *ibid.*, 623.

80.   a) F. A. Street-Perrott, Paleo-perspectives: Changes in terrestrial ecosystems, *Ambio*, **1994**, *23*, 37. b) R. Corfield, Close encounters with crystalline gas, *Chem. Br.*, **2002**, *38*, May, 22.

81.   a) M. Milankovitch, *Mathematische Klimalehre und Astronomische Theorie der Khmaschwankungen*, Gebruder Borntreger, Berlin, Germany, 1930. b) J. D. Hays, J. Imbrie and N. J. Shackleton, Variations in the Earth's orbit: Pacemaker of the ice ages, *Science*, 1976, *194*, 1121. c) J. Imbrie and K. P. Imbrie, *Ice Ages: Solving the mystery*, Harvard University Press, Cambridge, USA, 1979. d) C. Covey, The Earth's orbit and the ice ages, *Sci. Am.*, **1984**, *250*, Feb., 42. e) A. G. Dawson, *Ice Age Earth*, Routledge, London, UK, 1992. f) G. Henderson, Deep freeze, *New Sci.*, **1998**, 14 Feb., 28.

82.   W. S. Broecker and G. H. Denton, What drives glacial cycles? *Sci. Am.*, **1990**, *262*, Jan., 42.

83.   J. Croll, *Climate and Time*, Appleton, New York, USA, 1875.

84.   a) A. J. Weaver *et al.*, Simulated influence of carbon dioxide, orbital forcing and ice sheets on the climate of the last glacial maximum, *Nature*, **1998**, *394*, 847. b) M. E. Raymo, Glacial puzzles, *Science*, **1998**, *281*, 1467. c) R. A. Kerr, Why the ice ages don't keep time, *Science*, **1999**, *285*, 503. d) J. A. Rial, Pacemaking the ice ages by frequency modulation of Earth's orbital eccentricity, *ibid.*, 564. e) P. U. Clark, R. B. Alley and D. Pollard, Northern hemisphere ice-sheet influences on global climate change, *Science*, **1999**, *286*, 1104. f) G. M. Henderson and N. C. Slowey, Evidence from U-Th dating against northern hemisphere forcing of the penultimate deglaciation, *Nature*, **2000**, *404*, 61. g) D. P. Schrag, Of ice and elephants, *Nature*, **2000**, *404*, 23. h) D. B. Karner and R. A. Muller, A causuality problem for Milankovitch, *Science*, **2000**, *288*, 2143. i) R. A. Kerr, Ice, mud point to CO_2 role in glacial cycle, *Science*, **2000**, *289*, 1868. j) N. Shackleton, The 100,000-year ice-age cycle identified and found to lag temperature, carbon dioxide, and orbital eccentricity, *ibid.*, 1897. k) M. Khodri *et al.*, Simulating the amplification of orbital forcing by ocean feedbacks in the last glaciation, *Nature*, **2001**, *410*, 570. l) R. A. Kerr, An orbital confluence leaves its mark, *Science*, **2001**, *292*, 191. m) J. C. Zachos *et al.*, Climate responses to orbital forcing across the Oligocene-Miocene boundary, *ibid.*, 274.

85.   a) E. N. Parker, Sunny side of global warming, *Nature*, **1999**, *399*, 416. b) J. Lean and D. Rind, Earth's response to a variable sun, *Science*, **2001**, *292*, 234.

86.   a) W. Karlén, Global temperature forced by solar irradiation and greenhouse gases? *Ambio*, **2001**, *30*, 349. b) R. Rind, The sun's role in climate variations, *Science*, **2002**, *296*, 673.

87.   A. Berger and M. F. Loutre, An exceptionally long interglacial ahead, *Science*, **2002**, *297*, 1287.

88.   a) J. L. Sarmiento, Ocean carbon cycle, *Chem. Eng. News*, **1993**, *71*, 31 May, 30. b) P. P. Tans and P. S. Bakwin, Climate change and carbon dioxide, *Ambio*, **1995**, *24*, 374.

89.   a) K. Caldeira and M. E. Wickett, Anthropogenic carbon and ocean pH, *Nature*, **2003**, *425*, 365. b) Q. Schiermeier, Researchers seek to turn the tide on problem of acid seas, *Nature*, **2004**, *430*, 820. c) R. A. Feely *et al.*, Impact of anthropogenic CO_2 on the $CaCO_3$ system in the oceans, *Science*, **2004**, *305*, 362.

90.   a) P. G. Falkowski, R. T. Barber and V. Smetacek, Biogeochemical controls and feedbacks on ocean primary production, *Science*, **1998**, *281*, 200. b) C. Field *et al.*, Primary

production of the biosphere: integrating terrestrial and oceanic components, *ibid.*, 237. c) P. G. Falkowski, The ocean's invisible forest, *Sci. Am.*, **2002**, *287*, Aug., 38.

91.   a) S. C. Doney, The ocean's productive deserts, *Nature*, **1997**, *389*, 905. b) S. Emerson *et al.*, Experimental determination of the organic carbon flux from open-ocean surface waters, *ibid.*, 951.

92.   a) W. S. Broecker, Chaotic climate, *Sci. Am.*, **1995**, *273*, Nov., 44. b) S. Rahmstorf, Ice-cold in Paris, *New Sci.*, **1997**, 8 Feb., 26. c) S. Rahmstorf, Risk of sea-change in the Atlantic, *Nature*, **1997**, *388*, 826. d) T. F. Stocker and A. Schmittner, Influence of CO_2 emission rates on the stability of the thermohaline circulation, *ibid.*, 862. e) W. S. Broecker, Thermohaline circulation, the Achilles heel of our climate system: Will man-made CO_2 upset the current balance? *Science*, **1997**, *278*, 1582. f) R. A. Kerr, Warming's unpleasant surprise: shivering in the greenhouse? *Science*, **1998**, *281*, 156. g) D. Cromwell, Ocean circulation, *New Sci.*, *Inside Science 130*, **2000**, 20 May, 1. h) A. Ganachaud and C. Wunsch, Improved estimates of global ocean circulation, heat transport and mixing from hydrographic data, *Nature*, **2000**, *408*, 453. i) W. S Broecker and S. Hemming, Climate swings come into focus, *Science*, **2001**, *294*, 2308. j) S. Rahmsdorf, Ocean circulation and climate change during the past 120,000 years, *Nature*, **2002**, *419*, 207. k) T. F. Stocker, North-south connections, *Science*, **2002**, *297*, 1814. l) V. Morgan *et al.*, Relative timing of deglacial climate events in Antarctica and Greenland, *ibid.* 1862. m) R. A. Kerr, Mild winters mostly hot air, not gulf stream, *ibid.*, 2202. n) B. Hansen *et al.*, Already the day after tomorrow? *Science*, **2004**, *305*, 953. o) R. B. Alley, Abrupt climate change, *Sci. Am.*, **2004**, *291*, Nov., 40.

93.   a) F. Pearce, Will a sea change turn up the heat? *New Sci.*, **1996**, 30 Nov., 16. b) B. Dickson, From the Labrador Sea to global change, *Nature*, **1997**, *386*, 649. c) S. Lehman, Sudden end of an interglacial, *Nature*, **1997**, *390*, 117. d) J. F. Adkins *et al.*, Variability of the North Atlantic thermohaline circulation during the last interglacial period, *ibid.*, 154. e) W. H. Calvin, The great climate flip-flop, *The Atlantic Monthly*, **1998**, Jan., 47. f) T. F. Stocker, A glimpse of the glacial, *Nature*, **1998**, *391*, 338. g) F. Pearce, Quick change, *New Sci.*, **1998**, 14 Nov., 15. h) R. A. Kerr, The Little Ice Age - only the latest big chill, *Science*, **1999**, *284*, 2069. i) K. Taylor, Rapid climate change, **1999**, 87, 320. k) B. Edwards, Freezing future, *New Sci.*, **1999**, 27 Nov., 6. j) P. U. Clark *et al.*, The role of the thermohaline circulation in abrupt climate change, *Nature*, **2002**, *415*, 863. k) M. Visbeck, The Ocean's role in Atlantic climate variability, **2002**, *297*, 2223. l) W. S. Broecker, Does the trigger for abrubt climate change reside in the ocean or in the atmosphere? *Science*, **2003**, *300*, 1519.

94.   a) P. Williamson and H. Oeschger, Climate instability in a warmer world, *Ambio*, **1993**, *22*, 411. b) K. C. Taylor *et al.*, The Holocene-Younger Dryas transition recorded at Summit, Greenland, *Science*, **1997**, *278*, 825. c) J. P. Severinghaus *et al.*, Timing of abrupt climate change at the end of the Younger Dryas interval from thermally fractionated gases in polar ice, *Nature*, **1998**, *391*, 141.

95.   a) R. A. Kerr, Millennial climate oscillation spied, *Science*, **1996**, *271*, 146. b) R. A. Kerr, Sea floor records reveal interglacial climate cycles, *Science*, **1998**, *279*, 1304. c) D. Oppo, J. F. McManus and J. L. Cullen, Abrupt climate events 500,000 to 340,000 years ago: Evidence from subpolar North Atlantic sediments, *Science*, **1998**, *279*, 1335.

96.   L. Labeyrie, Glacial climate instability, *Science*, **2000**, *290*, 1905.

97.   a) R. A. Kerr, A variable sun paces millennial climate, *Science*, **2001**, *294*, 1431. b) G. Bond *et al.*, Persistent solar influence on North Atlantic climate during the Holocene, *ibid.*, 2130.

98.   a) W. S. Broecker, Massive iceberg discharges as triggers for global climate change, *Nature*, **1994**, *372*, 421. b) F. Pearce, Feel the pulse, *New Sci.*, **2000**, 2 Sept., 30.

99.   a) P. U. Clark *et al.*, Fresh water forcing of abrupt climate change during the last glaciation, *Science*, **2001**, *293*, 283. b) J. T. Teller, D. W. Leverington and J. D. Mann, Fresh water outbursts to the oceans from glacial Lake Agassiz and their role in climate change during

the last deglaciation, *Quat, Sci. Rev.*, **2002**, *21*, 879. c) S. M. Colman, A fresh look at glacial floods, *Science*, **2002**, *296*, 1251. b)

100. a) D. C. Barber *et al.*, Forcing of the cold event of 8,200 years ago by catastrophic drainage of Laurentide lakes, *Nature*, **1999**, *400*, 344. b) G. Clarke *et al.*, Superlakes, megafloods, and abrupt climate change, *Science*, **2003**, *301*, 922.

101. a) J. C. White and E. J. Steig, Timing is everything in a game of two hemispheres, *Nature*, **1998**, *394*, 717. b) T. Bunier *et al.*, Asynchrony of Antarctic and Greenland climate change during the last glacial period, *ibid.*, 739. c) N. Shackleton, Climate change across the hemispheres, *Science*, **2001**, *291*, 58. d) T. Blunier and E. J. Brook, Timing of millenial-scale climate change in Antarctica and Greenland during the last glacial period, *ibid.*, 109.

102. a) R. A. Kerr, Will the Arctic Ocean lose all of its ice? *Science*, **1999**, *286*, 1828. b) K. Y. Vinnikov *et al.*, Global warming and northern hemisphere sea ice extent, *ibid.*, 1934. c) O. M. Johannessen, E. V. Shalina and M. W. Miles, Satellite evidence for an Arctic sea ice cover in transformation, *ibid.*, 1937. d) D. MacKenzie, Arctic meltdown, *New Sci.*, **2002**, 2 March, 5. e) R. A. Kerr, A warmer Arctic means a change for all, *Science*, **2002**, *297*, 1490. f) S. Laxon, N. Peacock and D. Smith, High interannual variability of sea ice thickness in the Arctic region, *Nature*, **2003**, *425*, 947. g) R. A. Kerr, Scary Arctic ice loss? Blame the wind, *Science*, **2005**, *307*, 203. h) R. Curry and C. Mauritzen, Dilution of the northern North Atlantic Ocean in recent decades, *Science*, **2005**, *308*, 1772.

103. J. Overpeck *et al.*, Possible role of dust-induced regional warming in abrupt climate change during the last glacial period, *Nature*, **1996**, *386*, 447.

104. a) M. E. Katz *et al.*, The source and fate of massive carbon input during the latest Paleocene thermal maximum, *Science*, **1999**, *286*, 1531. b) J. P. Kennett *et al.*, Carbon isotope evidence for methane hydrate instability during Quarternary interstadials, *Science*, **2000**, *288*, 128.

105. a) R. S. Bradley *et al.*, ENSO signal in continental temperature and precipitation records, *Nature*, **1987**, *327*, 497. b) D. B. Enfield, Historical and prehistorical overview of El Niño/Southern Oscillation, in *El Niño: Historical and Paleoclimatic Aspects of the Southern Oscillation*, H. F. Diaz and V. Markgraf, eds., Cambridge University Press, Cambridge, UK, 1992. c) M. Golnaraghi and R. Kaul, Responding to ENSO, *Environment*, **1995**, 37, Jan./Feb., 16. d) M. Gantz, *Currents of Change: El Niño's Impact on Climate and Society*, Cambridge University Press, Cambridge, UK, 1996. e) R. Allan, J. Lindesay, and D. Parker, *El Niño Southern Oscillation and Climatic Variability*, CSIRO, Collingwood, Australia, 1996. f) P. J. Webster and J. A. Curry, The oceans and weather, *Sci. Am. Quart, The Oceans*, **1998**, *9*, Fall, 38. g) J. Gribbin and M. Gribbin, Big Weather, *New Sci.*, *Inside Science 120*, **1999**, 22 May, 1. h) F. Pearce, Weather warning, *New Sci.*, **1999**, 3 Oct., 36. i) J. Cole, A slow dance for El Niño, *Science*, **2001**, *291*, 1496. j) C. M. Moy *et al.*, Variability of El Niño/Southern Oscillation activity at millennial timescales during the Holocene epoch, *Nature*, **2002**, *420*, 162.

106. a) P. J. Webster and T. N. Palmer, The past and the future of El Niño, *Nature*, **1997**, *390*, 562. b) R.-H. Zhang, L.M. Rothstein and A. J. Busalacchi, Origin of upper-ocean warming and El Niño change on decadal scales in the tropical Pacific Ocean, *Nature*, **1998**, *391*, 879. c) M. J. McPhaden, Genesis and evolution of the 1997-98 El Niño, *Science*, **1999**, *283*, 950. d) P. J. Webster *et al.*, Coupled ocean-atmosphere dynamics in the Indian Ocean during 1997-98, *Nature*, **1999**, *401*, 356. e) F. P. Chaves *et al.*, Biological and chemical response of the equatorial Pacific Ocean to the 1997-1998 El Niño, *Science*, **1999**, *286*, 2126. f) R. A. Kerr, Signs of success in forecasting El Niño, *Science*, **2002**, *297*, 497. g) R. A. Kerr, A surprise La Niña, *Science*, **2003**, *300*, 1215. h) F. Pearce, The mother of all El Niños revealed, *New Sci.*, 18 Jan., **2003**, 4. i) F. P. Chavez *et al.*, From anchovies to sardines and back: Multidecadal change in the Pacific Ocean, *Science*, **2003**, *299*, 217.

107. a) K. S. Jayaraman, Monsoon prediction model, *Nature*, **1989**, *342*, 4. b) F. Pearce, Sneaky El Niño outwits weather forecasters, *New Sci.*, **1997**, 31 May, 6. c) E. Masood, El Niño forecast fails to convince sceptics, *Nature*, **1997**, *388*, 108. d) K. S. Jayaraman, Indian

monsoon 'progressing normally', *ibid.*, 108. e) C. Holden, Monsoons defy El Niño, *Science*, **1997**, *278*, 581.

108. a) R. A. Kerr, Big El Niños ride the back of slower climate change, *Science*, **1999**, *283*, 1108. b) A. Timmermann *et al.*, Increased El Niño frequency in a climate model forced by future greenhouse warming, *Nature*, **1999**, *398*, 694. c) J. Cole, A slow dance for El Niño, *Science*, **2001**, *291*, 1496.

109. C. D. Charles, D. E. Hunter and R. G. Fairbanks, Interaction between the ENSO and the Asian monsoon in a coral record of tropical climate, *Science*, **1997**, *277*, 925.

110. C. F. Ropelewski, Predicting El Niño, *Nature*, **1992**, *356*, 476.

111. a) M. McCartney, Is the ocean at the helm? *Nature*, **1997**, *388*, 521. b) R. T. Sutton and M. R. Allen, Decadal predictability of North Atlantic sea surface temperature and climate, *ibid.*, 563. c) R. A. Kerr, A new force in high-latitude climate, *Science*, **1999**, *284*, 241.

112. a) T. J. Goreau and R. L. Hayes, Coral bleaching and ocean "hot spots", *Ambio*, **1994**, *23*, 176. b) E. Pennisi, New threat seen from carbon dioxide, *Science*, **1998**, *279*, 989. c) C. Wilkinson *et al.*, Ecological and socioeconomic impacts of 1998 coral mortality in the Indian Ocean: An ENSO impact and a warning of future change? *Ambio*, **1999**, *28*, 188. d) P. Pockley, Global warming "could kill most coral reefs by 2100", *Nature*, **1999**, *400*, 98. e) J. A. Kleypas *et al.*, Geochemical consequences of increased atmospheric carbon dioxide on coral reefs, *Science*, **1999**, *284*, 118. f) B. E. Brown *et al.*, Bleaching patterns in reef corals, *Nature*, **2000**, *404*, 142. g) J.-P. Gattuso and R. W. Buddemeier, Calcification and CO_2, *Nature*, **2000**, *407*, 311. h) U. Riebesell *et al.*, Reduced calcification of marine plankton in response to increased atmospheric CO_2, *ibid.*, 364. i) P. Pockley, Global warming identified as main threat to coral reefs, *ibid.*, 932. j) D. Normile, Warmer waters more deadly to coral reefs than pollution, *Science*, **2000**, *290*, 682. k) F. Pearce, Grief on the reef, *New Sci.*, **2002**, 20 April, 11. l) E. Pennisis, Survey confirms coral reefs are in peril, *Science*, **2002**, *297*, 1622. m) R. Nowak, Great Barrier bluff, *New Sci.*, **2003**, 4 Jan., 8. n) E. Wolanski *et al.*, Mud, marine snow and coral reefs, *Am. Sci.*, **2003**, 91, 44. o) M. McCulloch *et al.*, Coral record of increased sediment flux to the inner Great Barrier Reef since European settlement, *Nature*, **2003**, *421*, 727. p) D. R. Bellwood *et al.*, Confronting the coral crisis, *Nature*, **2004**, *429*, 827.

113. C. C Johnson, The rise and fall of rudist reefs, *Am. Sci.*, **2002**, *90*, 149.

114. a) *United Nations Framework Convention on Climate Change*, United Nations, New York, USA, 1992. b) *Caring for Climate. A guide to the climate Change Convention and the Kyoto Protocol*, UNFCCC, Bonn, Germany, 2003.

115. a) F. Pearce, Chill winds at the summit, *New Sci.*, **1997**, 1 March, 12. b) P. Pockley, Australia plays hard to get, *New Sci.*, **1997**, 27 Nov., 323. c) K. Ramakrishna, The great debate on CO_2 emissions, *Nature*, **1997**, *390*, 227. d) M. Hulme and M. Parry, Whistling in the dark, *New Sci.*, **1997**, 6 Dec., 51.

116. a) B. Hileman, Kyoto climate conference, *Chem. Eng. News*, **1997**, *75*, 22 Dec., 20. b) B. Bolin, The Kyoto negotiations on climate change: A science perspective, *Science*, **1998**, *279*, 330.

117. F. Pearce, Pollution detectives add to greenhouse woes, *New Sci.*, **1996**, 24 Aug., 6.

118. a) C. M. Cooney, Nations seek "fair" greenhouse gas treaty in Kyoto, *Env. Sci. Technol.*, **1997**, *31*, 516 A. b) J. Lanchbery, Expectations for the climate talks in Buenos Aires, *Environment*, **1998**, *40*, Oct., 16. c) J. Reilly *et al.*, Multi-gas assessment of the Kyoto protocol, *Nature*, **1999**, *401*, 549. d) F. D'Evie and J. Taylor, Greenhouse gas emission abatement: Equitable burden sharing, *Ambio*, **1999**, *28*, 148. e) D. Dickson, Deadlock in The Hague, but hopes remain for spring climate deal, *Nature*, **2000**, *408*, 503. f) R. A. Kerr, Can the Kyoto climate treaty be saved from itself? *Science*, **2000**, *290*, 920. g) B. Hileman, Deadlock on climate change, *Chem. Eng. News*, **2000**, *78*, 4 Dec., 11. h) B. Hileman, Climate treaty stalemate, *Chem. Eng. News*, **2000**, *78*, 18 Dec., 17. i) T. Blundell, After The Hague, *Chem. Indust.*, **2001**, Jan., 10. j) B. Hileman, Bush changes course on CO_2, *Chem. Eng. News*, **2001**, *79*, 19 March, 10. k) B. Hileman, U.S. abandons Kyoto protocol, *ibid.*, 2 April, 17. l) F. Pearce, A real roasting, *New Sci.*, **2001**, 7 April, 11. m) J. Giles, 'Political

fix' saves Kyoto deal from collapse, *Nature* **2001**, *412*, 365. n) W. D. Nordhaus, Global warming economics, *Science*, **2001**, *294*, 1283. o) R. T. Watson, Climate change: The political situation, *Science*, **2003**, *302*, 1925.

119.   *Anon.*, Climate respite? *New Sci.*, **2005**, 19 Feb., 4.
120.   B. Hileman, How to reduce greenhouse gases, *Chem. Eng. News*, **2002**, *80*, 27 May, 37.
121.   F. Pearce, Kyoto won't stop climate change, *New Sci.,* **2004**, 9 Oct., 6.
122.   a) J. D. Mahlman, Uncertainties in projections of human-caused climate warming, *Science*, **1997**, *278*, 1416. b) K. Schmidt, Coming to grips with the world's greenhouse gases, *Science*, **1998**, *281*, 504. c) A. J. Weaver and F. W. Zwiers, Uncertainty in climate change, *Nature*, **2000**, *407*, 571. d) M. R. Allen *et al.*, Quantifying the uncertainty in forecasts of anthropogenic climate change, *ibid.*, 617. e) F. W. Zwiers and A. J. Weaver, The causes of 20th century warming, *Science*, **2000**, *290*, 2081. f) B. Hileman, Pace of global change quickens, *Chem. Eng. News*, **2001**, *79*, 29 Jan., 9. g) *Anon.*, IPCC says global warming is our own fault, *Chem. Indust.*, **2001**, 5 Feb., 66. h) S. Leviticus *et al.*, Anthropogenic warming of Earth's climate system, *Science,* **2001**, *292*, 267. i) S. H. Schneider, What is 'dangerous' climate change? *Nature*, **2001**, *411*, 17. j) K. E. Trenberth, Climate variability and global warming, *Science*, **2001**, *293*, 48. k) J. Reilly *et al.*, Uncertainity and climate change assessments, *ibid.*, 430. l) M. Allen, S. Raper and J. Mitchell, Uncertainty in the IPCC's third assessment report, *ibid.*, 430. m) T. R. Karl and K. E Trenberth, Modern global climate change, *Science*, **2003**, *302*, 1719. n) Q. Schiermeier, Modellers deplore 'short-termism' on climate, *Nature*, **2004**, *428*, 593.
123.   a) K. Hasselmann *et al.*, The challenge of long-term climate change, *Science*, **2003**, *302*, 1923. b) B. Hileman, Stark effects from global warming, *Chem. Eng. News*, **2005**, *83*, *21*, March, 47.

Chapter 9

The Ozone Layer: Earth's Stratospheric Defence

"Keeping the global environment livable in the face of the rapidly increasing global population and the universal desire for a higher standard of living presents mankind's greatest challenge for the 21st century."

F. Sherwood Rowland, 1995 Nobel Laureate, in *Ambio*, October 1990.

9.1. The Atmosphere and the Ozone Layer

The rarefied air of the stratosphere and beyond provides humanity and life generally with a vital protection from intense high energy solar radiation. An important part of this protection is provided by ozone, a surprisingly small component of the air in the stratosphere that is usually referred to as the ozone layer. From the late 1970s, the news media carried articles about the ozone layer and an ozone hole over Antarctica at frequent intervals and in some cases called this hole "a hole in the sky". Such a description suggested that something spectacular and worrying had occurred and this was indeed the case [1-5]. To gain an understanding of this it is necessary to first look at the atmosphere and its composition. Despite its apparent continuity and almost total transparency, the atmosphere consists of well-defined regions as shown in Fig. 9.1. The lowest region, the troposphere, that contains about eighty-five percent of the air, extends upwards to twelve kilometres in altitude in the mid latitudes, as high as fifteen kilometres above the equator and as low as ten kilometres above the poles. Because equatorial air is warmed more than that over the poles it rises and polar air flows towards the equator at lower altitudes to take its place. This circulating airflow is complex and mixes air both vertically and horizontally to make the troposphere a turbulent region in which clouds are formed and weather patterns are established.

As altitude increases the air thins and the troposphere cools but with sufficient warmth being retained close to Earth's surface to maintain biologically benign temperatures because of the warming greenhouse effect as discussed in Chapter 8. This warming is a consequence of the troposphere absorbing much of the infrared radiation emitted from Earth's surface rather than a direct warming by sunlight. As a result, the troposphere cools as altitude increases until a broad boundary layer is reached where the density of air is about one percent of that at sea level. This is the tropopause that separates the troposphere from the stratosphere. Now the cooling trend reverses as the comparatively tranquil stratosphere is entered. Here, the absorption of ultraviolet radiation from the sun by oxygen, O_2, and the consequent production of ozone, O_3, produces a warm layer of stratospheric air over the cold upper troposphere, or an inversion layer. This slows the mixing of air between the troposphere and stratosphere so that individual molecules can take a year or more to pass from one to the other.

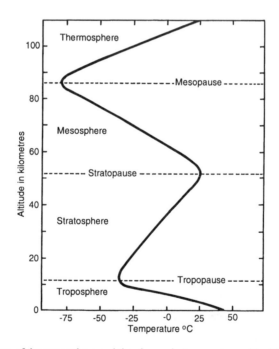

Fig. 9.1. The structure of the atmosphere and the change in temperature with altitude.

As the ascent through the rapidly thinning air of the stratosphere is continued, the temperature rises further until another broad boundary is reached, the stratopause, at about fifty kilometres altitude. Here, the air density is only about 0.01 percent of its density at sea level. As this boundary is passed the mesosphere is entered and the air cools rapidly with increase in altitude as the amount of oxygen and ozone decreases rapidly and the absorption of ultraviolet radiation diminishes. The next boundary encountered is the mesopause at about eighty-five kilometres altitude where the air density falls to about 0.001 percent of its value at sea level. Now the thermosphere is entered and temperature starts to rise again as very high energy ultraviolet-C radiation is absorbed to ionise, or remove an electron from, oxygen and nitrogen molecules. As altitude increases further the air continues to thin until the boundary with space is reached at about four hundred kilometres altitude.

It is apparent from this brief exploration of the atmosphere that Earth is enveloped in several distinct layers of air showing different temperature variations. Together, these layers act as both thermal insulation and as a thermostat as Earth basks in the sunlight. Dry air consists of 78.09 percent nitrogen, N_2, 20.94 percent oxygen, O_2, by volume and most of the remainder is unreactive argon, Ar, at 0.93 percent together with a very small, but increasing amount of carbon dioxide, CO_2, at about 0.038 percent. The atmosphere also contains a minuscule amount of ozone, so small that for every ten million molecules of air only three are ozone. Ninety percent of it is concentrated in the ozone layer in the lower stratosphere between fifteen to twenty-five kilometres in altitude. In total there are about 3.3 billion tonnes of ozone in the atmosphere that, although large in absolute terms, is a very small proportion of the atmosphere. This is readily appreciated when it is realized that ozone would be left as a three millimetre deep layer of gas covering Earth's surface if all other atmospheric gases where removed.

The triatomic ozone molecule is made up of three oxygen atoms linked in a bent chain. It is produced in the stratosphere by the action of high energy ultraviolet-C radiation on oxygen. Because the amount of oxygen is small above thirty kilometres altitude not much ozone is produced there, and while there is plenty of oxygen in the troposphere very little ultraviolet-C radiation penetrates there and so little ozone is produced. As a result most ozone is found in the lower stratosphere where it constitutes the ozone layer. Ozone is a pungent and very reactive gas that is injurious to health. It is routinely produced by high energy ultraviolet lamps and is sometimes used in water purification and in paper bleaching. Even so, this scarcely sounds like the description of an atmospheric component essential to the survival of the biosphere. It is not until ultraviolet induced reactions in the stratosphere are considered that the vital importance of the ozone layer emerges.

9.2. Ozone in the Stratosphere

The presence of ozone in the stratosphere and its life protecting role together represent one of the vital balances that make Earth a habitable planet. The absorption of sunlight by diatomic oxygen produces triatomic ozone. Sunlight consists of a wide band of radiation whose energy increases as its wavelength decreases. It contains radiation visible to the human eye that ranges from wavelengths of 400 to 700 nanometres, or from violet to red. (A nanometre is one billionth of a metre.) Much of this visible radiation is absorbed by chlorophyll in green plants to power photosynthesis that sustains most life by generating plant growth that also provides food for herbivorous animals. Sunlight also contains infrared radiation with wavelengths greater than 700 nanometres that is felt as warmth and ultraviolet radiation with wavelengths less than 400 nanometres where that of wavelengths less than 320 nanometres causes sunburn, particularly in light-skinned people. It is the interaction of oxygen and ozone with sunlight in the stratosphere that protects humanity from the dangerous ultraviolet wavelengths of sunlight shorter than 320 nanometres to a major extent and totally from the high energy wavelengths shorter than 290 nanometres. This protection arises because the relatively plentiful diatomic oxygen in the stratosphere and above absorbs ultraviolet-C radiation with wavelengths of 42 to 242 nanometres and breaks into two very reactive oxygen atoms, or radicals, that react with diatomic oxygen to form triatomic ozone as shown in Fig. 9.2. The major constituent of air, diatomic nitrogen, also absorbs ultraviolet radiation and contributes to this complete filtering out of ultraviolet-C radiation.

Ultraviolet-B radiation with wavelengths of 320 nanometres or less splits ozone into diatomic oxygen and atomic oxygen, or an oxygen radical. The newly produced atomic oxygen, O, then either reacts with diatomic oxygen to regenerate ozone or with another ozone molecule to produce two diatomic oxygen molecules. The overall effect is that a constant level of ozone is maintained in the stratosphere by the action of sunlight and the chemical reactions that it induces. In this way, ozone absorbs most of the very dangerous DNA damaging ultraviolet-B radiation in the 290 to 320 nanometres wavelength range, and wise people do not excessively expose their skin to the small portion that reaches Earth's surface. As a consequence, it is predominantly ultraviolet-A radiation in the range 320 to 400 nanometres that reaches the surface.

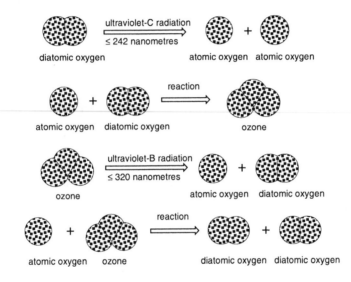

Fig. 9.2. The processes by which sunlight continuously produces ozone in the stratosphere and breaks it down to oxygen again in a cycle of reactions that maintains ozone at a constant level. Each chequered sphere represents an oxygen atom.

The absorption of ultraviolet-C radiation also causes warming of the stratosphere because the combination of atomic and diatomic oxygen forms an ozone molecule with excess energy that is transferred through collisions to the more plentiful diatomic oxygen and nitrogen molecules that vibrate and move more rapidly as a consequence. This is measured and felt as an increase in temperature. As has already been seen, this creates an inversion layer of warm stratospheric air above cold tropospheric air that encloses Earth and greatly slows the mixing of air between the warm stratosphere with the cold upper troposphere. Overall, the ozone layer is a vital component of Earth's ultraviolet screen, plays a significant part in the maintenance of Earth's temperature within habitable bounds and helps regulate the structure of the atmosphere.

9.3. Ozone Destruction

Other natural processes occurring in the stratosphere also affect the level of ozone and all would be well if they were unaffected by human activities. The natural chemistry of the stratosphere is complex and only its salient aspects are considered here. However, its importance cannot be overestimated and its study has absorbed the energies of large numbers of scientists, three of whom, Paul J. Crutzen, Mario L. Molina and F. Sherwood Rowland, were awarded Noble Prizes in 1995 [3-5]. Of the additional natural processes affecting the level of stratospheric ozone that they studied, the most important is the reaction of highly reactive atomic chlorine, Cl, or the chlorine radical, with ozone to produce diatomic oxygen and diatomic chlorine monoxide, ClO, and its subsequent reaction with atomic oxygen to regenerate atomic chlorine and diatomic oxygen as shown in Fig. 9.3. One ozone molecule is destroyed each time this cyclic process occurs. It is estimated that this cycle repeats itself about one hundred

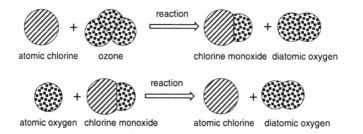

Fig. 9.3. The cyclic process through which atomic chlorine destroys ozone in the stratosphere. The chequered and hatched spheres represent oxygen and chlorine atoms, respectively.

thousand times so that a hundred thousand ozone molecules are destroyed by each atomic chlorine. Eventually, atomic chlorine is converted to hydrogen chloride, HCl, through other reactions and leaves the stratosphere to be washed out by rain in the troposphere as dilute hydrochloric acid. The natural source of atomic chlorine is biologically produced methyl chloride, CH_3Cl, that rises into the stratosphere where it is decomposed by the intense ultraviolet radiation. The extent of this effect of atomic chlorine on the natural ozone balance is now being greatly increased by additional atomic chlorine reaching the stratosphere because of the release of vast amounts of chlorofluorocarbons, or CFCs, into the atmosphere. This began to generate disquiet from the early 1970s onwards [6]. However, it was not until a decade later that a highly disconcerting event was to more sharply focus this disquiet, the discovery of the Antarctic ozone hole. (Nitric oxide and the hydroxyl radical are also important in natural ozone destruction but are not discussed here as their stratospheric levels have not been dangerously changed by human activities [2-4])

9.4. The Antarctic Ozone Hole

On 16 May 1985 there appeared in the journal Nature a letter revealing that the amount of ozone in the stratosphere above the British Antarctic Base at Halley Bay at 76°S had steadily declined each southern spring until in October 1985 it was down to about sixty-five percent of its 1957 level as seen in Fig. 9.4 [7]. The

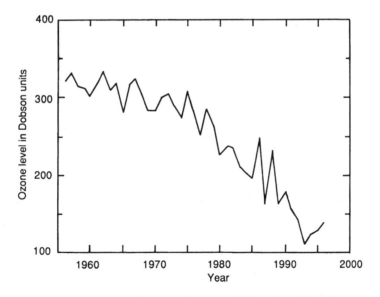

Fig. 9.4. The progressive depletion of the ozone layer over Halley Bay during the Antarctic spring. Data from [7] and [8].

scientists who made these observations, J. C. Franklin, B. G. Gardiner and J. D. Shanklin, further pointed out that this ozone depletion coincided with an increase in the amount of CFCs in the Antarctic air during the same period of time. However, from mid spring until the return of the continuous darkness of the Antarctic winter the levels of ozone built up again, but each subsequent year there was a trend of further depletion of ozone levels over Antarctica during the southern spring [7,8]. This was the first time that the opening of the "ozone hole" over Antarctica in the southern spring was reported and its extent has been continuously monitored from both satellites and ground stations ever since.

A Tiros N satellite map of the 1996 ozone hole and ozone levels in the Southern Hemisphere is shown in Fig. 9.5 and typifies this annual phenomenon.

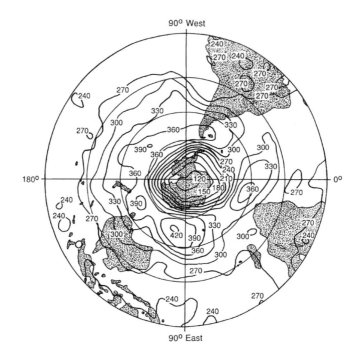

Fig. 9.5. The Antarctic ozone hole as seen on 6 October 1996 through total ozone vertical analysis from the Tiros N satellite. The ozone level contours are labelled in Dobson units. The region of low ozone resembles the Antarctic continent in shape and there is a ridge of high ozone levels over the southern parts of Africa, Australia and South America. The ozone level decreases as latitude decreases towards the tropical zone. The circles of increasing diameter outwards from the South Pole represent the latitudes 60°S, 40°S and 20°S, respectively. Based on data provided by the Climate Prediction Centre/NCEP/NWS/NOAA.

In the Antarctic spring of 1996, the ozone hole covered an area of twenty-two million square kilometres and was roughly the shape of the Antarctic continent and extended to about thirty kilometres altitude in the stratosphere. It had quite a dynamic character as its shape fluctuated through buffeting by external airflows. Because the process producing the ozone hole requires both the cold of winter and sunlight, its formation starts in winter at the lower latitude edge of Antarctica where in June and July there is sufficient sunlight during the short days for ozone destruction to begin slowly as shown by measurements made at the Faraday base at 65°S [9]. This ozone depletion accelerates and extends southwards as sunlight returns with the southern spring and the ozone hole reaches its maximum extent.

Ozone levels are often quoted in Dobson units named after G. M. B. Dobson who developed a method for measuring ozone levels in the atmosphere by comparing the amounts of different wavelengths of ultraviolet-A and -B radiation reaching Earth's surface. One Dobson unit represents a layer of pure ozone one hundredth of a millimetre thick when the temperature is 0°C and the pressure is 101.3 kilopascals, the normal atmospheric pressure at Earth's surface. Thus, when the amount of ozone measured is 400 Dobson units all of the ozone in a column of air of one square centimetre cross-section extending upwards to the boundary of space would make a layer four millimetres thick over that square centimetre if it was brought down to the surface. (This is the same as the unevenly distributed ozone in the column averaging to four parts per million by volume.) It is seen from this that the satellite map in Fig. 9.5 shows a deep hollow in the ozone layer over Antarctica, the ozone hole, surrounded by a high ridge of ozone over the southern parts of Africa, Australia and South America. As the tropics are approached the ozone layer thins again but nowhere near to the extent to which it thins in the Antarctic ozone hole.

9.5. The International Response and the Montreal Protocol

The 1985 report of the Antarctic ozone hole caused immediate scientific concern and quite quickly thereafter international concern at governmental level. However, there had been earlier concerns that the CFCs used in refrigerators, aerosol sprays and other applications could lead to a depletion of the ozone layer and a resulting increase in dangerous ultraviolet-B radiation reaching Earth's surface. At about the same time concerns had also been raised about the possibilities of depletion of the ozone layer resulting from the exhaust gases of the large numbers of stratospheric supersonic transports then planned [3-5]. The overall result was the first tentative governmental and industrial enquiries into the dangers arising from the release of CFCs and other gases into the atmosphere. The discovery of the Antarctic ozone hole and the correlation of its growth with the increasing levels of CFCs in the atmosphere were major catalysts for the unusually swift international response to the threat to the ozone layer that has been an impressive demonstration of the effective, if sometimes uneasy, combination of scientific investigation and political will to avert a major environmental threat [10].

The dominant concern was that should ozone depletion become a global stratospheric phenomenon it would have the effect of opening wider a window through which would pour increased amounts of ultraviolet-B radiation. That this was a justified concern was shown by ground measurements in Toronto where, in the period 1989-1993, levels of ultraviolet-B radiation increased by thirty-five percent a year in winter and seven percent a year in summer coincident with growing stratospheric ozone depletion [11]. Global ozone measurements from satellites over the period 1979-1993 found substantial increases in ultraviolet-B radiation levels at high and mid latitudes in both hemispheres and only small changes in the tropics [12]. Such radiation increases are likely to produce multiple environmental effects including an increase in skin cancer, eye damage, a weakening of the immune system in humans and other animals, damage to plants and phyto- and zooplankton, and result in an overall increase in disease and a decrease in food production [13-17]. An increase in ultraviolet radiation induced chemical reactions in the troposphere might also occur with difficult to predict consequences [18]. In addition to these effects, wood, plastic and natural and artificial fibres would become brittle more rapidly under the increased ultraviolet-B radiation intensity with significant economic loss resulting [19]. (Prolonged exposure to ultraviolet-A radiation also causes skin cancer but it is not absorbed by ozone and does not feature in these considerations of the ozone layer.)

It is estimated that global ozone depletion could have become as high as thirteen percent if the production of CFCs had not been curbed [20]. These fears resulted in the internationally binding Montreal Protocol on Substances that Deplete the Ozone Layer in 1987 and its subsequent amendments in 1990 (London), 1992 (Copenhagen) and 1996 (Vienna) [21-26]. These introduced increasingly stringent controls designed to protect the ozone layer and provide a blueprint for tackling other environmental problems given the political will. The initial 1987 agreement was for CFC production to be halved by 2000. In 1990, this was changed to a complete phase out by 2000. More recent amendments, agreed to by 130 nations, required developed nations to cease the use of all CFCs, along with carbon tetrachloride and 1,1,1-trichloroethane that also produce atomic chlorine in the stratosphere, by 1 January 1996 with the developing nations to cease such use by 2010. The Montreal Protocol signatories established a US$240 million fund in 1990 to assist developing nations with the CFC phase out and replacement with alternatives by 2006 and pledged a further US$510 million for this purpose in 1992.

9.6. A Closer Look at the Ozone Layer

To understand the driving force behind the Montreal Protocol a closer examination of the ozone layer is required. Ozone is unevenly distributed in the stratosphere and its level shows substantial seasonal variations as seen in Fig. 9.6 [2,3,27]. In the northern hemisphere measurements made at St Petersburg (Russian Federation) and Arosa (Switzerland) show that the highest ozone levels occur in March to April and the lowest levels occur in October to November. The seasonal variation in the southern hemisphere mirrors that of the north but is six months out of sequence as measurements at Aspendale (Australia) show. This is also true at Huancayo (Peru) where the ozone levels and their variations are much smaller as is typical of regions closer to the Equator. These variations arise from the latitudinal and seasonal changes of the intensity of solar ultraviolet radiation together with changes in stratospheric mixing patterns that arise from the seasonal warming and cooling of the hemispheres and the influence of Earth's rotation. Each year the patterns seen in Fig. 9.6 are broadly repeated.

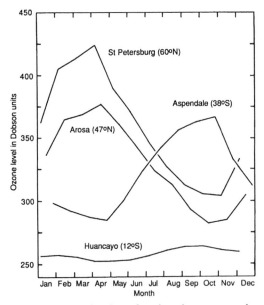

Fig. 9.6. Seasonal variations of ozone levels at four locations averaged over twenty-five years. Adapted from [3].

Most of the stratospheric ozone is produced in the tropics, where sunlight is most intense, and stratospheric movement subsequently distributes it globally. At the north and south polar latitudes sunlight strikes the atmosphere very obliquely so that the production of ozone is small compared with that in the tropics. Because of this, levels of atomic oxygen are also low and as a result the ozone destroying cycle involving atomic chlorine and chlorine monoxide shown in Fig. 9.3 is not as effective in the polar stratosphere as is the case in the lower latitudes. Clearly, something else occurs in the stratosphere above Antarctica to create the ozone hole.

9.7. The Antarctic Vortex

A very major annual event occurs over Antarctica that has not yet been taken into account. As the atmosphere cools with the approach of winter the air sinks and a westward rotating wind pattern builds up in the stratosphere above the South Pole under the influence of Earth's rotation, or Coriolis forces. This huge circulating wind pattern that reaches speeds increasing from zero to 300 kilometres per hour from surface level to fifty kilometres altitude, respectively, is the Antarctic vortex. It acts as an enormous doughnut shaped container that encompasses most of the atmosphere above Antarctica and largely prevents the almost stationary air at the vortex centre mixing with the air outside from mid autumn until mid spring [28-30]. This isolation of air inside the vortex has a profound influence on the formation of the Antarctic ozone hole.

Unlike the troposphere, the stratosphere is very dry and virtually cloudless. The amount of water it contains is similar to that of the scarce ozone. However, there is often a thin globally distributed aerosol layer present composed mainly of sulfuric acid, H_2SO_4, droplets. This is the result of biologically generated carbonyl sulfide, COS, reaching the stratosphere where it is oxidized to sulfur dioxide, SO_2, and then to sulfur trioxide, SO_3, that reacts with the scarce stratospheric water to form sulfuric acid over several months. An additional source of this aerosol is major volcanic eruptions, such as those of El Chichón in 1982 and Pinatubo in 1991, that eject huge amounts of sulfur dioxide directly into the stratosphere where it is also converted to sulfuric acid.

Clouds are unusual in the stratosphere but as the temperature drops in the Antarctic vortex in the long southern winter darkness polar stratospheric clouds, PSCs, of two types, I and II, form at twenty to twenty-five kilometres altitude where the stratospheric temperature is lowest. The type I ice clouds are formed by gases such as naturally occurring nitrous oxide, N_2O, nitric oxide, NO, and nitrogen dioxide, NO_2, and water freezing onto the sulfuric acid aerosol droplets at -83°C to -87°C. The type II ice clouds consist predominantly of water crystals and form below the water stratospheric frost point that is about -88.2°C to -90.2°C. They form either as pure water ice or through a build up of layers of water frozen onto Type I particles. Both types of clouds consist of tiny crystals that slowly grow in size during the winter and early spring and extend from ten to one hundred kilometres in length and are several kilometres thick. They show a beautiful iridescence and are sometimes called nacreous, or mother of pearl, clouds.

The effects of the formation of the polar stratospheric ice clouds are twofold. The first effect is that the cloud ice crystals may become so heavy that they fall from the stratosphere into the troposphere and thereby remove both sulfuric acid and oxides of nitrogen from the stratosphere. This decreases the amount of nitrogen dioxide present to react with chlorine monoxide to form chlorine nitrate, $ClONO_2$, that is not an ozone destroying molecule. As a consequence there is a higher level of chlorine monoxide (produced as shown in Fig. 9.3.) than usual inside the vortex and some of it combines to form chlorine peroxide, $ClOOCl$, in the southern winter darkness. In the early spring sunlight chlorine peroxide breaks down into atomic chlorine and chlorine dioxide that subsequently decomposes into atomic chlorine and diatomic oxygen as shown in Fig. 9.7. Thus, the removal of nitrogen dioxide by the polar stratospheric ice clouds has the effect of increasing the production of ozone destroying atomic chlorine inside the vortex.

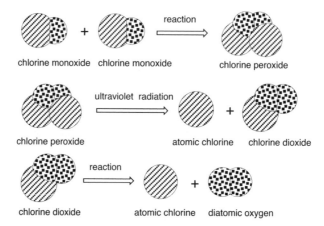

chlorine monoxide chlorine monoxide   chlorine peroxide

ultraviolet radiation

chlorine peroxide   atomic chlorine   chlorine dioxide

reaction

chlorine dioxide   atomic chlorine   diatomic oxygen

Fig. 9.7. The first of the routes through which ozone destroying atomic chlorine is produced in the Antarctic vortex. The chequered and hatched spheres represent oxygen and chlorine atoms, respectively.

The second effect is that hydrogen chloride, HCl, resulting from the reaction of atomic chorine and naturally occurring methane, CH_4, reacts with chlorine nitrate on the surface of the ice crystals to release diatomic chlorine, Cl_2, and nitric acid, HNO_3. In the darkness of the Antarctic winter the diatomic chlorine has little effect on ozone levels but the spring sunlight splits it into atomic chlorine as shown in Fig. 9.8. The overall result is that the amount of atomic chlorine available to deplete ozone within the Antarctic vortex is very high compared with the levels outside it. The Antarctic stratospheric ice clouds persist into the southern spring until the return of sunlight and the consequent increase in stratospheric temperature causes them to melt.

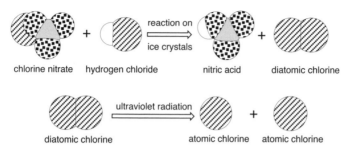

Fig. 9.8. The second of the routes through which ozone destroying atomic chlorine is produced in the Antarctic vortex. The chequered, hatched, shaded and clear spheres represent oxygen, chlorine, nitrogen and hydrogen atoms, respectively.

An Arctic vortex forms in the northern winter but it is less pronounced because the Arctic stratospheric temperatures are not as low as those in the Antarctic, and the formation of Arctic stratospheric ice clouds is correspondingly less [31]. It appears that the stratospheric sulfuric acid aerosol has a global effect similar to, but weaker than, that of the polar stratospheric ice clouds. However, the local ozone depletion caused is not so marked due to the continual stratospheric mixing occurring outside the polar vortices.

9.8. Flight into the Antarctic Vortex

To investigate changes in ozone levels with latitude a high flying ER2 reconnaissance aircraft was equipped to measure ozone as it flew through the rapidly rotating wall of the Antarctic vortex into its centre during the Airborne Antarctic Ozone Experiment, AAOE [29]. On the southern winter day of 23 August 1987, the ER2 flew from Puente Arenas, at the southern tip of Patagonia, into the Antarctic vortex at an altitude of eighteen kilometres. Only moderate changes in the ozone level were detected as the ER2 passed through the wall of the vortex but the chlorine monoxide level rose sharply as seen in Fig. 9.9. On a similar flight in the spring sunlight of the 16 September 1987, the pilot was greeted by a huge polar stratospheric ice cloud in the shape of an eye with a green pupil surrounded by a bright red iris as sunlight was diffracted by

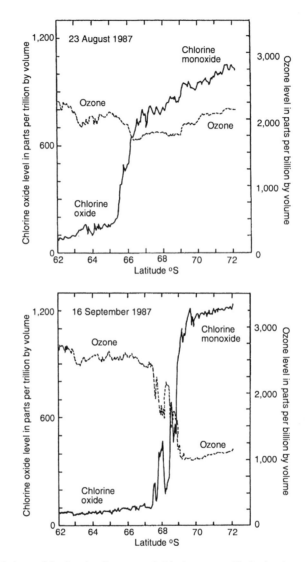

Fig. 9.9. The variations of the levels of ozone and chlorine monoxide in the Antarctic vortex on 23 August 1987 and 16 September 1987. Because the measurements of these gases were made at a single altitude, ozone and chlorine monoxide levels are quoted in terms of their number relative to that of the total number of air molecules at that altitude, i.e. parts per billion and parts per trillion by volume, respectively. Adapted from [28].

the ice crystals. A steep decline in ozone levels was detected as the vortex was entered coincident with a great increase in chlorine monoxide levels as seen in

Fig. 9.9. These two flights, and others like them, showed the development of the Antarctic ozone hole with the arrival of spring sunlight.

The variation in ozone levels above Antarctica with altitude, season and the sun's eleven year cycle has also been studied through the release of balloons carrying ozone measuring equipment from several research stations. The measurements of ozone made in this way are all similar and are typified by those made at the American McMurdo station during 1989. In the winter darkness of 23 August 1989 it was found that the ozone level increased rapidly from about ten kilometres altitude to a maximum in the region of fourteen to eighteen kilometres and then rapidly diminished as altitude increased [32]. This showed that most of the ozone was concentrated in the lower stratosphere and provided a basis for comparison with subsequent measurements made in the spring. A balloon flight on 20 October 1989 provided a very different picture of the ozone layer. Now ozone levels diminished sharply above eleven kilometres to very low levels between altitudes of fourteen to twenty-one kilometres.

Both the E2R and balloon ozone measurements are consistent with the polar stratospheric ice clouds causing the release of much higher levels of the ozone destroying combination of chlorine monoxide and atomic chlorine through the processes shown in Figs. 9.7 and 9.8 inside the Antarctic vortex than is the case outside it. The result is that the springtime ozone depletion in the isolated stratospheric air in the vortex is much greater than in the warmer air outside it as seen in Figs. 9.5 and 9.9.

A general picture of the natural ozone production, depletion and circulation processes emerges from the discussion so far and is shown in simplified form in Fig. 9.10. A continuous production of the major part of stratospheric ozone occurs above the tropics where sunlight is most intense. It is transported to higher latitudes, where ozone production is less, by stratospheric airflows generated by seasonal and rotational effects. In the Antarctic winter, stratospheric ice clouds form in the Antarctic vortex and with the return of the sun in spring reactions occur within these ice clouds that destroy ozone to produce the ozone hole. In summer the ice clouds melt and the Antarctic vortex breaks up to release ozone depleted air that drifts northwards. Ozone destruction occurs over the Arctic through a similar process in the northern spring but to a lesser extent compared with that over the colder Antarctic in the southern spring. It is thought that the stratospheric sulfuric acid aerosol that encircles Earth may also cause global ozone destruction but is less effective than the polar stratospheric ice clouds.

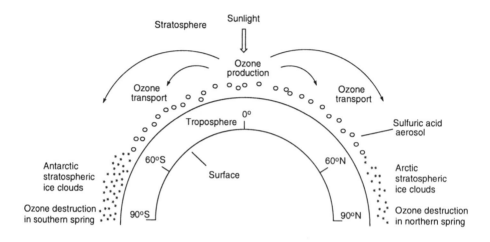

Fig. 9.10. An illustration (not to scale) of the natural ozone production, depletion and circulation processes.

9.9. Chlorofluorocarbons: Nemesis of the Ozone Layer

It is clear that any increase in atomic chlorine over its natural level is likely to result in a depletion of the ozone layer. Quite unintentionally, this is precisely what has been happening since the early 1970s as a result of technological innovation. Chlorofluorocarbons, CFCs, were first made in the 1920s and by 1986 global annual production had reached 1.2 million tonnes. The CFCs contain chlorine, fluorine and carbon and are not found in nature. They are very stable non toxic liquids that have a high vapour pressure so that they readily vaporize. This makes them ideal for use as refrigerants, propellants in spray cans, and as the foaming, or blowing, gas in the production of flexible open and rigid closed cell plastics used in vehicle dashboards, bedding and as thermal insulation. The combination of their ability to dissolve grease and their lack of corrosive properties resulted in their extensive use as cleaning agents for electronic components. Particularly widely used examples of CFCs were trichlorofluoromethane, $CFCl_3$ or CFC11, dichlorodifluoromethane, CF_2Cl_2 or CFC12, and 1,2-trichlorofluoroethane, $CF_2ClCFCl_2$ or CFC113. (Industry has devised a strange code to identify CFCs where the units digit is the number of fluorine atoms, the tens digit is the number of hydrogen atoms plus one and the hundreds digit is the number of carbon atoms minus one which digit is not shown if it is zero. The remaining atoms are chlorine atoms where the total number of atoms in the CFC is five if only one carbon atom is present and is eight if two carbons are present.)

The stability of the CFCs that made them industrially attractive also made them a major menace to the ozone layer. CFCs do not dissolve in water to a significant extent and so are not washed out of the troposphere by rain as are hydrogen chloride and sodium chloride particles generated by sea spray. Neither do they react appreciably with other gases in the troposphere. As a consequence they have long lifetimes in the troposphere. Although they are heavier than air, normal air movement carries the CFCs up to the tropopause where they slowly enter the stratosphere over a period of years. There, they are rapidly broken down by the intense stratospheric ultraviolet radiation. This releases ozone destroying atomic chlorine together with atomic fluorine that is a thousand times less effective in destroying ozone. This addition of atomic chlorine above the natural level hastens the depletion of the ozone layer with its most impressive manifestation being the Antarctic ozone hole.

While ninety-five percent of CFCs have been released in the northern hemisphere, they are now spread globally in the troposphere and the stratosphere. Their impact on the ozone layer in the Antarctic vortex was particularly marked and greatly increased the size of the ozone hole. However, a cooling in the stratosphere in high northern latitudes in the 1992-1993 and 1994-1995 winters also led to a substantial ozone depletion in the northern temperate regions approaching ten percent in winter and spring and five percent in summer and autumn [33]. In the winter 1995-1996 a particularly strong Arctic vortex developed and produced a large area of ozone depletion in March 1996 extending over Scandinavia, northern Canada, Alaska and the Russian Federation and as far south as the United Kingdom. In the latter case an unprecedentedly low ozone level of 195 Dobson units was recorded over Lerwick that was only fifty-three percent of the March average [34,35]. Prior to this, cynical northern observers might have thought that events high over the distant Antarctic had little relevance to them but, judging from media reaction, a heightening of interest occurred in the populous north. The popular science journal New Scientist was moved to editorialise with the headline "Disaster in the stratosphere" on 16 March 1996 [34].

Clearly major ozone depletion has occurred globally not only because of the effect due to CFCs but also because of a global cooling of the stratosphere [36]. This cooling arises from the depletion of the ozone layer and decreased warming through the absorption of ultraviolet-B radiation by ozone combined with an enhanced greenhouse effect that traps an increasing amount of the Earth's reradiated thermal infrared radiation in the troposphere (as is discussed in Chapter 8) and diminishes the amount of reradiated heat available to warm the stratosphere from below. Such stratospheric cooling favours the formation of deep vortices in winter at both poles and the increased formation of polar stratospheric ice clouds within them that increases ozone destruction. This is not an encouraging prospect as global ozone levels, currently estimated to be up to ten percent below normal levels, are unlikely to begin recovering as rapidly as anticipated from the phasing out of CFCs if stratospheric cooling continues [36]. However, this effect of cooling may be offset by increased turbulence of the warming troposphere that will cause greater mixing between it and the stratosphere at the tropopause so that CFCs enter the stratosphere more rapidly and are there destroyed faster than anticipated in the absence of tropospheric warming [37].

9.10. Changes in Stratospheric Chlorine and Chlorofluorocarbon Levels

The amount of chlorine in CFCs and other chlorine containing molecules in the atmosphere appeared to reach a maximum in 1997 at about 4.2 parts per billion by volume, five times its level in the 1950s and twice its level in the 1970s [2,22,23]. To that time, growth in stratospheric chlorine was paralleled by the increase in CFCs levels detected in the troposphere. Fortunately, the levels of CFCs and other organochlorine gases in the troposphere appear to have peaked as the Montreal Protocol and its subsequently strengthened restrictions take effect. However, as the time taken for mixing air between the troposphere and the lower stratosphere is several years, stratospheric chlorine levels will only slowly decline to the natural levels in a hundred years or so, as shown in Fig. 9.11 [38]. The CFCs and other organochlorines are broken down by ultraviolet radiation of about 230 nanometres wavelength but as none of this radiation enters

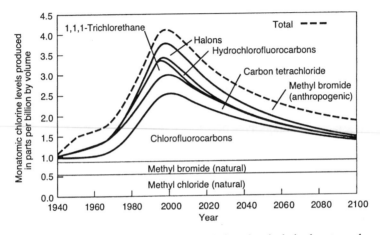

Fig. 9.11. Actual and predicted levels of ozone depleting chemicals in the atmosphere shown in terms of the amount of atomic chlorine they produce in the stratosphere in the case of the CFCs, HCFCs, methylchloride, trichlorethylene and carbon tetrachloride, and the amount of monatomic bromine expressed in terms of its effect equivalent to monatomic chlorine in the case of the halons and methyl bromide. Adapted from [38].

the troposphere they have to rise into the stratosphere for this to occur. The predicted levels shown in Fig. 9.11 are based on measured changes in chlorine and bromine levels and it is reassuring that the predicted peaking of atmospheric chlorine and bromine in the late 1990s has occurred [38]. The impact of the Montreal Protocol should show up strongly as the depth of the Antarctic ozone hole decreases as the mid twenty-first century is approached.

While the increased stratospheric chlorine level accounts for about three quarters of the ozone depletion, its close relative, bromine, accounts for the remaining one quarter of the depletion. Bromine destroys ozone through processes similar to those involving chlorine, and atom for atom is about forty times more effective in ozone destruction than is chlorine. Bromine reaches the stratosphere in organobromine molecules collectively known as halons that are used in fire-fighting equipment. Fortunately, the amount of halons produced has been much less than the amount of CFCs produced. Production of halons, H-1301 (CF_3Br), H-1211 (CF_2ClBr) and H-2402 (CF_2BrCF_2Br) stopped in 1994, but they are still being released from existing equipment. Another source of stratospheric bromine is methyl bromide, CH_3Br, that is produced biologically and is also manufactured for use as an agricultural fumigant. The Montreal Protocol proposes that this use should be phased out by developed and developing nations by 2005 and 2015, respectively, and its atmospheric level should slowly decline to its natural level thereafter [25,38].

The phasing out of CFCs, halons and related ozone depleting halocarbons clearly cannot bring to a standstill those activities in which they play a major role without economic loss and social disruption. Two basic approaches to this problem have been adopted. The first has been to adopt different methods such as roll-on applicators or small hand pumps for deodorant and perfume applications. The second has been a replacement program using markedly less ozone threatening replacements for CFCs such as the hydrochlorofluorocarbons, HCFCs, and hydrofluorocarbons, HFCs. These compounds, exemplified by HCFC-22 (CHF_2Cl), HCFC-142b (CH_3CF_2Cl), HCFC-141b (CH_3CFCl_2) and HFC-134a (CH_2FCF_2) are broken down relatively rapidly in the troposphere where they last for thirteen, twenty, nine and fourteen years, respectively. This compares with tropospheric lifetimes of fifty, one hundred and two and eighty-five years for CFC-11, CFC-12 and CFC-113 years, respectively. The shorter lived HCFCs do not enter the stratosphere to anywhere near the extent to which the CFCs do and therefore cause much less ozone depletion because of this and because they contain less chlorine. The only halogen released by HFCs in the stratosphere is fluorine that is much less reactive with ozone than is chlorine. The HCFCs are either to be replaced by the more ozone benign hydrocarbons such as cyclopentane and isobutane that have been adopted by some nations as blowing agents and coolants or to be made redundant by new technologies by 2030 [39].

9.11. Lingering Concerns

Clearly the threat to the ozone layer has been greatly reduced through the provisions of the Montreal Protocol. This is a rare example of an international environmental protection agreement that seems to be working. It is hopefully the precursor of a range of international agreements that will safeguard Earth's habitability. However, the banning of CFCs created an opportunity for criminal organizations to smuggle an estimated 30,000 tonnes of CFCs to unscrupulous users in Europe and North America annually at the turn of the century. Fortunately, governments and international agencies have acted to close down the illegal CFC production plants, albeit slowly [40]. Another problem persists with the search for a replacement for the ozone depleting pesticide methyl bromide that is being hindered by a combination of acceptable substitutes being less readily available than anticipated and politically motivated disagreements [25]. An example of this was the request by American millers in late 2002 for more time to find an alternative to methyl bromide. The agricultural use of methyl bromide by the United States dropped from 25,528 tonnes in 1991 to 7,674 tonnes in 2003. However, it was claimed that the United States would need to increase its use of methyl bromide to 9,445 tonnes in 2005 and 2006 that would constitute a very large share of the 14,750 tonnes allowed under the Montreal Protocol for the fifteen nations that sought permission for a limited use of methyl bromide in agriculture [26]. Perplexingly, requirements for food safety and the prevention of spreading of pests through increasing international trade may lead to an increase in methyl bromide use for fumigation in the short term at least. So far this has resulted in a probably underestimated global increase in the use of methyl bromide for this purpose from 11,000 tonnes in 2002 to 18,000 tonnes in 2004.

There are other human activities that pose potential threats to the ozone layer. While the building of large numbers of stratospheric transports seems unlikely at present, the oxides of nitrogen that their engines produce have significant ozone destroying properties and an increase in the level of these gases in the stratosphere could generate another round of ozone depletion [41]. There remains another concern about the oxides of nitrogen, the CFCs and their replacements.

This is the greenhouse character of these gases that causes an increased retention of Earth's thermal infrared radiation energy in the troposphere. Per molecule, the CFCs, HCFCs, halons and related molecules are up to twenty thousand times more effective in absorbing infrared radiation than is carbon dioxide. The global loss of HFC-134a to the troposphere from cooling systems was expected to be about 150,000 tonnes annually from 2000 onwards and would have a greenhouse effect as great as all of the carbon dioxide produced through fossil fuel combustion in the United Kingdom annually [39]. Thus, beneficial changes for the ozone layer may have undesirable consequences in terms of global warming and climate change in the short to medium term. Despite the slow exchange of air between the troposphere and the stratosphere, anthropogenic impacts on both are intimately related. It appears that this may be particularly so in Antarctica where the cooling of the stratosphere due to ozone depletion in the ozone hole may retain cold tropospheric air over the centre of the continent longer and cause abnormally strong and warm westerly air flows around its periphery to accelerate the break up of ice shelves [42]. Interactions such as these are but another illustration of how closely interwoven are all aspects of the biosphere and humanity's activities.

References

1.   a) J. Farman, What hope for the ozone layer now? *New Sci.*, **1987**, 12 Nov., 50. b) R. S. Stolarski, The Antarctic ozone hole, *Sci. Am.*, **1988**, *258*, Jan., 20. c) F. S. Rowland, Stratospheric ozone depletion, *Annu. Rev. Phys. Chem.*, **1991**, *42*, 731.

2.   P. S. Zurer, Ozone depletion's recurring surprises challenge atmospheric scientists, *Chem. Eng. News*, **1993**, *71*, May 24, 8.

3.   F. S. Rowland, Stratospheric ozone depletion by chlorofluorocarbons (Nobel Lecture), *Angew. Chem. Int. Edn. Engl.*, **1996**, *35*, 1787.

4.   P. J. Crutzen, My life with O_3, NO_x, and other YZO_x compounds (Nobel Lecture), *Angew. Chem. Int. Edn. Engl*, **1996**, *35*, 1759.

5.   M. J. Molina, Polar ozone depletion (Nobel Lecture), *Angew. Chem. Int. Edn. Engl.*, **1996**, *35*, 1779.

6.   a) J. E. Lovelock, R. J. Maggs and R. J. Wade, Halogenated hydrocarbons in and over the Atlantic, *Nature*, **1973**, *241*, 194. b) M. J. Molina and F. S. Rowland, Stratospheric sink for chlorofluoromethanes: chlorine atom-catalysed destruction of ozone, *Nature*, **1974**, *249*, 810. c) F. S. Rowland, Chlorofluorocarbons and the depletion of stratospheric ozone, *Am. Sci.*, **1989**, *77*, 36. d) R. Garcia, Causes of ozone depletion, *Physics World*, **1994**, April, 49. e) F. S. Rowland and M. J. Molina, Ozone depletion: 20 years after the alarm, *Chem. Eng. News*, **1994**, *721*, 15 Aug., 8.

7.   J. C. Farman, B. G. Gardiner and J. D. Shanklin, Large losses of ozone in Antarctica reveal seasonal ClO_x /NO_x interaction, *Nature*, **1985**, *315*, 207.

8.   The British Antarctic Survey, http://www.nbs.ac.uk/public/icd/jds/ozone/

9.   H. K. Roscoe, A. E. Jones and A. M. Lee, Midwinter start to Antarctic ozone depletion: evidence from observations and models, *Science*, **1997**, *278*, 93.

10. K. T. Litfin, *Ozone Discourses: Science and Politics in Global Environmental Cooperation*, Columbia University Press, New York, USA, 1994.

11. J. B. Kerr and C. T. McElroy, Evidence for large upward trends in ultraviolet-B radiation linked to ozone depletion, *Science*, **1993**, *262*, 1032.

12. S. Madronich, Changes in ultraviolet radiation reaching the Earth's surface, *Ambio*, **1995**, *24*, 143.

13. J. C. van der Leun, X. Tang and M. Tevini, Environmental effects of ozone depletion: 1994 assessment, *Ambio*, **1995**, *24*, 138.

14. a) J.-S. Taylor, DNA, sunlight, and skin cancer, *J. Chem. Ed.*, **1990**, *67*, 835. b) S. Mandronich and F. R. de Gruiji, Skin cancer and UV radiation, *Nature*, **1993**, *366*, 23. c) J. D. Longstreth, Effects of increased solar ultraviolet radiation on human health, *Ambio*, **1995**, *24*, 153. d) D. J. Leffell and D. E. Brash, Sunlight and skin cancer, *Sci. Am.*, **1996**, *275*, July, 38. e) H. Slaper, Estimates of ozone depletion and skin cancer incidence to examine the Vienna convention achievements, *Nature*, **1996**, *384*, 256.

15. a) A. R. Blaustein and D. W. Blake, The puzzle of declining amphibian populations, *Sci. Am.*, **1995**, *272*, April, 56. b) T. Halliday, A declining amphibian conundrum, *Nature*, **1998**, *394*, 418.

16. M. M. Caldwell, Effects of increased solar ultraviolet radiation on terrestrial plants, *Ambio*, **1995**, *24*, 166.

17. a) D.-P. Häder Effects of increased solar ultraviolet radiation on aquatic ecosystems, *Ambio*, **1995**, *24*, 174. b) R. G. Zepp, T. V. Callaghan and D. J. Erickson, Effects of increased solar ultraviolet radiation on biogeochemical cycles, *ibid.*, 181.

18. X. Tang and S. Madronich, Effects of increased solar ultraviolet radiation on tropospheric composition and air quality, *Ambio*, **1995**, *24*, 188.

19. A. L. Andrady, Effects of increased solar ultraviolet radiation on materials, *Ambio*, **1995**, *24*, 191.

20. F. S. Rowland, Stratospheric ozone depletion by chlorofluorocarbons, *Ambio*, **1990**, *19*, 281.

21. a) S. R. Seidel and D. P. Blank, The Montreal Protocol: pollution prevention on a global scale, *Ambio*, **1990**, *19*, 301. b) D. A. Wirth and D. A. Lashof, Beyond Vienna and Montreal - multilateral agreements on greenhouse gases, *ibid.*, 305. c) J. H. Chafee and S. J. Shimberg, Supplementing the Montreal Protocol: the need for domestic legislation, *Ambio*, **1990**, 19, 310. d) Á. Rosencranz and R. Milligan, CFC abatement: the needs of developing countries, *ibid.*, 312. e) G. D. Phillips, CFCs in the developing nations: a major economic development opportunity. Will the institutions help or hinder? *Ibid.* 316. f) C. A. Moore, Industry responses to the Montreal Protocol, *ibid.*, 320. g) H. F. French, Making environmental treaties work, *Sci. Am.*, **1994**, *271*, Dec., 62. h) E. A. Parson and O. Greene, The complex chemistry of the international ozone agreements, *Environment*, **1995**, *37*, March, 16.

22. V. Morell, Ozone destroying chlorine tops out, *Science*, **1996**, *271*, 32.

23.   a) S. A. Montzka Decline in the tropospheric abundance of halogen from halocarbons: implications for stratospheric ozone depletion, *Science*, **1996**, *272*, 1318. b) J. H. Butler, Better budgets for methyl halides? *Nature*, **2000**, *403*, 260. c) S. A. Montzka, Present and future trends in the atmospheric burden of ozone-depleting halogens, *ibid.*, 690.

24.   a) K. J. Holmes and J. H. Ellis, Potential environmental impacts of future halocarbon emissions, *Env. Sci. Technol.*, **1996**, *30*, 348 A. b) P. J. Fraser and M. J. Prather, Uncertain road to ozone recovery, *Nature*, **1999**, *398*, 663. c) E. Stokstad, A brighter outlook for good ozone, *Science*, **2002**, *297*, 1623.

25.   a) F. Pearce, Promising the Earth, *New Sci.*, **1997**, 30 Aug., 4. b) F. Pearce, US millers fight for banned pesticide, *New Sci.*, **2002**, 5 Oct., 11. c) *Anon.*, Fruit threat to ozone layer, *New Sci.*, **2004**, 4 Dec., 4.

26.   B. Hileman, Methyl bromide phaseout stymied, *Chem. Eng. News*, **2005**, *83*, 17 Jan., 30.

27.   D. W. Fahey and A. R. Ravishankara, Summer in the stratosphere, *Science*, **1999**, *285*, 208.

28.   J. G. Anderson, D. W. Toohey and W. H. Brune, Free radicals within the Antarctic vortex: the role of CFCs in Antarctic ozone loss, *Science*, **1991**, *251*, 39.

29.   M. R. Schoeberl and D. L. Hartmann, The dynamics of the stratospheric polar vortex and its relation to springtime ozone depletions, *Science*, **1991**, *251*, 46.

30.   a) O. B. Toon and R. P. Turco, Polar stratospheric clouds and ozone depletion, *Sci. Am.*, **1991**, *264*, June, 40. b) J. M. Rodriguez, Probing stratospheric ozone, *Science*, **1993**, *261*, 1128.

31.   a) C. R. Webster, Chlorine chemistry on polar stratospheric cloud particles in the Arctic winter, *Science*, **1993**, *261*, 1130. b) D. W. Toohey, The seasonal evolution of reactive chlorine in the northern hemisphere stratosphere, *ibid.*, 1134. c) M. L. Santee, Interhemispheric differences in polar stratospheric HNO_3, H_2O, ClO, and O_3, *Science*, **1995**, *267*, 849.

32.   T. Deshler, Ozone and temperature profiles over McMurdo station Antarctica in the spring of 1989, *Geophys. Res. Lett*, **1990**, *17*, 151.

33.   World Meteorological Organisation, *Global Ozone Research and Monitoring Project, Report No 37. Scientific Assessment of Ozone depletion: 1994*, WMO/NASA. 1995.

34.   *Anon.*, Disaster in the stratosphere, *New Sci.*, **1996**, 16 March, 3.

35.   a) F. Pearce, Big freeze digs a deeper hole in ozone layer, *New Sci.*, **1996**, 16 March, 7. b) *Anon.*, The hole where the ultraviolet comes in, *New Sci.*, **1996**, 4 May, 3. c) R. Müller, Severe chemical ozone loss in the Arctic during the winter of 1995-96, *Nature*, **1997**, *389*, 709. d) R. Stolarski, A bad winter for Arctic ozone, *Nature*, **1997**, *389*, 788. e) M. Rex, Prolonged, stratospheric ozone loss in the 1995-96 Arctic Winter, *Nature*, **1997**, *389*, 835. f) L. O. Björn, The problem of ozone depletion in northern Europe, *Ambio*, **1998**, *27*, 275.

36.   a) C. Brühl and P. J. Crutzen, Ozone and climate changes in the light of the Montreal Protocol: a model study, *Ambio*, **1990**, *19*, 293. b) R. J. Salawitch, A greenhouse warming connection, *Nature*, **1998**, *392*, 551. c) D. T. Shindell, D. Rind and P. Lonergan, Increased polar stratospheric ozone losses and delayed eventual recovery owing to increasing greenhouse gas concentrations, *Nature*, **1998**, *392*, 589. d) R. A. Kerr, Ozone loss, greenhouse gases linked, *Science*, **1998**, *280*, 202. e) P. Zurer, Greenhouse gases impeding ozone recovery, *Chem. Eng. News*, **1998**, *76*, 13 April, 12. f) R. A. Kerr, Deep chill triggers record ozone hole, *Science*, **1998**, *282*, 391. g) J. Hecht, Polar Alert. Climate change is taking the brakes off ozone depletion, *New Sci.*, **1999**, 12 June, 6. h) M. Schrope, Successes

in fight to save ozone layer could close holes by 2050, *Nature*, **2000**, *408*, 627. i) G. Walker, The hole story, *New Sci.*, **2000**, 25 March, 24.

37. N. Butchart and A. A. Scaife, Removal of chlorocarbons by increased mass exchange between the stratosphere and troposphere in a changing climate, *Nature*, **2001**, *410*, 799.

38. a) D. J. Wuebbles and J. M. Calm, An environmental rational for retention of endangered chemicals, *Science*, **1997**, *278*, 1090. b) C. Baird, *Environmental Chemistry*, 2nd Edn. Freeman, New York, USA, 1999, p. 61.

39. B. Hileman, A chilling battle, *Chem. Eng. News*, **1998**, *76*, 3 Aug., 33.

40. a) *Scientific Assessment of Ozone Depletion: 1994*. Report 37, Global ozone research and monitoring project, World Meteorological Organisation, Geneva, 1995. b) F. Pearce, The hole that will not mend, *New Sci.*, **1997**, 30 Aug., 16. c) D. Spurgeon, Ozone treaty "must tackle CFC smuggling", *Nature*, **1997**, *389*, 219. d) M. Smith and M. Vincent, Tanking a killer coolant, *Can. Geog.*, **1997**, Sept./Oct., 40.

41. S. Nemecek, Holes in ozone Science, *Sci. Am.*, **1995**, *272*, Jan., 26.

42. S. Simpson, A push from above, *Sci. Am.*, **2002**, *287*, Aug., 9.

Subject Index

photosynthesis, 3, 19, 79, 144-146, 160, 179, 219, 221,360, 361, 384, 405, 420, 428, 29, 431-433, 452, 462, 480., 486, 118, 119.
pituitary gland, 239.
plague, 269, 270-272, 283, 335-337.
plankton, 32, 116, 124, 179-181, 380, 430, 431, 452, 457.
plant, biological, 14, 19, 25, 45, 77, 79-86, 103, 106, 117, 118, 123, 144, 145, 146, 148, 152, 153, 155, 158, 160-162, 164, 166, 167, 171, 173, 174, 176, 180, 181-195, 197, 211-213, 215, 219, 221, 222, 224-228, 237, 238, 240-243, 246, 255, 360, 361, 386, 420, 425, 429, 431-434, 445, 446, 452, 480, 486.
plasmid, 212, 213, 226, 239, 241, 285, 287.
Plasmodium falciparum, 311-315.
Plasmodium ovale, 311.
Plasmodium malarea, 311.
Plasmodium vivax, 311, 313.
plutonium-239, 33, 99, 393, 398-402.
pneumonia, 284, 271, 274, 288, 291, 292, 297, 301, 317.
Pneumocystis, carinii, 301.
Poland, 8, 26, 47, 57, 58, 66, 68, 87. 123, 157, 191, 364, 372, 381, 429.
poliomyelitis, 117, 284, 293, 308, 316, 319, 333.
poliovirus, 240, 316.
pollen, 188, 190, 194, 211, 212.
pollination, 182, 189, 190, 193, 194, 211.
pollution, 49, 85, 88, 90, 106, 114-120, 124, 178, 321, 326, 387, 407, 437, 462.

polonium-214/218, 326.
polychlorinated biphenyls, PCBs 119.
polypeptide, 220, 223, 224, 239.
Polynesia, 4, 86.
population, animal, 240, 380, 434, 446, 452, 457, 463.
population change/growth, human, 15, 45, 85 46, 51, 57, 55, 56, 63, 128, 147, 156, 159, 385, 423.
population, fish, 172, 174, 176, 177, 179.
population, human, 1, 14, 50-52, 63, 83, 85-87, 95, 106, 111, 113, 114, 125, 126, 140-144, 147-149, 152, 155, 156, 157, 159, 161, 166, 170, 190, 191, 193-195, 232-234, 240, 255, 270, 271, 286, 293-295, 297, 302, 307, 308, 313, 316, 317, 319, 337, 359, 363, 365, 367, 386, 425.
population, plant, 187, 433, 434.
population pyramid, 63.
"Population Bomb, The" (the book), 155.
potassium, 154, 158, 159, 170.
potato, 149, 150, 152, 153, 166, 182, 188, 190, 228, 243, 285, 432.
potato blight, 149.
potato famine, Irish, 26, 149.
potato, sweet, 153, 182.
poverty, 70, 294, 303.
prion, 239, 276, 277.
Prince William Sound, 32.
prokaryote, 212, 213, 215, 223, 227, 243.
production, primary, 8, 9, 141.
proteome, 234, 235.
protein, 150, 160, 161, 164, 171, 177, 182, 183, 187, 189, 212-215,